Maximal Function Methods for Sobolev Spaces

Mathematical
Surveys
and
Monographs

Volume 257

Maximal Function Methods for Sobolev Spaces

Juha Kinnunen
Juha Lehrbäck
Antti Vähäkangas

AMS AMERICAN MATHEMATICAL SOCIETY
Providence, Rhode Island

2020 *Mathematics Subject Classification.* Primary 42B25, 46E35;
Secondary 26D10, 28A12, 31B15, 35A23, 35J92, 42B37.

For additional information and updates on this book, visit
www.ams.org/bookpages/surv-257

Library of Congress Cataloging-in-Publication Data

Names: Kinnunen, Juha, author. | Lehrbäck, Juha (Juha Tapio), 1979– author. | Vähäkangas, Antti V., author.

Title: Maximal function methods for Sobolev spaces / Juha Kinnunen, Juha Lehrbäck, Antti Vähäkangas.

Description: Providence, Rhode Island: American Mathematical Society, [2021] | Series: Mathematical surveys and monographs, 0076-5376; volume 257 | Includes bibliographical references and index.

Identifiers: LCCN 2021013318 | ISBN 9781470465759 (paperback) | ISBN 9781470466602 (ebook)

Subjects: LCSH: Sobolev spaces. | Maximal functions. | Inequalities (Mathematics) | AMS: Harmonic analysis on Euclidean spaces – Harmonic analysis in several variables – Maximal functions, Littlewood-Paley theory. | Functional analysis – Linear function spaces and their duals – Sobolev spaces and other spaces of "smooth" functions, embedding theorems, trace theorems. | Real functions – Inequalities – Inequalities involving derivatives and differential and integral operators. | Measure and integration – Classical measure theory – Contents, measures, outer measures, capacities. | Potential theory – Higher-dimensional theory – Potentials and capacities, extremal length and related notions in higher dimensions. | Partial differential equations – General topics in partial differential equations – Inequalities applied to PDEs involving derivatives, differential and integral operators, or integrals. | Partial differential equations – Elliptic equations and systems – Quasilinear elliptic equations with p-Laplacian. | Harmonic analysis on Euclidean spaces – Harmonic analysis in several variables – Harmonic analysis and PDE.

Classification: LCC QA323 .K56 2021 | DDC 515/.782–dc23

LC record available at https://lccn.loc.gov/2021013318

♾ The paper used in this book is acid-free and falls within the guidelines
established to ensure permanence and durability.
Visit the AMS home page at `https://www.ams.org/`

10 9 8 7 6 5 4 3 2 1   26 25 24 23 22 21

Contents

Preface

This book is about maximal functions and their applications in Sobolev spaces. There are many good texts on the use of maximal functions in harmonic analysis, but we feel that there is room for a source book gathering advances in maximal function methods related to Poincaré and Sobolev inequalities, pointwise estimates and approximation for Sobolev functions, Hardy's inequalities, and partial differential equations. A recurring theme throughout the book is self-improvement of uniform quantitative conditions. Our approach is partially motivated by the theory of analysis on metric measure spaces, but in order to avoid extra complications we restrict our attention only to prototypes in Euclidean space. Nevertheless, the methods applied in analysis on metric measure spaces are useful already in the Euclidean context.

Besides maximal functions and Sobolev spaces, we discuss several related concepts. Capacities are needed for the study of fine properties of Sobolev functions and characterization of Sobolev spaces with zero boundary values. The capacity density condition is applied for Hardy's inequalities and in partial differential equations. In addition to the Hausdorff dimension, we use the Assouad dimension and the lower dimension to characterize density conditions and to give sufficient and necessary conditions for Hardy's inequalities and their generalizations. The distance function appears in Hardy's inequality and also has applications in Sobolev spaces. We study the Muckenhoupt weight properties of distance functions and combine these with general weighted norm inequalities. At the end of the book we discuss the theory of weak solutions to the p-Laplace equation and show how maximal function techniques can be used in this context. The choice of topics is exclusive and reflects the research interests of the authors.

Our style is concise and brief. Instead of lengthy motivations, we give detailed proofs in order to make the arguments flexible and transparent. For this reason some of the proofs are relatively long. We demonstrate interesting techniques, with the idea that the methods will have a wider range of applications beyond the topics covered by this book. This is not always the most direct approach and causes some overlap. Part of the material is rather standard, while some results appear for the first time in book form. One of our goals was to gather material that has been scattered in research papers and make it accessible to a wider audience. Standard references and related research papers are mentioned in the notes at the end of each chapter. We hope that the list of references, which is long but not complete, will make it easier for the reader to further investigate the literature.

We have primarily aimed the book for graduate students, but we also believe that it can be used as a reference by researchers. Some parts of the material have been used by the authors in graduate courses. Most of the book is self-contained, only assuming knowledge of measure and integration theory, in particular the Lebesgue measure and L^p spaces, as well as some functional analysis. We have made an effort to arrange the material so that the chapters can be read relatively independently.

We would like to thank many colleagues for discussions and comments, in particular, David Cruz-Uribe, Sylvester Eriksson-Bique, Jonathan Fraser, Napoleon Freitas Paajanen, Tero Kilpeläinen, Peter Lindqvist, Juan Manfredi, Derek Robinson, Nageswari Shanmugalingam and Xiao Zhong. The authors have been supported by the Academy of Finland during the writing of this book.

Juha Kinnunen
Juha Lehrbäck
Antti Vähäkangas

Notation

We use the following notation for basic concepts throughout the text.

- $\mathbb{N} = \{1, 2, 3, \ldots\}$, $\mathbb{N}_0 = \mathbb{N} \cup \{0\}$, $\mathbb{Z} = \{\ldots, -2, -1, 0, 1, 2, \ldots\}$, $\mathbb{Q}$ is the set of rational numbers, $\mathbb{R}$ is the set of real numbers and $[-\infty, \infty] = \mathbb{R} \cup \{-\infty, \infty\}$ is the set of extended real numbers.
- $|x| = (x_1^2 + x_2^2 + \cdots + x_n^2)^{\frac{1}{2}}$ is the Euclidean norm and $x \cdot y = x_1 y_1 + x_2 y_2 + \cdots + x_n y_n$ is the inner product of $x = (x_1, x_2, \ldots, x_n) \in \mathbb{R}^n$ and $y = (y_1, y_2, \ldots, y_n) \in \mathbb{R}^n$.
- $e_j = (0, \ldots, 0, 1, 0, \ldots, 0)$, $j = 1, 2, \ldots, n$, is the jth coordinate unit vector in $\mathbb{R}^n$.
- $\operatorname{diam}(A) = \sup\{|a - b| : a, b \in A\}$ is the diameter of $A \subset \mathbb{R}^n$.
- $d(A, B) = \inf\{|a - b| : a \in A \text{ and } b \in B\}$ is the distance between $A, B \subset \mathbb{R}^n$ and $d(x, A) = d(\{x\}, A)$ is the distance between $x \in \mathbb{R}^n$ and $A \subset \mathbb{R}^n$.
- ∂A is the boundary, $\overline{A} = A \cup \partial A$ is the closure and $A^c = \mathbb{R}^n \setminus A$ is the complement of $A \subset \mathbb{R}^n$.
- A is compactly contained in $\Omega \subset \mathbb{R}^n$, denoted $A \Subset \Omega$, if $\overline{A}$ is a compact subset of Ω.
- $|A|$ is the n-dimensional Lebesgue outer measure of $A \subset \mathbb{R}^n$.
- $B(x, r) = \{y \in \mathbb{R}^n : |y - x| < r\}$ is an open ball with center $x \in \mathbb{R}^n$ and radius $r > 0$. The corresponding closed ball is $\overline{B}(x, r) = \{y \in \mathbb{R}^n : |y - x| \leq r\}$.
- $\omega_n = |B(0, 1)|$, where $B(0, 1) \subset \mathbb{R}^n$. Then $|B(x, r)| = \omega_n r^n$ for every $x \in \mathbb{R}^n$ and $r > 0$.
- $\sigma(\partial B(0, 1)) = n\omega_n$ is the $(n-1)$-dimensional surface measure of $\partial B(0, 1) \subset \mathbb{R}^n$. Here σ is a normalized Hausdorff measure; see Remark 3.1.
- $Q(x, r) = \{y \in \mathbb{R}^n : -r \leq y_j - x_j < r, j = 1, \ldots, n\}$ is a half-open cube with center $x \in \mathbb{R}^n$ and side length $2r$ with $r > 0$.
- χ_A is the characteristic function of a set $A \subset \mathbb{R}^n$, defined by $\chi_A(x) = 1$ for $x \in A$ and $\chi_A(x) = 0$ for $x \in A^c$.
- For a function $f\colon A \to \mathbb{R}$ and $B \subset A$, the restriction of f to B is $f|_B\colon B \to \mathbb{R}$, $f|_B(x) = f(x)$ for every $x \in B$.
- $f_+(x) = \max\{f(x), 0\}$ is the positive part and $f_-(x) = \max\{-f(x), 0\}$ is the negative part of a function $f\colon A \to \mathbb{R}$.
- $\operatorname{supp} f = \overline{\{x \in A : f(x) \neq 0\}}$ is the support of a function $f\colon A \to \mathbb{R}$. If $\operatorname{supp} f$ is a compact subset of $B \subset A$, then f is compactly supported in B.
- $\Omega \subset \mathbb{R}^n$ is an open set.
- $C(\Omega) = C^0(\Omega) = \{u : \Omega \to \mathbb{R} : u \text{ is continuous}\}$.

- $C^k(\Omega) = \{u : \Omega \to \mathbb{R} : u \text{ is } k \text{ times continuously differentiable}\}$, $k \in \mathbb{N}$.
- $C_0(\Omega) = \{u \in C(\Omega) : u \text{ is compactly supported in } \Omega\}$.
- $C_0^k(\Omega) = \{u \in C^k(\Omega) : u \text{ is compactly supported in } \Omega\}$, $k \in \mathbb{N}$.
- $C^\infty(\Omega) = \bigcap_{k=0}^\infty C^k(\Omega)$, $C_0^\infty(\Omega) = \bigcap_{k=0}^\infty C_0^k(\Omega)$.
- The integral mean value of an integrable function f over a Lebesgue measurable set $A \subset \mathbb{R}^n$ with $0 < |A| < \infty$ is

$$f_A = \fint_A f(x)\,dx = \frac{1}{|A|}\int_A f(x)\,dx.$$

CHAPTER 1

Maximal Functions

Maximal functions are standard tools in real and harmonic analysis, but they are also useful in connection with Sobolev spaces and partial differential equations. This chapter discusses definitions and basic properties of maximal functions. The Hardy–Littlewood–Wiener maximal function theorem is considered in Section 1.2, the Lebesgue differentiation theorem for locally integrable functions is proved in Section 1.3 and local integrability of the maximal function is discussed in Section 1.4. Bounds for the Riesz potentials in Section 1.6 are applied in the proofs of Sobolev–Poincaré inequalities in Chapter 3.

1.1. Hardy–Littlewood maximal function

The Hardy–Littlewood maximal function gives the maximal integral average of the absolute value of the function on balls centered at a point.

DEFINITION 1.1. Assume that $f \in L^1_{\mathrm{loc}}(\mathbb{R}^n)$, that is, f is Lebesgue measurable and $\int_K |f(x)|\,dx < \infty$ for every compact set $K \subset \mathbb{R}^n$. The centered Hardy–Littlewood maximal function $Mf\colon \mathbb{R}^n \to [0,\infty]$ is defined by

$$Mf(x) = \sup_{r>0} \fint_{B(x,r)} |f(y)|\,dy$$

for every $x \in \mathbb{R}^n$.

It is possible to consider the maximal function without the absolute values, but the definition above will suffice for our purposes. It follows from the Lebesgue differentiation theorem that the absolute value of a locally integrable function is bounded from above by the maximal function almost everywhere, see Corollary 1.23. This explains the terminology.

It is enough to assume that $f\colon \mathbb{R}^n \to [-\infty,\infty]$ is a measurable function in the definition above, but the assumption $f \in L^1_{\mathrm{loc}}(\mathbb{R}^n)$ guarantees that the integral averages are finite. If the function f is not locally integrable, then the maximal function is infinite everywhere: if there is a compact set K where f is not integrable, then at every point x there is a ball $B(x,r)$ that contains K, so $Mf(x) = \infty$. Observe that $Mf(x)$ is defined at every point $x \in \mathbb{R}^n$ even if f is only defined up to a set of measure zero. Moreover, if $f(x) = g(x)$ for almost every $x \in \mathbb{R}^n$, then $Mf(x) = Mg(x)$ for every $x \in \mathbb{R}^n$.

REMARK 1.2. There are several alternative definitions for maximal functions. The noncentered maximal function of $f \in L^1_{\mathrm{loc}}(\mathbb{R}^n)$ at $x \in \mathbb{R}^n$ is

$$M^*f(x) = \sup_{B \ni x} \fint_B |f(y)|\,dy,$$

where the supremum is taken over all open balls B with $x \in B$. Then $Mf(x) \leq M^*f(x)$ for every $x \in \mathbb{R}^n$. On the other hand, if $B = B(z,r) \ni x$, then $B(z,r) \subset B(x,2r)$ and

$$\fint_B |f(y)|\,dy \leq \frac{|B(x,2r)|}{|B(z,r)|} \fint_{B(x,2r)} |f(y)|\,dy = 2^n \fint_{B(x,2r)} |f(y)|\,dy \leq 2^n Mf(x).$$

This implies $M^*f(x) \leq 2^n Mf(x)$ and thus

$$Mf(x) \leq M^*f(x) \leq 2^n Mf(x)$$

for every $x \in \mathbb{R}^n$.

Instead of balls we will sometimes use cubes in the definition of the maximal function. Such maximal function is again comparable to Mf with two-sided estimates as above. Maximal functions defined on dyadic cubes will be discussed in Chapter 8.

Recall that a function $g\colon \mathbb{R}^n \to [-\infty,\infty]$ is lower semicontinuous if the distribution set $\{x \in \mathbb{R}^n : g(x) > t\}$ is open for every $t \in \mathbb{R}$. Next we show that Mf is always lower semicontinuous and thus measurable. This also indicates that the maximal operator improves the regularity of the function.

LEMMA 1.3. *Assume that $f \in L^1_{\mathrm{loc}}(\mathbb{R}^n)$. Then Mf is lower semicontinuous.*

PROOF. Let $E_t = \{x \in \mathbb{R}^n : Mf(x) > t\}$ with $t \in \mathbb{R}$. For a fixed $x \in E_t$, there exists $r > 0$ such that

$$\fint_{B(x,r)} |f(y)|\,dy > t.$$

By the properties of the Lebesgue measure, we have

$$\fint_{B(x,r)} |f(y)|\,dy = \lim_{R\to r} \frac{1}{|B(x,R)|} \int_{B(x,r)} |f(y)|\,dy,$$

from which we conclude that there exists $R > r$ with

$$\frac{1}{|B(x,R)|} \int_{B(x,r)} |f(y)|\,dy > t.$$

If $z \in \mathbb{R}^n$ with $|x-z| < R-r$, then

$$|y-z| \leq |y-x| + |x-z| < r + (R-r) = R$$

for every $y \in B(x,r)$, and thus $B(x,r) \subset B(z,R)$. This implies

$$\begin{aligned} t &< \frac{1}{|B(x,R)|} \int_{B(x,r)} |f(y)|\,dy \leq \frac{1}{|B(x,R)|} \int_{B(z,R)} |f(y)|\,dy \\ &= \frac{1}{|B(z,R)|} \int_{B(z,R)} |f(y)|\,dy \leq Mf(z), \end{aligned}$$

whenever $|x-z| < R-r$. This shows that $B(x,R-r) \subset E_t$ and thus E_t is open. □

From the following example we see that the maximal function can be discontinuous. However, this cannot happen if the function is continuous, see Lemma 1.7 below.

EXAMPLE 1.4. Let $R > 0$ and define $f\colon \mathbb{R}^n \to \mathbb{R}$, $f(x) = \chi_{B(0,R)}(x)$ for every $x \in \mathbb{R}^n$. Then $Mf(x) = 1$ for every $x \in B(0,R)$ and $Mf(x) < 1$ for every $x \in \partial B(0,R)$. The first claim holds since $B(0,R)$ is open, and the second follows from the fact that there exists a ball $B(y,\frac{r}{2}) \subset B(x,r) \setminus B(0,R)$ if $x \in \partial B(0,R)$ and $r > 0$. This shows that Mf is not continuous at the points $x \in \partial B(0,R)$.

A few elementary properties of the Hardy–Littlewood maximal operator are collected below.

LEMMA 1.5. *Assume that $f, g \in L^1_{\mathrm{loc}}(\mathbb{R}^n)$ and let $x \in \mathbb{R}^n$. The following assertions hold:*

(a) $Mf(x) \geq 0$,
(b) $M(f+g)(x) \leq Mf(x) + Mg(x)$,
(c) $|Mf(x) - Mg(x)| \leq M(f-g)(x)$ *if* $Mf(x) < \infty$ *and* $Mg(x) < \infty$,
(d) $M(af)(x) = |a|Mf(x)$, *for every* $a \in \mathbb{R}$,
(e) $M(\tau_h f)(x) = (\tau_h Mf)(x)$, *where* $\tau_h f(x) = f(x+h)$ *with* $h \in \mathbb{R}^n$,
(f) $M(\delta_a f)(x) = (\delta_a Mf)(x)$, *where* $\delta_a f(x) = f(ax)$ *with* $a > 0$.

PROOF. Assertion (a) is clear. To prove assertion (b), we observe that

$$\begin{aligned} M(f+g)(x) &= \sup_{r>0} \fint_{B(x,r)} |f(y) + g(y)|\,dy \\ &\leq \sup_{r>0} \left(\fint_{B(x,r)} |f(y)|\,dy + \fint_{B(x,r)} |g(y)|\,dy \right) \\ &\leq \sup_{r>0} \fint_{B(x,r)} |f(y)|\,dy + \sup_{r>0} \fint_{B(x,r)} |g(y)|\,dy \\ &= Mf(x) + Mg(x). \end{aligned}$$

For the proof of (c), let $x \in \mathbb{R}^n$ be such that $Mf(x) < \infty$ and $Mg(x) < \infty$. By property (b), we have

$$Mf(x) = M(f - g + g)(x) \leq M(f-g)(x) + Mg(x).$$

By switching the roles of f and g above, we conclude

$$|Mf(x) - Mg(x)| \leq M(f-g)(x).$$

Clearly

$$M(af)(x) = \sup_{r>0} \fint_{B(x,r)} |af(y)|\,dy = |a| \sup_{r>0} \fint_{B(x,r)} |f(y)|\,dy = |a|Mf(x),$$

for every $a \in \mathbb{R}$. This proves (d), with the convention $0 \cdot \infty = 0$.

Assertion (e) follows from a change of variables

$$\begin{aligned} M(\tau_h f)(x) &= \sup_{r>0} \fint_{B(x,r)} |f(y+h)|\,dy = \sup_{r>0} \frac{1}{|B(x,r)|} \int_{B(x+h,r)} |f(z)|\,dz \\ &= \sup_{r>0} \fint_{B(x+h,r)} |f(z)|\,dz = (\tau_h Mf)(x). \end{aligned}$$

Another change of variables gives

$$\begin{aligned}M(\delta_a f)(x) &= \sup_{r>0} \fint_{B(x,r)} |f(ay)|\,dy = \sup_{r>0} \frac{1}{|B(x,r)|} \int_{B(ax,ar)} |f(z)| a^{-n}\,dz \\ &= \sup_{r>0} \fint_{B(ax,ar)} |f(z)|\,dz = (\delta_a Mf)(x),\end{aligned}$$

where in the final step we used the fact that $\{ar : r > 0\} = \{r > 0\}$. This implies (f). $\square$

EXAMPLE 1.6. Let $0 < \alpha < n$ and define $f \colon \mathbb{R}^n \setminus \{0\} \to \mathbb{R}$, $f(x) = |x|^{-\alpha}$. Let $x \in \mathbb{R}^n \setminus \{0\}$ and write $z = \frac{x}{|x|}$. By Lemma 1.5 (f) and (d), we have

$$Mf(x) = Mf(|x|z) = (\delta_{|x|} Mf)(z) = M(\delta_{|x|} f)(z) = |x|^{-\alpha} Mf(z).$$

Thus $Mf(x) = Mf(z)|x|^{-\alpha}$ for every $x \in \mathbb{R}^n \setminus \{0\}$, where $z \in \partial B(0,1)$. Since f is radial, the value $Mf(z)$ is independent of $z \in \partial B(0,1)$. Moreover,

$$\begin{aligned}Mf(z) &= \sup_{r>0} \fint_{B(z,r)} |f(y)|\,dy \le \sup_{0<r<\frac12} \fint_{B(z,r)} |f(y)|\,dy + \sup_{r\ge\frac12} \fint_{B(z,r)} |f(y)|\,dy \\ &\le \left(\frac12\right)^{-\alpha} + C(n) \sup_{r\ge\frac12} \fint_{B(0,r+1)} |y|^{-\alpha}\,dy \\ &\le 2^\alpha + C(n) \sup_{r\ge\frac12} \frac{1}{(r+1)^n} \int_0^{r+1} \rho^{-\alpha} \rho^{n-1}\,d\rho \\ &\le 2^\alpha + C(n,\alpha) \sup_{r\ge\frac12} (r+1)^{-\alpha} < \infty.\end{aligned}$$

This shows that Mf is a constant multiple of f. This is closely related to the fact that f is a Muckenhoupt A_1-weight, see Chapter 8 and Example 10.27.

In contrast to Example 1.4, the maximal function of a continuous function is continuous under a finiteness assumption. This property will be further strengthened in Remark 1.18.

LEMMA 1.7. *Assume that f is a continuous function in $\mathbb{R}^n$. If $Mf(x) < \infty$ for every $x \in \mathbb{R}^n$, then Mf is continuous in $\mathbb{R}^n$.*

PROOF. Let $x, h \in \mathbb{R}^n$. By the finiteness assumption and properties (c), (b) and (e) in Lemma 1.5, we have

$$|M(\tau_h f)(x) - Mf(x)| \le M(\tau_h f - f)(x) \le Mf(x+h) + Mf(x) < \infty.$$

Let $\varepsilon > 0$. Then there exists a ball $B(x,r)$ such that

$$M(\tau_h f - f)(x) \le \fint_{B(x,r)} |\tau_h f(y) - f(y)|\,dy + \frac{\varepsilon}{2}.$$

Since f is uniformly continuous in $B(x, r+1)$, there is $0 < \delta < 1$ such that

$$|\tau_h f(y) - f(y)| = |f(y+h) - f(y)| < \frac{\varepsilon}{2}$$

whenever $y \in B(x,r)$ and $|h| < \delta$. This implies

$$|M(\tau_h f)(x) - Mf(x)| \le M(\tau_h f - f)(x) \le \fint_{B(x,r)} |\tau_h f(y) - f(y)|\,dy + \frac{\varepsilon}{2} < \varepsilon$$

whenever $|h| < \delta$, and by Lemma 1.5 (e) we obtain

$$|Mf(x+h) - Mf(x)| = |(\tau_h Mf)(x) - Mf(x)| = |M(\tau_h f)(x) - Mf(x)| < \varepsilon$$

whenever $|h| < \delta$. This shows that Mf is continuous. $\square$

EXAMPLE 1.8. The function $f(x) = |x|$ is Lipschitz continuous in $\mathbb{R}^n$, see Definition 2.1, but $Mf(x) = \infty$ for every $x \in \mathbb{R}^n$. Thus the maximal function may be infinite everywhere, even if the function is finite everywhere. This explains the finiteness assumption in Lemma 1.7.

We conclude this section by giving a two-sided pointwise estimate for the maximal function of the characteristic function of a ball.

EXAMPLE 1.9. Let $R > 0$. We will show that

$$\frac{R^n}{(|x|+R)^n} \le M(\chi_{B(0,R)})(x) \le \frac{C(n)R^n}{(|x|+R)^n} \tag{1.1}$$

for every $x \in \mathbb{R}^n$. In order to prove (1.1), let $x \in \mathbb{R}^n$. Lemma 1.5 (f) implies

$$M(\chi_{B(0,R)})(x) = M(\delta_{R^{-1}}\chi_{B(0,1)})(x) = (\delta_{R^{-1}} M\chi_{B(0,1)})(x).$$

Therefore, it suffices to prove (1.1) for $R = 1$.

First assume that $|x| > 2$ and let $y \in B(0,1)$. Then $|y - x| \le |y| + |x| < 1 + |x|$, and consequently $B(0,1) \subset B(x, |x|+1)$. Moreover, if $|x - y| < |x| - 1$, then $|y| \ge |x| - |x-y| \ge 1$ and $B(x, |x|-1) \cap B(0,1) = \emptyset$. From this we conclude that

$$\begin{aligned}
\frac{|B(0,1)|}{|B(x,|x|+1)|} &= \fint_{B(x,|x|+1)} \chi_{B(0,1)}(y)\,dy \le M(\chi_{B(0,1)})(x) \\
&\le \sup_{\rho > |x|-1} \fint_{B(x,\rho)} \chi_{B(0,1)}(y)\,dy + \sup_{|x|-1 \ge \rho > 0} \fint_{B(x,\rho)} \chi_{B(0,1)}(y)\,dy \\
&= \sup_{\rho > |x|-1} \frac{1}{|B(x,\rho)|} \int_{B(x,\rho)} \chi_{B(0,1)}(y)\,dy \\
&\le \sup_{\rho > |x|-1} \frac{|B(0,1)|}{|B(x,\rho)|} = \frac{|B(0,1)|}{|B(x,|x|-1)|}.
\end{aligned}$$

Since $|x| > 2$, we have

$$|x| + 1 = |x| + 4 - 3 < 3|x| - 3 = 3(|x| - 1),$$

which implies

$$\frac{|B(x,|x|+1)|}{|B(x,|x|-1)|} = \left(\frac{|x|+1}{|x|-1}\right)^n \le 3^n.$$

Thus

$$\frac{|B(0,1)|}{|B(x,|x|+1)|} \le M(\chi_{B(0,1)})(x) \le \frac{|B(0,1)|}{|B(x,|x|-1)|} \le 3^n \frac{|B(0,1)|}{|B(x,|x|+1)|},$$

and this proves (1.1) in the case $|x| > 2$.

On the other hand, if $|x| \le 2$, then

$$\frac{|B(0,1)|}{|B(x,|x|+1)|} \ge \frac{|B(0,1)|}{|B(x,3)|} = 3^{-n}.$$

Thus by Lemma 1.10

$$\frac{|B(0,1)|}{|B(x,|x|+1)|} \leq M(\chi_{B(0,1)})(x) \leq 1 \leq 3^n \frac{|B(0,1)|}{|B(x,|x|+1)|},$$

and this completes the proof of (1.1) with $C(n) = 3^n$.

1.2. Hardy–Littlewood–Wiener maximal function theorem

This section discusses bounds for the maximal function on $L^p(\mathbb{R}^n)$. Recall that for a Lebesgue measurable set $A \subset \mathbb{R}^n$ and $1 \leq p \leq \infty$, the space $L^p(A)$ consists of Lebesgue measurable functions $f\colon A \to [-\infty, \infty]$ with $\|f\|_{L^p(A)} < \infty$, where

$$\|f\|_p = \|f\|_{L^p(A)} = \left(\int_A |f(x)|^p \, dx\right)^{\frac{1}{p}}, \qquad \text{for } 1 \leq p < \infty,$$

and

$$\|f\|_\infty = \|f\|_{L^\infty(A)} = \operatorname*{ess\,sup}_{x\in A} |f(x)|.$$

The space $L^p(A)$ becomes a Banach space when we identify two functions that differ on a set of measure zero. In addition, $L^p_{\mathrm{loc}}(A)$ is the space of functions $f\colon A \to [-\infty, \infty]$ with $f \in L^p(K)$ for every compact set $K \subset A$.

First we show that the maximal function is essentially bounded if the function is essentially bounded. Intuitively this is clear since the integral averages in the maximal function cannot be larger than the essential supremum of the function.

Lemma 1.10. *Assume that $f \in L^\infty(\mathbb{R}^n)$. Then $Mf \in L^\infty(\mathbb{R}^n)$ and*

$$Mf(x) \leq \|f\|_{L^\infty(\mathbb{R}^n)}$$

for every $x \in \mathbb{R}^n$. In particular,

$$\|Mf\|_{L^\infty(\mathbb{R}^n)} \leq \|f\|_{L^\infty(\mathbb{R}^n)}.$$

Proof. For every $x \in \mathbb{R}^n$ and $r > 0$, we have

$$\fint_{B(x,r)} |f(y)| \, dy \leq \|f\|_{L^\infty(\mathbb{R}^n)}.$$

Thus $Mf(x) \leq \|f\|_{L^\infty(\mathbb{R}^n)}$ for every $x \in \mathbb{R}^n$. □

Lemma 1.10 asserts that $M\colon L^\infty(\mathbb{R}^n) \to L^\infty(\mathbb{R}^n)$, $f \mapsto Mf$, is a bounded operator. In contrast, $f \in L^1(\mathbb{R})$ does not imply $Mf \in L^1(\mathbb{R})$.

Example 1.11. Example 1.9, with $R = 1$, shows that

$$M(\chi_{B(0,1)})(x) \geq \frac{1}{(|x|+1)^n}$$

for every $x \in \mathbb{R}^n$. This implies $M(\chi_{B(0,1)}) \notin L^1(\mathbb{R}^n)$. Thus the Hardy–Littlewood maximal operator is not bounded on $L^1(\mathbb{R}^n)$.

Remark 1.12. In fact, the maximal function is in $L^1(\mathbb{R}^n)$ if and only if the function is zero almost everywhere. In other words, $Mf \notin L^1(\mathbb{R}^n)$ for every nonzero $f \in L^1_{\mathrm{loc}}(\mathbb{R}^n)$. To see this, assume first that f is not zero almost everywhere and choose $R > 0$ to be large enough that

$$\int_{B(0,R)} |f(y)| \, dy > 0.$$

For $x \in \mathbb{R}^n$ with $|x| \geq R$ we have

$$\begin{aligned} Mf(x) &\geq \fint_{B(x,2|x|)} |f(y)|\,dy \geq \frac{1}{|B(x,2|x|)|}\int_{B(0,R)} |f(y)|\,dy \\ &= \frac{C(n)}{|x|^n}\int_{B(0,R)} |f(y)|\,dy, \end{aligned}$$

and thus $Mf \notin L^1(\mathbb{R}^n)$. On the other hand, if $f(x) = 0$ for almost every $x \in \mathbb{R}^n$, then $Mf(x) = 0$ for every $x \in \mathbb{R}^n$ and thus $Mf \in L^1(\mathbb{R}^n)$.

The following Vitali-type $5r$-covering lemma will be used several times throughout this book, for instance in the proof of the maximal function theorem, Theorem 1.15. For a proof, see Mattila [**315**, p. 23].

LEMMA 1.13. *Let $\mathcal{B}$ be a collection of balls $B(x, r_x) \subset \mathbb{R}^n$, with the radius $r_x > 0$ possibly depending on x. Assume that the balls in $\mathcal{B}$ have uniformly bounded radii. Then there exists a countable subcollection of pairwise disjoint balls $B(x_i, r_i) \in \mathcal{B}$, $i \in \mathbb{N}$, such that*

$$\bigcup_{B\in\mathcal{B}} B \subset \bigcup_{i=1}^{\infty} B(x_i, 5r_i).$$

We remark that in Lemma 1.13 and in many corresponding statements the countable subcollection may be finite.

We discuss briefly how the $5r$-covering lemma is applied in practice. Let $\mathcal{B}$ be a collection of balls $B(x, r_x)$ with $0 < r_x < R$. We would like to extract a countable subcollection of pairwise disjoint balls that covers the union of the original balls. In general, this is not possible without expanding the balls as in Lemma 1.13. Let $B(x_i, r_i) \in \mathcal{B}$, $i \in \mathbb{N}$, be a countable subcollection of pairwise disjoint balls given by Lemma 1.13. Then

$$\begin{aligned} \Big|\bigcup_{B\in\mathcal{B}} B\Big| &\leq \Big|\bigcup_{i=1}^{\infty} B(x_i, 5r_i)\Big| \leq \sum_{i=1}^{\infty} |B(x_i, 5r_i)| \\ &= 5^n \sum_{i=1}^{\infty} |B(x_i, r_i)| = 5^n \Big|\bigcup_{i=1}^{\infty} B(x_i, r_i)\Big| \leq 5^n \Big|\bigcup_{B\in\mathcal{B}} B\Big|. \end{aligned}$$

Thus the measures of $\bigcup_{B\in\mathcal{B}} B$, $\bigcup_{i=1}^{\infty} B(x_i, 5r_i)$ and $\bigcup_{i=1}^{\infty} B(x_i, r_i)$ are comparable, and if one of these measures is infinite, then all of them are infinite. Moreover, if $x \in B \in \mathcal{B}$, then there exists $i \in \mathbb{N}$ such that $x \in B(x_i, 5r_i)$.

Also the following Cavalieri's principle and its proof will be applied several times in the book, see Hewitt and Stromberg [**191**, Theorem 21.71].

LEMMA 1.14. *Let $0 < p < \infty$. Assume that $A \subset \mathbb{R}^n$ is a measurable set and that $f\colon A \to [-\infty, \infty]$ is a measurable function. Then*

$$\int_A |f(x)|^p\,dx = p\int_0^{\infty} t^{p-1}|\{x \in A : |f(x)| > t\}|\,dt.$$

PROOF. By Fubini's theorem

$$\begin{aligned}\int_A |f(x)|^p\,dx &= \int_{\mathbb{R}^n} \chi_A(x) p \int_0^{|f(x)|} t^{p-1}\,dt\,dx \\ &= p\int_{\mathbb{R}^n}\int_0^\infty \chi_A(x)\chi_{[0,|f(x)|)}(t)t^{p-1}\,dt\,dx \\ &= p\int_0^\infty\int_{\mathbb{R}^n} \chi_A(x)\chi_{[0,|f(x)|)}(t)t^{p-1}\,dx\,dt \\ &= p\int_0^\infty t^{p-1}\int_{\mathbb{R}^n} \chi_A(x)\chi_{\{x\in\mathbb{R}^n:|f(x)|>t\}}(x)\,dx\,dt \\ &= p\int_0^\infty t^{p-1}|\{x\in A:|f(x)|>t\}|\,dt. \qquad \square\end{aligned}$$

Next we state the Hardy–Littlewood–Wiener maximal function theorem, which asserts that $M\colon L^p(\mathbb{R}^n)\to L^p(\mathbb{R}^n)$, $f\mapsto Mf$, is a bounded operator for $1<p\le\infty$. The corresponding result does not hold for $p=1$, but in this case we have a weak type estimate. The proof contains similar elements as the proof of the Marcinkiewicz interpolation theorem.

THEOREM 1.15. *Let $1\le p\le\infty$ and assume that $f\in L^p(\mathbb{R}^n)$.*

(a) *Then Mf is finite almost everywhere.*

(b) *If $p=1$, there exists a constant $C=C(n)$ such that*

$$|\{x\in\mathbb{R}^n : Mf(x)>t\}| \le \frac{C}{t}\int_{\mathbb{R}^n}|f(x)|\,dx$$

for every $t>0$.

(c) *If $1<p\le\infty$, then $Mf\in L^p(\mathbb{R}^n)$ and there exists a constant $C=C(n,p)$ such that*

$$\|Mf\|_{L^p(\mathbb{R}^n)} \le C\|f\|_{L^p(\mathbb{R}^n)}.$$

PROOF. Assume that $f\in L^1(\mathbb{R}^n)$ and let $E_t=\{x\in\mathbb{R}^n : Mf(x)>t\}$ for $t>0$. By the definition of Mf, for every $x\in E_t$ there exists $r_x>0$ such that

$$\int_{B(x,r_x)}|f(y)|\,dy > t|B(x,r_x)| = t\omega_n r_x^n. \tag{1.2}$$

This implies

$$r_x^n \le \frac{1}{\omega_n t}\int_{B(x,r_x)}|f(y)|\,dy \le \frac{1}{\omega_n t}\int_{\mathbb{R}^n}|f(y)|\,dy<\infty.$$

Thus $\mathcal{B}=\{B(x,r_x): x\in E_t\}$ is a collection of balls with uniformly bounded radii and $E_t\subset\bigcup_{B\in\mathcal{B}}B$. The $5r$-covering lemma, Lemma 1.13, gives pairwise disjoint balls $B(x_i,r_i)\in\mathcal{B}$, $i\in\mathbb{N}$, such that

$$|E_t|\le\sum_{i=1}^\infty|B(x_i,5r_i)| = 5^n\sum_{i=1}^\infty|B(x_i,r_i)|.$$

Thus

$$|E_t|\le 5^n\sum_{i=1}^\infty|B(x_i,r_i)| \le \frac{5^n}{t}\sum_{i=1}^\infty\int_{B(x_i,r_i)}|f(y)|\,dy \le \frac{5^n}{t}\int_{\mathbb{R}^n}|f(y)|\,dy.$$

This proves assertion (b) with $C(n) = 5^n$, and assertion (a) for $p = 1$. The latter claim follows by taking $t \to \infty$.

Next we prove assertion (a) and assertion (c) for $1 < p \le \infty$. The case $p = \infty$ follows from Lemma 1.10, and thus we may assume that $1 < p < \infty$. Let $f \in L^p(\mathbb{R}^n)$ and $t > 0$, and define

$$\widetilde{f}(x) = \begin{cases} f(x), & \text{if } |f(x)| \ge \frac{t}{2}, \\ 0, & \text{otherwise.} \end{cases}$$

Then $|f| \le |\widetilde{f}| + \frac{t}{2}$, which implies by Lemma 1.5 (b) that $Mf(x) \le M\widetilde{f}(x) + \frac{t}{2}$ for every $x \in \mathbb{R}^n$. Consequently,

$$\{x \in \mathbb{R}^n : Mf(x) > t\} \subset \Big\{x \in \mathbb{R}^n : M\widetilde{f}(x) > \frac{t}{2}\Big\}.$$

Observe that

$$|\widetilde{f}(x)| = \chi_{\{y \in \mathbb{R}^n :\, \widetilde{f}(y) \neq 0\}}(x) |\widetilde{f}(x)|^{1-p} |\widetilde{f}(x)|^p \le \left(\frac{t}{2}\right)^{1-p} |f(x)|^p,$$

which implies $\widetilde{f} \in L^1(\mathbb{R}^n)$. By (b) we have

$$\begin{aligned} |\{x \in \mathbb{R}^n : Mf(x) > t\}| &\le \frac{2 \cdot 5^n}{t} \|\widetilde{f}\|_{L^1(\mathbb{R}^n)} \\ &= \frac{2 \cdot 5^n}{t} \int_{\{x \in \mathbb{R}^n :\, |f(x)| \ge \frac{t}{2}\}} |f(x)|\, dx. \end{aligned} \tag{1.3}$$

By Lemma 1.14 and its proof, we obtain

$$\begin{aligned} \int_{\mathbb{R}^n} (Mf(x))^p\, dx &= p \int_0^\infty t^{p-1} \big|\{x \in \mathbb{R}^n : Mf(x) > t\}\big|\, dt \\ &\le p 2 \cdot 5^n \int_0^\infty t^{p-2} \int_{\{x \in \mathbb{R}^n :\, |f(x)| \ge \frac{t}{2}\}} |f(x)|\, dx\, dt \\ &= p 2 \cdot 5^n \int_{\mathbb{R}^n} |f(x)| \int_0^{2|f(x)|} t^{p-2}\, dt\, dx \\ &= p 2 \cdot 5^n \frac{2^{p-1}}{p-1} \int_{\mathbb{R}^n} |f(x)| |f(x)|^{p-1}\, dt\, dx. \end{aligned} \tag{1.4}$$

This proves assertion (c) with $1 < p < \infty$ and

$$C(n,p) = 2 \cdot \left(\frac{5^n p}{p-1}\right)^{\frac{1}{p}}.$$

Finally, assertion (a) for $1 < p < \infty$ holds since $Mf \in L^p(\mathbb{R}^n)$. □

REMARK 1.16. The centered maximal function M is a nonlinear operator. In general, the boundedness of a nonlinear operator does not imply its continuity. However, since M is a sublinear operator which is bounded on $L^p(\mathbb{R}^n)$, it is also continuous on $L^p(\mathbb{R}^n)$ for $1 < p \le \infty$. To see this, let $f, g \in L^p(\mathbb{R}^n)$. By Lemma 1.5 (c) and Theorem 1.15 (a), we have

$$|Mf(x) - Mg(x)| \le M(f-g)(x)$$

for almost every $x \in \mathbb{R}^n$. Theorem 1.15 (c) implies

$$\|Mf - Mg\|_{L^p(\mathbb{R}^n)} \le \|M(f-g)\|_{L^p(\mathbb{R}^n)} \le C(n,p) \|f-g\|_{L^p(\mathbb{R}^n)}.$$

The continuity of M on $L^p(\mathbb{R}^n)$ is a consequence of this estimate. When $p = 1$, the argument above shows that

$$\begin{aligned}\big|\{x \in \mathbb{R}^n : |Mf(x) - Mg(x)| > t\}\big| &\le \big|\{x \in \mathbb{R}^n : M(f-g)(x) > t\}\big| \\ &\le \frac{C(n)}{t} \int_{\mathbb{R}^n} |f(x) - g(x)|\, dx\end{aligned}$$

whenever $t > 0$ and $f, g \in L^1(\mathbb{R}^n)$. In particular, it follows that if $(f_i)_{i\in\mathbb{N}}$ converges to f in $L^1(\mathbb{R}^n)$, then Mf_i converges to Mf in measure as $i \to \infty$.

It is clear that $f(x) = 0$ for almost every $x \in \mathbb{R}^n$ if and only if there exists a point $x_0 \in \mathbb{R}^n$ such that $Mf(x_0) = 0$. Consequently, $Mf(x) = 0$ for every $x \in \mathbb{R}^n$ if $Mf(x_0) = 0$ for some $x_0 \in \mathbb{R}^n$. Next we show that if the maximal function of a locally integrable function is finite at one point, then it is finite almost everywhere.

LEMMA 1.17. *Assume that $f \in L^1_{\mathrm{loc}}(\mathbb{R}^n)$. If there exists a point $x_0 \in \mathbb{R}^n$ such that $Mf(x_0) < \infty$, then $Mf(x) < \infty$ for almost every $x \in \mathbb{R}^n$.*

PROOF. Let $f_k = f\chi_{B(0,k)}$ for $k \in \mathbb{N}$. Then $f_k \in L^1(\mathbb{R}^n)$, and it follows from Theorem 1.15 (a) that for every $k \in \mathbb{N}$ we have $Mf_k(x) < \infty$ for almost every $x \in \mathbb{R}^n$. Hence if

$$E = \bigcup_{k=1}^{\infty} \{x \in \mathbb{R}^n : Mf_k(x) = \infty\},$$

then $|E| = 0$.

We claim that $Mf(x) < \infty$ for every $x \in \mathbb{R}^n \setminus E$. Let $x \in \mathbb{R}^n \setminus E$ and choose $k \in \mathbb{N}$ large enough that $B(x,1) \subset B(0,k)$. If $0 < r \le 1$, then

$$\fint_{B(x,r)} |f(y)|\, dy = \fint_{B(x,r)} |f_k(y)|\, dy \le Mf_k(x) < \infty.$$

Then assume that $r > 1$. Since $B(x,r) \subset B(x_0,r_0)$ with $r_0 = r + |x - x_0|$, we have

$$\begin{aligned}\fint_{B(x,r)} |f(y)|\, dy &\le \frac{|B(x_0,r_0)|}{|B(x,r)|} \fint_{B(x_0,r_0)} |f(y)|\, dy \le \frac{|B(x_0,r_0)|}{|B(x,r)|} Mf(x_0) \\ &= \left(\frac{r + |x - x_0|}{r}\right)^n Mf(x_0) \le (1 + |x - x_0|)^n Mf(x_0) < \infty.\end{aligned}$$

This implies

$$Mf(x) = \sup_{r>0} \fint_{B(x,r)} |f(y)|\, dy \le Mf_k(x) + (1 + |x - x_0|)^n Mf(x_0) < \infty,$$

which shows that $Mf(x) < \infty$ for every $x \in \mathbb{R}^n \setminus E$. □

REMARK 1.18. Assume that f is a continuous function in $\mathbb{R}^n$. If there exists a point $x_0 \in \mathbb{R}^n$ with $Mf(x_0) < \infty$, then the proof of Lemma 1.17 shows that $Mf(x) < \infty$ for every $x \in \mathbb{R}^n$. Thus Mf is a continuous function in $\mathbb{R}^n$, by Lemma 1.7.

1.3. Lebesgue differentiation theorem

We discuss the Lebesgue differentiation theorem as an application of the maximal function theorem. The core of the argument is that the weak type estimate implies almost everywhere convergence of integral averages. This is based on the fact that such convergence holds for a dense class of continuous functions.

THEOREM 1.19. *Let $1 \le p < \infty$ and assume that $f \in L^p_{\mathrm{loc}}(\mathbb{R}^n)$. Then*

$$\lim_{r\to 0} \fint_{B(x,r)} |f(y) - f(x)|^p \, dy = 0 \tag{1.5}$$

for almost every $x \in \mathbb{R}^n$.

PROOF. By considering the functions $f\chi_{B(0,k)}$, for $k \in \mathbb{N}$, we may assume that $f \in L^p(\mathbb{R}^n)$. Let

$$E_t = \Big\{ x \in \mathbb{R}^n : \limsup_{r\to 0} \fint_{B(x,r)} |f(y) - f(x)|^p \, dy > 2 \cdot 3^p t \Big\},$$

for $t > 0$. Then

$$\Big\{ x \in \mathbb{R}^n : \limsup_{r\to 0} \fint_{B(x,r)} |f(y) - f(x)|^p \, dy > 0 \Big\} = \bigcup_{j=1}^{\infty} E_{\frac{1}{j}}.$$

Hence, it suffices to show that $|E_t| = 0$ for every $t > 0$.

Let $t > 0$, $x \in E_t$ and $\varepsilon > 0$, and let $g \colon \mathbb{R}^n \to \mathbb{R}$ be a continuous function satisfying $\|f - g\|_{L^p(\mathbb{R}^n)} < \varepsilon$. Since g is continuous, we have

$$\limsup_{r\to 0} \fint_{B(x,r)} |g(y) - g(x)|^p \, dy = 0,$$

and thus

$$\begin{aligned}
2 \cdot 3^p t &< \limsup_{r\to 0} \fint_{B(x,r)} |f(y) - f(x)|^p \, dy \\
&\le 3^p \limsup_{r\to 0} \Big(\fint_{B(x,r)} |f(y) - g(y)|^p \, dy + \fint_{B(x,r)} |g(y) - g(x)|^p \, dy \Big) \\
&\quad + 3^p |g(x) - f(x)|^p \\
&= 3^p \limsup_{r\to 0} \fint_{B(x,r)} |f(y) - g(y)|^p \, dy + 3^p |f(x) - g(x)|^p \\
&\le 3^p M(|f - g|^p)(x) + 3^p |f(x) - g(x)|^p.
\end{aligned}$$

It follows that if $x \in E_t$, then

$$M(|f - g|^p)(x) > t \quad \text{or} \quad |f(x) - g(x)|^p > t.$$

By Theorem 1.15 (b) and Chebyshev's inequality, we obtain

$$\begin{aligned}
|E_t| &\le |\{x \in \mathbb{R}^n : M(|f - g|^p)(x) > t\}| + |\{x \in \mathbb{R}^n : |f(x) - g(x)|^p > t\}| \\
&\le \frac{C(n)}{t} \int_{\mathbb{R}^n} |f(x) - g(x)|^p \, dx + \frac{1}{t} \int_{\mathbb{R}^n} |f(x) - g(x)|^p \, dx \le \frac{C(n) + 1}{t} \varepsilon^p.
\end{aligned}$$

Letting $\varepsilon \to 0$ shows that $|E_t| = 0$. $\square$

DEFINITION 1.20. We say that $x \in \mathbb{R}^n$ is a Lebesgue point of $f \in L^1_{\mathrm{loc}}(\mathbb{R}^n)$ if the limit

$$f^*(x) = \lim_{r\to 0} \fint_{B(x,r)} f(y) \, dy \tag{1.6}$$

exists as a real number and

$$\lim_{r\to 0} \fint_{B(x,r)} |f(y) - f^*(x)| \, dy = 0.$$

The set of Lebesgue points of f is denoted by $\mathrm{Leb}(f)$.

In (1.6) one might be tempted to write $f(x)$ instead of $f^*(x)$, but the equivalence class of $f \in L^1_{\mathrm{loc}}(\mathbb{R}^n)$ is only defined up to a set of measure zero. If $f = g$ almost everywhere, then f and g have the same set of Lebesgue points and $f^*(x) = g^*(x)$ for every Lebesgue point x of f and g. Thus the notion of Lebesgue point is independent of the representative in the equivalence class in $L^1_{\mathrm{loc}}(\mathbb{R}^n)$. Whether x is a Lebesgue point of f is completely independent of the value $f(x)$, and the function f does not even need to be defined at x.

The following Lebesgue differentiation theorem asserts that a locally integrable function has Lebesgue points almost everywhere. It allows us to pass from information on averages to pointwise information outside a set of measure zero.

THEOREM 1.21. *Assume that* $f \in L^1_{\mathrm{loc}}(\mathbb{R}^n)$. *Then* $|\mathbb{R}^n \setminus \mathrm{Leb}(f)| = 0$. *Moreover, there exists a set* $E \subset \mathrm{Leb}(f)$ *such that* $|\mathbb{R}^n \setminus E| = 0$ *and* $f^*(x) = f(x)$ *for every* $x \in E$.

PROOF. Let

$$E = \left\{ x \in \mathbb{R}^n : \lim_{r\to 0} \fint_{B(x,r)} |f(y) - f(x)|\,dy = 0 \right\}.$$

Theorem 1.19 implies $|\mathbb{R}^n \setminus E| = 0$. If $x \in E$, then

$$\begin{aligned} 0 &\le \left| \fint_{B(x,r)} f(y)\,dy - f(x) \right| = \left| \fint_{B(x,r)} (f(y) - f(x))\,dy \right| \\ &\le \fint_{B(x,r)} |f(y) - f(x)|\,dy \xrightarrow{r\to 0} 0. \end{aligned}$$

This implies

$$f^*(x) = \lim_{r\to 0} \fint_{B(x,r)} f(y)\,dy = f(x)$$

and

$$\fint_{B(x,r)} |f(y) - f^*(x)|\,dy = \fint_{B(x,r)} |f(y) - f(x)|\,dy \xrightarrow{r\to 0} 0,$$

for every $x \in E$. It follows that $E \subset \mathrm{Leb}(f)$, and thus $|\mathbb{R}^n \setminus \mathrm{Leb}(f)| \le |\mathbb{R}^n \setminus E| = 0$. This completes the proof. $\square$

If f is a pointwise defined function in an equivalence class in $L^1_{\mathrm{loc}}(\mathbb{R}^n)$, then by the Lebesgue differentiation theorem the limit

$$f^*(x) = \lim_{r\to 0} \fint_{B(x,r)} f(y)\,dy$$

equals $f(x)$ for almost every $x \in \mathbb{R}^n$. It follows that f^* belongs to the same equivalence class as f in $L^1_{\mathrm{loc}}(\mathbb{R}^n)$. In arguments related to Lebesgue points we use for $f \in L^1_{\mathrm{loc}}(\mathbb{R}^n)$ the representative f^*, unless otherwise specified.

EXAMPLE 1.22. Let $f\colon \mathbb{R} \to \mathbb{R}$ be the Heaviside function

$$f(x) = \begin{cases} 1, & x > 0, \\ \frac{1}{2}, & x = 0, \\ 0, & x < 0. \end{cases}$$

Then

$$\lim_{r\to 0} \frac{1}{2r} \int_{x-r}^{x+r} f(y)\,dy = f(x)$$

for every $x \in \mathbb{R}$, but $x = 0$ is not a Lebesgue point of f. Indeed, we have

$$\frac{1}{2r}\int_{-r}^{r}|f(y)-a|\,dy = \frac{1}{2r}\int_{-r}^{0}|a|\,dy + \frac{1}{2r}\int_{0}^{r}|1-a|\,dy = \frac{1}{2}|a| + \frac{1}{2}|1-a| \neq 0$$

for every $a \in \mathbb{R}$ and $r > 0$.

As a simple corollary of the Lebesgue differentiation theorem we obtain pointwise estimate for a function by its maximal function.

COROLLARY 1.23. *Assume that $f \in L^1_{\mathrm{loc}}(\mathbb{R}^n)$. Then $|f(x)| \le Mf(x)$ for almost every $x \in \mathbb{R}^n$.*

PROOF. Theorem 1.21 implies

$$|f(x)| = |f^*(x)| = \lim_{r\to 0}\left|\fint_{B(x,r)} f(y)\,dy\right| \le \sup_{r>0}\fint_{B(x,r)}|f(y)|\,dy = Mf(x),$$

for almost every $x \in \mathbb{R}^n$. □

Another consequence of Theorem 1.21 is the following Lebesgue density theorem.

THEOREM 1.24. *Assume that $A \subset \mathbb{R}^n$ is a measurable set. Then*

$$\lim_{r\to 0}\frac{|A\cap B(x,r)|}{|B(x,r)|} = \begin{cases} 1, & \text{for a.e. } x \in A,\\ 0, & \text{for a.e. } x \notin A.\end{cases}$$

PROOF. Consider the function $f(x) = \chi_A(x)$ for every $x \in \mathbb{R}^n$, and let $E \subset \mathbb{R}^n$ be the corresponding set from Theorem 1.21. Then we have, for every $x \in A \cap E$,

$$1 = f(x) = f^*(x) = \lim_{r\to 0}\fint_{B(x,r)} f(y)\,dy = \lim_{r\to 0}\frac{|A\cap B(x,r)|}{|B(x,r)|}.$$

Similar reasoning shows that the limit is 0 for every $x \in A^c \cap E$, and this proves the claim. □

1.4. A theorem of Stein

The following lemma gives a reverse weak type estimate, which is a converse of Theorem 1.15 (b). Reverse weak type estimates play a central role in self-improvement of reverse Hölder inequalities in Chapter 8.

LEMMA 1.25. *Assume that $f \in L^1(\mathbb{R}^n)$. There exists a constant $C = C(n)$ such that*

$$\frac{1}{t}\int_{\{x\in\mathbb{R}^n : Mf(x)>t\}}|f(x)|\,dx \le C|\{x\in\mathbb{R}^n : Mf(x) > t\}|$$

for every $t > 0$.

PROOF. Let $t > 0$. By Lemma 1.3, the set $E_t = \{x \in \mathbb{R}^n : Mf(x) > t\}$ is open. Theorem 1.15 (b) implies that $|E_t| < \infty$, and thus $\mathbb{R}^n \setminus E_t \neq \emptyset$. Let $K \subset E_t$ be a compact set and define $r_x = d(x, \partial E_t) > 0$ for $x \in K$. Then $\mathcal{B} = \{B(x, r_x) : x \in K\}$ is a cover of K. For every $x \in K$ we have $B(x, r_x) \subset E_t$ and

$$0 < r_x \le \operatorname{diam}(K) + d(K, \partial E_t) < \infty.$$

Let $x \in K$. Since $B(x, 2r_x) \cap (\mathbb{R}^n \setminus E_t) \neq \emptyset$, there exists a point $z \in B(x, 2r_x) \setminus E_t$ with $Mf(z) \leq t$. Thus

$$\fint_{B(z,r)} |f(y)| \, dy \leq t$$

for every $r > 0$ and, in particular,

$$\fint_{B(z,7r_x)} |f(y)| \, dy \leq t.$$

Since $B(x, 5r_x) \subset B(z, 7r_x)$, we have

$$\fint_{B(x,5r_x)} |f(y)| \, dy \leq C(n) \fint_{B(z,7r_x)} |f(y)| \, dy \leq C(n)t$$

with $C(n) = (\frac{7}{5})^n$. Lemma 1.13 gives pairwise disjoint balls $B(x_i, r_i) \in \mathcal{B}$, $i \in \mathbb{N}$, satisfying $K \subset \bigcup_{i=1}^{\infty} B(x_i, 5r_i)$. This implies

$$\begin{aligned} \int_K |f(y)| \, dy &\leq \sum_{i=1}^{\infty} \int_{B(x_i,5r_i)} |f(y)| \, dy \leq C(n)t \sum_{i=1}^{\infty} |B(x_i, 5r_i)| \\ &= C(n)t \sum_{i=1}^{\infty} |B(x_i, r_i)| \leq C(n)t|E_t|. \end{aligned}$$

The claim follows since

$$\int_{E_t} |f(y)| \, dy = \sup_{K \subset E_t} \int_K |f(y)| \, dy \leq C(n)t|E_t|$$

with $C(n) = 7^n$, where the supremum is taken over all compact subsets K of E_t. □

By Remark 1.12, $Mf \in L^1(\mathbb{R}^n)$ only if $f \in L^1_{\mathrm{loc}}(\mathbb{R}^n)$ is zero almost everywhere. However, the maximal function Mf can be locally integrable even if f is not zero almost everywhere. Next we characterize the local integrability of Mf for compactly supported functions $f \in L^1(\mathbb{R}^n)$ by Zygmund's class $L \log^+ L(\mathbb{R}^n)$. We say that a measurable function $f \colon \mathbb{R}^n \to [-\infty, \infty]$ belongs to $L \log^+ L(\mathbb{R}^n)$ if

$$\int_{\mathbb{R}^n} |f(x)| \log^+ |f(x)| \, dx < \infty.$$

Here $\log^+ |f| = \chi_{\{x \in \mathbb{R}^n : |f(x)| > 1\}} \log |f|$.

THEOREM 1.26. *Assume that $f \in L^1(\mathbb{R}^n)$ is a compactly supported function. Then $Mf \in L^1_{\mathrm{loc}}(\mathbb{R}^n)$ if and only if $f \in L \log^+ L(\mathbb{R}^n)$.*

PROOF. First assume that $f \in L\log^+ L(\mathbb{R}^n)$ and let $B \subset \mathbb{R}^n$ be a ball. Lemma 1.14 and the estimate in (1.3) give

$$\begin{aligned}\int_B Mf(x)\,dx &= 2\int_0^\infty |\{x \in B : Mf(x) > 2t\}|\,dt \\ &\le 2|B| + 2\int_1^\infty |\{x \in B : Mf(x) > 2t\}|\,dt \\ &\le 2|B| + C(n)\int_1^\infty \frac{1}{t}\int_{\{x\in\mathbb{R}^n:|f(x)|\ge t\}} |f(x)|\,dx\,dt \\ &\le 2|B| + C(n)\int_{\mathbb{R}^n} |f(x)|\int_1^{\max\{|f(x)|,1\}} \frac{1}{t}\,dt\,dx \\ &= 2|B| + C(n)\int_{\mathbb{R}^n} |f(x)|\log^+|f(x)|\,dx < \infty.\end{aligned}$$

Since B is an arbitrary ball and $Mf \in L^1(B)$, we conclude that $Mf \in L^1_{\mathrm{loc}}(\mathbb{R}^n)$.

Then assume that $Mf \in L^1_{\mathrm{loc}}(\mathbb{R}^n)$. Since the support of f is compact and $f \in L^1(\mathbb{R}^n)$, the maximal function $Mf(x)$ decays to zero as $|x| \to \infty$ and thus $E = \{x \in \mathbb{R}^n : Mf(x) > 1\}$ is a bounded set. Corollary 1.23 and Lemma 1.25 imply

$$\begin{aligned}\int_{\mathbb{R}^n} |f(x)|\log^+|f(x)|\,dx &= \int_{\mathbb{R}^n} |f(x)|\int_1^{\max\{|f(x)|,1\}} \frac{1}{t}\,dt\,dx \\ &\le \int_1^\infty \frac{1}{t}\int_{\{x\in\mathbb{R}^n:|f(x)|>t\}} |f(x)|\,dx\,dt \\ &\le \int_1^\infty \frac{1}{t}\int_{\{x\in\mathbb{R}^n:Mf(x)>t\}} |f(x)|\,dx\,dt \\ &\le C(n)\int_1^\infty |\{x \in \mathbb{R}^n : Mf(x) > t\}|\,dt \\ &= C(n)\int_1^\infty |\{x \in E : Mf(x) > t\}|\,dt \\ &\le C(n)\int_E Mf(x)\,dx < \infty,\end{aligned}$$

since Mf is locally integrable. This shows that $f \in L\log^+ L(\mathbb{R}^n)$. □

1.5. Restricted maximal function

We apply maximal functions with a restricted collection of radii in many arguments.

DEFINITION 1.27. Let $R\colon \mathbb{R}^n \to [0,\infty]$ be a function and assume that $f \in L^1_{\mathrm{loc}}(\mathbb{R}^n)$. The restricted centered maximal function $M_R f\colon \mathbb{R}^n \to [0,\infty]$ is

$$M_R f(x) = \begin{cases} |f(x)|, & \text{if } R(x) = 0, \\ \sup \fint_{B(x,r)} |f(y)|\,dy, & \text{if } R(x) > 0, \end{cases} \tag{1.7}$$

where the supremum is taken over all radii r with $0 < r < R(x)$. If $R \ge 0$ is a real number, we use the convention that $R(x) = R$ for every $x \in \mathbb{R}^n$.

Let $f \in L^1_{\mathrm{loc}}(\mathbb{R}^n)$. As in Corollary 1.23, we have

$$|f(x)| \le M_R f(x) \le Mf(x)$$

for almost every $x \in \mathbb{R}^n$. Thus the maximal function theorem, see Theorem 1.15, also holds for the restricted maximal function, assuming that $M_R f$ is measurable whenever $f \in L^1_{\mathrm{loc}}(\mathbb{R}^n)$. If $R\colon \mathbb{R}^n \to [0,\infty)$ is continuous, then the proof of Lemma 1.3 can be adapted to show that $M_R f$ is measurable for every $f \in L^1_{\mathrm{loc}}(\mathbb{R}^n)$.

The following result is a variant of Lemma 1.7. Under the assumptions of Lemma 1.28, we have $M_R f(x) < \infty$ for every $x \in \mathbb{R}^n$.

LEMMA 1.28. *Assume that $R\colon \mathbb{R}^n \to [0,\infty)$ and $f\colon \mathbb{R}^n \to \mathbb{R}$ are continuous. Then $M_R f$ is a continuous function in $\mathbb{R}^n$.*

PROOF. We observe that

$$F(x,r) = \begin{cases} |f(x)|, & \text{if } r = 0, \\ \fint_{B(x,r)} |f(y)|\,dy, & \text{if } r > 0, \end{cases}$$

is a continuous function in $\mathbb{R}^n \times [0,\infty)$. Let $x \in \mathbb{R}^n$ and $\varepsilon > 0$. By the uniform continuity of F in $\overline{B}(x,1) \times [0, R(x)+1]$, there exists $0 < \eta < 1$ such that

$$|F(y,s) - F(x,t)| < \varepsilon$$

whenever $|y-x| + |s-t| < \eta$ and $0 \le s,t \le R(x)+1$. By the continuity of R at x, there exists $0 < \delta < \frac{\eta}{2}$ such that

$$|R(x) - R(y)| < \frac{\eta}{2}$$

whenever $|x-y| < \delta$. To prove the continuity of $M_R f$ at the point x, let $y \in \mathbb{R}^n$ be such that $|x-y| < \delta$. Then it holds for some $0 \le r(y) \le R(y)$ that

$$M_R(f)(y) = F(y, r(y)) \le F(x, \min\{r(y), R(x)\}) + \varepsilon \le M_R f(x) + \varepsilon,$$

since $|y-x| + |r(y) - \min\{r(y), R(x)\}| < \eta$ and $0 \le r(y) \le R(y) \le R(x)+1$. On the other hand, for some $0 \le r(x) \le R(x)$, we have

$$M_R f(x) = F(x, r(x)) \le F(y, \min\{r(x), R(y)\}) + \varepsilon \le M_R f(y) + \varepsilon.$$

The continuity of $M_R f$ at x follows by combining the estimates above. □

REMARK 1.29. Let $1 \le p < \infty$ and $R\colon \mathbb{R}^n \to [0,\infty)$, and assume that $f \in L^1_{\mathrm{loc}}(\mathbb{R}^n)$. By Hölder's inequality

$$\fint_{B(x,r)} |f(y)|\,dy \le \left(\fint_{B(x,r)} |f(y)|^p\,dy \right)^{\frac{1}{p}},$$

for every $x \in \mathbb{R}^n$ and $r > 0$, and thus

$$M_R f(x) \le \left(M_R |f|^p(x) \right)^{\frac{1}{p}}$$

for every $x \in \mathbb{R}^n$.

The following maximal function with respect to an open set will be studied in Section 4.5.

DEFINITION 1.30. Let $\Omega \subsetneq \mathbb{R}^n$ be an open set and assume that $f \in L^1_{\mathrm{loc}}(\Omega)$. The maximal function $M_\Omega f\colon \Omega \to [0,\infty]$ is defined by

$$M_\Omega f(x) = \sup \fint_{B(x,r)} |f(y)|\, dy, \tag{1.8}$$

where the supremum is taken over all radii r with $0 < r < d(x, \partial\Omega)$.

We remark that $M_\Omega f$ is measurable if $f \in L^1_{\mathrm{loc}}(\Omega)$. The maximal function theorem, see Theorem 1.15 (c), implies that the maximal operator M_Ω is bounded on $L^p(\Omega)$ for $1 < p \leq \infty$, that is

$$\|M_\Omega f\|_{L^p(\Omega)} \leq \|M(f\chi_\Omega)\|_{L^p(\mathbb{R}^n)} \leq C(n,p)\|f\chi_\Omega\|_{L^p(\mathbb{R}^n)} = C(n,p)\|f\|_{L^p(\Omega)}, \tag{1.9}$$

for every $f \in L^p(\Omega)$. The first inequality above follows since $B(x,r) \subset \Omega$ if $x \in \Omega$ and $0 < r < d(x, \partial\Omega)$.

LEMMA 1.31. *Let $\Omega \subsetneq \mathbb{R}^n$ be an open set and assume that $f\colon \Omega \to \mathbb{R}$ has a continuous extension to $\mathbb{R}^n$. Then $M_\Omega f$ is continuous in Ω.*

PROOF. Let $\widetilde{f}\colon \mathbb{R}^n \to \mathbb{R}$ be a continuous function such that $\widetilde{f}(x) = f(x)$ for every $x \in \Omega$. Write $R(x) = d(x, \partial\Omega)$ for every $x \in \mathbb{R}^n$. Then

$$M_\Omega f(x) = M_\Omega(f\chi_\Omega)(x) = M_\Omega(\widetilde{f}\chi_\Omega)(x) = M_R(\widetilde{f}\chi_\Omega)(x) = M_R\widetilde{f}(x)$$

for every $x \in \Omega$. The claim follows from Lemma 1.28. □

1.6. Riesz potential

Standard applications of Riesz potentials are in potential theory and harmonic analysis. We apply them in connection with Sobolev–Poincaré type inequalities in Chapter 3.

DEFINITION 1.32. Let $0 < \alpha < n$ and assume that $f \in L^1_{\mathrm{loc}}(\mathbb{R}^n)$ is a nonnegative function. The Riesz potential $I_\alpha f\colon \mathbb{R}^n \to [0,\infty]$ is defined by

$$I_\alpha f(x) = \int_{\mathbb{R}^n} \frac{f(y)}{|x-y|^{n-\alpha}}\, dy,$$

for every $x \in \mathbb{R}^n$.

It is possible to consider Riesz potentials also for signed functions, but nonnegative functions suffice for our purposes. We begin with a bound for the Riesz potential of a characteristic function of a measurable set with finite measure.

LEMMA 1.33. *Assume that $E \subset \mathbb{R}^n$ is a measurable set with $|E| < \infty$. There exists a constant $C = C(n)$ such that*

$$I_1\chi_E(x) = \int_E \frac{1}{|x-y|^{n-1}}\, dy \leq C|E|^{\frac{1}{n}}$$

for every $x \in \mathbb{R}^n$.

PROOF. The claim is trivial if $|E| = 0$, and hence we may assume that $|E| > 0$. Let $x \in \mathbb{R}^n$ and let $B = B(x,r)$ be a ball with $|B| = |E|$. Then

$$|E \setminus B| = |E| - |E \cap B| = |B| - |B \cap E| = |B \setminus E|$$

and thus

$$\int_{E\setminus B} \frac{1}{|x-y|^{n-1}}\, dy \leq \frac{|E\setminus B|}{r^{n-1}} = \frac{|B\setminus E|}{r^{n-1}} \leq \int_{B\setminus E} \frac{1}{|x-y|^{n-1}}\, dy.$$

This implies

$$\begin{aligned}\int_E \frac{1}{|x-y|^{n-1}}\,dy &= \int_{E\setminus B} \frac{1}{|x-y|^{n-1}}\,dy + \int_{E\cap B} \frac{1}{|x-y|^{n-1}}\,dy\\ &\le \int_{B\setminus E} \frac{1}{|x-y|^{n-1}}\,dy + \int_{E\cap B} \frac{1}{|x-y|^{n-1}}\,dy\\ &= \int_B \frac{1}{|x-y|^{n-1}}\,dy = C(n)r = C(n)|B|^{\frac{1}{n}} = C(n)|E|^{\frac{1}{n}}. \qquad \square\end{aligned}$$

We also have a norm estimate for the Riesz potential.

LEMMA 1.34. *Let $1 \le p < \infty$. Assume that $\Omega \subset \mathbb{R}^n$ is an open set with $|\Omega| < \infty$ and let $f \in L^p(\Omega)$. There exists a constant $C = C(n,p)$ such that*

$$\big\|I_1(|f|\chi_\Omega)\big\|_{L^p(\Omega)} \le C|\Omega|^{\frac{1}{n}}\|f\|_{L^p(\Omega)}.$$

PROOF. Let $x \in \Omega$. If $1 < p < \infty$ and $p' = \frac{p}{p-1}$, then Hölder's inequality and Lemma 1.33 give

$$\begin{aligned}\int_\Omega \frac{|f(y)|}{|x-y|^{n-1}}\,dy &= \int_\Omega \frac{|f(y)|}{|x-y|^{\frac{1}{p}(n-1)}}\frac{1}{|x-y|^{\frac{1}{p'}(n-1)}}\,dy\\ &\le \Big(\int_\Omega \frac{|f(y)|^p}{|x-y|^{n-1}}\,dy\Big)^{\frac{1}{p}}\Big(\int_\Omega \frac{1}{|x-y|^{n-1}}\,dy\Big)^{\frac{1}{p'}}\\ &\le C(n,p)|\Omega|^{\frac{1}{np'}}\Big(\int_\Omega \frac{|f(y)|^p}{|x-y|^{n-1}}\,dy\Big)^{\frac{1}{p}}\\ &= C(n,p)|\Omega|^{\frac{p-1}{np}}\Big(\int_\Omega \frac{|f(y)|^p}{|x-y|^{n-1}}\,dy\Big)^{\frac{1}{p}}.\end{aligned}$$

Note that the inequality above also holds for $p = 1$. By Fubini's theorem and Lemma 1.33, we obtain

$$\begin{aligned}\int_\Omega |I_1(|f|\chi_\Omega)(x)|^p\,dx &\le C(n,p)|\Omega|^{\frac{p-1}{n}}\int_\Omega\int_\Omega \frac{|f(y)|^p}{|x-y|^{n-1}}\,dy\,dx\\ &\le C(n,p)|\Omega|^{\frac{p-1}{n}}|\Omega|^{\frac{1}{n}}\int_\Omega |f(y)|^p\,dy. \qquad \square\end{aligned}$$

Next we consider a pointwise upper bound for the Riesz potential in terms of the maximal function.

LEMMA 1.35. *Let $0 < \alpha < n$ and assume that $f \in L^1_{\mathrm{loc}}(\mathbb{R}^n)$ is a nonnegative function. There exists a constant $C = C(n,\alpha)$ such that*

$$\int_{B(x,r)} \frac{f(y)}{|x-y|^{n-\alpha}}\,dy \le Cr^\alpha Mf(x)$$

for every $x \in \mathbb{R}^n$ and $r > 0$.

PROOF. Let $B_j = B(x, 2^{-j}r)$ and $A_j = B_j \setminus B_{j+1}$ for $j \in \mathbb{N}_0$. Since $|x-y| \geq 2^{-j-1}r$ for every $y \in A_j$ and $|B_j| = \omega_n(2^{-j}r)^n$, we obtain

$$\begin{aligned}
\int_{B(x,r)} \frac{f(y)}{|x-y|^{n-\alpha}}\,dy &= \sum_{j=0}^{\infty} \int_{A_j} \frac{f(y)}{|x-y|^{n-\alpha}}\,dy \\
&\leq \sum_{j=0}^{\infty} (2^{-j-1}r)^{\alpha-n} \int_{A_j} f(y)\,dy \\
&\leq 2^{n-\alpha} \sum_{j=0}^{\infty} (2^{-j}r)^{\alpha}(2^{-j}r)^{-n} \int_{B_j} f(y)\,dy \\
&= \omega_n 2^{n-\alpha} r^{\alpha} \sum_{j=0}^{\infty} 2^{-\alpha j} \fint_{B_j} f(y)\,dy \\
&\leq C(n,\alpha) r^{\alpha} Mf(x) \sum_{j=0}^{\infty} 2^{-\alpha j} \\
&\leq C(n,\alpha) r^{\alpha} Mf(x).
\end{aligned}$$

□

The following Hardy–Littlewood–Sobolev theorem for the Riesz potential will be applied in Chapter 3. The exponent p^* below is called the Sobolev conjugate of p. Observe that $\frac{1}{p} - \frac{1}{p^*} = \frac{\alpha}{n}$.

THEOREM 1.36. *Let $1 < p < \infty$, $0 < \alpha < \frac{n}{p}$ and $p^* = \frac{np}{n-\alpha p}$. Assume that $f \in L^p(\mathbb{R}^n)$ is a nonnegative function. There exists a constant $C = C(n,p,\alpha)$ such that*

$$\|I_\alpha f\|_{L^{p^*}(\mathbb{R}^n)} \leq C\|f\|_{L^p(\mathbb{R}^n)}.$$

PROOF. If $f = 0$ almost everywhere, there is nothing to prove, and thus we may assume that $f > 0$ on a set of positive measure. This implies $Mf(x) > 0$ for every $x \in \mathbb{R}^n$. Let $x \in \mathbb{R}^n$, $r > 0$ and $p' = \frac{p}{p-1}$. By Hölder's inequality

$$\begin{aligned}
\int_{\mathbb{R}^n\setminus B(x,r)} \frac{f(y)}{|x-y|^{n-\alpha}}\,dy &\leq \left(\int_{\mathbb{R}^n\setminus B(x,r)} f(y)^p\,dy\right)^{\frac{1}{p}} \left(\int_{\mathbb{R}^n\setminus B(x,r)} |x-y|^{(\alpha-n)p'}\,dy\right)^{\frac{1}{p'}} \\
&\leq C(n,p,\alpha) r^{\alpha-\frac{n}{p}} \|f\|_{L^p(\mathbb{R}^n)}.
\end{aligned}$$

Lemma 1.35 implies

$$\begin{aligned}
I_\alpha f(x) &= \int_{\mathbb{R}^n} \frac{f(y)}{|x-y|^{n-\alpha}}\,dy = \int_{B(x,r)} \frac{f(y)}{|x-y|^{n-\alpha}}\,dy + \int_{\mathbb{R}^n\setminus B(x,r)} \frac{f(y)}{|x-y|^{n-\alpha}}\,dy \\
&\leq C(n,p,\alpha)\big(r^{\alpha} Mf(x) + r^{\alpha-\frac{n}{p}} \|f\|_{L^p(\mathbb{R}^n)}\big).
\end{aligned}$$

By choosing

$$r = \left(\frac{Mf(x)}{\|f\|_{L^p(\mathbb{R}^n)}}\right)^{-\frac{p}{n}} > 0,$$

we obtain

$$I_\alpha f(x) \leq C(n,p,\alpha) Mf(x)^{1-\frac{\alpha p}{n}} \|f\|_{L^p(\mathbb{R}^n)}^{\frac{\alpha p}{n}} \tag{1.10}$$

for every $x \in \mathbb{R}^n$. Theorem 1.15 (c) gives

$$\begin{aligned}\int_{\mathbb{R}^n} |I_\alpha f(x)|^{p^*}\,dy &\le C(n,p,\alpha)\|f\|_{L^p(\mathbb{R}^n)}^{\frac{\alpha p}{n}p^*} \int_{\mathbb{R}^n} (Mf(x))^p\,dx \\ &\le C(n,p,\alpha)\|f\|_{L^p(\mathbb{R}^n)}^{\frac{\alpha p}{n}p^*} \|f\|_{L^p(\mathbb{R}^n)}^p,\end{aligned}$$

which shows that

$$\|I_\alpha f\|_{L^{p^*}(\mathbb{R}^n)} \le C(n,p,\alpha)\|f\|_{L^p(\mathbb{R}^n)}^{\frac{\alpha p}{n}+\frac{p}{p^*}} = C(n,p,\alpha)\|f\|_{L^p(\mathbb{R}^n)}. \qquad \square$$

From the proof of the previous theorem we also obtain a weak type estimate for $p = 1$.

LEMMA 1.37. *Let $0 < \alpha < n$ and assume that $f \in L^1(\mathbb{R}^n)$ is a nonnegative function. There exists a constant $C = C(n,\alpha)$ such that*

$$|\{x \in \mathbb{R}^n : I_\alpha f(x) > t\}| \le Ct^{-\frac{n}{n-\alpha}}\|f\|_{L^1(\mathbb{R}^n)}^{\frac{n}{n-\alpha}}$$

for every $t > 0$.

PROOF. We may assume that $\|f\|_{L^1(\mathbb{R}^n)} > 0$ since otherwise $I_\alpha f(x) = 0$ for every $x \in \mathbb{R}^n$. By a straightforward modification of (1.10) with $p = 1$, we have

$$I_\alpha f(x) \le C(n,\alpha)Mf(x)^{1-\frac{\alpha}{n}}\|f\|_{L^1(\mathbb{R}^n)}^{\frac{\alpha}{n}}$$

for every $x \in \mathbb{R}^n$. Theorem 1.15 (b) implies

$$\begin{aligned}|\{x \in \mathbb{R}^n : I_\alpha f(x) > t\}| &\le \left|\left\{x \in \mathbb{R}^n : C(n,\alpha)Mf(x)^{\frac{n-\alpha}{n}}\|f\|_{L^1(\mathbb{R}^n)}^{\frac{\alpha}{n}} > t\right\}\right| \\ &\le \left|\left\{x \in \mathbb{R}^n : Mf(x) > C(n,\alpha)t^{\frac{n}{n-\alpha}}\|f\|_{L^1(\mathbb{R}^n)}^{-\frac{\alpha}{n}\frac{n}{n-\alpha}}\right\}\right| \\ &\le C(n,\alpha)t^{-\frac{n}{n-\alpha}}\|f\|_{L^1(\mathbb{R}^n)}^{\frac{\alpha}{n-\alpha}}\|f\|_{L^1(\mathbb{R}^n)} \\ &= C(n,\alpha)t^{-\frac{n}{n-\alpha}}\|f\|_{L^1(\mathbb{R}^n)}^{\frac{n}{n-\alpha}}\end{aligned}$$

for every $t > 0$. $\square$

1.7. Fractional maximal function

The fractional maximal function has many applications in potential theory and partial differential equations.

DEFINITION 1.38. Let $0 \le \alpha \le n$ and $R > 0$, and assume that $f \in L^1_{\mathrm{loc}}(\mathbb{R}^n)$. The centered fractional maximal function $M_{\alpha,R}f\colon \mathbb{R}^n \to [0,\infty]$ is defined by

$$M_{\alpha,R}f(x) = \sup_{0<r<R} r^\alpha \fint_{B(x,r)} |f(y)|\,dy.$$

For $R = \infty$ we write $M_{\alpha,\infty} = M_\alpha$. If $\alpha = 0$ and $R = \infty$, we get the usual Hardy–Littlewood maximal function, that is, $M_0 = M$. A similar argument as in the proof of Lemma 1.3 shows that the fractional maximal function is lower semicontinuous and thus measurable.

REMARK 1.39. The range $0 \le \alpha \le n$ is natural for M_α. For $\alpha < 0$, we have $M_\alpha f(x) = \infty$ for every $x \in \mathbb{R}^n$ with

$$|f(x)| = |f^*(x)| = \lim_{r\to 0}\left|\fint_{B(x,r)} f(y)\,dy\right| > 0.$$

On the other hand, for $\alpha > n$, we have $M_\alpha f(x) = \infty$ for every $x \in \mathbb{R}^n$ if $f \neq 0$ on a set of positive measure.

When $0 < \alpha < n$, there is a close connection between the fractional maximal function and the Riesz potential.

LEMMA 1.40. *Let $0 < \alpha < n$ and assume that $f \in L^1_{\mathrm{loc}}(\mathbb{R}^n)$. There exists a constant $C = C(n)$ such that*

$$M_\alpha f(x) \le C I_\alpha |f|(x) \tag{1.11}$$

for every $x \in \mathbb{R}^n$.

PROOF. Let $x \in \mathbb{R}^n$ and $r > 0$. Then

$$\begin{aligned} r^\alpha \fint_{B(x,r)} |f(y)|\,dy &= \frac{r^{\alpha-n}}{\omega_n} \int_{B(x,r)} |f(y)|\,dy \\ &\le \frac{1}{\omega_n} \int_{B(x,r)} \frac{|f(y)|}{|x-y|^{n-\alpha}}\,dy \le \frac{1}{\omega_n} \int_{\mathbb{R}^n} \frac{|f(y)|}{|x-y|^{n-\alpha}}\,dy. \end{aligned}$$

Taking supremum over radii $r > 0$ proves the claim with $C(n) = \frac{1}{\omega_n}$. □

REMARK 1.41. It is not possible to obtain a pointwise inequality in the reverse direction in (1.11). However, such an inequality holds in average by a result of Muckenhoupt and Wheeden. That is, for every $1 < p < \infty$ there exists a constant $C = C(n, p, \alpha)$ such that

$$C^{-1} \|M_\alpha f\|_{L^p(\mathbb{R}^n)} \le \big\|I_\alpha |f|\big\|_{L^p(\mathbb{R}^n)} \le C \|M_\alpha f\|_{L^p(\mathbb{R}^n)}$$

for every $f \in L^1_{\mathrm{loc}}(\mathbb{R}^n)$, see Muckenhoupt and Wheeden [**335**] and Adams and Hedberg [**4**, Theorem 3.6.1].

The following version of the Hardy–Littlewood–Sobolev theorem, see Theorem 1.36 and Lemma 1.37, will be useful in Section 4.6.

THEOREM 1.42. *Let $1 \le p < \infty$, $0 < \alpha < \frac{n}{p}$ and $p^* = \frac{np}{n-\alpha p}$.*

(a) *If $p > 1$, there exists a constant $C = C(n, p, \alpha)$ such that*

$$\|M_\alpha f\|_{L^{p^*}(\mathbb{R}^n)} \le C \|f\|_{L^p(\mathbb{R}^n)}$$

for every $f \in L^p(\mathbb{R}^n)$.

(b) *If $p = 1$, there exists a constant $C = C(n, \alpha)$ such that*

$$|\{x \in \mathbb{R}^n : M_\alpha f(x) > t\}| \le C t^{-\frac{n}{n-\alpha}} \|f\|_{L^1(\mathbb{R}^n)}^{\frac{n}{n-\alpha}}$$

for every $f \in L^1(\mathbb{R}^n)$ and $t > 0$.

PROOF. If $f \in L^p(\mathbb{R}^n)$, $p > 1$ and $0 < \alpha < \frac{n}{p}$, then (1.11) and Theorem 1.36 imply

$$\|M_\alpha f\|_{L^{p^*}(\mathbb{R}^n)} \le C(n) \big\|I_\alpha |f|\big\|_{L^{p^*}(\mathbb{R}^n)} \le C(n, p, \alpha) \|f\|_{L^p(\mathbb{R}^n)}.$$

On the other hand, if $f \in L^1(\mathbb{R}^n)$ and $0 < \alpha < n$, then (1.11) and Lemma 1.37 give

$$\begin{aligned} |\{x \in \mathbb{R}^n : M_\alpha f(x) > t\}| &\le |\{x \in \mathbb{R}^n : C(n) I_\alpha |f|(x) > t\}| \\ &\le C(n, \alpha) t^{-\frac{n}{n-\alpha}} \|f\|_{L^1(\mathbb{R}^n)}^{\frac{n}{n-\alpha}} \end{aligned}$$

for every $t > 0$. □

The size of the distribution set $\{x \in \mathbb{R}^n : M_\alpha f(x) > t\}$ can be estimated using the Hausdorff content. We recall here also the definition of Hausdorff measures, which are natural lower dimensional measures on $\mathbb{R}^n$.

DEFINITION 1.43. Let $E \subset \mathbb{R}^n$ and $\lambda \geq 0$. For $0 < \delta \leq \infty$, the λ-dimensional Hausdorff δ-content of E is

$$\mathcal{H}^\lambda_\delta(E) = \inf\left\{\sum_{i=1}^\infty r_i^\lambda : E \subset \bigcup_{i=1}^\infty B(x_i, r_i),\ 0 < r_i \leq \delta\right\}.$$

In the case $\lambda = 0$ we also allow finite summations. The (spherical) λ-dimensional Hausdorff measure of E is

$$\mathcal{H}^\lambda(E) = \lim_{\delta \to 0} \mathcal{H}^\lambda_\delta(E) = \sup_{\delta > 0} \mathcal{H}^\lambda_\delta(E).$$

Both $\mathcal{H}^\lambda_\delta$ and $\mathcal{H}^\lambda$ are outer measures and Borel sets in $\mathbb{R}^n$ are $\mathcal{H}^\lambda$-measurable. For every $E \subset \mathbb{R}^n$, we have

$$\mathcal{H}^\lambda_\infty(E) \leq \mathcal{H}^\lambda_\delta(E) \leq \mathcal{H}^\lambda(E)$$

whenever $0 < \delta \leq \infty$. On the other hand, $\mathcal{H}^\lambda(E) = 0$ if and only if $\mathcal{H}^\lambda_\delta(E) = 0$ for every (or some) $0 < \delta \leq \infty$.

DEFINITION 1.44. The Hausdorff dimension of $E \subset \mathbb{R}^n$ is

$$\dim_{\mathrm{H}}(E) = \inf\{\lambda \geq 0 : \mathcal{H}^\lambda(E) = 0\} = \inf\{\lambda \geq 0 : \mathcal{H}^\lambda_\infty(E) = 0\}.$$

We refer to Evans and Gariepy [**122**, Chapter 2] and Mattila [**315**, Chapter 4] for these and other basic properties of Hausdorff measures and Hausdorff dimension.

A standard covering argument, as in the proof of Theorem 1.15, gives the following weak type inequality for the fractional maximal function.

LEMMA 1.45. *Let $0 \leq \alpha < n$ and assume that $f \in L^1(\mathbb{R}^n)$. There exists a constant $C = C(n, \alpha)$ such that*

$$\mathcal{H}^{n-\alpha}_\infty(\{x \in \mathbb{R}^n : M_\alpha f(x) > t\}) \leq \frac{C}{t}\int_{\mathbb{R}^n} |f(x)|\,dx$$

for every $t > 0$.

PROOF. Let $E_t = \{x \in \mathbb{R}^n : M_\alpha f(x) > t\}$ for $t > 0$. By the definition of $M_\alpha f$, for every $x \in E_t$ there exists $r_x > 0$ such that

$$r_x^\alpha \fint_{B(x, r_x)} |f(y)|\,dy > t.$$

This implies

$$r_x^{n-\alpha} \leq \frac{1}{\omega_n t}\int_{B(x,r_x)} |f(y)|\,dy \leq \frac{1}{\omega_n t}\int_{\mathbb{R}^n} |f(y)|\,dy < \infty.$$

Thus $\mathcal{B} = \{B(x, r_x) : x \in E_t\}$ is a collection of balls with uniformly bounded radii and $E_t \subset \bigcup_{B \in \mathcal{B}} B$. Lemma 1.13 gives pairwise disjoint balls $B(x_i, r_i) \in \mathcal{B}$, $i \in \mathbb{N}$, such that $E_t \subset \bigcup_{i=1}^\infty B(x_i, 5r_i)$. It follows that

$$\mathcal{H}^{n-\alpha}_\infty(E_t) \leq \sum_{i=1}^\infty (5r_i)^{n-\alpha} \leq \frac{5^{n-\alpha}}{\omega_n t}\sum_{i=1}^\infty \int_{B(x_i,r_i)} |f(y)|\,dy \leq \frac{5^{n-\alpha}}{\omega_n t}\int_{\mathbb{R}^n} |f(y)|\,dy,$$

and this proves the claim with $C(n, \alpha) = \frac{5^{n-\alpha}}{w_n}$. □

1.8. Notes

The maximal function was introduced by Hardy and Littlewood [**179**] in the one-dimensional case and by Wiener [**397**] in the higher dimensional case. References for maximal functions include Bennet and Sharpley [**28**], DeVore and Sharpley [**100**], Duoandikoetxea [**107**], García-Cuerva and Rubio de Francia [**139**], Grafakos [**156, 157**], de Guzmán [**98, 99**], Journé [**216**], Lu, Ding and Yan [**286**], Sadosky [**356**], Stein [**369, 371**] and Torchinsky [**383**].

The maximal function theorem, Theorem 1.15, holds for the centered maximal function with respect to more general measures than the Lebesgue measure by the Besicovitch covering theorem, see Mattila [**315**, p. 40]. Sjögren [**362**] showed that the noncentered maximal function for more general measures than the Lebesgue measure may fail to satisfy the weak type estimate. Theorem 1.17 is by Fiorenza and Krbec [**129**]. See also Wik [**399**]. For Hausdorff content estimates, see Orobitg and Verdera [**341**]. Boundedness of maximal operators on weighted $L^p(\mathbb{R}^n)$ spaces is discussed in Chapter 8 and Chapter 9.

Stein [**370**] showed that the $L^p(\mathbb{R}^n)$ norm of the centered Hardy–Littlewood maximal operator with $1 < p \le \infty$ is bounded from above by a dimension-free constant, see also Stein and Strömberg [**372**]. For the noncentered maximal function the constant depends on the dimension, see Grafakos and Montgomery-Smith [**159**]. The best constant in the weak type estimate for the centered Hardy–Littlewood maximal operator was obtained by Melas [**323**]. In the one-dimensional case best constants have been investigated, for example, by Aldaz [**10**], Bernal [**29**], Grafakos and Kinnunen [**158**] and Grafakos and Montgomery-Smith [**159**].

The class $L\log^+ L(\mathbb{R}^n)$ in Theorem 1.26 was introduced by Zygmund and the necessity of this condition in Theorem 1.26 was observed by Stein [**368**]. Reverse weak type estimates as in Theorem 1.25 with more general measures have been investigated by Andersen and Young [**15, 16**].

For the Riesz potentials we refer to Adams and Hedberg [**4**, Chapter 3], Aikawa and Essén [**8**], Landkof [**254**, Chapter 1], Mizuta [**329**], Riesz [**355**], Stein [**369**, Section V.1] and Ziemer [**407**, Section 2.6]. The proof of Theorem 1.36 is from Hedberg [**181**]. Maximal function estimates for potentials are discussed, for example, in Adams and Hedberg [**4**, Chapter 3], Malý and Ziemer [**303**, Section 1.2], Maz′ya [**321**, Chapter 12], Turesson [**388**, Chapter 3] and Ziemer [**407**, Section 2.8]. For the Riesz potentials and the fractional maximal function, see also Adams [**3**].

CHAPTER 2

Lipschitz and Sobolev Functions

Sobolev spaces are natural function spaces in the theory of partial differential equations and variational problems. The definition of Sobolev function involves weak derivatives, but Sobolev spaces are also obtained by completion from smooth functions. In addition, we consider the flexible class of Lipschitz continuous functions. The definition and properties of Sobolev spaces are reviewed in Section 2.2 and Section 2.3. Sobolev spaces with zero boundary values are considered in Section 2.4 and the topic will be developed further in Section 5.5. This is the natural function space associated with Hardy's inequalities in Chapter 6 and it enables us to assign and compare boundary values of functions in Chapter 11. Weak convergence methods in Sobolev spaces are considered in Section 2.5 and they are applied in Section 4.4, Section 4.5 and Section 12.2.

2.1. Lipschitz functions

We begin by recalling the definition of a Lipschitz continuous function.

DEFINITION 2.1. Let $A \subset \mathbb{R}^n$ and $0 \le L < \infty$. A function $u\colon A \to \mathbb{R}$ is L-Lipschitz if

$$|u(x) - u(y)| \le L|x - y|$$

for every $x, y \in A$. The number L is called a Lipschitz constant of u. We say that $u\colon A \to \mathbb{R}$ is Lipschitz in A if it is L-Lipschitz for some L. The space of all Lipschitz functions in A is denoted by $\operatorname{Lip}(A)$. A function $u\colon \Omega \to \mathbb{R}$ is locally Lipschitz in an open set $\Omega \subset \mathbb{R}^n$ if $u \in \operatorname{Lip}(K)$ for every compact $K \subset \Omega$. The space of all locally Lipschitz functions in Ω is denoted by $\operatorname{Lip}_{\mathrm{loc}}(\Omega)$.

Observe that the Lipschitz constant of a function in $\operatorname{Lip}_{\mathrm{loc}}(\Omega)$ may depend on the compact subset $K \subset \Omega$.

EXAMPLE 2.2. Let $y \in \mathbb{R}^n$ and let $E \subset \mathbb{R}^n$ with $E \neq \emptyset$. The triangle inequality implies that the distance functions $x \mapsto |x - y|$ and $x \mapsto d(x, E)$ are 1-Lipschitz functions in $\mathbb{R}^n$. Note that these functions are not, in general, differentiable everywhere. However, by Rademacher's theorem, see Theorem 2.9, they are differentiable almost everywhere.

EXAMPLE 2.3. Consider the slit domain

$$\Omega = \{x \in \mathbb{R}^2 : |x| < 1\} \setminus \{(x_1, 0) \in \mathbb{R}^2 : x_1 \ge 0\}.$$

Let $u\colon \Omega \to \mathbb{R}$, $u(x) = \theta$, where θ is the argument of x in polar coordinates with $0 < \theta < 2\pi$. It is clear that u is not Lipschitz in Ω. However, this function u is locally Lipschitz in Ω.

Functions with zero boundary values will be of particular importance for us. For Lipschitz functions, it turns out to be convenient to make the further distinction between compactly supported functions and functions that vanish in the complement and have a compact support.

DEFINITION 2.4. Let $A \subset \mathbb{R}^n$.

(a) The space of all compactly supported Lipschitz functions in A is

$$\operatorname{Lip}_c(A) = \{u \in \operatorname{Lip}(\mathbb{R}^n) : \operatorname{supp} u \text{ is a compact subset of } A\}.$$

(b) The space of all compactly supported Lipschitz functions that vanish in the complement of A is

$$\operatorname{Lip}_0(A) = \{u \in \operatorname{Lip}_c(\mathbb{R}^n) : u = 0 \text{ in } \mathbb{R}^n \setminus A\}.$$

REMARK 2.5. It is clear that $\operatorname{Lip}_c(A) \subset \operatorname{Lip}_0(A)$. For a bounded set $A \subset \mathbb{R}^n$, we have

$$\operatorname{Lip}_0(A) = \{u \in \operatorname{Lip}(\mathbb{R}^n) : u = 0 \text{ in } \mathbb{R}^n \setminus A\}.$$

Moreover, if $\Omega \subset \mathbb{R}^n$ is an open set, then

$$C_0^\infty(\Omega) \subset C_0^k(\Omega) \subset \operatorname{Lip}_c(\Omega)$$

for every $k \in \mathbb{N}$. Indeed, by considering the zero extension of $u \in C_0^1(\Omega)$, we may assume that $u \in C_0^1(\mathbb{R}^n)$. Let $x, y \in \mathbb{R}^n$, $x \neq y$. By the mean value theorem, there exists z in the line-segment between x and y such that

$$|u(x) - u(y)| = |\nabla u(z) \cdot (x - y)| \leq \|\nabla u\|_{L^\infty(\mathbb{R}^n)} |x - y|.$$

This shows that u is L-Lipschitz with $L = \|\nabla u\|_{L^\infty(\mathbb{R}^n)}$.

One useful property of Lipschitz functions, not shared by smooth functions, is that the pointwise minimum and maximum of L-Lipschitz functions are L-Lipschitz. This is also true for pointwise infimum and supremum of L-Lipschitz functions if the infimum and supremum are finite at one point. In particular, if u is an L-Lipschitz function in $A \subset \mathbb{R}^n$ and $t \in \mathbb{R}$, then the truncations $\max\{u, t\}$ and $\min\{u, t\}$ are L-Lipschitz in A.

EXAMPLE 2.6. Let $x \in \mathbb{R}^n$ and $r > 0$. Define

$$u(y) = \max\{0, 1 - \tfrac{1}{r} d(y, B(x, r))\},$$

for $y \in \mathbb{R}^n$. The function u is $\frac{1}{r}$-Lipschitz in $\mathbb{R}^n$, $u = 1$ in $\overline{B}(x, r)$, and $u = 0$ in $\mathbb{R}^n \setminus B(x, 2r)$. This kind of function is used as a cutoff to localize estimates.

The following result gives a method to extend L-Lipschitz functions from a subset $A \subset \mathbb{R}^n$ to L-Lipschitz functions in the entire $\mathbb{R}^n$.

THEOREM 2.7. *Let $A \subset \mathbb{R}^n$ and assume that $u \colon A \to \mathbb{R}$ is an L-Lipschitz function with $0 \leq L < \infty$. There exists an L-Lipschitz function $\widetilde{u} \colon \mathbb{R}^n \to \mathbb{R}$ such that $\widetilde{u}(x) = u(x)$ for every $x \in A$.*

PROOF. Define

$$\widetilde{u}(x) = \inf_{a \in A} \{u(a) + L|x - a|\}, \tag{2.1}$$

for $x \in \mathbb{R}^n$. We claim that $\widetilde{u}(x) = u(x)$ for every $x \in A$. To see this we fix $x \in A$ and observe that, for every $a \in A$,

$$u(x) - u(a) \leq |u(x) - u(a)| \leq L|x - a|,$$

which implies $u(x) \le u(a) + L|x-a|$ for every $a \in A$. By taking infimum over $a \in A$, we obtain $u(x) \le \widetilde{u}(x)$. On the other hand, by the definition $\widetilde{u}(x) \le u(x)$, and thus $\widetilde{u}(x) = u(x)$ for every $x \in A$.

Then we show that $\widetilde{u}$ is L-Lipschitz in $\mathbb{R}^n$. In particular, the proof below also shows that $\widetilde{u}$ is real-valued since $\widetilde{u} = u$ in A. Let $x, y \in \mathbb{R}^n$. Then

$$\begin{aligned}\widetilde{u}(x) &= \inf_{a\in A}\{u(a) + L|x-a|\} \le \inf_{a\in A}\{u(a) + L|y-a| + L|x-y|\}\\ &= \inf_{a\in A}\{u(a) + L|y-a|\} + L|x-y| = \widetilde{u}(y) + L|x-y|.\end{aligned}$$

By switching the roles of x and y, we obtain $\widetilde{u}(y) \le \widetilde{u}(x) + L|x-y|$. Thus

$$-L|x-y| \le \widetilde{u}(x) - \widetilde{u}(y) \le L|x-y|,$$

which shows that $\widetilde{u}$ is L-Lipschitz in $\mathbb{R}^n$. □

REMARK 2.8. The function $\widetilde{u}$ in (2.1) is called the McShane extension of u. A similar argument as used in the proof of Theorem 2.7 shows that the function

$$\sup_{a\in A}\{u(a) - L|x-a|\}, \tag{2.2}$$

for $x \in \mathbb{R}^n$, is also an L-Lipschitz extension of u. The McShane extension $\widetilde{u}$ in (2.1) is the largest L-Lipschitz extension of u, in the sense that if $v\colon \mathbb{R}^n \to \mathbb{R}$ is L-Lipschitz and $v|_A = u$, then $v \le \widetilde{u}$. Correspondingly, the function defined by (2.2) is the smallest L-Lipschitz extension of u.

Let $\Omega \subset \mathbb{R}^n$ be an open set. A function $u\colon \Omega \to \mathbb{R}$ is differentiable at $x \in \Omega$ if there exists a linear map $Du\colon \mathbb{R}^n \to \mathbb{R}$, called the derivative of u, such that

$$\lim_{y\to x}\frac{|u(y) - u(x) - Du(y-x)|}{|y-x|} = 0. \tag{2.3}$$

If the derivative Du exists, it is unique and satisfies

$$Du(y-x) = \nabla u(x)\cdot(y-x)$$

for every $y \in \mathbb{R}^n$, where

$$\nabla u(x) = \left(\frac{\partial u}{\partial x_1}(x), \dots, \frac{\partial u}{\partial x_n}(x)\right)$$

is the pointwise gradient of u at x.

We recall the classical Rademacher theorem. See Evans and Gariepy [**122**, Section 3.1.2] for a proof.

THEOREM 2.9. *Let $\Omega \subset \mathbb{R}^n$ be an open set and assume that $u \in \operatorname{Lip}(\Omega)$. Then u is differentiable almost everywhere in Ω.*

As a corollary of Theorem 2.9, we obtain differentiation rules for Lipschitz functions.

COROLLARY 2.10. *Let $\Omega \subset \mathbb{R}^n$ be an open set. If $u, v \in \operatorname{Lip}(\Omega)$, then the Leibniz rule*

$$\nabla(uv)(x) = \nabla u(x)v(x) + u(x)\nabla v(x) \tag{2.4}$$

holds for almost every $x \in \Omega$. Moreover, if $f \in C^1(\mathbb{R})$ and $u \in \operatorname{Lip}(\Omega)$, then the chain rule

$$\nabla(f\circ u)(x) = f'(u(x))\nabla u(x) \tag{2.5}$$

is valid for almost every $x \in \Omega$.

Instead of the gradient ∇u, we are often interested in $|\nabla u|$, for which we have the following representation.

LEMMA 2.11. *Let $\Omega \subset \mathbb{R}^n$ be an open set and assume that $u \in \operatorname{Lip}(\Omega)$. Then*

$$|\nabla u(x)| = \lim_{r\to 0} \sup_{y\in B(x,r)} \frac{|u(y)-u(x)|}{r} \tag{2.6}$$

for almost every $x \in \Omega$.

PROOF. By Theorem 2.7, we may assume that $u \in \operatorname{Lip}(\mathbb{R}^n)$, and by Theorem 2.9 u is differentiable at almost every $x \in \Omega$. Let $x \in \Omega$ be such a point and let $r > 0$. Then

$$\sup_{y\in B(x,r)} \frac{|\nabla u(x)\cdot(y-x)|}{r} \le \sup_{y\in B(x,r)} \frac{|\nabla u(x)||y-x|}{r} \le |\nabla u(x)|.$$

On the other hand, we choose $z = x + r\frac{\nabla u(x)}{|\nabla u(x)|}$ if $\nabla u(x) \neq 0$, and $z = x$ if $\nabla u(x) = 0$. Since $z \in \overline{B}(x,r)$, we have

$$\begin{aligned}\sup_{y\in B(x,r)} \frac{|\nabla u(x)\cdot(y-x)|}{r} &= \sup_{y\in \overline{B}(x,r)} \frac{|\nabla u(x)\cdot(y-x)|}{r} \\ &\ge \frac{|\nabla u(x)\cdot(z-x)|}{r} = |\nabla u(x)|.\end{aligned}$$

Combining the estimates above we obtain

$$|\nabla u(x)| = \sup_{y\in B(x,r)} \frac{|\nabla u(x)\cdot(y-x)|}{r}, \tag{2.7}$$

for every $r > 0$.

Since u is differentiable at x, it follows from (2.3) that

$$\lim_{r\to 0} \sup_{y\in B(x,r)} \frac{|u(y)-u(x)-\nabla u(x)\cdot(y-x)|}{r} = 0. \tag{2.8}$$

Let $r > 0$ and $y \in B(x,r)$, and write

$$\begin{aligned} a(y,r) &= \frac{\nabla u(x)\cdot(y-x)}{r}, \\ b(y,r) &= \frac{u(y)-u(x)-\nabla u(x)\cdot(y-x)}{r}, \\ c(y,r) &= \frac{u(y)-u(x)}{r}. \end{aligned}$$

Then $c(y,r) = a(y,r) + b(y,r)$ and hence

$$\begin{aligned}|a(y,r)| - \sup_{z\in B(x,r)} |b(z,r)| &\le |a(y,r)| - |b(y,r)| \le |c(y,r)| \\ &\le |a(y,r)| + |b(y,r)| \le |a(y,r)| + \sup_{z\in B(x,r)} |b(z,r)|.\end{aligned}$$

By taking supremums over all $y \in B(x,r)$, we obtain

$$\begin{aligned}\sup_{y\in B(x,r)} |a(y,r)| - \sup_{y\in B(x,r)} |b(y,r)| &\le \sup_{y\in B(x,r)} |c(y,r)| \\ &\le \sup_{y\in B(x,r)} |a(y,r)| + \sup_{y\in B(x,r)} |b(y,r)|.\end{aligned}$$

The claim follows by taking $r \to 0$ and using (2.7) and (2.8). $\square$

REMARK 2.12. From (2.6) we see that if $u \in \operatorname{Lip}(\Omega)$ is an L-Lipschitz function, then $|\nabla u(x)| \le L$ for almost every $x \in \Omega$.

The following locality property is a useful consequence of (2.6).

LEMMA 2.13. *Let $\Omega \subset \mathbb{R}^n$ be an open set and assume that $u \in \operatorname{Lip}(\Omega)$ and $t \in \mathbb{R}$. Then $\nabla u = 0$ almost everywhere in the set $\{x \in \Omega : u(x) = t\}$.*

PROOF. By Theorem 2.7, we may assume that $\Omega = \mathbb{R}^n$ and $u \in \operatorname{Lip}(\mathbb{R}^n)$. Let $A = \{x \in \Omega : u(x) = t\}$ and let $x \in A$ be such that (2.6) holds. By the Lebesgue density theorem, see Theorem 1.24, we may assume that

$$\lim_{r\to 0} \frac{|A \cap B(x,r)|}{|B(x,r)|} = 1. \tag{2.9}$$

Define

$$d(r) = \sup_{y \in B(x,r)} d(y,A) + r^2 > 0$$

for $r > 0$. By (2.9) we have

$$\lim_{r\to 0} \frac{d(r)}{r} = 0.$$

Let $r > 0$ and $y \in B(x,r)$. There exists a point $z \in A \cap B(y, d(r))$. Since $u \in \operatorname{Lip}(\mathbb{R}^n)$ is an L-Lipschitz function for some constant $L > 0$, we have

$$\frac{|u(y) - u(x)|}{r} = \frac{|u(y) - u(z)|}{r} \le \frac{L|y - z|}{r} \le \frac{Ld(r)}{r}.$$

This implies

$$|\nabla u(x)| = \lim_{r\to 0} \sup_{y \in B(x,r)} \frac{|u(y) - u(x)|}{r} \le \lim_{r\to 0} \frac{Ld(r)}{r} = 0,$$

and the proof is complete. □

2.2. Sobolev spaces

Let $\Omega \subset \mathbb{R}^n$ be an open set and let $u \in \operatorname{Lip}_{\mathrm{loc}}(\Omega)$. Then u is absolutely continuous on all closed line segments that are contained in Ω. Using integration by parts on lines parallel to the coordinate axes and Fubini's theorem, we obtain

$$\int_\Omega u(x) \frac{\partial \varphi}{\partial x_i}(x)\, dx = -\int_\Omega \frac{\partial u}{\partial x_i}(x) \varphi(x)\, dx, \tag{2.10}$$

for every test function $\varphi \in C_0^\infty(\Omega)$ and $i = 1, \dots, n$. Weak derivatives are defined as functions that satisfy the integration by parts formula (2.10).

DEFINITION 2.14. Let $\Omega \subset \mathbb{R}^n$ be an open set and assume that $u \in L^1_{\mathrm{loc}}(\Omega)$. A function $v \in L^1_{\mathrm{loc}}(\Omega)$ is the ith weak partial derivative of u in Ω, for $i = 1, \dots, n$, if

$$\int_\Omega u(x) \frac{\partial \varphi}{\partial x_i}(x)\, dx = -\int_\Omega v(x) \varphi(x)\, dx$$

for every $\varphi \in C_0^\infty(\Omega)$. We write $\frac{\partial u}{\partial x_i} = v$. If all weak partial derivatives $\frac{\partial u}{\partial x_i}$, with $i = 1, \dots, n$, exist, then

$$\nabla u(x) = \left(\frac{\partial u}{\partial x_1}(x), \dots, \frac{\partial u}{\partial x_n}(x) \right),$$

for $x \in \Omega$, is the weak gradient of u in Ω.

REMARK 2.15. A weak gradient is unique up to a set of measure zero. To see this, let f and g be weak gradients of $u \in L^1_{\mathrm{loc}}(\Omega)$ in Ω. Then

$$\int_\Omega f(x)\varphi(x)\,dx = -\int_\Omega u(x)\nabla\varphi(x)\,dx = \int_\Omega g(x)\varphi(x)\,dx,$$

for every $\varphi \in C_0^\infty(\Omega)$, where the integrals are taken componentwise. From this it follows that $f = g$ almost everywhere in Ω. Moreover, we have $f(x) = g(x)$ at every point $x \in \Omega$ that is a Lebesgue point of all the components of both f and g. It is also clear that changing a function on a set of measure zero does not affect its weak derivatives.

REMARK 2.16. Let $u \in \mathrm{Lip}_{\mathrm{loc}}(\Omega)$ and $i = 1, \dots, n$. Then the ith classical partial derivative $\frac{\partial u}{\partial x_i}$ exists almost everywhere in Ω by Rademacher's theorem, Theorem 2.9. By (2.10), the ith classical partial derivative coincides almost everywhere in Ω with the ith weak partial derivative $\frac{\partial u}{\partial x_i}$. We use the same notation for the weak and pointwise derivatives of functions, and the interpretation will be clear from the context.

DEFINITION 2.17. Let $1 \le p \le \infty$ and let $\Omega \subset \mathbb{R}^n$ be an open set. The Sobolev space $W^{1,p}(\Omega)$ consists of functions $u \in L^p(\Omega)$ whose weak partial derivatives $\frac{\partial u}{\partial x_i}$, $i = 1, \dots, n$, exist and belong to $L^p(\Omega)$. For $1 \le p < \infty$, the space $W^{1,p}(\Omega)$ is equipped with the norm

$$\|u\|_{W^{1,p}(\Omega)} = \left(\|u\|^p_{L^p(\Omega)} + \|\nabla u\|^p_{L^p(\Omega)}\right)^{\frac{1}{p}}, \tag{2.11}$$

and the space $W^{1,\infty}(\Omega)$ is equipped with the norm

$$\|u\|_{W^{1,\infty}(\Omega)} = \|u\|_{L^\infty(\Omega)} + \|\nabla u\|_{L^\infty(\Omega)}, \tag{2.12}$$

where $\|\nabla u\|_{L^p(\Omega)} = \big\||\nabla u|\big\|_{L^p(\Omega)}$ for every $1 \le p \le \infty$. A function u in Ω belongs to the local Sobolev space $W^{1,p}_{\mathrm{loc}}(\Omega)$ if $u \in W^{1,p}(\Omega')$ for every open set $\Omega' \Subset \Omega$.

REMARK 2.18. In some cases, it is convenient to use different norms on the Sobolev space.

(a) For $1 \le p < \infty$, the norms

$$\left(\|u\|^p_{L^p(\Omega)} + \sum_{i=1}^n \left\|\frac{\partial u}{\partial x_i}\right\|^p_{L^p(\Omega)}\right)^{\frac{1}{p}}, \quad \|u\|_{L^p(\Omega)} + \sum_{i=1}^n \left\|\frac{\partial u}{\partial x_i}\right\|_{L^p(\Omega)},$$

and $\|u\|_{L^p(\Omega)} + \|\nabla u\|_{L^p(\Omega)}$ are all equivalent to (2.11).

(b) For $p = \infty$, the norm

$$\max\left\{\|u\|_{L^\infty(\Omega)}, \left\|\frac{\partial u}{\partial x_1}\right\|_{L^\infty(\Omega)}, \dots, \left\|\frac{\partial u}{\partial x_n}\right\|_{L^\infty(\Omega)}\right\}$$

is equivalent to (2.12).

The spaces $W^{1,p}(\Omega)$, $1 \le p \le \infty$, are Banach spaces with the above norms. Moreover, the spaces $W^{1,p}(\Omega)$ are reflexive when $1 < p < \infty$. As in $L^p(\Omega)$ spaces we identify functions in $W^{1,p}(\Omega)$ that are equal almost everywhere.

2.3. Approximation and calculus in Sobolev spaces

In this section we review some basic properties of the first-order Sobolev spaces. Instead of discussing all details, we give standard references. We illustrate the well-known density properties of smooth functions in Sobolev spaces by considering the case of $W^{1,p}(\mathbb{R}^n)$ with $1 < p < \infty$, see Evans and Gariepy [**122**, Section 4.2.1]. The argument is based on convolution approximation and maximal functions.

Recall that for measurable functions $f, g \colon \mathbb{R}^n \to [-\infty, \infty]$ the convolution $f * g$ is defined by

$$(f * g)(x) = \int_{\mathbb{R}^n} f(y) g(x - y)\, dy$$

for every $x \in \mathbb{R}^n$ for which the integral is defined. If $1 \le p \le \infty$, $f \in L^p(\mathbb{R}^n)$ and $g \in L^1(\mathbb{R}^n)$, then $(f * g)(x)$ exists for almost every $x \in \mathbb{R}^n$ and

$$\|f * g\|_{L^p(\mathbb{R}^n)} \le \|f\|_{L^p(\mathbb{R}^n)} \|g\|_{L^1(\mathbb{R}^n)}, \tag{2.13}$$

see Grafakos [**156**, Theorem 1.2.12]. This is Young's convolution inequality.

Let $\varphi \colon \mathbb{R}^n \to \mathbb{R}$,

$$\varphi(x) = \begin{cases} a \exp\left(\dfrac{1}{|x|^2 - 1}\right), & x \in B(0,1), \\ 0, & x \in \mathbb{R}^n \setminus B(0,1), \end{cases}$$

where the constant a is chosen so that $\int_{\mathbb{R}^n} \varphi(x)\, dx = 1$. For $x \in \mathbb{R}^n$, $\varepsilon > 0$ and $f \in L^1_{\mathrm{loc}}(\mathbb{R}^n)$, we define

$$\varphi_\varepsilon(x) = \frac{1}{\varepsilon^n} \varphi\left(\frac{x}{\varepsilon}\right). \tag{2.14}$$

The convolution

$$(\varphi_\varepsilon * f)(x) = \int_{\mathbb{R}^n} \varphi_\varepsilon(y) f(x - y)\, dy$$

can be considered as a weighted mean value of the function f in $B(x, \varepsilon)$ and it is clearly independent of the representative of f. Let $u \in W^{1,p}(\mathbb{R}^n)$ and $\varepsilon > 0$. Since $\varphi \in C_0^\infty(\mathbb{R}^n)$, the dominated convergence theorem implies that $\varphi_\varepsilon * u \in C^\infty(\mathbb{R}^n) \cap W^{1,p}(\mathbb{R}^n)$ and

$$\frac{\partial(\varphi_\varepsilon * u)}{\partial x_i} = \varphi_\varepsilon * \frac{\partial u}{\partial x_i} = \frac{\partial \varphi_\varepsilon}{\partial x_i} * u \tag{2.15}$$

for every $i = 1, \dots, n$.

The Hardy–Littlewood maximal function gives uniform pointwise control for the convolution approximations.

LEMMA 2.19. *Assume that $f \in L^1_{\mathrm{loc}}(\mathbb{R}^n)$. Then*

$$\sup_{\varepsilon > 0} |(\varphi_\varepsilon * f)(x)| \le M f(x)$$

for every $x \in \mathbb{R}^n$.

PROOF. First consider a function ψ of the form

$$\psi(x) = \sum_{j=1}^{N} a_j \chi_{B(0, r_j)}(x),$$

where $x \in \mathbb{R}^n$, $N \in \mathbb{N}$ and $a_j > 0$ for every $j = 1, \ldots, N$. Then

$$\|\psi\|_{L^1(\mathbb{R}^n)} = \sum_{j=1}^{N} a_j |B(0, r_j)|.$$

Let $x \in \mathbb{R}^n$ and $\varepsilon > 0$. A change of variables $z = \frac{y}{\varepsilon}$ gives

$$\begin{aligned}
|(\psi_\varepsilon * f)(x)| &= \left| \int_{\mathbb{R}^n} \psi_\varepsilon(y) f(x-y)\, dy \right| = \left| \frac{1}{\varepsilon^n} \int_{\mathbb{R}^n} \psi\Big(\frac{y}{\varepsilon}\Big) f(x-y)\, dy \right| \\
&= \left| \int_{\mathbb{R}^n} \psi(z) f(x - \varepsilon z)\, dz \right| = \left| \sum_{j=1}^{N} a_j \int_{B(0,r_j)} f(x-\varepsilon z)\, dz \right| \\
&\leq \sum_{j=1}^{N} a_j |B(0,r_j)| \fint_{B(0,r_j)} |f(x - \varepsilon z)|\, dz.
\end{aligned}$$

With $y = x - \varepsilon z$ we have

$$\begin{aligned}
\fint_{B(0,r_j)} |f(x-\varepsilon z)|\, dz &= \frac{1}{\varepsilon^n |B(0,r_j)|} \int_{B(x,\varepsilon r_j)} |f(y)|\, dy \\
&= \frac{1}{|B(x,\varepsilon r_j)|} \int_{B(x,\varepsilon r_j)} |f(y)|\, dy \leq Mf(x).
\end{aligned}$$

Thus

$$|(\psi_\varepsilon * f)(x)| \leq \sum_{j=1}^{N} a_j |B(0,r_j)| Mf(x) = \|\psi\|_{L^1(\mathbb{R}^n)} Mf(x)$$

for every $x \in \mathbb{R}^n$.

To conclude the proof of the lemma, consider an increasing sequence of functions ψ_k as above such that $\psi_k(x) \to \varphi(x)$ for every $x \in \mathbb{R}^n$ as $k \to \infty$. This is possible since φ is nonnegative, radial and radially decreasing. By the monotone convergence theorem we have

$$\begin{aligned}
|(\varphi_\varepsilon * f)(x)| &\leq \int_{\mathbb{R}^n} \varphi_\varepsilon(y) |f(x-y)|\, dy = \int_{\mathbb{R}^n} \lim_{k\to\infty} (\psi_k)_\varepsilon(y) |f(x-y)|\, dy \\
&= \lim_{k\to\infty} \int_{\mathbb{R}^n} (\psi_k)_\varepsilon(y) |f(x-y)|\, dy = \lim_{k\to\infty} \big| ((\psi_k)_\varepsilon * |f|)(x) \big| \\
&\leq \lim_{k\to\infty} \|\psi_k\|_{L^1(\mathbb{R}^n)} Mf(x) = \|\varphi\|_{L^1(\mathbb{R}^n)} Mf(x) = Mf(x)
\end{aligned}$$

for every $\varepsilon > 0$ and $x \in \mathbb{R}^n$, and the claim follows. □

LEMMA 2.20. *Let $1 < p < \infty$ and assume that $u \in W^{1,p}(\mathbb{R}^n)$. Then $\varphi_\varepsilon * u \to u$ in $W^{1,p}(\mathbb{R}^n)$ as $\varepsilon \to 0$. In particular, $C^\infty(\mathbb{R}^n) \cap W^{1,p}(\mathbb{R}^n)$ is dense in $W^{1,p}(\mathbb{R}^n)$.*

PROOF. By (2.15), it suffices to show that if $f \in L^p(\mathbb{R}^n)$, then $\varphi_\varepsilon * f \to f$ in $L^p(\mathbb{R}^n)$ as $\varepsilon \to 0$. To this end, let $\varepsilon > 0$ and $f \in L^p(\mathbb{R}^n)$ with $f = f^*$, and let x be a Lebesgue point of f. By the Lebesgue differentiation theorem, Theorem 1.21, we obtain

$$\begin{aligned}|(\varphi_\varepsilon * f)(x) - f(x)| &= \left|\int_{\mathbb{R}^n} \varphi_\varepsilon(y)(f(x-y) - f(x))\,dy\right| \\ &\le \frac{\|\varphi\|_{L^\infty(\mathbb{R}^n)}}{\varepsilon^n} \int_{B(0,\varepsilon)} |f(x-y) - f(x)|\,dy \\ &= C(n)\|\varphi\|_{L^\infty(\mathbb{R}^n)} \fint_{B(x,\varepsilon)} |f(y) - f(x)|\,dy \xrightarrow{\varepsilon\to 0+} 0.\end{aligned}$$

By Lemma 2.19, we have

$$|(\varphi_\varepsilon * f)(x) - f(x)| \le |(\varphi_\varepsilon * f)(x)| + |f(x)| \le Mf(x) + |f(x)|.$$

Theorem 1.21 implies that almost every point $x \in \mathbb{R}^n$ is a Lebesgue point of f. Since $Mf + |f| \in L^p(\mathbb{R}^n)$ by Theorem 1.15 (c), we can apply the dominated convergence theorem to obtain

$$\lim_{\varepsilon\to 0} \int_{\mathbb{R}^n} |(\varphi_\varepsilon * f)(x) - f(x)|^p\,dx = \int_{\mathbb{R}^n} \lim_{\varepsilon\to 0} |(\varphi_\varepsilon * f)(x) - f(x)|^p\,dx = 0,$$

and the proof is complete. □

Lemma 2.20 also holds for $p = 1$, see Evans and Gariepy [**122**, Section 4.2.1]. Note that the proof of Lemma 2.20 shows that $C^\infty(\mathbb{R}^n)$ is dense in $L^p(\mathbb{R}^n)$, for every $1 < p < \infty$.

As another application of convolution approximation, we identify the space of bounded Lipschitz continuous functions with the space $W^{1,\infty}(\mathbb{R}^n)$.

THEOREM 2.21. *Assume that* $u \in L^\infty(\mathbb{R}^n)$.

(a) *Then* $u \in W^{1,\infty}(\mathbb{R}^n)$ *if and only if* u *has a representative in* $\operatorname{Lip}(\mathbb{R}^n)$.

(b) *If* $u \in W^{1,\infty}(\mathbb{R}^n)$, *then there exists a set* $E \subset \mathbb{R}^n$ *such that* $|E| = 0$ *and*

$$|u(x) - u(y)| \le \|\nabla u\|_{L^\infty(\mathbb{R}^n)}|x - y|$$

for every $x, y \in \mathbb{R}^n \setminus E$.

(c) *If* u *is* L*-Lipschitz in* $\mathbb{R}^n$, *then* $\|\nabla u\|_{L^\infty(\mathbb{R}^n)} \le L$.

PROOF. Assume first that $u \in W^{1,\infty}(\mathbb{R}^n)$ and define $u_\varepsilon = \varphi_\varepsilon * u$ for every $\varepsilon > 0$. Let $\varepsilon > 0$. By Young's convolution inequality, we have

$$\|\nabla u_\varepsilon\|_{L^\infty(\mathbb{R}^n)} = \|\varphi_\varepsilon * \nabla u\|_{L^\infty(\mathbb{R}^n)} \le \big\|\varphi_\varepsilon * |\nabla u|\big\|_{L^\infty(\mathbb{R}^n)} \le \|\nabla u\|_{L^\infty(\mathbb{R}^n)},$$

where the first convolution is taken componentwise. Since $u_\varepsilon \in C^\infty(\mathbb{R}^n)$, we have

$$\begin{aligned}|u_\varepsilon(x) - u_\varepsilon(y)| &= \left|\int_0^1 \nabla u_\varepsilon(y + t(x-y)) \cdot (x-y)\,dt\right| \\ &\le \|\nabla u_\varepsilon\|_{L^\infty(\mathbb{R}^n)}|x-y| \le \|\nabla u\|_{L^\infty(\mathbb{R}^n)}|x-y|\end{aligned}$$

for every $x, y \in \mathbb{R}^n$. By the proof of Theorem 2.20, the convolutions u_ε converge to u pointwise almost everywhere in $\mathbb{R}^n$ as $\varepsilon \to 0$. Hence, there is a set $E \subset \mathbb{R}^n$ such that $|E| = 0$ and

$$|u(x) - u(y)| \le \|\nabla u\|_{L^\infty(\mathbb{R}^n)}|x-y| \tag{2.16}$$

for every $x, y \in \mathbb{R}^n \setminus E$. This proves (b). By uniform continuity, u can be redefined on the set E of measure zero in such a way that (2.16) holds for every $x, y \in \mathbb{R}^n$. In particular, u has a representative in $\operatorname{Lip}(\mathbb{R}^n)$.

Then assume that $u \in L^\infty(\mathbb{R}^n)$ is L-Lipschitz in $\mathbb{R}^n$. Using integration by parts as in connection with (2.10), we find that the classical gradient of u, which exists almost everywhere, is the weak gradient of u. Moreover, by Theorem 2.9 and (2.6) we see that $|\nabla u(x)| \le L$ for almost every $x \in \mathbb{R}^n$. Thus $u \in W^{1,\infty}(\mathbb{R}^n)$ and $\|\nabla u\|_\infty \le L$, and this implies the remaining assertions. □

REMARK 2.22. In addition to Lemma 2.20, we have the following density properties for smooth functions in Sobolev spaces.

(a) The subspace $C_0^\infty(\mathbb{R}^n)$ is dense in $W^{1,p}(\mathbb{R}^n)$ with $1 \le p < \infty$. For $p > 1$, this follows from Lemma 2.20 and a truncation argument. Since $C_0^\infty(\mathbb{R}^n)$ functions are Lipschitz continuous, we conclude that $\operatorname{Lip}_c(\mathbb{R}^n)$ is dense in $W^{1,p}(\mathbb{R}^n)$ with $1 \le p < \infty$.

(b) If $1 \le p < \infty$ and $\Omega \subset \mathbb{R}^n$ is an open set, then $W^{1,p}(\Omega) \cap C^\infty(\Omega)$ is dense in $W^{1,p}(\Omega)$. This follows from a standard partition of unity and convolution argument, see Evans and Gariepy [**122**, Section 4.2.1]. Moreover, if $u \in W^{1,p}(\Omega)$ is nonnegative, the approximating functions can be taken to be nonnegative. However, Example 2.3 shows that Lipschitz continuous functions are not dense in $W^{1,p}(\Omega)$ for every open set $\Omega \subset \mathbb{R}^n$.

The density properties above allow us to transfer results for classical derivatives to the weak derivatives, see Evans and Gariepy [**122**, Section 4.2.2] and Heinonen, Kilpeläinen and Martio [**187**, Theorem 1.24].

LEMMA 2.23. *Let $1 \le p < \infty$ and write $p' = \frac{p}{p-1}$, for $1 < p < \infty$, and $p' = \infty$, for $p = 1$. Assume that $\Omega \subset \mathbb{R}^n$ is an open set and that $u \in W^{1,p}_{\mathrm{loc}}(\Omega)$ and $v \in W^{1,p'}_{\mathrm{loc}}(\Omega)$. Then*

$$\nabla(uv)(x) = \nabla u(x)v(x) + u(x)\nabla v(x) \tag{2.17}$$

for almost every $x \in \Omega$.

Another useful result is the chain rule for weak derivatives, see Evans and Gariepy [**122**, Section 4.2.2] and Heinonen, Kilpeläinen and Martio [**187**, Theorem 1.18].

LEMMA 2.24. *Let $\Omega \subset \mathbb{R}^n$ be an open set. Assume that $u \in W^{1,p}(\Omega)$ and $f \in C^1(\mathbb{R})$ with $f' \in L^\infty(\mathbb{R})$ and $f(0) = 0$. Then $f \circ u \in W^{1,p}(\Omega)$ and*

$$\nabla(f \circ u)(x) = f'(u(x))\nabla u(x),$$

for almost every $x \in \Omega$.

The space of Lipschitz functions is closed under taking absolute values, minima and maxima. These useful properties are inherited by the Sobolev spaces $W^{1,p}(\Omega)$ with $1 \le p < \infty$, see Evans and Gariepy [**122**, 4.2.2] and Heinonen, Kilpeläinen and Martio [**187**, Theorem 1.20].

THEOREM 2.25. *Let $1 \le p < \infty$ and let $\Omega \subset \mathbb{R}^n$ be an open set.*

(a) *If $u \in W^{1,p}(\Omega)$, then $|u| \in W^{1,p}(\Omega)$ and*

$$\nabla|u| = \begin{cases} \nabla u, & \text{a.e. in } \{x \in \Omega : u(x) > 0\}, \\ 0, & \text{a.e. in } \{x \in \Omega : u(x) = 0\}, \\ -\nabla u, & \text{a.e. in } \{x \in \Omega : u(x) < 0\}. \end{cases}$$

(b) *If* $u \in W^{1,p}(\Omega)$ *and* $t \in \mathbb{R}$, *then* $\nabla u = 0$ *almost everywhere in the set*

$$\{x \in \Omega : u(x) = t\}.$$

(c) *If* $u, v \in W^{1,p}(\Omega)$, *then* $\max\{u, v\} \in W^{1,p}(\Omega)$ *and* $\min\{u, v\} \in W^{1,p}(\Omega)$. *Moreover,*

$$\nabla \max\{u, v\} = \begin{cases} \nabla u, & \text{a.e. in } \{x \in \Omega : u(x) \geq v(x)\}, \\ \nabla v, & \text{a.e. in } \{x \in \Omega : u(x) \leq v(x)\}, \end{cases}$$

and

$$\nabla \min\{u, v\} = \begin{cases} \nabla u, & \text{a.e. in } \{x \in \Omega : u(x) \leq v(x)\}, \\ \nabla v, & \text{a.e. in } \{x \in \Omega : u(x) \geq v(x)\}. \end{cases}$$

Corresponding claims as in (a)–(c) *also hold for functions in the local Sobolev space* $W^{1,p}_{\text{loc}}(\Omega)$.

REMARK 2.26. Theorem 2.25 implies that we can cut a Sobolev function at a certain level and the truncated function belongs to the same Sobolev space. More precisely, let $1 \leq p < \infty$ and $t \geq 0$, and assume that $u \in W^{1,p}(\Omega)$. Then $\min\{u, t\} \in W^{1,p}(\Omega)$ and

$$\nabla \min\{u, t\} = \begin{cases} \nabla u, & \text{a.e. in } \{x \in \Omega : u(x) < t\}, \\ 0, & \text{a.e. in } \{x \in \Omega : u(x) \geq t\}. \end{cases}$$

A similar claim also holds for $\max\{u, -t\}$ for $t \geq 0$. The restriction $t \geq 0$ is necessary if Ω is unbounded, since neither $\min\{u, t\}$ nor $\max\{u, -t\}$ belongs to $W^{1,p}(\Omega)$ if $t < 0$ and Ω is unbounded. Moreover, the truncated functions

$$u_t = \max\{-t, \min\{u, t\}\} = \begin{cases} t, & \text{in } \{x \in \Omega : u(x) \geq t\}, \\ u, & \text{in } \{x \in \Omega : -t < u(x) < t\}, \\ -t, & \text{in } \{x \in \Omega : u(x) \leq -t\}, \end{cases}$$

converge to $u \in W^{1,p}(\Omega)$ in $W^{1,p}(\Omega)$ as $t \to \infty$. Thus bounded $W^{1,p}(\Omega)$ functions are dense in $W^{1,p}(\Omega)$ when $1 \leq p < \infty$.

2.4. Sobolev spaces with zero boundary values

Roughly speaking, a Sobolev function belongs to the space $W^{1,p}_0(\Omega)$ if it vanishes on the boundary $\partial\Omega$. However, the meaning of boundary values is delicate since Sobolev functions are only defined up to Lebesgue measure zero. We discuss a pointwise approach to this question in Section 5.5.

DEFINITION 2.27. Let $1 \leq p < \infty$ and let $\Omega \subset \mathbb{R}^n$ be an open set. The Sobolev space with zero boundary values $W^{1,p}_0(\Omega)$ is the completion of $C^\infty_0(\Omega)$ with respect to the Sobolev norm (2.11). The space $W^{1,p}_0(\Omega)$ is equipped with the norm (2.11). We say that functions $u, v \in W^{1,p}(\Omega)$ have the same boundary values in the Sobolev sense if $u - v \in W^{1,p}_0(\Omega)$.

By definition, $u \in W^{1,p}_0(\Omega)$ if and only if there exist functions $u_i \in C^\infty_0(\Omega)$, $i \in \mathbb{N}$, such that $u_i \to u$ in $W^{1,p}(\Omega)$ as $i \to \infty$. The only difference compared to $W^{1,p}(\Omega)$ is that functions in $W^{1,p}_0(\Omega)$ can be approximated by $C^\infty_0(\Omega)$ functions instead of more general $C^\infty(\Omega)$ functions. That is, for $1 \leq p < \infty$, we have

$$W^{1,p}(\Omega) = \overline{W^{1,p}(\Omega) \cap C^\infty(\Omega)} \quad \text{and} \quad W^{1,p}_0(\Omega) = \overline{C^\infty_0(\Omega)},$$

where the completions are taken with respect to the Sobolev norm. It follows that $W_0^{1,p}(\Omega)$ is a closed subspace of $W^{1,p}(\Omega)$. Moreover,

$$C_0^\infty(\Omega) \subset \mathrm{Lip}_c(\Omega) \subset \mathrm{Lip}_0(\Omega) \subset W_0^{1,p}(\Omega) \subset W^{1,p}(\Omega) \subset L^p(\Omega)$$

for every open set $\Omega \subset \mathbb{R}^n$. The third inclusion above is proved in Lemma 2.29 and the other inclusions are clear.

REMARK 2.28. Let $1 \le p < \infty$ and let $\Omega \subset \mathbb{R}^n$ be an open set.

(a) If $u \in W^{1,p}(\Omega)$ has a compact support in Ω, then $u \in W_0^{1,p}(\Omega)$.

(b) If $u \in W^{1,p}(\Omega)$ and $\varphi \in \mathrm{Lip}_c(\Omega)$, then $u\varphi \in W_0^{1,p}(\Omega)$.

The proofs can be found in Heinonen, Kilpeläinen and Martio [**187**, Theorem 1.24 and Lemma 1.25].

LEMMA 2.29. *Let* $1 \le p < \infty$ *and let* $\Omega \subset \mathbb{R}^n$ *be an open set. Then* $\mathrm{Lip}_0(\Omega) \subset W_0^{1,p}(\Omega)$.

PROOF. By Remark 2.28 (b), $\mathrm{Lip}_c(\Omega) \subset W_0^{1,p}(\Omega)$. Hence, it suffices to show that functions in $\mathrm{Lip}_0(\Omega)$ can be approximated in $W^{1,p}(\Omega)$ by functions in $\mathrm{Lip}_c(\Omega)$. Let $u \in \mathrm{Lip}_0(\Omega)$. Define $v = u_+ \in \mathrm{Lip}_0(\Omega)$ and

$$v_\varepsilon(x) = \max\{v(x) - \varepsilon, 0\},$$

for $x \in \mathbb{R}^n$ and $0 < \varepsilon < 1$. Since $v(x) = 0$ for every $x \in \mathbb{R}^n \setminus \Omega$ and the support of v is bounded, it follows from the Lipschitz continuity of v that $v_\varepsilon \in \mathrm{Lip}_c(\Omega)$ for every $0 < \varepsilon < 1$.

By Theorem 2.25 (c) we have

$$\nabla v_\varepsilon = \begin{cases} \nabla v, & \text{a.e. in } \{x \in \Omega : v(x) \ge \varepsilon\}, \\ 0, & \text{a.e. in } \{x \in \Omega : v(x) \le \varepsilon\}, \end{cases}$$

and hence

$$\|\nabla v - \nabla v_\varepsilon\|_{L^p(\Omega)} \le \|\chi_{\{0 \le v \le \varepsilon\}} \nabla v\|_{L^p(\Omega)}.$$

By the dominated convergence theorem and Theorem 2.25 (b), we obtain

$$\begin{aligned} \lim_{\varepsilon \to 0} \|\chi_{\{0 \le v \le \varepsilon\}} \nabla v\|_{L^p(\Omega)} &= \left(\lim_{\varepsilon \to 0} \int_\Omega \chi_{\{0 \le v \le \varepsilon\}}(x) |\nabla v(x)|^p \, dx \right)^{\frac{1}{p}} \\ &= \left(\int_\Omega \lim_{\varepsilon \to 0} \chi_{\{0 \le v \le \varepsilon\}}(x) |\nabla v(x)|^p \, dx \right)^{\frac{1}{p}} \\ &= \left(\int_\Omega \chi_{\{v = 0\}}(x) |\nabla v(x)|^p \, dx \right)^{\frac{1}{p}} = 0. \end{aligned}$$

Here $\chi_{\{0 \le v \le \varepsilon\}} |\nabla v|^p \le |\nabla v|^p$ for every $0 < \varepsilon < 1$, and so $|\nabla v|^p \in L^1(\Omega)$ may be used as an integrable majorant. Since $0 \le v - v_\varepsilon \le \varepsilon$ and $\mathrm{supp}(v - v_\varepsilon) \subset \mathrm{supp}\, v$, we conclude that

$$\begin{aligned} \|v - v_\varepsilon\|_{W^{1,p}(\Omega)} &\le \|v - v_\varepsilon\|_{L^p(\Omega)} + \|\nabla v - \nabla v_\varepsilon\|_{L^p(\Omega)} \\ &\le \varepsilon \|\chi_{\mathrm{supp}\, v}\|_{L^p(\Omega)} + \|\chi_{\{0 \le v \le \varepsilon\}} \nabla v\|_{L^p(\Omega)} \longrightarrow 0, \end{aligned}$$

as $\varepsilon \to 0$. The function $u_- \in \mathrm{Lip}_0(\Omega)$ can be approximated with $\mathrm{Lip}_c(\Omega)$ functions in a similar manner, and the claim follows since $u = u_+ - u_-$. □

We have the following version of Theorem 2.25 for $W_0^{1,p}(\Omega)$, see Heinonen, Kilpeläinen and Martio [**187**, Lemma 1.23].

THEOREM 2.30. *Let $1 \le p < \infty$ and let $\Omega \subset \mathbb{R}^n$ be an open set. If $u, v \in W_0^{1,p}(\Omega)$, then $|u| \in W_0^{1,p}(\Omega)$, $\max\{u, v\} \in W_0^{1,p}(\Omega)$, and $\min\{u, v\} \in W_0^{1,p}(\Omega)$.*

It is easy to extend functions in Sobolev spaces with zero boundary values.

THEOREM 2.31. *Let $1 \le p < \infty$ and let $\Omega \subset \mathbb{R}^n$ be an open set. Assume that $u \in W_0^{1,p}(\Omega)$ and let u_0 be the zero extension of u, that is,*

$$u_0(x) = \begin{cases} u(x), & x \in \Omega, \\ 0, & x \in \mathbb{R}^n \setminus \Omega. \end{cases}$$

Then $u_0 \in W^{1,p}(\mathbb{R}^n)$ and

$$\nabla u_0 = \begin{cases} \nabla u, & a.e.\ in\ \Omega, \\ 0, & a.e.\ in\ \mathbb{R}^n \setminus \Omega. \end{cases}$$

In particular, $\|u_0\|_{W^{1,p}(\mathbb{R}^n)} = \|u\|_{W^{1,p}(\Omega)}$.

PROOF. Let ∇u be the weak gradient of u with respect to Ω, and let g be the componentwise zero extension of ∇u. By definition, there exist functions $u_i \in C_0^\infty(\Omega)$, $i \in \mathbb{N}$, such that $u_i \to u$ in $W^{1,p}(\Omega)$ as $i \to \infty$. If $\varphi \in C_0^\infty(\mathbb{R}^n)$, then

$$-\int_{\mathbb{R}^n} u_0(x)\nabla\varphi(x)\,dx = -\int_\Omega u(x)\nabla\varphi(x)\,dx = -\lim_{i\to\infty}\int_\Omega u_i(x)\nabla\varphi(x)\,dx.$$

Since the weak gradient of φ coincides with the classical gradient of φ in Ω, the last term can be written as

$$\lim_{i\to\infty}\int_\Omega \nabla u_i(x)\varphi(x)\,dx = \int_\Omega \nabla u(x)\varphi(x)\,dx = \int_{\mathbb{R}^n} g(x)\varphi(x)\,dx.$$

By the uniqueness of weak gradients, see Remark 2.15, we conclude that the weak gradient ∇u_0 of u_0 with respect to $\mathbb{R}^n$ coincides almost everywhere with g. The remaining conclusions follow from this. □

2.5. Weak convergence and Sobolev spaces

Let $1 \le p < \infty$ and $m \in \mathbb{N}$, and let $A \subset \mathbb{R}^n$ be a measurable set. Then $L^p(A; \mathbb{R}^m)$ is the space of functions $f = (f_1, \dots, f_m) \colon A \to \mathbb{R}^m$ with $f_i \in L^p(A)$ for every $i = 1, \dots, m$. Integration of such functions is done componentwise. The space $L^p(A; \mathbb{R}^m)$ is equipped with the norm

$$\|f\|_{L^p(A;\mathbb{R}^m)} = \left(\int_A |f(x)|^p\,dx\right)^{\frac{1}{p}},$$

where $|f(x)|$ is the Euclidean norm of $f(x) \in \mathbb{R}^m$. We will often abbreviate $L^p(A; \mathbb{R}^m) = L^p(A)$, but sometimes use $L^p(A; \mathbb{R}^m)$ to emphasize the difference.

The space $L^p(A; \mathbb{R}^m)$ is a Banach space if we identify two functions that differ on a set of measure zero. For $1 < p < \infty$, the dual space of $L^p(A; \mathbb{R}^m)$ can be identified with $L^{p'}(A; \mathbb{R}^m)$, where $p' = \frac{p}{p-1}$ is the conjugate exponent of p. The duality between $f \in L^p(A; \mathbb{R}^m)$ and $g \in L^{p'}(A; \mathbb{R}^m)$ is given by

$$\langle g, f\rangle = \int_A f(x) \cdot g(x)\,dx.$$

In particular, the space $L^p(A; \mathbb{R}^m)$ is reflexive for $1 < p < \infty$.

In this section we discuss weak convergence techniques for $L^p(\Omega;\mathbb{R}^m)$, where $\Omega \subset \mathbb{R}^n$ is an open set. Most of the results hold for more general Banach spaces as well.

DEFINITION 2.32. Let $1 < p < \infty$ and $m \in \mathbb{N}$, and let $\Omega \subset \mathbb{R}^n$ be an open set. A sequence $(f_i)_{i\in\mathbb{N}}$ of functions in $L^p(\Omega;\mathbb{R}^m)$ converges weakly in $L^p(\Omega;\mathbb{R}^m)$ to a function $f \in L^p(\Omega;\mathbb{R}^m)$, if

$$\lim_{i\to\infty} \int_\Omega f_i(x) \cdot g(x)\, dx = \int_\Omega f(x) \cdot g(x)\, dx$$

for every $g \in L^{p'}(\Omega;\mathbb{R}^m)$ with $p' = \frac{p}{p-1}$.

We remark that a weak limit is unique in $L^p(\Omega;\mathbb{R}^m)$ and hence almost everywhere. By the Cauchy–Schwarz and Hölder inequalities, convergent sequences in $L^p(\Omega;\mathbb{R}^m)$ are weakly convergent, but the converse is not true.

EXAMPLE 2.33. Let $1 < p < \infty$ and $x_i = (i, 0, \dots, 0)$, for $i \in \mathbb{N}$. We claim that the functions $f_i = \chi_{B(x_i,1)}$ converge weakly to zero in $L^p(\mathbb{R}^n) = L^p(\mathbb{R}^n;\mathbb{R}^1)$ as $i \to \infty$. Note that these functions do not converge to zero in $L^p(\mathbb{R}^n)$ since

$$\|f_i\|_{L^p(\mathbb{R}^n)} = \|f_1\|_{L^p(\mathbb{R}^n)} > 0$$

for every $i \in \mathbb{N}$.

In order to show the weak convergence to zero, let $g \in L^{p'}(\mathbb{R}^n)$ and $\varepsilon > 0$, and choose $\varphi \in L^{p'}(\mathbb{R}^n)$ with compact support satisfying $\|g - \varphi\|_{L^{p'}(\mathbb{R}^n)} \le \varepsilon$. Then

$$\begin{aligned}\limsup_{i\to\infty} \left| \int_{\mathbb{R}^n} f_i(x) g(x)\, dx \right| &= \limsup_{i\to\infty} \left| \int_{\mathbb{R}^n} f_i(x)(g(x) - \varphi(x))\, dx \right| \\ &\le \limsup_{i\to\infty} \|f_i\|_{L^p(\mathbb{R}^n)} \|g - \varphi\|_{L^{p'}(\mathbb{R}^n)} \le \varepsilon \|f_1\|_{L^p(\mathbb{R}^n)}.\end{aligned}$$

Since this holds for every $\varepsilon > 0$, we have

$$\lim_{i\to\infty} \int_{\mathbb{R}^n} f_i(x) g(x)\, dx = 0,$$

and thus the sequence $(f_i)_{i\in\mathbb{N}}$ converges weakly to zero in $L^p(\mathbb{R}^n)$.

Next we show that weakly convergent sequences are bounded and that the L^p norm is lower semicontinuous with respect to the weak convergence.

LEMMA 2.34. *Let $1 < p < \infty$ and $m \in \mathbb{N}$, and let $\Omega \subset \mathbb{R}^n$ be an open set. If a sequence $(f_i)_{i\in\mathbb{N}}$ converges to f weakly in $L^p(\Omega;\mathbb{R}^m)$, then $(f_i)_{i\in\mathbb{N}}$ is bounded in $L^p(\Omega;\mathbb{R}^m)$. Moreover, we have*

$$\|f\|_{L^p(\Omega;\mathbb{R}^m)} \le \liminf_{i\to\infty} \|f_i\|_{L^p(\Omega;\mathbb{R}^m)}. \tag{2.18}$$

PROOF. The claim

$$\sup_i \|f_i\|_{L^p(\Omega;\mathbb{R}^m)} < \infty$$

follows from the uniform boundedness principle in Wojtaszczyk [**400**, I.A.7] or the closed graph theorem in Wojtaszczyk [**400**, II.A.3]. In order to prove (2.18), let $g \in L^{p'}(\Omega;\mathbb{R}^m)$ with $\|g\|_{L^{p'}(\Omega;\mathbb{R}^m)} = 1$ and

$$\|f\|_{L^p(\Omega;\mathbb{R}^m)} = \int_\Omega f(x) \cdot g(x)\, dx.$$

The Cauchy–Schwarz and Hölder inequalities imply

$$\begin{aligned}\int_\Omega f(x)\cdot g(x)\,dx &= \lim_{i\to\infty}\int_\Omega f_i(x)\cdot g(x)\,dx\\ &\le \liminf_{i\to\infty}\|f_i\|_{L^p(\Omega;\mathbb{R}^m)}\|g\|_{L^{p'}(\Omega;\mathbb{R}^m)} = \liminf_{i\to\infty}\|f_i\|_{L^p(\Omega;\mathbb{R}^m)},\end{aligned}$$

and the desired inequality follows. □

Pointwise uniform bounds are preserved under weak convergence.

THEOREM 2.35. *Let $1 < p < \infty$ and $m \in \mathbb{N}$, and let $\Omega \subset \mathbb{R}^n$ be an open set. Assume that sequences $(f_i)_{i\in\mathbb{N}}$ and $(g_i)_{i\in\mathbb{N}}$ are such that f_i converges to f weakly in $L^p(\Omega;\mathbb{R}^m)$ and g_i converges to g weakly in $L^p(\Omega)$ as $i\to\infty$. If $|f_i(x)| \le g_i(x)$ for almost every $x \in \Omega$, then $|f(x)| \le g(x)$ for almost every $x\in\Omega$.*

PROOF. Let $x \in \Omega$ be a Lebesgue point of g and all components of f. Let $0 < r < d(x,\partial\Omega)$ and assume that $\int_{B(x,r)} f(y)\,dy \neq 0$. Define

$$z = \left|\fint_{B(x,r)} f(y)\,dy\right|^{-1}\fint_{B(x,r)} f(y)\,dy \in \mathbb{R}^m$$

and

$$\psi = z|B(x,r)|^{-1}\chi_{B(x,r)} \in L^{p'}(\Omega;\mathbb{R}^m).$$

By the Cauchy–Schwarz inequality and the assumptions, we have

$$\begin{aligned}\left|\fint_{B(x,r)} f(y)\,dy\right| &= z\cdot\fint_{B(x,r)} f(y)\,dy = \int_\Omega f(y)\cdot\psi(y)\,dy\\ &= \lim_{i\to\infty}\int_\Omega f_i(y)\cdot\psi(y)\,dy \le \liminf_{i\to\infty}\int_\Omega |f_i(y)||\psi(y)|\,dy\\ &\le \liminf_{i\to\infty}\int_\Omega g_i(y)|\psi(y)|\,dy = \fint_{B(x,r)} g(y)\,dy.\end{aligned}$$

Hence,

$$\left|\fint_{B(x,r)} f(y)\,dy\right| \le \fint_{B(x,r)} g(y)\,dy,$$

which clearly also holds if $\int_{B(x,r)} f(y)\,dy = 0$. The claim follows by taking $r\to 0$ on both sides since almost every point $x\in\Omega$ is a Lebesgue point of g and all components of f, by Theorem 1.21. □

A bounded sequence in $L^p(\Omega;\mathbb{R}^m)$ need not have a convergent subsequence. However, the following Theorem 2.36 shows that it always has a weakly convergent subsequence if $1 < p < \infty$. This will be important in our applications of weak convergence. Since $L^p(\Omega;\mathbb{R}^m)$ is reflexive and separable when $1 < p < \infty$, Theorem 2.36 follows from Wojtaszczyk [**400**, II.A.14 and II.C.3]. This theorem does not hold when $p = 1$, which can be seen by considering the sequence $(\varphi_{\frac{1}{k}})_{k\in\mathbb{N}}$ of functions from (2.14) approximating Dirac's delta.

THEOREM 2.36. *Let $1 < p < \infty$ and $m \in \mathbb{N}$, and let $\Omega \subset \mathbb{R}^n$ be an open set. Assume that $(f_i)_{i\in\mathbb{N}}$ is a bounded sequence in $L^p(\Omega;\mathbb{R}^m)$. Then there exists a subsequence $(f_{i_k})_{k\in\mathbb{N}}$ and a function $f \in L^p(\Omega;\mathbb{R}^m)$ such that $f_{i_k} \to f$ weakly in $L^p(\Omega;\mathbb{R}^m)$ as $k\to\infty$.*

Weak convergence is often too weak a mode of convergence and we need tools to upgrade it to stronger modes of convergence. We begin with the following result, which is related to Lemma 2.34. Since $L^p(\Omega;\mathbb{R}^m)$ is a uniformly convex Banach space, see Smith and Turett [**363**], the next lemma follows from Brezis [**56**, Proposition 3.32] or Hewitt and Stromberg [**191**, 15.17].

LEMMA 2.37. *Let $1 < p < \infty$ and $m \in \mathbb{N}$, and let $\Omega \subset \mathbb{R}^n$ be an open set. Assume that a sequence $(f_i)_{i\in\mathbb{N}}$ converges to f weakly in $L^p(\Omega;\mathbb{R}^m)$ and*

$$\limsup_{i\to\infty} \|f_i\|_{L^p(\Omega;\mathbb{R}^m)} \le \|f\|_{L^p(\Omega;\mathbb{R}^m)}. \tag{2.19}$$

Then $f_i \to f$ in $L^p(\Omega;\mathbb{R}^m)$, as $i \to \infty$.

Under the assumptions in Lemma 2.37 we have, by (2.18) and (2.19),

$$\|f\|_{L^p(\Omega;\mathbb{R}^m)} \le \liminf_{i\to\infty} \|f_i\|_{L^p(\Omega;\mathbb{R}^m)} \le \limsup_{i\to\infty} \|f_i\|_{L^p(\Omega;\mathbb{R}^m)} \le \|f\|_{L^p(\Omega;\mathbb{R}^m)},$$

which implies

$$\lim_{i\to\infty} \|f_i\|_{L^p(\Omega;\mathbb{R}^m)} = \|f\|_{L^p(\Omega;\mathbb{R}^m)}.$$

This means that the limit exists with an equality in (2.19).

For a proof of the following version of Mazur's lemma, see Yosida [**401**, pp. 120–121] and Björn and Björn [**31**, Section 6.1].

THEOREM 2.38. *Assume that X is a normed linear space and that $(x_i)_{i\in\mathbb{N}}$ is a sequence in X such that $x_i \to x$ weakly in X as $i \to \infty$. Then for all $i \in \mathbb{N}$ there exists $m_i > i$ and a convex combination $y_i = \sum_{j=i}^{m_i} a_{i,j}x_j$, with $a_{i,j} \ge 0$ and $\sum_{j=i}^{m_i} a_{i,j} = 1$, such that $y_i \to x$ in the norm of X as $i \to \infty$.*

Mazur's lemma asserts that for every weakly converging sequence, there is a sequence of convex combinations that converges strongly. Thus weak convergence is upgraded to strong convergence for a sequence of convex combinations. Observe that the convex combination is essentially for a subsequence since some of the coefficients $a_{i,j}$ may be zero. The feature that the convex combinations only involve the tail of the original sequence is applied to identify the weak limit in Theorem 2.40.

THEOREM 2.39. *Let $1 < p < \infty$ and let $\Omega \subset \mathbb{R}^n$ be an open set. Assume that $(u_i)_{i\in\mathbb{N}}$ is a bounded sequence in $W^{1,p}(\Omega)$. Then there exists a subsequence $(u_{i_k})_{k\in\mathbb{N}}$ and $u \in W^{1,p}(\Omega)$ such that $(u_{i_k}, \nabla u_{i_k}) \to (u, \nabla u)$ weakly in $L^p(\Omega;\mathbb{R}^{n+1})$ as $k \to \infty$. Moreover, if $u_i \in W_0^{1,p}(\Omega)$ for every $i \in \mathbb{N}$, then $u \in W_0^{1,p}(\Omega)$.*

PROOF. Let $f_i = (u_i, \nabla u_i) \in L^p(\Omega;\mathbb{R}^{n+1})$ for every $i \in \mathbb{N}$. Then $(f_i)_{i\in\mathbb{N}}$ is a bounded sequence in $L^p(\Omega;\mathbb{R}^{n+1})$. By Theorem 2.36, there exists a subsequence $(f_{i_k})_{k\in\mathbb{N}}$ that converges weakly to some f in $L^p(\Omega;\mathbb{R}^{n+1})$ as $k \to \infty$. Write $f = (u, v)$ with $u \in L^p(\Omega)$ and $v \in L^p(\Omega;\mathbb{R}^n)$. We show that $u \in W^{1,p}(\Omega)$ and that $(u_{i_k}, \nabla u_{i_k})$ converges to $(u, \nabla u)$ weakly in $L^p(\Omega;\mathbb{R}^{n+1})$ as $k \to \infty$. It suffices to prove that v is the weak gradient of u in Ω.

By using test functions of the form $(g_1, 0, \dots, 0)$ or $(0, g_2, \dots, g_{n+1})$, we conclude that $u_{i_k} \to u$ weakly in $L^p(\Omega)$ and $\nabla u_{i_k} \to v$ weakly in $L^p(\Omega;\mathbb{R}^n)$ as $k \to \infty$. Let $\varphi \in C_0^\infty(\Omega)$ and $j = 1, \dots, n$. Then

$$\int_\Omega \frac{\partial u_{i_k}}{\partial x_j}(x)\varphi(x)\,dx = -\int_\Omega u_{i_k}(x)\frac{\partial \varphi}{\partial x_j}(x)\,dx \xrightarrow{k\to\infty} -\int_\Omega u(x)\frac{\partial \varphi}{\partial x_j}(x)\,dx.$$

On the other hand, since $\nabla u_{i_k} \to v$ weakly in $L^p(\Omega;\mathbb{R}^n)$, by testing with function $(0,\dots,\varphi,\dots,0) \in L^{p'}(\Omega;\mathbb{R}^n)$, where φ is in the jth position, we see that

$$\int_\Omega \frac{\partial u_{i_k}}{\partial x_j}(x)\varphi(x)\,dx \xrightarrow{k\to\infty} \int_\Omega v_j(x)\varphi(x)\,dx.$$

This implies

$$\int_\Omega u(x)\frac{\partial \varphi}{\partial x_j}(x)\,dx = -\int_\Omega v_j(x)\varphi(x)\,dx$$

for every $\varphi \in C_0^\infty(\Omega)$, and thus $v = \nabla u$ in Ω.

For the last claim, we assume that $u_i \in W_0^{1,p}(\Omega)$ for every $i \in \mathbb{N}$ and that the sequence

$$(f_{i_k})_{k\in\mathbb{N}} = ((u_{i_k}, \nabla u_{i_k}))_{k\in\mathbb{N}}$$

converges weakly to $f = (u, \nabla u)$ in $L^p(\Omega;\mathbb{R}^{n+1})$. By Theorem 2.38, there exists a sequence of convex combinations

$$h_k = \sum_{j=k}^{m_k} a_{k,j} f_{i_j} = \left(\sum_{j=k}^{m_k} a_{k,j} u_{i_j}, \nabla\left(\sum_{j=k}^{m_k} a_{k,j} u_{i_j}\right)\right)$$

that converges to $f = (u, \nabla u)$ in $L^p(\Omega;\mathbb{R}^{n+1})$ as $k \to \infty$. In particular, the first component

$$h_{k,1} = \sum_{j=k}^{m_k} a_{k,j} u_{i_j} \in W_0^{1,p}(\Omega)$$

converges to u in $W^{1,p}(\Omega)$ as $k \to \infty$. Since $W_0^{1,p}(\Omega)$ is a closed subspace of $W^{1,p}(\Omega)$, it follows that $u \in W_0^{1,p}(\Omega)$, and the proof is complete. □

The following variant of Theorem 2.39 will be useful. It incorporates an additional mechanism for identifying the weak limit and also ensures that the original sequence converges weakly to this limit. We remark that Theorems 2.39 and 2.40 do not hold when $p = 1$.

THEOREM 2.40. *Let $1 < p < \infty$ and let $\Omega \subset \mathbb{R}^n$ be an open set. Assume that $(u_i)_{i\in\mathbb{N}}$ is a bounded sequence in $W^{1,p}(\Omega)$ such that either $u_i \to u$ weakly in $L^p(\Omega)$, as $i \to \infty$, or $u_i \to u$ almost everywhere in Ω as $i \to \infty$.*

(a) *Then $u \in W^{1,p}(\Omega)$, $u_i \to u$ weakly in $L^p(\Omega)$, and $\nabla u_i \to \nabla u$ weakly in $L^p(\Omega;\mathbb{R}^n)$ as $i \to \infty$.*
(b) *If $u_i \in W_0^{1,p}(\Omega)$ for every $i \in \mathbb{N}$, then $u \in W_0^{1,p}(\Omega)$.*

PROOF. Assertion (b) follows from (a) and Theorem 2.39. For (a) it suffices to show that $u \in W^{1,p}(\Omega)$ and $(u_i, \nabla u_i) \to (u, \nabla u)$ weakly in $L^p(\Omega;\mathbb{R}^{n+1})$ as $i \to \infty$. We prove the latter claim by showing that each subsequence $(u_{i_k})_{k\in\mathbb{N}}$ has a further subsequence, also denoted by $(u_{i_k})_{k\in\mathbb{N}}$, such that

$$(u_{i_k}, \nabla u_{i_k}) \to (u, \nabla u) \tag{2.20}$$

weakly in $L^p(\Omega;\mathbb{R}^{n+1})$ as $k \to \infty$. Indeed, if $g \in L^{p'}(\Omega;\mathbb{R}^{n+1})$, then

$$a_i = \int_\Omega (u_i(x), \nabla u_i(x)) \cdot g(x)\,dx \xrightarrow{i\to\infty} \int_\Omega (u(x), \nabla u(x)) \cdot g(x)\,dx = a,$$

since otherwise the definition of convergent real-valued sequences implies that $(a_i)_{i\in\mathbb{N}}$ has a subsequence all of whose subsequences fail to converge to a. This is a contradiction with respect to (2.20) when tested with g.

Let $(u_{i_k})_{k\in\mathbb{N}}$ be a subsequence of $(u_i)_{i\in\mathbb{N}}$. By Theorem 2.39, there exists a subsequence, also denoted by $(u_{i_k})_{k\in\mathbb{N}}$, and a function $v \in W^{1,p}(\Omega)$ such that $(u_{i_k}, \nabla u_{i_k}) \to (v, \nabla v)$ weakly in $L^p(\Omega;\mathbb{R}^{n+1})$ as $k \to \infty$. It suffices to show that $u = v$ almost everywhere. If $u_i \to u$ weakly in $L^p(\Omega)$, then in particular $u_{i_k} \to u$ weakly in $L^p(\Omega)$ and $u = v$ almost everywhere by the uniqueness of weak limits. Hence we may assume that $u_i \to u$ almost everywhere in Ω as $i \to \infty$.

By Theorem 2.38, there exists a sequence of convex combinations

$$h_k = \sum_{j=k}^{m_k} a_{k,j}(u_{i_j}, \nabla u_{i_j})$$

that converges to $(v, \nabla v)$ in $L^p(\Omega;\mathbb{R}^{n+1})$ as $k \to \infty$. In particular, the first component $h_{k,1} = \sum_{j=k}^{m_k} a_{k,j}u_{i_j}$ converges to v in $L^p(\Omega)$, and therefore some subsequence of $(h_{k,1})_{k\in\mathbb{N}}$ converges to v almost everywhere in Ω. On the other hand, since $u_i \to u$ almost everywhere in Ω as $i \to \infty$, we have

$$h_{k,1} = \sum_{j=k}^{m_k} a_{k,j}u_{i_j} \xrightarrow{k\to\infty} u \tag{2.21}$$

almost everywhere in Ω. Hence $u = v$ almost everywhere in Ω, from which we conclude that $u \in W^{1,p}(\Omega)$ and that (2.20) holds. $\square$

REMARK 2.41. The fact that $u_i \to u$ almost everywhere in Ω, as $i \to \infty$, does not imply for general convex combinations that

$$\sum_{j=1}^{m_k} a_{k,j}u_{i_j} \xrightarrow{k\to\infty} u$$

almost everywhere in Ω. However, for the tail convex combinations in (2.21) the convergence holds.

In order to demonstrate weak convergence techniques, we prove continuity of the operator $u \mapsto |u|$ on Sobolev spaces. This operator is easily shown to be bounded on Sobolev spaces, but continuity is not a consequence of boundedness since the operator is not linear.

THEOREM 2.42. *Let $1 < p < \infty$ and let $\Omega \subset \mathbb{R}^n$ be an open set. Assume that $(u_i)_{i\in\mathbb{N}}$ is a sequence in $W^{1,p}(\Omega)$ that converges to u in $W^{1,p}(\Omega)$. Then $|u_i| \to |u|$ in $W^{1,p}(\Omega)$ as $i \to \infty$.*

PROOF. Since $u_i \to u$ in $L^p(\Omega)$ and

$$\big||u_i(x)| - |u(x)|\big| \le |u_i(x) - u(x)|$$

for every $x \in \Omega$, we see that $|u_i| \to |u|$ in $L^p(\Omega)$ as $i \to \infty$. By Theorem 2.25 (a) and (b),

$$\big|\nabla|u_i|(x)\big| = |\nabla u_i(x)| \quad \text{and} \quad \big|\nabla|u|(x)\big| = |\nabla u(x)| \tag{2.22}$$

for almost every $x \in \Omega$. Since $(u_i)_{i\in\mathbb{N}}$ converges in $W^{1,p}(\Omega)$, we have by (2.22) that $(|u_i|)_{i\in\mathbb{N}}$ is a bounded sequence in $W^{1,p}(\Omega)$. Since $|u_i| \to |u|$ weakly in $L^p(\Omega)$ as $i \to \infty$, Theorem 2.40 (a) implies $\nabla|u_i| \to \nabla|u|$ weakly in $L^p(\Omega;\mathbb{R}^n)$ as $i \to \infty$.

From (2.22) we obtain

$$\lim_{i\to\infty}\int_{\Omega}\left|\nabla|u_i|(x)\right|^p dx = \lim_{i\to\infty}\int_{\Omega}|\nabla u_i(x)|^p dx = \int_{\Omega}|\nabla u(x)|^p dx = \int_{\Omega}\left|\nabla|u|(x)\right|^p dx. \tag{2.23}$$

Lemma 2.37 implies that $\nabla|u_i|$ converges to $\nabla|u|$ in $L^p(\Omega;\mathbb{R}^n)$, as $i\to\infty$. This shows that $|u_i|\to|u|$ in $W^{1,p}(\Omega)$ as $i\to\infty$. □

2.6. Difference quotients

This section gives a characterization of Sobolev spaces in terms of difference quotients. The proof is based on weak convergence methods from Section 2.5.

DEFINITION 2.43. Let $\Omega\subset\mathbb{R}^n$ be an open set. Assume that $u\in L^1_{\text{loc}}(\Omega)$ and let $\Omega'\Subset\Omega$ be an open set. The jth difference quotient, with $j=1,\dots,n$, of u at $x\in\Omega'$ is

$$D_j^h u(x)=\frac{u(x+he_j)-u(x)}{h},$$

for $h\in\mathbb{R}$ with $0<|h|<d(\Omega',\partial\Omega)$. Here e_j is the jth coordinate unit vector in $\mathbb{R}^n$. We also write

$$D^h u=(D_1^h u,\dots,D_n^h u).$$

REMARK 2.44. The definition of difference quotients makes sense at every $x\in\Omega$ whenever $0<|h|<d(x,\partial\Omega)$. If $\Omega=\mathbb{R}^n$, then the definition makes sense for every $h\neq 0$. We interpret $d(\Omega',\partial\mathbb{R}^n)=\infty$ in Definition 2.43 and Theorem 2.45.

Pointwise derivatives are defined as limits of difference quotients. Theorem 2.45 shows that Sobolev spaces can be characterized by integrated difference quotients, bypassing this limiting process. We remark that assertions (b) and (c) in Theorem 2.45 do not hold for $p=1$.

THEOREM 2.45. *Let $\Omega\subset\mathbb{R}^n$ and $\Omega'\Subset\Omega$ be open sets.*

(a) *Let $1\le p<\infty$ and assume that $u\in W^{1,p}(\Omega)$. There exists a constant $C=C(n)$ such that*

$$\|D^h u\|_{L^p(\Omega';\mathbb{R}^n)}\le C\|\nabla u\|_{L^p(\Omega;\mathbb{R}^n)} \tag{2.24}$$

whenever $0<|h|<d(\Omega',\partial\Omega)$.

(b) *Let $1<p<\infty$. Assume that $u\in L^p(\Omega)$ and that there exists a constant C such that*

$$\|D^h u\|_{L^p(\Omega';\mathbb{R}^n)}\le C$$

whenever $0<|h|<d(\Omega',\partial\Omega)$. Then the weak gradient ∇u exists in Ω', $u\in W^{1,p}(\Omega')$ and $\|\nabla u\|_{L^p(\Omega';\mathbb{R}^n)}\le C$.

(c) *Let $1<p<\infty$. Assume that $u\in L^p(\mathbb{R}^n)$ and that there exists a constant C such that*

$$\|D^h u\|_{L^p(\mathbb{R}^n;\mathbb{R}^n)}\le C \tag{2.25}$$

for every $h\neq 0$. Then the weak gradient ∇u with respect to $\mathbb{R}^n$ exists, $u\in W^{1,p}(\mathbb{R}^n)$ and $\|\nabla u\|_{L^p(\mathbb{R}^n;\mathbb{R}^n)}\le C$.

PROOF. To prove part (a), we first assume that $u \in C^\infty(\Omega) \cap W^{1,p}(\Omega)$. Let $x \in \Omega'$ and let $h \in \mathbb{R}$ be such that $0 < |h| < d(\Omega', \partial\Omega)$. Then, for every $j = 1, \dots, n$,

$$\begin{aligned} u(x + he_j) - u(x) &= \int_0^h \frac{\partial}{\partial t}(u(x + te_j))\,dt = \int_0^h \nabla u(x + te_j) \cdot e_j\,dt \\ &= \int_0^h \frac{\partial u}{\partial x_j}(x + te_j)\,dt. \end{aligned}$$

By Hölder's inequality

$$\begin{aligned} |D_j^h u(x)| &= \left|\frac{u(x + he_j) - u(x)}{h}\right| \le \frac{1}{|h|}\int_{-|h|}^{|h|}\left|\frac{\partial u}{\partial x_j}(x + te_j)\right| dt \\ &\le \frac{1}{|h|}\left(\int_{-|h|}^{|h|}\left|\frac{\partial u}{\partial x_j}(x + te_j)\right|^p dt\right)^{\frac{1}{p}} |2h|^{1-\frac{1}{p}}, \end{aligned}$$

which implies

$$|D_j^h u(x)|^p \le \frac{2^{p-1}}{|h|}\int_{-|h|}^{|h|}\left|\frac{\partial u}{\partial x_j}(x + te_j)\right|^p dt.$$

Next we integrate over Ω' and switch the order of integration by Fubini's theorem as follows

$$\begin{aligned} \int_{\Omega'} |D_j^h u(x)|^p\,dx &\le \frac{2^{p-1}}{|h|}\int_{\Omega'}\int_{-|h|}^{|h|}\left|\frac{\partial u}{\partial x_j}(x + te_j)\right|^p dt\,dx \\ &= \frac{2^{p-1}}{|h|}\int_{-|h|}^{|h|}\int_{\Omega'}\left|\frac{\partial u}{\partial x_j}(x + te_j)\right|^p dx\,dt \le 2^p \int_\Omega \left|\frac{\partial u}{\partial x_j}(x)\right|^p dx. \end{aligned}$$

The final step above follows from the assumption $0 < |h| < d(\Omega', \partial\Omega)$, since if $|t| \le |h|$, then

$$\int_{\Omega'}\left|\frac{\partial u}{\partial x_j}(x + te_j)\right|^p dx \le \int_\Omega \left|\frac{\partial u}{\partial x_j}(x)\right|^p dx.$$

Using the elementary inequality $(a_1 + \cdots + a_n)^\alpha \le n^\alpha(a_1^\alpha + \cdots + a_n^\alpha)$, where $a_i \ge 0$, $\alpha > 0$, we obtain

$$\begin{aligned} \int_{\Omega'} |D^h u(x)|^p\,dx &= \int_{\Omega'}\left(\sum_{j=1}^n |D_j^h u(x)|^2\right)^{\frac{p}{2}} dx \le n^{\frac{p}{2}}\int_{\Omega'}\sum_{j=1}^n |D_j^h u(x)|^p\,dx \\ &= n^{\frac{p}{2}}\sum_{j=1}^n\int_{\Omega'} |D_j^h u(x)|^p\,dx \le 2^p n^{\frac{p}{2}}\sum_{j=1}^n\int_\Omega\left|\frac{\partial u}{\partial x_j}(x)\right|^p dx \\ &\le 2^p n^{1+\frac{p}{2}}\int_\Omega |\nabla u(x)|^p\,dx. \end{aligned}$$

The case $u \in C^\infty(\Omega) \cap W^{1,p}(\Omega)$ of inequality (2.24) follows with $C = C(n) = 2n^{\frac{3}{2}}$.

Then we consider the general case $u \in W^{1,p}(\Omega)$. We choose functions $u_i \in C^\infty(\Omega) \cap W^{1,p}(\Omega)$, $i \in \mathbb{N}$, such that $u_i \to u$ in $W^{1,p}(\Omega)$ as $i \to \infty$. By passing to a subsequence, if necessary, we may also assume that $u_i \to u$ pointwise almost everywhere in Ω as $i \to \infty$. Assume that $0 < |h| < d(\Omega', \partial\Omega)$. Then $D^h u_i(x) \to D^h u(x)$ for almost every $x \in \Omega'$ as $i \to \infty$. By Fatou's lemma and inequality (2.24), we obtain

$$\begin{aligned}\int_{\Omega'}|D^h u(x)|^p\,dx &\le \liminf_{i\to\infty}\int_{\Omega'}|D^h u_i(x)|^p\,dx\\ &\le C(n)\liminf_{i\to\infty}\int_{\Omega}|\nabla u_i(x)|^p\,dx = C(n)\int_{\Omega}|\nabla u(x)|^p\,dx,\end{aligned}$$

and this proves part (a).

For the proof of (b), we let $\varphi\in C_0^\infty(\Omega')$ and $j=1,\dots,n$, and assume that $0<|h|<d(\operatorname{supp}\varphi,\partial\Omega')$. A change of variables gives

$$\int_{\Omega'}u(x)\frac{\varphi(x+he_j)-\varphi(x)}{h}\,dx = -\int_{\Omega'}\frac{u(x-he_j)-u(x)}{-h}\varphi(x)\,dx.$$

This shows that

$$\int_{\Omega'}u(x)D_j^h\varphi(x)\,dx = -\int_{\Omega'}D_j^{-h}u(x)\varphi(x)\,dx \tag{2.26}$$

whenever $0<|h|<d(\operatorname{supp}\varphi,\partial\Omega')$. On the other hand, by assumption

$$\sup_{0<|h|<d(\Omega',\partial\Omega)}\|D^{-h}u\|_{L^p(\Omega';\mathbb{R}^n)}\le C.$$

Since $1<p<\infty$, Theorem 2.36 implies that there exist $f\in L^p(\Omega';\mathbb{R}^n)$ and a sequence $(h_i)_{i\in\mathbb{N}}$ converging to zero such that $D^{-h_i}u\to f$ weakly in $L^p(\Omega';\mathbb{R}^n)$ as $i\to\infty$. By (2.26), we have

$$\begin{aligned}\int_{\Omega'}u(x)\frac{\partial\varphi}{\partial x_j}(x)\,dx &= \int_{\Omega'}u(x)\Big(\lim_{i\to\infty}D_j^{h_i}\varphi(x)\Big)\,dx = \lim_{i\to\infty}\int_{\Omega'}u(x)D_j^{h_i}\varphi(x)\,dx\\ &= -\lim_{i\to\infty}\int_{\Omega'}D_j^{-h_i}u(x)\varphi(x)\,dx = -\int_{\Omega'}f_j(x)\varphi(x)\,dx\end{aligned}$$

for every $\varphi\in C_0^\infty(\Omega')$. Here the second equality follows from the dominated convergence theorem and the facts that Ω' is bounded and $u\in L^p(\Omega)$, and the final equality is the weak convergence tested with $g=(0,\dots,\varphi,\dots,0)$, where φ is in the jth position. It follows that $\nabla u=f$ in the weak sense with respect to Ω', and hence $u\in W^{1,p}(\Omega')$. By inequality (2.18),

$$\|\nabla u\|_{L^p(\Omega';\mathbb{R}^n)} = \|f\|_{L^p(\Omega';\mathbb{R}^n)}\le\liminf_{i\to\infty}\|D^{-h_i}u\|_{L^p(\Omega';\mathbb{R}^n)}\le C,$$

and this concludes the proof of (b).

Recall that in part (c) we assume that (2.25) holds for $u\in L^p(\mathbb{R}^n)$ whenever $h\ne 0$. Let $\Omega_i=B(0,2i)$ and $\Omega_i'=B(0,i)$ for every $i\in\mathbb{N}$. Assertion (b) and the assumption (2.25) imply that $u_i=u|_{\Omega_i}$, $i\in\mathbb{N}$, has a weak gradient ∇u_i in Ω_i' and $\|\nabla u_i\|_{L^p(\Omega_i';\mathbb{R}^n)}\le C$. Since $\nabla u_{i+1}=\nabla u_i$ almost everywhere in Ω_i', we see that the limit

$$f(x)=\lim_{i\to\infty}\chi_{\Omega_i'}(x)\nabla u_i(x)$$

exists for almost every $x\in\mathbb{R}^n$. The weak gradient of u with respect to $\mathbb{R}^n$ coincides with $f\in L^1_{\mathrm{loc}}(\mathbb{R}^n;\mathbb{R}^n)$ and Fatou's lemma implies

$$\begin{aligned}\|\nabla u\|_{L^p(\mathbb{R}^n;\mathbb{R}^n)} = \|f\|_{L^p(\mathbb{R}^n;\mathbb{R}^n)} &= \left(\int_{\mathbb{R}^n}\lim_{i\to\infty}\left|\chi_{\Omega_i'}(x)\nabla u_i(x)\right|^p dx\right)^{\frac1p}\\ &\le\liminf_{i\to\infty}\left(\int_{\Omega_i'}|\nabla u_i(x)|^p\,dx\right)^{\frac1p}\le C.\end{aligned}$$

From this it also follows that $u\in W^{1,p}(\mathbb{R}^n)$, and the proof is complete. □

2.7. Notes

References for Sobolev spaces include Adams and Hedberg [**4**], Adams and Fournier [**5**], Björn and Björn [**31**], Diening, Harjulehto, Hästö and Růžička [**104**], Edmunds and Evans [**116**], Evans [**121**], Evans and Gariepy [**122**], Goldstein and Reshetnyak [**154**], Gilbarg and Trudinger [**152**], Heinonen, Kilpeläinen and Martio [**187**], Heinonen, Koskela, Shanmugalingam and Tyson [**190**], Kufner [**248**], Leoni [**266**], Maz′ya [**318, 321**], Reshetnyak [**354**], Sobolev [**365**], Stein [**369**], Triebel [**384**] and Ziemer [**407**].

For Lipschitz functions, we refer to Evans and Gariepy [**122**, Chapter 3] and Heinonen [**186**]. Theorem 2.7 is from [**122**, 3.1.1]. Lemma 2.19 can be found in García-Cuerva and Rubio de Francia [**139**, Theorem 4.13] and Grafakos [**156**, Section 2.1.2]. For Sobolev spaces with zero boundary values, we refer to Adams and Hedberg [**4**], Diening, Harjulehto, Hästö and Růžička [**104**], Heinonen, Kilpeläinen and Martio [**187**] and Maz′ya [**318, 321**]. Characterizations for Sobolev spaces with zero boundary values will be obtained in Section 5.5. Traces and extensions are discussed, for instance, in Brudnyi and Brudnyi [**59, 60**], Ihnatsyeva and Vähäkangas [**202**], Jonsson and Wallin [**215**], Shvartsman [**361**] and Triebel [**385**].

For the general theory of weak convergence we refer to Brezis [**56**], Hewitt and Stromberg [**191**], Wojtaszczyk [**400**] and Yosida [**401**]. Weak convergence techniques for Sobolev spaces are discussed in Heinonen, Kilpeläinen and Martio [**187**, Chapter 1] and Heinonen, Koskela, Shanmugalingam and Tyson [**190**]. For a general theory of Banach space valued L^p functions, we refer to Hytönen, van Neerven, Veraar and Weis [**200**]. For Theorem 2.39 see Heinonen, Kilpeläinen and Martio [**187**, Theorem 1.31] and for Theorem 2.40 see [**187**, Theorem 1.32]. The proof of Theorem 2.40 is inspired by Jakszto [**210**]. Theorem 2.42 also holds when $p = 1$, see Marcus and Mizel [**304**]. Theorem 2.45 can be found in Brezis [**56**, Proposition 9.3], Evans [**121**, 5.8.2] and Gilbarg and Trudinger [**152**, Section 7.11].

CHAPTER 3

Sobolev and Poincaré Inequalities

This chapter studies several versions of Sobolev and Poincaré inequalities, which are among the most important tools in the theory of partial differential equations and Sobolev spaces. As we shall see later in Theorem 4.16, even the definition of Sobolev spaces can be based on a Poincaré inequality. Pointwise estimates and maximal functions are applied in the proofs. The inequalities are first proved for Lipschitz functions and then extended to Sobolev functions by approximation. Lipschitz truncation method is applied in Theorem 3.6 and Theorem 3.13 to cover the borderline case $p = 1$.

3.1. Pointwise estimates for Lipschitz functions

We discuss pointwise estimates for Lipschitz and Sobolev functions in terms of the Riesz potential and the Hardy–Littlewood maximal function of the gradient. In particular, these results hold for $C_0^\infty(\mathbb{R}^n)$ functions.

REMARK 3.1. We consider a slightly different version of the Hausdorff measure than in Definition 1.43. Let

$$H_\delta^\lambda(E) = \frac{\alpha(\lambda)}{2^\lambda} \inf\left\{ \sum_{i=1}^\infty \operatorname{diam}(A_i)^\lambda : E \subset \bigcup_{i=1}^\infty A_i,\ 0 < \operatorname{diam}(A_i) \le \delta \right\},$$

where $\alpha(\lambda) = \frac{\pi^{\frac{\lambda}{2}}}{\Gamma(\frac{\lambda}{2}+1)}$ and Γ is the gamma function. The normalized Hausdorff measure is

$$H^\lambda(E) = \lim_{\delta \to 0} H_\delta^\lambda(E).$$

It is easy to see that there exists a constant $C = C(\lambda) \ge 1$ such that

$$\frac{1}{C} H^\lambda(E) \le \mathcal{H}^\lambda(E) \le C H^\lambda(E)$$

for every $E \subset \mathbb{R}^n$, where $\mathcal{H}^\lambda$ is the Hausdorff measure in Definition 1.43. Note that $\alpha(n) = \omega_n$ for every $n \in \mathbb{N}$. The main advantage of H^λ compared to $\mathcal{H}^\lambda$ is that $H^n(E) = |E|$ for every measurable $E \subset \mathbb{R}^n$, see Evans and Gariepy [**122**, Section 2.2]. With this normalization the natural $(n-1)$-dimensional surface measure of $E \subset \mathbb{R}^n$ is given by $\sigma(E) = H^{n-1}(E)$. An integration over spheres formula for $f \in L^1(\mathbb{R}^n)$ is

$$\int_{\mathbb{R}^n} f(x)\,dx = \int_0^\infty \int_{\partial B(0,r)} f(x)\,d\sigma(x)\,dr, \tag{3.1}$$

see Evans and Gariepy [**122**, Proposition 1, p. 118].

We begin with a local pointwise estimate for the oscillation of a Lipschitz function in terms of the Riesz potential of the gradient, see Definition 1.32.

LEMMA 3.2. *Assume that $u \in \operatorname{Lip}(\mathbb{R}^n)$ and let $B = B(x_0, r) \subset \mathbb{R}^n$ be a ball. Then*

$$|u(x) - u_B| \le \frac{2^n}{n\omega_n} I_1(|\nabla u|\chi_B)(x) \tag{3.2}$$

for every $x \in B$.

PROOF. Let $x \in B$. Since Lipschitz functions on $\mathbb{R}$ are absolutely continuous, we have

$$u(x) - u(y) = -\int_0^{|x-y|} \frac{\partial}{\partial t}\left(u\left(x + t\frac{y-x}{|y-x|}\right)\right) dt,$$

for every $y \in B$. Since the line-segment between x and $y \in B$ is a subset of B, we obtain

$$\begin{aligned} u(x) - u_B &= \fint_B (u(x) - u(y))\, dy \\ &= -\frac{1}{|B|}\int_B \int_0^{|x-y|} \chi_B\left(x + t\frac{y-x}{|y-x|}\right) \frac{\partial}{\partial t}\left(u\left(x + t\frac{y-x}{|y-x|}\right)\right) dt\, dy. \end{aligned}$$

By writing $y = x + \rho z$ with $0 < \rho < 2r$ and $|z| = 1$, and integrating over spheres,

$$\begin{aligned} |u(x) - u_B| &\le \frac{1}{|B|}\int_0^{2r} \int_{\partial B(0,1)} \int_0^{2r} \chi_B(x+tz)\left|\frac{\partial}{\partial t}(u(x+tz))\right| dt\, d\sigma(z)\, \rho^{n-1}\, d\rho \\ &= \frac{(2r)^n}{n|B|}\int_{\partial B(0,1)} \int_0^{2r} \chi_B(x+tz)\left|\frac{\partial}{\partial t}(u(x+tz))\right| dt\, d\sigma(z). \end{aligned}$$

Let E be the set of those $z \in \partial B(0,1)$ for which u is $\mathbb{R}^n$-differentiable at H^1-almost every point on the line-segment from x and $x + 2rz$. By Rademacher's theorem, see Theorem 2.9, followed by integration over spheres and Fubini's theorem, it holds that $H^{n-1}(\partial B(0,1) \setminus E) = 0$. By the chain rule, we get

$$\int_0^{2r} \chi_B(x+tz)\left|\frac{\partial}{\partial t}(u(x+tz))\right| dt = \int_0^{2r} \chi_B(x+tz)\left|\nabla u(x+tz) \cdot z\right| dt$$

for every $z \in E$. Fubini's theorem and integration over spheres imply

$$\begin{aligned} |u(x) - u_B| &\le \frac{(2r)^n}{n|B|}\int_{\partial B(0,1)} \int_0^{2r} \frac{\chi_B(x+tz)|\nabla u(x+tz)|}{|x-(x+tz)|^{n-1}} t^{n-1}\, dt\, d\sigma(z) \\ &\le \frac{(2r)^n}{n|B|}\int_B \frac{|\nabla u(y)|}{|x-y|^{n-1}}\, dy = \frac{2^n}{n\omega_n} I_1(|\nabla u|\chi_B)(x). \end{aligned}$$ □

There is a global version of Lemma 3.2 for compactly supported Lipschitz functions.

LEMMA 3.3. *Assume that $u \in \operatorname{Lip}_c(\mathbb{R}^n)$. Then*

$$|u(x)| \le \frac{2^n}{n\omega_n} I_1|\nabla u|(x)$$

for every $x \in \mathbb{R}^n$.

PROOF. Let $B_j = B(0, j) \subset \mathbb{R}^n$ for $j \in \mathbb{N}$. We have

$$|u_{B_j}| \le \fint_{B_j} |u(x)|\, dx \le |B_j|^{-1}\|u\|_{L^1(\mathbb{R}^n)} \xrightarrow{j\to\infty} 0.$$

Thus $\lim_{j\to\infty} u_{B_j} = 0$. Since $x \in B_j$ for $j > |x|$, by Lemma 3.2 we obtain

$$|u(x)| = \lim_{j\to\infty} |u(x) - u_{B_j}| \le \frac{2^n}{n\omega_n} \limsup_{j\to\infty} I_1(|\nabla u|\chi_{B_j})(x) \le \frac{2^n}{n\omega_n} I_1|\nabla u|(x)$$

for every $x \in \mathbb{R}^n$. □

Lemma 3.2 also implies a useful pointwise estimate for the oscillation of a function in terms of the maximal function of the gradient. This can be considered as a pointwise version of the Poincaré inequality, see Lemma 3.9 below.

THEOREM 3.4. *Assume that $u \in \operatorname{Lip}(\mathbb{R}^n)$ and let $B = B(x_0, r) \subset \mathbb{R}^n$ be a ball. There exists a constant $C = C(n)$ such that*

$$|u(x) - u_B| \le CrM(|\nabla u|\chi_B)(x)$$

for every $x \in B$.

PROOF. Let $x \in B$. Then $B \subset B(x, 2r)$, and we obtain from Lemma 3.2 and Lemma 1.35 that

$$\begin{aligned} |u(x) - u_B| &\le C(n) I_1(|\nabla u|\chi_B)(x) = C(n) \int_{B(x,2r)} \frac{|\nabla u(y)|\chi_B(y)}{|x-y|^{n-1}}\, dy \\ &\le C(n) r M(|\nabla u|\chi_B)(x). \end{aligned}$$

□

Next we show how the pointwise estimates above can be generalized to Sobolev functions.

THEOREM 3.5. *Let $1 \le p < \infty$. Assume that $u \in W^{1,p}_{\mathrm{loc}}(\mathbb{R}^n)$ and let $B = B(x_0, r) \subset \mathbb{R}^n$ be a ball. Then*

$$|u(x) - u_B| \le C(n) I_1(|\nabla u|\chi_B)(x) \le C(n) r M(|\nabla u|\chi_B)(x) \tag{3.3}$$

for almost every $x \in B$.

PROOF. By Remark 2.22, functions in $W^{1,p}(B(x_0, 2r)) \cap C^\infty(B(x_0, 2r))$ are dense in $W^{1,p}(B(x_0, 2r))$, and they are Lipschitz continuous in $B = B(x_0, r)$. Hence there exist functions $u_i \in \operatorname{Lip}(B)$, $i \in \mathbb{N}$, such that $u_i \to u$ in $W^{1,p}(B)$ as $i \to \infty$. By Theorem 2.7, we may assume that $u_i \in \operatorname{Lip}(\mathbb{R}^n)$ for every $i \in \mathbb{N}$. By passing to a subsequence, if necessary, we obtain a set $E_1 \subset B$ with $|E_1| = 0$ such that

$$\lim_{i\to\infty} u_i(x) = u(x) < \infty$$

for every $x \in B \setminus E_1$. Lemma 1.37 implies

$$\begin{aligned} &\left|\{x \in \mathbb{R}^n : \left|I_1(|\nabla u_i|\chi_B)(x) - I_1(|\nabla u|\chi_B)(x)\right| > t\}\right| \\ &\le \left|\{x \in \mathbb{R}^n : I_1\big(\big||\nabla u_i| - |\nabla u|\big|\chi_B\big)(x) > t\}\right| \\ &\le C(n) t^{-\frac{n}{n-1}} \|\nabla u_i - \nabla u\|_{L^1(B)}^{\frac{n}{n-1}} \end{aligned}$$

for every $t > 0$. Thus, Hölder's inequality shows that

$$I_1(|\nabla u_i|\chi_B) \to I_1(|\nabla u|\chi_B)$$

in measure as $i \to \infty$. By passing again to a subsequence, if necessary, we obtain a set $E_2 \subset B$ with $|E_2| = 0$ such that

$$\lim_{i\to\infty} I_1(|\nabla u_i|\chi_B)(x) = I_1(|\nabla u|\chi_B)(x) < \infty$$

for every $x \in B \setminus E_2$.

Moreover, it follows from Hölder's inequality and L^p convergence that

$$(3.4)\qquad |(u_i)_B - u_B| \le \fint_B |u_i(x) - u(x)|\,dx \le \Big(\fint_B |u_i(x) - u(x)|^p\,dx\Big)^{\frac{1}{p}} \xrightarrow{i\to\infty} 0,$$

and thus $(u_i)_B \to u_B$ as $i \to \infty$. By applying (3.2) and arguing as in the proof of Theorem 3.4, we obtain

$$\begin{aligned} |u(x) - u_B| &= \lim_{i\to\infty} |u_i(x) - (u_i)_B| \le C(n) \lim_{i\to\infty} I_1(|\nabla u_i|\chi_B)(x) \\ &= C(n) I_1(|\nabla u|\chi_B)(x) \le C(n) r M(|\nabla u|\chi_B)(x), \end{aligned}$$

for every $x \in B \setminus (E_1 \cup E_2)$, where $|E_1 \cup E_2| = 0$. □

3.2. Sobolev–Gagliardo–Nirenberg inequality

Pointwise estimates in Section 3.1 imply the Sobolev–Gagliardo–Nirenberg inequality. The proof of the case $p > 1$ is a consequence of Lemma 3.3 and the Hardy–Littlewood–Sobolev inequality for the Riesz potential in Theorem 1.36. The argument for $p = 1$ is based on a truncation method and the weak type estimate for the Riesz potential in Lemma 1.37.

THEOREM 3.6. *Let $1 \le p < n$ and $p^* = \frac{np}{n-p}$, and assume that $u \in W^{1,p}(\mathbb{R}^n)$. There exists a constant $C = C(n,p)$ such that*

$$\|u\|_{L^{p^*}(\mathbb{R}^n)} \le C \|\nabla u\|_{L^p(\mathbb{R}^n)}.$$

PROOF. Assume first that $u \in \operatorname{Lip}_c(\mathbb{R}^n)$. By Lemma 3.3,

$$|u(x)| \le \frac{2^n}{n\omega_n} I_1|\nabla u|(x),$$

for every $x \in \mathbb{R}^n$. If $1 < p < n$, then Theorem 1.36 implies

$$\|u\|_{L^{p^*}(\mathbb{R}^n)} \le C(n) \big\| I_1|\nabla u| \big\|_{L^{p^*}(\mathbb{R}^n)} \le C(n,p) \|\nabla u\|_{L^p(\mathbb{R}^n)},$$

and this proves the claim for $u \in \operatorname{Lip}_c(\mathbb{R}^n)$ when $1 < p < n$.

Then we consider the case $p = 1$ for $u \in \operatorname{Lip}_c(\mathbb{R}^n)$. Let

$$A_j = \big\{x \in \mathbb{R}^n : 2^j < |u(x)| \le 2^{j+1}\big\},$$

with $j \in \mathbb{Z}$. Define $\varphi(t) = \max\{0, \min\{t, 1\}\}$, for $t \in \mathbb{R}$, and

$$u_j(x) = \varphi(2^{1-j}|u(x)| - 1) = \begin{cases} 0, & \text{when } |u(x)| \le 2^{j-1}, \\ 2^{1-j}|u(x)| - 1, & \text{when } 2^{j-1} < |u(x)| \le 2^j, \\ 1, & \text{when } |u(x)| > 2^j, \end{cases}$$

for $x \in \mathbb{R}^n$ and $j \in \mathbb{Z}$. Then $u_j \in \operatorname{Lip}_c(\mathbb{R}^n)$ for every $j \in \mathbb{Z}$ and by Lemma 2.13 we have $\nabla u_j = 0$ almost everywhere in $\mathbb{R}^n \setminus A_{j-1}$. Lemma 3.3 and Lemma 1.37 imply

$$\begin{aligned} |A_j| &\le \big|\{x \in \mathbb{R}^n : |u(x)| > 2^j\}\big| = \big|\{x \in \mathbb{R}^n : u_j(x) = 1\}\big| \\ &\le \Big|\Big\{x \in \mathbb{R}^n : I_1|\nabla u_j|(x) > \frac{n\omega_n}{2^{n+1}}\Big\}\Big| \le C(n) \Big(\int_{A_{j-1}} |\nabla u_j(x)|\,dx\Big)^{\frac{n}{n-1}} \\ &\le C(n) \Big(\int_{A_{j-1}} 2^{1-j} |\nabla u(x)|\,dx\Big)^{\frac{n}{n-1}} \le C(n) 2^{-j\frac{n}{n-1}} \Big(\int_{A_{j-1}} |\nabla u(x)|\,dx\Big)^{\frac{n}{n-1}}, \end{aligned}$$

for every $j \in \mathbb{Z}$. Since the sets A_j, with $j \in \mathbb{Z}$, are pairwise disjoint and their union is $\{x \in \mathbb{R}^n : u(x) \neq 0\}$, by summing over all $j \in \mathbb{Z}$ we obtain

$$\begin{aligned}\int_{\mathbb{R}^n} |u(x)|^{\frac{n}{n-1}}\,dx &= \sum_{j=-\infty}^{\infty} \int_{A_j} |u(x)|^{\frac{n}{n-1}}\,dx \le \sum_{j=-\infty}^{\infty} 2^{(j+1)\frac{n}{n-1}}|A_j| \\ &\le C(n) \sum_{j=-\infty}^{\infty} \Big(\int_{A_{j-1}} |\nabla u(x)|\,dx\Big)^{\frac{n}{n-1}} \\ &\le C(n)\Big(\sum_{j=-\infty}^{\infty} \int_{A_{j-1}} |\nabla u(x)|\,dx\Big)^{\frac{n}{n-1}} \\ &= C(n)\Big(\int_{\mathbb{R}^n} |\nabla u(x)|\,dx\Big)^{\frac{n}{n-1}}.\end{aligned}$$

This proves the case $p = 1$ for $u \in \operatorname{Lip}_c(\mathbb{R}^n)$.

Assume then that $u \in W^{1,p}(\mathbb{R}^n)$. By Remark 2.22 (a), there exist functions $u_i \in \operatorname{Lip}_c(\mathbb{R}^n)$, $i \in \mathbb{N}$, such that $u_i \to u$ in $W^{1,p}(\mathbb{R}^n)$ as $i \to \infty$. In particular, $u_i \to u$ in $L^p(\mathbb{R}^n)$ as $i \to \infty$, and hence, by passing to a subsequence if necessary, we may in addition assume that $u_i \to u$ almost everywhere in $\mathbb{R}^n$. Fatou's lemma and the claim for Lipschitz functions u_i, $i \in \mathbb{N}$, imply

$$\begin{aligned}\Big(\int_{\mathbb{R}^n} |u(x)|^{\frac{np}{n-p}}\,dx\Big)^{\frac{n-p}{np}} &\le \liminf_{i\to\infty}\Big(\int_{\mathbb{R}^n} |u_i(x)|^{\frac{np}{n-p}}\,dx\Big)^{\frac{n-p}{np}} \\ &\le C(n,p)\liminf_{i\to\infty}\Big(\int_{\mathbb{R}^n} |\nabla u_i(x)|^p\,dx\Big)^{\frac{1}{p}} \\ &= C(n,p)\Big(\int_{\mathbb{R}^n} |\nabla u(x)|^p\,dx\Big)^{\frac{1}{p}},\end{aligned}$$

where the final equality holds since $\|\nabla u_i\|_{L^p(\mathbb{R}^n)} \to \|\nabla u\|_{L^p(\mathbb{R}^n)}$ as $i \to \infty$. □

The proof above shows that in this case a weak type estimate implies a strong type estimate for $p = 1$. Observe from Theorem 1.15 (b) and Example 1.11 that this does not hold in general for $p = 1$. The reason why this is possible in the proof of Theorem 3.6 is that we consider gradients, which vanish almost everywhere on the set where the function is constant.

From the Sobolev–Gagliardo–Nirenberg inequality we obtain the following estimate for Sobolev functions with zero boundary values.

Corollary 3.7. *Let $1 \le p < \infty$, let $\Omega \subset \mathbb{R}^n$ be an open set with $|\Omega| < \infty$, and assume that $u \in W_0^{1,p}(\Omega)$. Let $1 \le q \le p^* = \frac{np}{n-p}$ for $1 \le p < n$, and $1 \le q < \infty$ for $n \le p < \infty$. There exists a constant $C = C(n,p,q)$ such that*

$$\Big(\int_\Omega |u(x)|^q\,dx\Big)^{\frac{1}{q}} \le C|\Omega|^{\frac{1}{n}-\frac{1}{p}+\frac{1}{q}}\Big(\int_\Omega |\nabla u(x)|^p\,dx\Big)^{\frac{1}{p}}. \tag{3.5}$$

Proof. Extend u as zero outside Ω. By Theorem 2.31 we then have $\nabla u(x) = 0$ for almost every $x \in \mathbb{R}^n \setminus \Omega$. Assume first that $1 \le p < n$. Hölder's inequality and

Theorem 3.6 imply

$$\begin{aligned}\left(\int_\Omega |u(x)|^q\,dx\right)^{\frac{1}{q}} &\le |\Omega|^{\frac{1}{n}-\frac{1}{p}+\frac{1}{q}}\left(\int_\Omega |u(x)|^{\frac{np}{n-p}}\,dx\right)^{\frac{n-p}{np}}\\ &\le C(n,p)|\Omega|^{\frac{1}{n}-\frac{1}{p}+\frac{1}{q}}\left(\int_\Omega |\nabla u(x)|^p\,dx\right)^{\frac{1}{p}}.\end{aligned}$$

Assume then that $n \le p < \infty$. If $q > p$, choose $1 < s < n$ satisfying $q = \frac{ns}{n-s}$. By the first part of the proof and Hölder's inequality, we obtain

$$\begin{aligned}\left(\int_\Omega |u(x)|^q\,dx\right)^{\frac{1}{q}} &\le C(n,p,q)|\Omega|^{\frac{1}{n}-\frac{1}{s}+\frac{1}{q}}\left(\int_\Omega |\nabla u(x)|^s\,dx\right)^{\frac{1}{s}}\\ &\le C(n,p,q)|\Omega|^{\frac{1}{n}-\frac{1}{p}+\frac{1}{q}}\left(\int_\Omega |\nabla u(x)|^p\,dx\right)^{\frac{1}{p}}.\end{aligned}$$

Finally, for $q \le p$, the claim follows by applying Hölder's inequality and the previous case with some $t > q$ on the left-hand side. □

REMARK 3.8. Let $1 \le p < n$ and let $\Omega \subset \mathbb{R}^n$ be an open set. The proof of Corollary 3.7 shows that the Sobolev inequality

$$\left(\int_\Omega |u(x)|^{\frac{np}{n-p}}\,dx\right)^{\frac{n-p}{np}} \le C(n,p)\left(\int_\Omega |\nabla u(x)|^p\,dx\right)^{\frac{1}{p}} \tag{3.6}$$

holds for every $u \in W_0^{1,p}(\Omega)$.

3.3. Sobolev–Poincaré inequalities

Sobolev–Poincaré inequalities for $1 < p < \infty$ are straightforward consequences of the pointwise estimates in Section 3.1 and the Hardy–Littlewood–Wiener maximal function theorem and the Hardy–Littlewood–Sobolev inequality for the Riesz potential, see Theorem 1.15 and Theorem 1.36 respectively. The argument for $p = 1$ is based on a Lipschitz truncation method and weak type estimates, as in the proof of Theorem 3.6. We begin with a p-Poincaré inequality for Lipschitz functions in the range $1 < p < \infty$.

LEMMA 3.9. *Let $1 < p < \infty$ and assume that $u \in \operatorname{Lip}(\mathbb{R}^n)$. There exists a constant $C = C(n,p)$ such that*

$$\int_B |u(x) - u_B|^p\,dx \le Cr^p\int_B |\nabla u(x)|^p\,dx,$$

for every ball $B = B(x_0,r) \subset \mathbb{R}^n$.

PROOF. By Theorem 3.4, we have

$$|u(x) - u_B|^p \le C(n,p)r^p\big(M(|\nabla u|\chi_B)(x)\big)^p$$

for every $x \in B$. Theorem 1.15 (c) implies

$$\begin{aligned}\int_B |u(x) - u_B|^p\,dx &\le C(n,p)r^p\int_{\mathbb{R}^n}\big(M(|\nabla u|\chi_B)(x)\big)^p\,dx\\ &\le C(n,p)r^p\int_{\mathbb{R}^n}\big(|\nabla u(x)|\chi_B(x)\big)^p\,dx\\ &= C(n,p)r^p\int_B |\nabla u(x)|^p\,dx.\end{aligned}$$

□

In the range $1 < p < n$ we have a Sobolev–Poincaré inequality for Lipschitz functions.

LEMMA 3.10. *Let $1 < p < n$ and $p^* = \frac{np}{n-p}$, and assume that $u \in \operatorname{Lip}(\mathbb{R}^n)$. There exists a constant $C = C(n,p)$ such that*

$$\Big(\fint_B |u(x) - u_B|^{p^*}\,dx\Big)^{\frac{1}{p^*}} \le C\Big(\fint_B |\nabla u(x)|^p\,dx\Big)^{\frac{1}{p}}$$

for every ball $B = B(x_0, r) \subset \mathbb{R}^n$.

PROOF. By Lemma 3.2, we have

$$|u(x) - u_B| \le C(n) I_1(|\nabla u|\chi_B)(x)$$

for every $x \in B$. Theorem 1.36, with $\alpha = 1$, implies

$$\begin{aligned}\Big(\int_B |u(x) - u_B|^{p^*}\,dx\Big)^{\frac{1}{p^*}} &\le C(n)\Big(\int_{\mathbb{R}^n} (I_1(|\nabla u|\chi_B)(x))^{p^*}\,dx\Big)^{\frac{1}{p^*}}\\ &\le C(n,p)\Big(\int_{\mathbb{R}^n} (|\nabla u(x)|\chi_B(x))^p\,dx\Big)^{\frac{1}{p}}\\ &= C(n,p)\Big(\int_B |\nabla u(x)|^p\,dx\Big)^{\frac{1}{p}}.\end{aligned}$$

□

Poincaré and Sobolev–Poincaré inequalities also hold for $p = 1$, and by approximation they can be extended for Sobolev functions, as well; see Theorem 3.14. The following two auxiliary results will be needed in the proof of the case $p = 1$ in Theorem 3.13.

LEMMA 3.11. *Assume that $E \subset \mathbb{R}^n$ is a measurable set and that $v \ge 0$ is a measurable function on E with*

$$\big|\{x \in E : v(x) = 0\}\big| \ge \frac{|E|}{2}.$$

Then

$$\big|\{x \in E : v(x) > t\}\big| \le \Big|\Big\{x \in E : |v(x) - a| > \frac{t}{2}\Big\}\Big| \tag{3.7}$$

for every $a \in \mathbb{R}$ and $t > 0$.

PROOF. First assume that $|a| \le \frac{t}{2}$. If $x \in E$ satisfies $v(x) > t$, then

$$|v(x) - a| \ge v(x) - |a| > \frac{t}{2},$$

and thus

$$\big|\{x \in E : v(x) > t\}\big| \le \Big|\Big\{x \in E : |v(x) - a| > \frac{t}{2}\Big\}\Big|.$$

Then assume that $|a| > \frac{t}{2}$. If $x \in E$ and $v(x) = 0$, then

$$|v(x) - a| = |a| > \frac{t}{2}.$$

Hence, in the case $|E| = \infty$ inequality (3.7) follows readily from the assumptions. On the other hand, if $|E| < \infty$, then the assumptions imply

$$\begin{aligned}\big|\{x \in E : v(x) > t\}\big| &\le |E| - \big|\{x \in E : v(x) = 0\}\big| \le \big|\{x \in E : v(x) = 0\}\big| \\ &\le \left|\left\{x \in E : |v(x) - a| > \frac{t}{2}\right\}\right|,\end{aligned}$$

and this proves the claim. $\square$

LEMMA 3.12. *Assume that $u \in \operatorname{Lip}(\mathbb{R}^n)$ and let $B \subset \mathbb{R}^n$ be a ball. There exists $b \in \mathbb{R}$ such that*

$$\big|\{x \in B : u(x) \ge b\}\big| \ge \frac{|B|}{2} \quad \text{and} \quad \big|\{x \in B : u(x) \le b\}\big| \ge \frac{|B|}{2}.$$

PROOF. Let $E_t = \{x \in B : u(x) \ge t\}$ for $t \in \mathbb{R}$, and let

$$b = \sup\left\{t \in \mathbb{R} : |E_t| \ge \frac{|B|}{2}\right\}.$$

Then $|b| \le \|u\|_{L^\infty(B)} < \infty$ and there exists an increasing sequence $(b_j)_{j\in\mathbb{N}}$ such that $b_j \to b$ as $j \to \infty$ and

$$|E_{b_j}| \ge \frac{|B|}{2}$$

for every $j \in \mathbb{N}$. Since $E_b = \bigcap_{j=1}^\infty E_{b_j}$ and the sequence $(E_{b_j})_{j\in\mathbb{N}}$ is decreasing, we have

$$\big|\{x \in B : u(x) \ge b\}\big| = |E_b| = \lim_{j\to\infty} |E_{b_j}| \ge \frac{|B|}{2}.$$

A similar argument shows that also the second inequality holds for b since

$$\big|\{x \in B : u(x) \le t\}\big| \ge |B| - |E_t| > \frac{|B|}{2}$$

whenever $t > b$. $\square$

We are ready to prove the Sobolev–Poincaré inequality for $p = 1$.

THEOREM 3.13. *Assume that $u \in \operatorname{Lip}(\mathbb{R}^n)$. There exists a constant $C = C(n)$ such that*

$$\left(\fint_B |u(x) - u_B|^{\frac{n}{n-1}}\,dx\right)^{\frac{n-1}{n}} \le C \int_B |\nabla u(x)|\,dx$$

for every ball $B = B(x_0, r) \subset \mathbb{R}^n$.

PROOF. By Lemma 3.12, there is a number $b \in \mathbb{R}$ satisfying

$$\big|\{x \in B : u(x) \ge b\}\big| \ge \frac{|B|}{2} \quad \text{and} \quad \big|\{x \in B : u(x) \le b\}\big| \ge \frac{|B|}{2}.$$

Define $v_+ = \max\{u - b, 0\}$ and $v_- = -\min\{u - b, 0\}$. Both of these functions belong to $\operatorname{Lip}(\mathbb{R}^n)$. In the sequel v denotes either v_+ or v_-; all statements are valid in both cases.

Let

$$A_j = \{x \in B : 2^j < v(x) \le 2^{j+1}\},$$

with $j \in \mathbb{Z}$. Define $\varphi(t) = \max\{0, \min\{t, 1\}\}$, for every $t \in \mathbb{R}$, and

$$v_j(x) = \varphi(2^{1-j} v(x) - 1),$$

for every $x \in \mathbb{R}^n$ and $j \in \mathbb{Z}$. Then $v_j \in \operatorname{Lip}(\mathbb{R}^n)$ for every $j \in \mathbb{Z}$. Let $j \in \mathbb{Z}$. By Lemma 3.2, we have

$$|v_j(x) - (v_j)_B| \le C(n) I_1(|\nabla v_j| \chi_B)(x)$$

for every $x \in B$. Hence we obtain from Lemma 3.11, with $t = \frac{1}{2}$ and $a = (v_j)_B$, that

$$\begin{aligned}|A_j| &\le \left|\{x \in B : 2^j < v(x)\}\right| \le \left|\left\{x \in B : v_j(x) > \frac{1}{2}\right\}\right| \\ &\le \left|\left\{x \in B : |v_j(x) - (v_j)_B| > \frac{1}{4}\right\}\right| \le \left|\left\{x \in \mathbb{R}^n : I_1(|\nabla v_j| \chi_B)(x) > C(n)\right\}\right|.\end{aligned}$$

The right-hand side can be estimated using the weak type estimate for the Riesz potential in Lemma 1.37 and the fact that

$$|\nabla v_j(x)| = 2^{1-j} |\nabla v(x)| \chi_{A_{j-1}}(x)$$

for almost every $x \in B$, see Lemma 2.13. It follows that

$$\begin{aligned}|A_j| &\le \left|\left\{x \in \mathbb{R}^n : I_1(|\nabla v_j| \chi_B)(x) > C(n)\right\}\right| \le C(n) \big\| |\nabla v_j| \chi_B \big\|_{L^1(\mathbb{R}^n)}^{\frac{n}{n-1}} \\ &\le C(n) 2^{-\frac{jn}{n-1}} \big\| |\nabla v| \chi_{A_{j-1} \cap B} \big\|_{L^1(\mathbb{R}^n)}^{\frac{n}{n-1}}.\end{aligned}$$

The sets A_j, for $j \in \mathbb{Z}$, are pairwise disjoint, and thus we obtain, using the estimate above,

$$\begin{aligned}\int_B v(x)^{\frac{n}{n-1}}\,dx &= \sum_{j\in\mathbb{Z}} \int_{A_j} v(x)^{\frac{n}{n-1}}\,dx \le \sum_{j\in\mathbb{Z}} 2^{\frac{(j+1)n}{n-1}} |A_j| \\ &\le C(n) \sum_{j\in\mathbb{Z}} 2^{\frac{(j+1)n}{n-1}} 2^{-\frac{jn}{n-1}} \big\| |\nabla v| \chi_{A_{j-1}\cap B} \big\|_{L^1(\mathbb{R}^n)}^{\frac{n}{n-1}} \\ &\le C(n) \Big\| \sum_{j\in\mathbb{Z}} |\nabla v| \chi_{A_{j-1}\cap B} \Big\|_{L^1(\mathbb{R}^n)}^{\frac{n}{n-1}} \le C(n) \big\| |\nabla u| \chi_B \big\|_{L^1(\mathbb{R}^n)}^{\frac{n}{n-1}}.\end{aligned}$$

Since $|u - b| = v_+ + v_-$, we arrive at

$$\begin{aligned}\left(\int_B |u(x) - u_B|^{\frac{n}{n-1}}\,dx\right)^{\frac{n-1}{n}} &\le 2\left(\int_B |u(x) - b|^{\frac{n}{n-1}}\,dx\right)^{\frac{n-1}{n}} \\ &\le 2\left(\int_B v_+(x)^{\frac{n}{n-1}}\,dx\right)^{\frac{n-1}{n}} + 2\left(\int_B v_-(x)^{\frac{n}{n-1}}\,dx\right)^{\frac{n-1}{n}} \\ &\le C(n) \big\| |\nabla u| \chi_B \big\|_{L^1(\mathbb{R}^n)} = C(n) \int_B |\nabla u(x)|\,dx.\end{aligned}$$

□

The next theorem gives a general Sobolev–Poincaré inequality for Sobolev functions. In particular, this shows that Lemma 3.9, Lemma 3.10 and Theorem 3.13 do not only hold for Lipschitz functions but also for functions in the corresponding Sobolev space. We refer to (3.8) below as the (q, p)-Poincaré inequality.

THEOREM 3.14. *Let $1 \le p < \infty$, let $\Omega \subset \mathbb{R}^n$ be an open set, and assume that $u \in W^{1,p}_{\mathrm{loc}}(\Omega)$. Let $1 \le q \le p^* = \frac{np}{n-p}$ for $1 \le p < n$, and $1 \le q < \infty$ for $n \le p < \infty$. There exists a constant $C = C(n, p, q)$ such that*

$$\left(\fint_B |u(x) - u_B|^q\,dx\right)^{\frac{1}{q}} \le Cr \left(\fint_B |\nabla u(x)|^p\,dx\right)^{\frac{1}{p}} \tag{3.8}$$

for every ball $B = B(x_0, r) \Subset \Omega$.

PROOF. Assume first that $u \in \operatorname{Lip}(\mathbb{R}^n)$. By Lemma 3.10, for $1 < p < n$, and Theorem 3.13, for $p = 1$, we obtain

$$\begin{aligned}\left(\fint_B |u(x) - u_B|^{\frac{np}{n-p}}\,dx\right)^{\frac{n-p}{np}} &= C(n,p)r^{-\frac{n-p}{p}}\left(\int_B |u(x) - u_B|^{\frac{np}{n-p}}\,dx\right)^{\frac{n-p}{np}} \\ &\le C(n,p)r^{1-\frac{n}{p}}\left(\int_B |\nabla u(x)|^p\,dx\right)^{\frac{1}{p}} \\ &= C(n,p)r\left(\fint_B |\nabla u(x)|^p\,dx\right)^{\frac{1}{p}}.\end{aligned} \tag{3.9}$$

For $1 \le p < n$, inequality (3.8) follows from (3.9) and Hölder's inequality on the left-hand side.

In the case $p \ge n$ we proceed as in the proof of Corollary 3.7. For $q > p \ge n$, there exists $1 < s < n$ such that $q = \frac{ns}{n-s}$, and (3.8) follows from (3.9) with exponent s and an application of Hölder's inequality on the right-hand side. For $q \le p$, the claim follows from the previous case and Hölder's inequality on the left-hand side.

For $u \in W^{1,p}_{\mathrm{loc}}(\Omega)$, we obtain (3.8) by approximation. By considering $R > r$ such that $B \subset B(x_0, R) \Subset \Omega$ and applying the McShane extension, see Theorem 2.7, we find functions $u_i \in \operatorname{Lip}(\mathbb{R}^n)$, $i \in \mathbb{N}$, satisfying $u_i \to u$ in $W^{1,p}(B)$ as $i \to \infty$; compare to the proof of Theorem 3.5. By passing to a subsequence, if necessary, we may in addition assume that $u_i \to u$ almost everywhere in B. As in (3.4), we have $(u_i)_B \to u_B$ as $i \to \infty$. Fatou's lemma and inequality (3.8) for the Lipschitz functions u_i give

$$\begin{aligned}\left(\fint_B |u(x) - u_B|^q\,dx\right)^{\frac{1}{q}} &\le \liminf_{i\to\infty}\left(\fint_B |u_i(x) - (u_i)_B|^q\,dx\right)^{\frac{1}{q}} \\ &\le \liminf_{i\to\infty} C(n,p,q)r\left(\fint_B |\nabla u_i(x)|^p\,dx\right)^{\frac{1}{p}} \\ &\le C(n,p,q)r\left(\fint_B |\nabla u(x)|^p\,dx\right)^{\frac{1}{p}},\end{aligned}$$

and the proof is complete. □

The following consequence of the Poincaré inequality will be applied in Chapter 11 and Chapter 12.

LEMMA 3.15. *Let $1 \le p < \infty$ and let $\Omega \subset \mathbb{R}^n$ be an open set. Assume that $u \in W^{1,p}_0(\Omega)$ is such that $|\nabla u| = 0$ almost everywhere in Ω. Then $u = 0$ almost everywhere in Ω.*

PROOF. Extend u as zero outside Ω. By Theorem 2.31 we then have $|\nabla u| = 0$ almost everywhere in $\mathbb{R}^n$. Define $B_j = B(0, j) \subset \mathbb{R}^n$ for $j \in \mathbb{N}$. By the $(1,p)$-Poincaré inequality in Theorem 3.14 we obtain

$$\fint_{B_j} |u(x) - u_{B_j}|\,dx \le C(n,p)r\left(\fint_{B_j} |\nabla u(x)|^p\,dx\right)^{\frac{1}{p}} = 0,$$

for every $j \in \mathbb{N}$. Hence, for every $j \in \mathbb{N}$ there is a set $E_j \subset B_j$ with $|E_j| = 0$ such that $u(x) = u_{B_j}$ for every $x \in B_j \setminus E_j$. Hölder's inequality implies

$$|u_{B_j}| \le \left(\fint_{B_j} |u(x)|^p \, dx \right)^{\frac{1}{p}} \le |B_j|^{-\frac{1}{p}} \|u\|_{L^p(\mathbb{R}^n)} \xrightarrow{j\to\infty} 0. \tag{3.10}$$

Thus $u(x) = \lim_{j\to\infty} u_{B_j} \chi_{B_j}(x) = 0$ for every $x \in \mathbb{R}^n \setminus E$, where $E = \bigcup_{j=1}^\infty E_j$ has zero measure. □

REMARK 3.16. The Sobolev–Poincaré inequality in Theorem 3.14 gives an alternative proof for the Sobolev–Gagliardo–Nirenberg inequality in Theorem 3.6. In order to show this, let $1 \le p < n$, assume that $u \in W^{1,p}(\mathbb{R}^n)$, and define $B_j = B(0,j)$, for $j \in \mathbb{N}$. By (3.10) we have $u_{B_j} \to 0$ as $j \to \infty$, and thus

$$|u(x)|^{\frac{np}{n-p}} = \liminf_{j\to\infty} |u(x) - u_{B_j}|^{\frac{np}{n-p}} \chi_{B_j}(x)$$

for every $x \in \mathbb{R}^n$. Using Fatou's lemma and Theorem 3.14, with $q = p^*$, we obtain

$$\begin{aligned}
\left(\int_{\mathbb{R}^n} |u(x)|^{\frac{np}{n-p}} \, dx \right)^{\frac{n-p}{np}} &\le \liminf_{j\to\infty} \left(\int_{B_j} |u(x) - u_{B_j}|^{\frac{np}{n-p}} \, dx \right)^{\frac{n-p}{np}} \\
&\le C(n,p) \liminf_{j\to\infty} \left(\int_{B_j} |\nabla u(x)|^p \, dx \right)^{\frac{1}{p}} \\
&\le C(n,p) \left(\int_{\mathbb{R}^n} |\nabla u(x)|^p \, dx \right)^{\frac{1}{p}}.
\end{aligned}$$

3.4. Poincaré inequalities for zero boundary values

If a Lipschitz or a Sobolev function u is zero outside a ball $B \subset \mathbb{R}^n$, then we can remove the mean value u_B from the left-hand side of the (q,p)-Poincaré inequality (3.8). This result follows from Corollary 3.7, but we give an alternative proof which is based on the (q,p)-Poincaré inequality. This technique can be adapted to many other situations as well, which is illustrated in Remark 3.18.

THEOREM 3.17. *Let $1 \le q \le p^* = \frac{np}{n-p}$ for $1 \le p < n$, and $1 \le q < \infty$ for $n \le p < \infty$. There exists a constant $C = C(n,p,q)$ such that*

$$\left(\fint_B |u(x)|^q \, dx \right)^{\frac{1}{q}} \le Cr \left(\fint_B |\nabla u(x)|^p \, dx \right)^{\frac{1}{p}} \tag{3.11}$$

for every ball $B = B(x_0, r) \subset \mathbb{R}^n$ and every $u \in W_0^{1,p}(B)$.

PROOF. We may assume that $q > 1$ since the claim for $q = 1$ follows from Hölder's inequality. Consider first the case $u \in \operatorname{Lip}_0(B)$. Then $u = 0$ in $B(x_0, 2r) \setminus B(x_0, r)$. By Hölder's inequality

$$\begin{aligned}
|u_{B(x_0,2r)}| &\le \fint_{B(x_0,2r)} |u(x)| \chi_{B(x_0,r)}(x) \, dx \\
&\le \left(\frac{|B(x_0,r)|}{|B(x_0,2r)|} \right)^{1-\frac{1}{q}} \left(\fint_{B(x_0,2r)} |u(x)|^q \, dx \right)^{\frac{1}{q}} \qquad (3.12) \\
&= (2^{-n})^{1-\frac{1}{q}} \left(\fint_{B(x_0,2r)} |u(x)|^q \, dx \right)^{\frac{1}{q}}.
\end{aligned}$$

Using Minkowski's inequality, the (q,p)-Poincaré inequality from Theorem 3.14 for $B(x_0,2r)$, and estimate (3.12), we obtain

$$\begin{aligned}\Big(\fint_{B(x_0,2r)}|u(x)|^q\,dx\Big)^{\frac1q} &\le \Big(\fint_{B(x_0,2r)}|u(x)-u_{B(x_0,2r)}|^q\,dx\Big)^{\frac1q}+|u_{B(x_0,2r)}|\\ &\le C(n,p,q)r\Big(\fint_{B(x_0,2r)}|\nabla u(x)|^p\,dx\Big)^{\frac1p}+(2^{-n})^{1-\frac1q}\Big(\fint_{B(x_0,2r)}|u(x)|^q\,dx\Big)^{\frac1q}<\infty.\end{aligned}$$

Since $(2^{-n})^{1-\frac1q}<1$, the second term on the right-hand side can be absorbed to the left-hand side, and thus

$$\Big(\fint_{B(x_0,2r)}|u(x)|^q\,dx\Big)^{\frac1q}\le C(n,p,q)r\Big(\fint_{B(x_0,2r)}|\nabla u(x)|^p\,dx\Big)^{\frac1p}.$$

Finally, the mean value integrals on both sides can be taken with respect to $B(x_0,r)$ since $u=0$ and $\nabla u=0$ in $B(x_0,2r)\setminus\overline{B}(x_0,r)$.

In the general case $u\in W_0^{1,p}(B)$ the claim follows by approximation since $\operatorname{Lip}_0(B)$ is dense in $W_0^{1,p}(B)$. The details are essentially same as in the proof of Theorem 3.6. □

REMARK 3.18. With minor modifications to the proof above, we can generalize the statement of Theorem 3.17 as follows. Let n, p and $q>1$ be as in Theorem 3.17. Assume that $u\in\operatorname{Lip}(\mathbb{R}^n)$ is such that $u=0$ in a set $E\subset B=B(x_0,r)$ satisfying $|E|\ge\gamma|B|$ with $0<\gamma\le1$. As in (3.12), we obtain

$$\begin{aligned}|u_B|&\le\fint_B|u(x)|\chi_{B\setminus E}(x)\,dx\le\Big(\frac{|B\setminus E|}{|B|}\Big)^{1-\frac1q}\Big(\fint_B|u(x)|^q\,dx\Big)^{\frac1q}\\ &\le(1-\gamma)^{1-\frac1q}\Big(\fint_B|u(x)|^q\,dx\Big)^{\frac1q}.\end{aligned}$$

Since $0\le(1-\gamma)^{1-\frac1q}<1$, we conclude that there exists a constant $C=C(n,p,q,\gamma)$ such that

$$\Big(\fint_B|u(x)|^q\,dx\Big)^{\frac1q}\le Cr\Big(\fint_B|\nabla u(x)|^p\,dx\Big)^{\frac1p}.$$

These ideas will be taken one step further in Section 5.9 with the concept of variational capacity; compare to Theorem 5.47 and also to Theorem 6.22.

3.5. Morrey's inequality

For $n<p<\infty$, we obtain Morrey's inequality, which can be regarded as a pointwise version of the Sobolev–Poincaré inequality in Section 3.3. The limiting case $p=n$ is studied in Section 9.7 together with the space of bounded mean oscillation.

THEOREM 3.19. *Let $n<p<\infty$ and assume that $u\in\operatorname{Lip}(\mathbb{R}^n)$. There exists a constant $C=C(n,p)$ such that*

$$|u(x)-u(y)|\le Cr\Big(\fint_B|\nabla u(z)|^p\,dz\Big)^{\frac1p}\tag{3.13}$$

for every ball $B=B(x_0,r)\subset\mathbb{R}^n$ and every $x,y\in B$.

PROOF. Let $B_0 = B = B(x_0, r)$ and choose a ball $B_1 = B(x_1, 2^{-1}r)$ such that $x \in B_1 \subset B_0$. For $j = 2, 3, \ldots$, choose recursively a ball $B_j = B(x_j, 2^{-j}r)$ such that $x \in B_j \subset B_{j-1}$. We apply a telescoping argument with the balls B_j. First, using the continuity of u, the triangle inequality, and the facts that $x \in B_j \subset B_{j-1}$ and $|B_{j-1}| = 2^n|B_j|$ for every $j \in \mathbb{N}$, we obtain

$$
\begin{aligned}
|u(x) - u_{B_0}| &= \lim_{j\to\infty} |u_{B_j} - u_{B_0}| \le \sum_{j=1}^{\infty} |u_{B_j} - u_{B_{j-1}}| \le \sum_{j=1}^{\infty} \fint_{B_j} |u(z) - u_{B_{j-1}}|\, dz \\
&\le \sum_{j=1}^{\infty} \frac{|B_{j-1}|}{|B_j|} \fint_{B_{j-1}} |u(z) - u_{B_{j-1}}|\, dz = 2^n \sum_{j=0}^{\infty} \fint_{B_j} |u(z) - u_{B_j}|\, dz.
\end{aligned}
$$

The $(1,p)$-Poincaré inequality, see Theorem 3.14, for each B_j, implies

$$
\begin{aligned}
|u(x) - u_{B_0}| &\le C(n,p) \sum_{j=0}^{\infty} 2^{-j} r \left(\fint_{B_j} |\nabla u(z)|^p\, dz \right)^{\frac{1}{p}} \\
&= C(n,p) \sum_{j=0}^{\infty} 2^{-j} r (2^{-j} r)^{-\frac{n}{p}} \left(\int_{B_j} |\nabla u(z)|^p\, dz \right)^{\frac{1}{p}} \\
&\le C(n,p) r^{1-\frac{n}{p}} \left(\int_{B_0} |\nabla u(z)|^p\, dz \right)^{\frac{1}{p}} \sum_{j=0}^{\infty} (2^{-j})^{1-\frac{n}{p}}.
\end{aligned}
$$

The geometric series converges since $p > n$, and so

$$
|u(x) - u_{B_0}| \le C(n,p) r^{1-\frac{n}{p}} \left(\int_{B_0} |\nabla u(z)|^p\, dz \right)^{\frac{1}{p}} = C(n,p) r \left(\fint_{B_0} |\nabla u(z)|^p\, dz \right)^{\frac{1}{p}}.
$$

The same estimate holds for $y \in B$ as well, and thus

$$
|u(x) - u(y)| \le |u(x) - u_B| + |u_B - u(y)| \le C(n,p) r \left(\fint_B |\nabla u(z)|^p\, dz \right)^{\frac{1}{p}}
$$

for every $x, y \in B$. □

REMARK 3.20. We consider an alternative proof of Morrey's inequality (3.13) through pointwise estimates and the Riesz potential. Let $n < p < \infty$, assume that $u \in \operatorname{Lip}(\mathbb{R}^n)$, and let $x, y \in B = B(x_0, r)$. By Lemma 3.2, we have

$$
\begin{aligned}
|u(x) - u(y)| &\le |u(x) - u_B| + |u_B - u(y)| \\
&\le C(n) \int_B \frac{|\nabla u(z)|}{|x - z|^{n-1}}\, dz + C(n) \int_B \frac{|\nabla u(z)|}{|y - z|^{n-1}}\, dz.
\end{aligned}
$$

Hölder's inequality gives

$$
\int_B \frac{|\nabla u(z)|}{|x - z|^{n-1}}\, dz \le \left(\int_B |\nabla u(z)|^p\, dz \right)^{\frac{1}{p}} \left(\int_B |x - z|^{(1-n)p'}\, dz \right)^{\frac{1}{p'}},
$$

where $p' = \frac{p}{p-1} < \frac{n}{n-1}$ is the Hölder conjugate of p and

$$
\int_B |x - z|^{(1-n)p'}\, dz \le \int_{B(x,2r)} |x - z|^{(1-n)p'}\, dz = C(n,p) r^{n-(n-1)p'}.
$$

Thus

$$
\int_B \frac{|\nabla u(z)|}{|x - z|^{n-1}}\, dz \le C(n,p) r^{1-\frac{n}{p}} \left(\int_B |\nabla u(z)|^p\, dz \right)^{\frac{1}{p}}.
$$

The same argument applies to the other term as well, and we conclude that

$$|u(x) - u(y)| \le C(n,p) r \Big(\fint_B |\nabla u(z)|^p \, dz \Big)^{\frac{1}{p}}.$$

By approximation, we obtain Morrey's inequality for Sobolev functions.

COROLLARY 3.21. *Let $n < p < \infty$, let $B = B(x_0, r) \subset \mathbb{R}^n$ be a ball, and assume that $u \in W^{1,p}_{\text{loc}}(\mathbb{R}^n)$. There exist a constant $C = C(n,p)$ and a set $E \subset B$ with $|E| = 0$ such that*

$$|u(x) - u(y)| \le Cr \Big(\fint_B |\nabla u(z)|^p \, dz \Big)^{\frac{1}{p}} \tag{3.14}$$

for every $x, y \in B \setminus E$.

PROOF. As in the proof of Theorem 3.5, we find functions $u_i \in \operatorname{Lip}(\mathbb{R}^n)$, $i \in \mathbb{N}$, such that $u_i \to u$ in $W^{1,p}(B)$ as $i \to \infty$. By passing to a subsequence, if necessary, we obtain a set $E \subset B$ such that $|E| = 0$ and

$$\lim_{i \to \infty} u_i(x) = u(x) < \infty$$

for every $x \in B \setminus E$. Theorem 3.19 implies

$$|u_i(x) - u_i(y)| \le C(n,p) r \Big(\fint_B |\nabla u_i(z)|^p \, dz \Big)^{\frac{1}{p}},$$

for every $x, y \in B \setminus E$, and the claim follows by passing to the limit on both sides. □

REMARK 3.22. Morrey's inequality in Corollary 3.21 gives an alternative proof for the (q,p)-Poincaré inequality (3.8) in the case $n < p < \infty$. Indeed, if $1 \le q < \infty$ and $u \in W^{1,p}_{\text{loc}}(\mathbb{R}^n)$, then using Hölder's and Morrey's inequalities we obtain

$$\begin{aligned} \Big(\fint_B |u(x) - u_B|^q \, dx \Big)^{\frac{1}{q}} &\le \Big(\fint_B \fint_B |u(x) - u(y)|^q \, dy \, dx \Big)^{\frac{1}{q}} \\ &\le C(n,p) r \Big(\fint_B |\nabla u(x)|^p \, dx \Big)^{\frac{1}{p}} \end{aligned}$$

for every ball $B \subset \mathbb{R}^n$.

The following consequence of Morrey's inequality can be seen as a limiting case $q = \infty$ of the (q,p)-Poincaré inequality in Theorem 3.14. It shows that a function in $W^{1,p}(\mathbb{R}^n)$, with $n < p < \infty$, has a Hölder continuous representative.

THEOREM 3.23. *Let $n < p < \infty$ and assume that $u \in W^{1,p}(\mathbb{R}^n)$. There exist a constant $C = C(n,p)$ and a set $E \subset \mathbb{R}^n$ with $|E| = 0$ such that*

$$|u(x) - u(y)| \le C \|\nabla u\|_{L^p(\mathbb{R}^n)} |x - y|^{1 - \frac{n}{p}} \tag{3.15}$$

for every $x, y \in \mathbb{R}^n \setminus E$. Moreover, the function u can be redefined in the set E in such a way that inequality (3.15) holds for every $x, y \in \mathbb{R}^n$, and the redefined function is Hölder continuous with the exponent $1 - \frac{n}{p}$.

PROOF. Assume first that $u \in \operatorname{Lip}_c(\mathbb{R}^n)$. Let $x, y \in \mathbb{R}^n$ and $B = B(x_0, |x-y|)$ be a ball with $x, y \in B$. By Morrey's inequality (3.13) we have

$$\begin{aligned} |u(x) - u(y)| &\leq C(n,p)|x-y| \left(\fint_B |\nabla u(z)|^p \, dz \right)^{\frac{1}{p}} \\ &\leq C(n,p) \|\nabla u\|_{L^p(\mathbb{R}^n)} |x-y|^{1-\frac{n}{p}}, \end{aligned}$$

and this proves the claim for $u \in \operatorname{Lip}_c(\mathbb{R}^n)$ with $E = \emptyset$.

Then assume that $u \in W^{1,p}(\mathbb{R}^n)$. By Remark 2.22 (a), there exist functions $u_i \in \operatorname{Lip}_c(\mathbb{R}^n)$, $i \in \mathbb{N}$, such that $u_i \to u$ in $W^{1,p}(\mathbb{R}^n)$ as $i \to \infty$. By passing to a subsequence, if necessary, we obtain a set $E \subset \mathbb{R}^n$ with $|E| = 0$ such that

$$\lim_{i \to \infty} u_i(x) = u(x) < \infty$$

for every $x \in \mathbb{R}^n \setminus E$. Let $x, y \in \mathbb{R}^n \setminus E$. We have already shown that the claim holds for the functions $u_i \in \operatorname{Lip}_c(\mathbb{R}^n)$ at all points of $\mathbb{R}^n$, and thus

$$|u_i(x) - u_i(y)| \leq C(n,p) \|\nabla u_i\|_{L^p(\mathbb{R}^n)} |x-y|^{1-\frac{n}{p}},$$

for every $i \in \mathbb{N}$. By passing to the limit on both sides we conclude that the desired inequality (3.15) holds for every $x, y \in \mathbb{R}^n \setminus E$.

The second claim, for $u \in W^{1,p}(\mathbb{R}^n)$, follows from the fact that $v = u|_{\mathbb{R}^n \setminus E}$ is uniformly continuous in the dense set $\mathbb{R}^n \setminus E \subset \mathbb{R}^n$, and thus v can be extended as a continuous function to the whole $\mathbb{R}^n$. By continuity, this extension satisfies (3.15) for every $x, y \in \mathbb{R}^n$. □

REMARK 3.24. A straightforward adaptation of the proof of Theorem 3.23 gives the following local version of the Hölder continuity. Let $n < p < \infty$ and let $\Omega \subset \mathbb{R}^n$ be an open set. Assume that $u \in W^{1,p}_{\text{loc}}(\Omega)$ and let $B \Subset \Omega$ be a ball. Then there is a set $E \subset B$ with $|E| = 0$ such that

$$|u(x) - u(y)| \leq C(n,p) \|\nabla u\|_{L^p(B)} |x-y|^{1-\frac{n}{p}}$$

for every $x, y \in B \setminus E$.

REMARK 3.25. Let $n < p < \infty$ and assume that $u \in \operatorname{Lip}(\mathbb{R}^n) \cap W^{1,p}(\mathbb{R}^n)$. Then

$$\sup_{x \in \mathbb{R}^n} |u(x)| \leq C(n,p) \|u\|_{L^p(\mathbb{R}^n)}^{1-\frac{n}{p}} \|\nabla u\|_{L^p(\mathbb{R}^n)}^{\frac{n}{p}} \leq C(n,p) \|u\|_{W^{1,p}(\mathbb{R}^n)}. \tag{3.16}$$

The second inequality in (3.16) is trivial. In order to prove the first inequality, we may assume, by Theorem 3.23, that $\|\nabla u\|_{L^p(\mathbb{R}^n)} > 0$. Let $x \in \mathbb{R}^n$ and $B = B(x, R)$, where $R > 0$ is to be chosen later. By Theorem 3.19, we have

$$|u(x) - u_B| \leq \fint_B |u(x) - u(y)| \, dy \leq C(n,p) \|\nabla u\|_{L^p(\mathbb{R}^n)} R^{1-\frac{n}{p}}.$$

This implies

$$\begin{aligned} |u(x)| &\leq |u(x) - u_B| + |u_B| \leq C(n,p) \|\nabla u\|_{L^p(\mathbb{R}^n)} R^{1-\frac{n}{p}} + |B|^{-\frac{1}{p}} \|u\|_{L^p(\mathbb{R}^n)} \\ &= C(n,p) R^{1-\frac{n}{p}} \|\nabla u\|_{L^p(\mathbb{R}^n)} + C(n,p) R^{-\frac{n}{p}} \|u\|_{L^p(\mathbb{R}^n)}. \end{aligned}$$

Choosing $R = \|u\|_{L^p(\mathbb{R}^n)} \|\nabla u\|_{L^p(\mathbb{R}^n)}^{-1} > 0$ concludes the proof of (3.16).

As another application of Morrey's inequality we show that all functions in $W^{1,p}_{\text{loc}}(\mathbb{R}^n)$, with $p > n$, are differentiable almost everywhere.

THEOREM 3.26. *Let $n < p \le \infty$ and assume that $u \in W^{1,p}_{\mathrm{loc}}(\mathbb{R}^n)$. Then u has a continuous representative which is differentiable almost everywhere. Moreover, the pointwise derivative of this representative equals the weak derivative of u almost everywhere.*

PROOF. Since $W^{1,\infty}_{\mathrm{loc}}(\mathbb{R}^n) \subset W^{1,p}_{\mathrm{loc}}(\mathbb{R}^n)$, we may assume $n < p < \infty$. Let $\varphi_j \in \mathrm{Lip}_c(\mathbb{R}^n)$ be such that $\varphi_j(x) = 1$ for every $x \in B(0, j)$, with $j \in \mathbb{N}$. Applying Theorem 3.23 to the functions $u\varphi_j \in W^{1,p}(\mathbb{R}^n)$ and taking $j \to \infty$, we see that u has a continuous representative in $\mathbb{R}^n$. From now on, we consider this continuous representative, still denoted by u. Moreover, all the derivatives below are weak, unless otherwise specified.

By Theorem 1.19, we have

$$\lim_{r\to 0} \fint_{B(x,r)} |\nabla u(z) - \nabla u(x)|^p \, dz = 0 \tag{3.17}$$

for almost every $x \in \mathbb{R}^n$. Let $x \in \mathbb{R}^n$ be such a point and define

$$v(y) = u(y) - u(x) - \nabla u(x) \cdot (y - x),$$

for every $y \in \mathbb{R}^n$. Then $v \in W^{1,p}_{\mathrm{loc}}(\mathbb{R}^n) \cap C(\mathbb{R}^n)$. By Morrey's inequality in Corollary 3.21, we have

$$|v(y) - v(x)| \le C(n,p)|y - x| \left(\fint_{B(x,2|y-x|)} |\nabla v(z)|^p \, dz \right)^{\frac{1}{p}},$$

for every $y \in \mathbb{R}^n$. Here we also use the fact that v is continuous. Since $v(x) = 0$ and $\nabla v(z) = \nabla u(z) - \nabla u(x)$ for almost every $z \in \mathbb{R}^n$, we obtain

$$\frac{|u(y) - u(x) - \nabla u(x) \cdot (y - x)|}{|y - x|} \le C(n,p) \left(\fint_{B(x,2|y-x|)} |\nabla u(z) - \nabla u(x)|^p \, dz \right)^{\frac{1}{p}}.$$

By (3.17) the right-hand side tends to zero, as $y \to x$. This shows that u is differentiable at x and that the weak and pointwise derivatives coincide at x. □

3.6. Notes

Sobolev–Poincaré inequalities are discussed in Adams and Hedberg [**4**], Adams and Fournier [**5**], Björn and Björn [**31**], Edmunds and Evans [**116**], Evans [**121**], Evans and Gariepy [**122**], Gilbarg and Trudinger [**152**], Hajłasz and Koskela [**171**], Heinonen [**185**], Heinonen, Koskela, Shanmugalingam and Tyson [**190**], Maz′ya [**318, 321**] and Ziemer [**407**]. The proof of Lemma 3.2 is similar to Edmunds and Evans [**116**, Theorem 3.18], see also Evans and Gariepy [**122**, 4.5.2] and Gilbarg and Trudinger [**152**, Lemma 7.15]. A version of Theorem 3.4 can be found in Bennett and Sharpley [**28**, Theorem V.5.6]. Theorem 3.6 was proved by Sobolev for $1 < p < n$ and by Gagliardo and Nirenberg for $p = 1$. For Lemma 3.3 and Theorem 3.6, for $1 < p < n$, we refer to Ziemer [**407**, Remark 2.8.6]. For the truncation method applied in the proof of Theorem 3.6, we refer to Maz′ya [**318**, 2.3.1, p. 110], [**321**, 2.3.1, p. 155]. See also Hajłasz [**167**] and Hajłasz and Koskela [**171**]. The proofs of Theorem 3.17 and Remark 3.18 are similar to Evans and Gariepy [**122**, 5.6.1]. Lemma 3.11 can be found in Hajłasz, [**167**, Lemma 5]. Theorem 3.26 is from Evans and Gariepy [**122**, 6.2] and Malý and Ziemer [**303**, Theorem 1.71].

CHAPTER 4

Pointwise Inequalities for Sobolev Functions

This chapter continues discussion on pointwise inequalities for Sobolev functions by the maximal function of the gradient. A Lipschitz truncation method in Section 4.2 gives an approximation of a Sobolev function with a Lipschitz continuous function which coincides with the original function outside a set of arbitrarily small measure. This result is applied in Section 12.3. A pointwise characterization for Sobolev functions is provided in Section 4.3. Regularity properties of maximal operators, with pointwise estimates for the gradient, are discussed in Section 4.4, Section 4.5 and Section 4.6.

4.1. Pointwise characterization of Sobolev spaces

The result below is closely related to Theorem 3.4. For the definition of the restricted maximal function, see Definition 1.27.

THEOREM 4.1. *Assume that* $u \in \operatorname{Lip}(\mathbb{R}^n)$. *There exists a constant* $C = C(n)$ *such that*

$$|u(x) - u(y)| \leq C|x-y|\big(M_{2|x-y|}|\nabla u|(x) + M_{2|x-y|}|\nabla u|(y)\big)$$

for every $x, y \in \mathbb{R}^n$.

PROOF. Let $x, y \in \mathbb{R}^n$. The claim is clear if $x = y$, and thus we may assume that $x \neq y$. Let $B = B(x_0, |x-y|)$ be a ball with $x, y \in B$. Then

$$B \subset B(x, 2|x-y|) \cap B(y, 2|x-y|),$$

and Theorem 3.4 with $R = 2|x-y|$ gives

$$\begin{aligned} |u(x) - u(y)| &\leq |u(x) - u_B| + |u_B - u(y)| \\ &\leq C(n)|x-y|\big(M(|\nabla u|\chi_B)(x) + M(|\nabla u|\chi_B)(y)\big) \\ &\leq C(n)|x-y|\big(M_{2|x-y|}|\nabla u|(x) + M_{2|x-y|}|\nabla u|(y)\big). \end{aligned}$$

The final step above follows from the general inequality

$$M(f\chi_B)(z) \leq M(f\chi_{B(z,R)})(z) \leq M_R f(z), \tag{4.1}$$

where $R > 0$ and $B \subset \mathbb{R}^n$ is a ball with $z \in B \subset B(z, R)$ and $f \in L^1(B(z,R))$. In (4.1) we also use the fact that $f\chi_{B(z,R)}$ vanishes in $\mathbb{R}^n \setminus B(z,R)$. $\square$

Theorem 4.1 has a counterpart for Sobolev functions. This generalization is particularly interesting since it can be used to give a characterization for Sobolev functions, see Theorem 4.6.

THEOREM 4.2. *Let $1 \le p < \infty$ and assume that $u \in W^{1,p}(\mathbb{R}^n)$. There exist a constant $C = C(n)$ and a set $E \subset \mathbb{R}^n$ with $|E| = 0$ such that*

$$|u(x) - u(y)| \le C|x-y|\big(M|\nabla u|(x) + M|\nabla u|(y)\big)$$

for every $x, y \in \mathbb{R}^n \setminus E$.

PROOF. Since compactly supported Lipschitz functions are dense in $W^{1,p}(\mathbb{R}^n)$, see Remark 2.22 (a), there exist functions $u_i \in \operatorname{Lip}_c(\mathbb{R}^n)$, $i \in \mathbb{N}$, such that $u_i \to u$ in $W^{1,p}(\mathbb{R}^n)$ as $i \to \infty$. By the maximal function theorem, see Theorem 1.15, and Remark 1.16, $(M|\nabla u_i|)_{i\in\mathbb{N}}$ converges to $M|\nabla u|$ in $L^p(\mathbb{R}^n)$ if $1 < p < \infty$, and in measure if $p = 1$. By passing twice to a subsequence, if necessary, we obtain a set $E \subset \mathbb{R}^n$ with $|E| = 0$ such that

$$\lim_{i\to\infty} u_i(x) = u(x) < \infty$$

and

$$\lim_{i\to\infty} M|\nabla u_i|(x) = M|\nabla u|(x) < \infty$$

for every $x \in \mathbb{R}^n \setminus E$. Theorem 4.1 implies

$$\begin{aligned}|u(x) - u(y)| &= \lim_{i\to\infty} |u_i(x) - u_i(y)| \\ &\le C(n) \limsup_{i\to\infty} |x-y|\big(M|\nabla u_i|(x) + M|\nabla u_i|(y)\big) \\ &= C(n)|x-y|\big(M|\nabla u|(x) + M|\nabla u|(y)\big)\end{aligned}$$

for every $x, y \in \mathbb{R}^n \setminus E$. □

REMARK 4.3. Let $1 \le p < \infty$, assume that $u \in W^{1,p}_{\mathrm{loc}}(\mathbb{R}^n)$, and let $B = B(x_0, R) \subset \mathbb{R}^n$ be a ball. By Theorem 3.5, inequality (4.1) and Remark 1.29, there exists a set $E_B \subset B$ with $|E_B| = 0$ such that

$$\begin{aligned}|u(x) - u_B| &\le C(n)RM(|\nabla u|\chi_B)(x) \le C(n)RM_{2R}|\nabla u|(x) \\ &\le C(n)R\big(M_{2R}|\nabla u|^p(x)\big)^{\frac{1}{p}}\end{aligned} \tag{4.2}$$

for every $x \in B \setminus E_B$.

Let $E = \bigcup_B E_B$, where the union is taken over all balls with center in $\mathbb{Q}^n$ and rational radii. Since the union is countable, we have $|E| = 0$. Moreover, for every $x, y \in \mathbb{R}^n$, $x \ne y$, there exists a ball $B = B(x_0, r)$ from the union with $x, y \in B$ and $r \le |x-y|$. By arguing as in the proof of Theorem 4.1 and applying (4.2), we obtain

$$|u(x) - u(y)| \le C(n)|x-y|\Big(\big(M_{2|x-y|}|\nabla u|^p(x)\big)^{\frac{1}{p}} + \big(M_{2|x-y|}|\nabla u|^p(y)\big)^{\frac{1}{p}}\Big)$$

for every $x, y \in \mathbb{R}^n \setminus E$.

REMARK 4.4. Assume that $u \in W^{1,\infty}(\mathbb{R}^n)$. Lemma 1.10 implies

$$M|\nabla u|(x) \le \|\nabla u\|_{L^\infty(\mathbb{R}^n)}$$

for every $x \in \mathbb{R}^n$. Since $u \in W^{1,\infty}(\mathbb{R}^n) \subset W^{1,1}_{\mathrm{loc}}(\mathbb{R}^n)$, by Remark 4.3 there exists a set $E \subset \mathbb{R}^n$ with $|E| = 0$ such that

$$|u(x) - u(y)| \le C(n)\|\nabla u\|_{L^\infty(\mathbb{R}^n)}|x-y|$$

for every $x, y \in \mathbb{R}^n \setminus E$. This gives another proof for the fact that every function $u \in W^{1,\infty}(\mathbb{R}^n)$ has a Lipschitz continuous representative; compare to Theorem 2.21.

Motivated by Theorem 4.2, we introduce the notion of a maximal gradient and the Hajłasz–Sobolev space.

DEFINITION 4.5. Let $1 < p < \infty$ and assume that $u \in L^p(\mathbb{R}^n)$. We write $g \in \mathcal{D}(u)$ for a measurable function $g\colon \mathbb{R}^n \to [0,\infty]$ if there exists a set $E \subset \mathbb{R}^n$ with $|E| = 0$ such that

$$|u(x) - u(y)| \le |x-y|\big(g(x)+g(y)\big) \tag{4.3}$$

for every $x, y \in \mathbb{R}^n \setminus E$. The Hajłasz–Sobolev space $M^{1,p}(\mathbb{R}^n)$ consists of functions $u \in L^p(\mathbb{R}^n)$ with $\mathcal{D}(u) \cap L^p(\mathbb{R}^n) \neq \emptyset$. This space is equipped with the norm

$$\|u\|_{M^{1,p}(\mathbb{R}^n)} = \|u\|_{L^p(\mathbb{R}^n)} + \inf_{g\in\mathcal{D}(u)} \|g\|_{L^p(\mathbb{R}^n)}.$$

THEOREM 4.6. *Let $1 < p < \infty$. Then $M^{1,p}(\mathbb{R}^n) = W^{1,p}(\mathbb{R}^n)$ and the associated norms are equivalent, that is, there exists a constant $C = C(n,p) \ge 1$ such that*

$$C^{-1}\|u\|_{W^{1,p}(\mathbb{R}^n)} \le \|u\|_{M^{1,p}(\mathbb{R}^n)} \le C\|u\|_{W^{1,p}(\mathbb{R}^n)}$$

for every $u \in W^{1,p}(\mathbb{R}^n)$.

PROOF. First we discuss the inclusion $W^{1,p}(\mathbb{R}^n) \subset M^{1,p}(\mathbb{R}^n)$. Let $u \in W^{1,p}(\mathbb{R}^n)$ and let $C(n)$ be the constant in Theorem 4.2. Theorem 1.15 (c) implies

$$C(n)M|\nabla u| \in \mathcal{D}(u) \cap L^p(\mathbb{R}^n)$$

and

$$\|u\|_{M^{1,p}(\mathbb{R}^n)} \le \|u\|_{L^p(\mathbb{R}^n)} + C(n)\big\|M|\nabla u|\big\|_{L^p(\mathbb{R}^n)} \le C(n,p)\|u\|_{W^{1,p}(\mathbb{R}^n)}.$$

Assume then that $u \in M^{1,p}(\mathbb{R}^n)$ with $g \in \mathcal{D}(u) \cap L^p(\mathbb{R}^n)$. Let $h \in \mathbb{R}\setminus\{0\}$ and $j = 1, \dots, n$. By the assumption we have

$$|u(x+he_j) - u(x)| \le |h|\big(g(x+he_j) + g(x)\big)$$

for almost every $x \in \mathbb{R}^n$, from which it follows that

$$\int_{\mathbb{R}^n} |u(x+he_j) - u(x)|^p\,dx \le |h|^p \int_{\mathbb{R}^n} \big(g(x+he_j)+g(x)\big)^p\,dx \le 2^p\|g\|^p_{L^p(\mathbb{R}^n)}|h|^p.$$

By the characterization of the Sobolev space with difference quotients, see Theorem 2.45 (c), we conclude that $u \in W^{1,p}(\mathbb{R}^n)$ and

$$\|u\|_{W^{1,p}(\mathbb{R}^n)} \le \|u\|_{L^p(\mathbb{R}^n)} + C(n,p)\|g\|_{L^p(\mathbb{R}^n)} < \infty.$$

The inequality

$$\|u\|_{W^{1,p}(\mathbb{R}^n)} \le C(n,p)\|u\|_{M^{1,p}(\mathbb{R}^n)}$$

follows by taking infimum over all $g \in \mathcal{D}(u) \cap L^p(\mathbb{R}^n)$. □

REMARK 4.7. The pointwise characterization in Theorem 4.6 can be used to give alternative proofs for many properties of Sobolev spaces. For example, if $1 < p < \infty$, $u \in M^{1,p}(\mathbb{R}^n)$ and $g \in \mathcal{D}(u) \cap L^p(\mathbb{R}^n)$, then by the triangle inequality

$$\big||u(x)| - |u(y)|\big| \le |u(x) - u(y)| \le |x-y|\big(g(x)+g(y)\big)$$

whenever $x, y \in \mathbb{R}^n \setminus E$ with $|E| = 0$. Thus $g \in \mathcal{D}(|u|) \cap L^p(\mathbb{R}^n)$ and consequently $|u| \in M^{1,p}(\mathbb{R}^n)$. By Theorem 4.6 we conclude that if $u \in W^{1,p}(\mathbb{R}^n)$, then also $|u| \in W^{1,p}(\mathbb{R}^n)$; compare to Theorem 2.25 (a).

The next result is similar to Theorem 2.40, but the proof is based on pointwise inequalities.

THEOREM 4.8. *Let $1 < p < \infty$ and assume that $u \in L^p(\mathbb{R}^n)$. Then $u \in W^{1,p}(\mathbb{R}^n)$ if and only if there exist functions $u_i \in L^p(\mathbb{R}^n)$ and $g_i \in \mathcal{D}(u_i) \cap L^p(\mathbb{R}^n)$, $i \in \mathbb{N}$, such that $u_i \to u$ almost everywhere and $g_i \to g \in L^p(\mathbb{R}^n)$ almost everywhere, as $i \to \infty$.*

PROOF. For $u \in W^{1,p}(\mathbb{R}^n)$ the claim follows from Theorem 4.6. To prove the reverse claim, assume that $u_i, g \in L^p(\mathbb{R}^n)$, $g_i \in \mathcal{D}(u_i) \cap L^p(\mathbb{R}^n)$, and $u_i \to u$ almost everywhere and $g_i \to g$ almost everywhere, as $i \to \infty$. Then for every $i \in \mathbb{N}$ there exists a set $E_i \subset \mathbb{R}^n$ with $|E_i| = 0$ such that

$$|u_i(x) - u_i(y)| \le |x - y|\big(g_i(x) + g_i(y)\big) \tag{4.4}$$

for every $x, y \in \mathbb{R}^n \setminus E_i$. Let $A \subset \mathbb{R}^n$ with $|A| = 0$ be such that

$$\lim_{i\to\infty} u_i(x) = u(x) < \infty \quad \text{and} \quad \lim_{i\to\infty} g_i(x) = g(x) < \infty$$

for every $x \in \mathbb{R}^n \setminus A$. Define $E = A \cup \bigcup_{i=1}^\infty E_i$ and observe that $|E| = 0$. Let $x, y \in \mathbb{R}^n \setminus E$ with $x \ne y$. By taking the limits in (4.4) we obtain

$$\begin{aligned} |u(x) - u(y)| &= \lim_{i\to\infty} |u_i(x) - u_i(y)| \le \limsup_{i\to\infty} |x - y|\big(g_i(x) + g_i(y)\big) \\ &= |x - y|\big(g(x) + g(y)\big), \end{aligned}$$

and thus $g \in \mathcal{D}(u) \cap L^p(\mathbb{R}^n)$. This implies $u \in M^{1,p}(\mathbb{R}^n)$ and $u \in W^{1,p}(\mathbb{R}^n)$ by Theorem 4.6. □

4.2. Lipschitz truncation of Sobolev functions

Theorem 4.2 implies that a Sobolev function is Lipschitz continuous in the set where the Hardy–Littlewood maximal function of the gradient is bounded. The size of the set where the maximal function of the gradient is large can be estimated by maximal function arguments. In contrast to the standard convolution approximation, see Remark 2.22, this approach gives an approximation of $u \in W^{1,p}(\mathbb{R}^n)$ with a Lipschitz continuous function which coincides with u outside a set of arbitrarily small measure.

THEOREM 4.9. *Let $1 \le p < \infty$ and assume that $u \in W^{1,p}(\mathbb{R}^n)$. For every $\varepsilon > 0$ there exists $\varphi_\varepsilon \in \operatorname{Lip}(\mathbb{R}^n)$ such that*

$$\|u - \varphi_\varepsilon\|_{W^{1,p}(\mathbb{R}^n)} < \varepsilon \quad \textit{and} \quad \big|\{x \in \mathbb{R}^n : u(x) \ne \varphi_\varepsilon(x)\}\big| < \varepsilon.$$

PROOF. By Theorem 4.2 there exists $E \subset \mathbb{R}^n$ with $|E| = 0$ such that

$$|u(x) - u(y)| \le C(n)|x - y|\big(M|\nabla u|(x)| + M|\nabla u|(y)\big)$$

for every $x, y \in \mathbb{R}^n \setminus E$. Let

$$F_t = \big\{x \in \mathbb{R}^n \setminus E : M|\nabla u|(x) \le t \text{ and } |u(x)| \le t\big\}$$

for $t > 0$. Since $|E| = 0$, we have

$$|\mathbb{R}^n \setminus F_t| = |(\mathbb{R}^n \setminus E) \setminus F_t| = \big|\{x \in \mathbb{R}^n \setminus E : M|\nabla u|(x) > t \text{ or } |u(x)| > t\}\big|.$$

If $x, y \in F_t$, then

$$|u(x) - u(y)| \le C(n)|x - y|\big(M|\nabla u|(x) + M|\nabla u|(y)\big) \le 2C(n)t|x - y|,$$

and thus the restriction $u|_{F_t}$ is a $2C(n)t$-Lipschitz function. Theorem 2.7 allows us to extend $u|_{F_t}$ from F_t to a $2C(n)t$-Lipschitz function $v_t\colon \mathbb{R}^n \to \mathbb{R}$. By truncation we obtain a $2C(n)t$-Lipschitz function

$$u_t = \max\{-t, \min\{v_t, t\}\}.$$

By inequality (1.3)

$$\begin{aligned}\big|\{x \in \mathbb{R}^n : M|\nabla u|(x) > t\}\big| &\le \frac{C(n)}{t \cdot t^{p-1}} \int_{\{x\in\mathbb{R}^n : |\nabla u(x)|\ge \frac{t}{2}\}} t^{p-1}|\nabla u(x)|\,dx \\ &\le \frac{C(n,p)}{t^p} \int_{\{x\in\mathbb{R}^n : |\nabla u(x)|\ge \frac{t}{2}\}} |\nabla u(x)|^p\,dx.\end{aligned}$$

Hence, by Chebyshev's inequality, we obtain

$$\begin{aligned}t^p|\mathbb{R}^n \setminus F_t| &\le t^p\big(\big|\{x \in \mathbb{R}^n : M|\nabla u|(x) > t\}\big| + \big|\{x \in \mathbb{R}^n : |u(x)| > t\}\big|\big) \\ &\le C(n,p) \int_{\{x\in\mathbb{R}^n : |\nabla u(x)|\ge \frac{t}{2}\}} |\nabla u(x)|^p\,dx + \int_{\{x\in\mathbb{R}^n : |u(x)|>t\}} |u(x)|^p\,dx.\end{aligned}$$

The dominated convergence theorem implies

$$\lim_{t\to\infty} t^p|\mathbb{R}^n \setminus F_t| = 0 \tag{4.5}$$

and consequently $|\mathbb{R}^n \setminus F_t| \to 0$ as $t \to \infty$. Observe that

$$\lim_{t\to\infty} |\{x \in \mathbb{R}^n : u(x) \ne u_t(x)\}| \le \lim_{t\to\infty} |\mathbb{R}^n \setminus F_t| = 0, \tag{4.6}$$

since $u = u_t$ in F_t for every $t > 0$.

Then we consider an estimate for $\|u - u_t\|_{W^{1,p}(\mathbb{R}^n)}$. Since $u = u_t$ in F_t and $|u_t| \le t$, the measure estimate in (4.5) implies

$$\begin{aligned}\|u_t - u\|^p_{L^p(\mathbb{R}^n)} &= \int_{\mathbb{R}^n\setminus F_t} |u_t(x) - u(x)|^p\,dx \\ &\le 2^p\left(\int_{\mathbb{R}^n\setminus F_t} |u_t(x)|^p\,dx + \int_{\mathbb{R}^n\setminus F_t} |u(x)|^p\,dx\right) \\ &\le 2^p\left(t^p|\mathbb{R}^n \setminus F_t| + \int_{\mathbb{R}^n\setminus F_t} |u(x)|^p\,dx\right) \xrightarrow{t\to\infty} 0.\end{aligned} \tag{4.7}$$

By Theorem 2.25 (b), we obtain

$$\nabla(u_t - u)(x) = \chi_{\mathbb{R}^n\setminus F_t}(x)\nabla(u_t - u)(x) = \chi_{\mathbb{R}^n\setminus F_t}(x)\nabla u_t(x) - \chi_{\mathbb{R}^n\setminus F_t}(x)\nabla u(x)$$

for almost every $x \in \mathbb{R}^n$. Since u_t is $2C(n)t$-Lipschitz in $\mathbb{R}^n$, we have $|\nabla u_t| \le 2C(n)t$ almost everywhere in $\mathbb{R}^n$, and by (4.5),

$$\begin{aligned}\|\nabla(u_t - u)\|^p_{L^p(\mathbb{R}^n)} &\le 2^p\left(\int_{\mathbb{R}^n\setminus F_t} |\nabla u_t(x)|^p\,dx + \int_{\mathbb{R}^n\setminus F_t} |\nabla u(x)|^p\,dx\right) \\ &\le 2^p\left((2C(n)t)^p|\mathbb{R}^n \setminus F_t| + \int_{\mathbb{R}^n\setminus F_t} |\nabla u(x)|^p\,dx\right) \xrightarrow{t\to\infty} 0.\end{aligned}$$

Together with (4.7) this implies

$$\lim_{t\to\infty} \|u - u_t\|_{W^{1,p}(\mathbb{R}^n)} = 0.$$

By (4.6) we conclude that, for a given $\varepsilon > 0$, we may choose $\varphi_\varepsilon = u_t$ for a sufficiently large $t > 0$. □

There are close connections between Sobolev spaces with zero boundary values and inequalities involving the distance to the boundary. Similar phenomena appear in connection with Hardy's inequalities, see Section 6.6.

LEMMA 4.10. *Let $1 < p < \infty$, let $\Omega \subsetneq \mathbb{R}^n$ be an open set, and assume that $u \in W^{1,p}(\Omega)$ satisfies*

$$\int_\Omega \frac{|u(x)|^p}{d(x,\partial\Omega)^p}\,dx < \infty. \tag{4.8}$$

Then $u \in W_0^{1,p}(\Omega)$.

PROOF. By considering the positive part u_+ and the negative part u_- separately, we may assume that $u \geq 0$. For every $t > 0$ we define $u_t\colon \Omega \to [0,\infty)$ by

$$u_t(x) = \max\bigl\{\min\{u(x), t d(x,\partial\Omega)\} - \tfrac{1}{t}, 0\bigr\}.$$

We show that $\{u_t : t > 0\}$ is a bounded collection of functions in $W_0^{1,p}(\Omega)$. Clearly $0 \leq u_t \leq u$ in Ω and thus

$$\int_\Omega u_t(x)^p\,dx \leq \int_\Omega u(x)^p\,dx < \infty.$$

To estimate the gradient we define

$$F_t = \{x \in \Omega : u(x) > t d(x,\partial\Omega)\},$$

for $t > 0$. Theorem 2.25 (c) implies

$$\begin{aligned}\int_\Omega |\nabla u_t(x)|^p\,dx &\leq \int_{\Omega\setminus F_t} |\nabla u(x)|^p\,dx + t^p \int_{F_t} |\nabla d(x,\partial\Omega)|^p\,dx \\ &\leq \int_\Omega |\nabla u(x)|^p\,dx + t^p|F_t|,\end{aligned}$$

where we also used the fact that $|\nabla d(x,\partial\Omega)| \leq 1$ for almost every $x \in \Omega$. Since

$$t^p|F_t| \leq \int_\Omega \frac{u(x)^p}{d(x,\partial\Omega)^p}\,dx < \infty,$$

for every $t > 0$, and $u_t(x) = 0$ if $d(x,\partial\Omega) < t^{-2}$, the estimates above give the uniform boundedness for the functions u_t in $W_0^{1,p}(\Omega)$. Since $u_t \to u$ almost everywhere in Ω, as $t \to \infty$, we conclude from Theorem 2.40 (b) that $u \in W_0^{1,p}(\Omega)$. □

Theorem 4.9, for $1 < p < \infty$, and Lemma 4.10 are special cases of the following general result. The key point is that the distance function gives an effective control for the zero extension of a Lipschitz function. This idea will be also used in many subsequent arguments.

THEOREM 4.11. *Let $1 < p < \infty$, let $\Omega \subsetneq \mathbb{R}^n$ be an open set, and assume that $u \in W^{1,p}(\Omega)$ satisfies*

$$\int_\Omega \frac{|u(x)|^p}{d(x,\partial\Omega)^p}\,dx < \infty. \tag{4.9}$$

Then $u \in W_0^{1,p}(\Omega)$ and for every $\varepsilon > 0$ there exists $\varphi_\varepsilon \in \operatorname{Lip}(\mathbb{R}^n)$ such that

(a) *$\varphi_\varepsilon(x) = 0$ for every $x \in \Omega^c = \mathbb{R}^n \setminus \Omega$,*
(b) *$\bigl|\{x \in \Omega : u(x) \neq \varphi_\varepsilon(x)\}\bigr| < \varepsilon$,*
(c) *$\|u - \varphi_\varepsilon\|_{W^{1,p}(\Omega)} < \varepsilon$.*

PROOF. Lemma 4.10 implies $u \in W_0^{1,p}(\Omega)$. By Theorem 2.31, we may assume that $u \in W^{1,p}(\mathbb{R}^n)$ and $u(x) = 0$ for every $x \in \Omega^c$. By Theorem 4.6 we have $u \in M^{1,p}(\mathbb{R}^n)$, and thus there exist $g \in \mathcal{D}(u) \cap L^p(\mathbb{R}^n)$ and $E \subset \mathbb{R}^n$ with $|E| = 0$ such that

$$|u(x) - u(y)| \le |x - y|\big(g(x) + g(y)\big) \tag{4.10}$$

for every $x, y \in \mathbb{R}^n \setminus E$. Let

$$F_t = \big\{x \in \Omega \setminus E : |u(x)| \le t,\ |u(x)|d(x, \partial\Omega)^{-1} \le t \text{ and } g(x) \le t\big\}$$

for $t > 0$. Since $|E| = 0$, we have

$$\begin{aligned}|\Omega \setminus F_t| \le \big|\{x \in \Omega \setminus E : |u(x)| > t\}\big| + \big|\{x \in \Omega \setminus E : |u(x)|d(x, \partial\Omega)^{-1} > t\}\big| \\ + \big|\{x \in \Omega \setminus E : g(x) > t\}\big|.\end{aligned}$$

Chebyshev's inequality implies

$$t^p\big|\{x \in \Omega \setminus E : |u(x)| > t\}\big| \le \int_{\{x\in\Omega\setminus E:|u(x)|>t\}} |u(x)|^p\,dx \xrightarrow{t\to\infty} 0.$$

Similar estimates for $|u|d(x, \partial\Omega)^{-1} \in L^p(\Omega)$ and $g \in L^p(\mathbb{R}^n)$ show that

$$\lim_{t\to\infty} t^p|\Omega \setminus F_t| = 0. \tag{4.11}$$

By (4.10) we have

$$|u(x) - u(y)| \le |x - y|\big(g(x) + g(y)\big) \le 2t|x - y|$$

for every $x, y \in F_t$. On the other hand, if $x \in F_t$ and $y \in \Omega^c$, then

$$|u(x) - u(y)| = |u(x)| \le td(x, \partial\Omega) = td(x, \Omega^c) \le t|x - y|.$$

These estimates and the fact that $u = 0$ in Ω^c show that the restriction $u|_{F_t\cup\Omega^c}$ is a $2t$-Lipschitz function. Theorem 2.7 allows us to extend $u|_{F_t\cup\Omega^c}$ to a $2t$-Lipschitz function $v_t : \mathbb{R}^n \to \mathbb{R}$. By truncation we obtain a $2t$-Lipschitz function

$$u_t = \max\{-t, \min\{v_t, t\}\}.$$

Clearly $u_t = 0$ in Ω^c. Since $u = v_t$ and $|u| \le t$ in F_t, we have $u_t = v_t = u$ in F_t. By (4.11) we obtain

$$\lim_{t\to\infty} |\{x \in \Omega : u(x) \ne u_t(x)\}| \le \lim_{t\to\infty} |\Omega \setminus F_t| = 0. \tag{4.12}$$

Proceeding as in the proof of Theorem 4.9, we conclude that

$$\lim_{t\to\infty} \|u_t - u\|_{W^{1,p}(\Omega)} = 0.$$

For a given $\varepsilon > 0$, we can choose $\varphi_\varepsilon = u_t$ for a sufficiently large $t > 0$, and this completes the proof. □

4.3. Campanato and Morrey approaches to Sobolev spaces

We study a Campanato-type approach based on the fractional sharp maximal function, which controls the oscillation of a locally integrable function.

DEFINITION 4.12. Let $0 \le \beta < \infty$ and $R > 0$, and assume that $f \in L^1_{\mathrm{loc}}(\mathbb{R}^n)$. The restricted fractional sharp maximal function $M^\sharp_{\beta,R} f : \mathbb{R}^n \to [0, \infty]$ is defined as

$$M^\sharp_{\beta,R} f(x) = \sup_{0<r<R} r^{-\beta} \fint_{B(x,r)} |f(y) - f_{B(x,r)}|\,dy,$$

for every $x \in \mathbb{R}^n$. For $R = \infty$, we write $M^\sharp_{\beta,\infty} f(x) = M^\sharp_\beta f(x)$ for every $x \in \mathbb{R}^n$.

The advantage of the following pointwise inequality compared to Theorem 4.1 and Theorem 4.2 is that it does not refer to the gradient of a function.

LEMMA 4.13. *Let $0 < \beta < \infty$ and assume that $f \in L^1_{\mathrm{loc}}(\mathbb{R}^n)$. There exist a constant $C = C(n, \beta)$ and a set $E \subset \mathbb{R}^n$ with $|E| = 0$ such that*

$$|f(x) - f(y)| \le C|x-y|^\beta \big(M^\sharp_{\beta,4|x-y|} f(x) + M^\sharp_{\beta,4|x-y|} f(y)\big)$$

for every $x, y \in \mathbb{R}^n \setminus E$. Moreover, the set E can be chosen to be the complement of the set of Lebesgue points of f.

PROOF. Let $E = \mathbb{R}^n \setminus \mathrm{Leb}(f)$ be the complement of the set of Lebesgue points of f. Theorem 1.21 implies $|E| = 0$. Let $x \in \mathbb{R}^n \setminus E$ and $r > 0$, and define $B_i = B(x, 2^{-i}r)$ for every $i \in \mathbb{N}_0$. Since x is a Lebesgue point of f, a telescoping argument, as in the proof of Theorem 3.19, gives

$$\begin{aligned} |f(x) - f_{B(x,r)}| &\le \sum_{i=0}^\infty |f_{B_{i+1}} - f_{B_i}| \le \sum_{i=0}^\infty \frac{|B_i|}{|B_{i+1}|} \fint_{B_i} |f(z) - f_{B_i}|\,dz \\ &\le C(n) \sum_{i=0}^\infty (2^{-i}r)^\beta (2^{-i}r)^{-\beta} \fint_{B_i} |f(z) - f_{B_i}|\,dz \\ &\le C(n,\beta) r^\beta M^\sharp_{\beta,2r} f(x). \end{aligned} \tag{4.13}$$

Let $y \in \mathbb{R}^n \setminus E$ and choose $r = |x-y|$. If $x = y$, the claim is clear, and thus we may assume that $x \ne y$. Since $B(x,r) \subset B(y,2r)$, the estimate in (4.13), in terms of y and $2r$, implies

$$\begin{aligned} |f(y) - f_{B(x,r)}| &\le |f(y) - f_{B(y,2r)}| + |f_{B(y,2r)} - f_{B(x,r)}| \\ &\le C(n,\beta) r^\beta M^\sharp_{\beta,4r} f(y) + \fint_{B(x,r)} |f(z) - f_{B(y,2r)}|\,dz \\ &\le C(n,\beta) r^\beta M^\sharp_{\beta,4r} f(y) + C(n) \fint_{B(y,2r)} |f(z) - f_{B(y,2r)}|\,dz \\ &\le C(n,\beta) r^\beta M^\sharp_{\beta,4r} f(y). \end{aligned}$$

By combining the estimates above we obtain

$$\begin{aligned} |f(x) - f(y)| &\le |f(x) - f_{B(x,r)}| + |f(y) - f_{B(x,r)}| \\ &\le C(n,\beta)|x-y|^\beta \big(M^\sharp_{\beta,4|x-y|} f(x) + M^\sharp_{\beta,4|x-y|} f(y)\big). \end{aligned}$$ □

REMARK 4.14. Lemma 4.13 gives a Campanato-type characterization of Hölder continuity. Assume first that $f \in L^1_{\mathrm{loc}}(\mathbb{R}^n)$ and let $0 < \beta \le 1$. By Lemma 4.13 there exists a set $E \subset \mathbb{R}^n$ with $|E| = 0$ such that

$$|f(x) - f(y)| \le C(n,\beta)|x-y|^\beta \big(M^\sharp_\beta f(x) + M^\sharp_\beta f(y)\big)$$

for every $x, y \in \mathbb{R}^n \setminus E$. If $M^\sharp_\beta f \in L^\infty(\mathbb{R}^n)$, then by uniform continuity f can be redefined on a set $E \cup F$ of measure zero, where F is an exceptional set for the essential supremum. Hence f becomes Hölder continuous in $\mathbb{R}^n$ with exponent β.

Conversely, if $r > 0$ and f is Hölder continuous in $\mathbb{R}^n$, with exponent β and constant C, then

$$|f(y) - f_{B(x,r)}| = \left| f(y) - \fint_{B(x,r)} f(z)\,dz \right| \le \fint_{B(x,r)} |f(y) - f(z)|\,dz \le C(2r)^\beta$$

for every $x \in \mathbb{R}^n$ and $y \in B(x,r)$. Thus

$$M_\beta^\sharp f(x) = \sup_{r>0} r^{-\beta} \fint_{B(x,r)} |f(y) - f_{B(x,r)}|\,dy \le 2^\beta C < \infty$$

for every $x \in \mathbb{R}^n$. This shows that $M_\beta^\sharp f$ is bounded in $\mathbb{R}^n$ by $2^\beta C$.

Since $M_\beta^\sharp f$ is invariant under redefinition of f on a set of measure zero, we conclude that $M_\beta^\sharp f \in L^\infty(\mathbb{R}^n)$ if and only if $f \in L^1_{\mathrm{loc}}(\mathbb{R}^n)$ can be redefined on a set of measure zero in such a way that f becomes Hölder continuous in $\mathbb{R}^n$ with exponent β.

REMARK 4.15. In the limiting case $\beta = 0$ we obtain the space of bounded mean oscillation $\mathrm{BMO}(\mathbb{R}^n)$, which consists of functions $f \in L^1_{\mathrm{loc}}(\mathbb{R}^n)$ satisfying $M_0^\sharp f \in L^\infty(\mathbb{R}^n)$. Functions of bounded mean oscillation are discussed in Section 9.7.

The following characterization of Sobolev spaces extends Theorem 4.6.

THEOREM 4.16. *Let $1 < p < \infty$. The following conditions are equivalent.*

(a) $u \in W^{1,p}(\mathbb{R}^n)$.

(b) $u \in L^p(\mathbb{R}^n)$ *and there exist a nonnegative function* $g \in L^p(\mathbb{R}^n)$ *and a set* $E \subset \mathbb{R}^n$ *with* $|E| = 0$ *such that*

$$|u(x) - u(y)| \le |x-y|\big(g(x) + g(y)\big)$$

for every $x, y \in \mathbb{R}^n \setminus E$.

(c) $u \in L^p(\mathbb{R}^n)$ *and there exists a nonnegative function* $g \in L^p(\mathbb{R}^n)$ *such that the Poincaré inequality*

$$\fint_{B(x,r)} |u(y) - u_{B(x,r)}|\,dy \le r \fint_{B(x,r)} g(y)\,dy$$

holds for every $x \in \mathbb{R}^n$ *and* $r > 0$.

(d) $u \in L^p(\mathbb{R}^n)$ *and* $M_1^\sharp u \in L^p(\mathbb{R}^n)$.

PROOF. Theorem 4.6 gives the equivalence of (a) and (b).

Then we prove that (b) implies (c). Let $y \in B(x,r) \setminus E$. We integrate the pointwise inequality in (b) twice over the ball $B(x,r)$. This gives

$$\begin{aligned} |u(y) - u_{B(x,r)}| &= \left| u(y) - \fint_{B(x,r)} u(z)\,dz \right| \le \fint_{B(x,r)} |u(y) - u(z)|\,dz \\ &\le 2r\left(g(y) + \fint_{B(x,r)} g(z)\,dz \right) \end{aligned}$$

and

$$\begin{aligned} \fint_{B(x,r)} |u(y) - u_{B(x,r)}|\,dy &\le 2r\left(\fint_{B(x,r)} g(y)\,dy + \fint_{B(x,r)} g(z)\,dz \right) \\ &\le r \fint_{B(x,r)} 4g(y)\,dy. \end{aligned}$$

Thus (c) holds with the function $4g \in L^p(\mathbb{R}^n)$.

To show that (c) implies (d), we observe that

$$0 \le M_1^\sharp u(x) = \sup_{r>0} \frac{1}{r} \fint_{B(x,r)} |u(y) - u_{B(x,r)}|\,dy \le \sup_{r>0} \fint_{B(x,r)} g(y)\,dy = Mg(x)$$

for every $x \in \mathbb{R}^n$. By Theorem 1.15 (c), we have $Mg \in L^p(\mathbb{R}^n)$, and thus $M_1^\sharp u \in L^p(\mathbb{R}^n)$.

Finally, we show that (d) implies (b). By Lemma 4.13, with $\beta = 1$, there exists $E \subset \mathbb{R}^n$ such that $|E| = 0$ and

$$|u(x) - u(y)| \le C(n)|x - y|\big(M_1^\sharp u(x) + M_1^\sharp u(y)\big)$$

for every $x, y \in \mathbb{R}^n \setminus E$. This gives (b) with $g = C(n)M_1^\sharp u \in L^p(\mathbb{R}^n)$. □

Next we consider a Morrey-type approach to pointwise inequalities. Compared to the Campanato approach above, here the sharp maximal function is replaced with a fractional maximal function of the gradient, see Definition 1.38.

THEOREM 4.17. *Let $0 \le \alpha < 1$ and assume that $u \in W^{1,1}_{\text{loc}}(\mathbb{R}^n)$. There exist a constant $C = C(n, \alpha)$ and a set $E \subset \mathbb{R}^n$ with $|E| = 0$ such that*

$$|u(x) - u(y)| \le C|x - y|^{1-\alpha}\big(M_{\alpha,4|x-y|}|\nabla u|(x) + M_{\alpha,4|x-y|}|\nabla u|(y)\big)$$

for every $x, y \in \mathbb{R}^n \setminus E$.

PROOF. Let $x \in \mathbb{R}^n$, $R > 0$ and $0 < r < R$. By the Poincaré inequality in Theorem 3.14, we have

$$\fint_{B(x,r)} |u(y) - u_{B(x,r)}|\, dy \le C(n) r \fint_{B(x,r)} |\nabla u(y)|\, dy.$$

This implies

$$r^{\alpha-1} \fint_{B(x,r)} |u(y) - u_{B(x,r)}|\, dy \le C(n) r^\alpha \fint_{B(x,r)} |\nabla u(y)|\, dy,$$

and consequently

$$M^\sharp_{1-\alpha,R} u(x) \le C(n) M_{\alpha,R}|\nabla u|(x).$$

The claim follows from Lemma 4.13 with $\beta = 1 - \alpha > 0$. □

REMARK 4.18. Theorem 4.17 gives a Morrey-type condition for Hölder continuity; compare to Remark 4.14, where Hölder continuity was characterized by a Campanato approach. Let $0 \le \alpha < 1$ and assume that $u \in W^{1,1}_{\text{loc}}(\mathbb{R}^n)$ and $M_\alpha|\nabla u| \in L^\infty(\mathbb{R}^n)$. By Theorem 4.17 there exists $E \subset \mathbb{R}^n$ with $|E| = 0$ such that

$$|u(x) - u(y)| \le C(n, \alpha)|x - y|^{1-\alpha}\big(M_\alpha|\nabla u|(x) + M_\alpha|\nabla u|(y)\big) \tag{4.14}$$

for every $x, y \in \mathbb{R}^n \setminus E$. By (4.14) we have

$$|u(x) - u(y)| \le C(n, \alpha)\big\|M_\alpha|\nabla u|\big\|_{L^\infty(\mathbb{R}^n)}|x - y|^{1-\alpha},$$

for every $x, y \in \mathbb{R}^n \setminus (E \cup F)$, where F is an exceptional set for essential supremum with $|F| = 0$. This shows that u is Hölder continuous with the exponent $1 - \alpha$, after a possible redefinition on a set of measure zero.

REMARK 4.19. From (4.14) we recover Theorem 3.23, which was obtained as a consequence of Morrey's inequality in Section 3.5. To see this, assume that $u \in W^{1,p}(\mathbb{R}^n)$ with $n < p < \infty$. By Hölder's inequality we have

$$M_{\frac{n}{p}}|\nabla u|(x) \le C(n)\big(M_n|\nabla u|^p(x)\big)^{\frac{1}{p}} \le C(n)\|\nabla u\|_{L^p(\mathbb{R}^n)} < \infty,$$

for every $x \in \mathbb{R}^n$. Thus (4.14), with $\alpha = \frac{n}{p}$, implies

$$|u(x) - u(y)| \le C(n, p)\|\nabla u\|_{L^p(\mathbb{R}^n)}|x - y|^{1-\frac{n}{p}}$$

for every $x, y \in \mathbb{R}^n \setminus E$. This shows that u is Hölder continuous with the exponent $1 - \frac{n}{p}$ after a possible redefinition on a set of measure zero.

4.4. Maximal operator on Sobolev spaces

We show that the Hardy–Littlewood maximal operator is not only bounded on $L^p(\mathbb{R}^n)$ but also on $W^{1,p}(\mathbb{R}^n)$ for $1 < p \le \infty$. We begin with the limiting case $p = \infty$. Recall that $W^{1,\infty}(\mathbb{R}^n)$ essentially coincides with the space of bounded Lipschitz continuous functions, see Theorem 2.21.

THEOREM 4.20. *Assume that $u \in W^{1,\infty}(\mathbb{R}^n)$. Then $Mu \in W^{1,\infty}(\mathbb{R}^n)$ and*

$$\|Mu\|_{W^{1,\infty}(\mathbb{R}^n)} \le \|u\|_{W^{1,\infty}(\mathbb{R}^n)}.$$

PROOF. Theorem 2.21 implies that the function $u \in W^{1,\infty}(\mathbb{R}^n)$, perhaps after redefinition on a set of measure zero, is bounded and Lipschitz continuous with constant $\|\nabla u\|_{L^\infty(\mathbb{R}^n)}$. Hence

$$|\tau_h u(y) - u(y)| = |u(y+h) - u(y)| \le \|\nabla u\|_{L^\infty(\mathbb{R}^n)}|h|$$

for every $y, h \in \mathbb{R}^n$, where $\tau_h u(y) = u(y+h)$. On the other hand, it follows from Lemma 1.10 that $Mu(y) \le \|u\|_{L^\infty(\mathbb{R}^n)} < \infty$ for every $y \in \mathbb{R}^n$. Since the maximal function commutes with translations and the maximal operator is sublinear, see Lemma 1.5, we have

$$\begin{aligned}|Mu(x+h) - Mu(x)| &= |(\tau_h Mu)(x) - Mu(x)| = |M(\tau_h u)(x) - Mu(x)| \\ &\le M(\tau_h u - u)(x) = \sup_{r>0} \fint_{B(x,r)} |\tau_h u(y) - u(y)|\,dy \\ &\le \|\nabla u\|_{L^\infty(\mathbb{R}^n)}|h|,\end{aligned}$$

for every $x, h \in \mathbb{R}^n$. This shows that Mu is Lipschitz continuous in $\mathbb{R}^n$ with constant $\|\nabla u\|_{L^\infty(\mathbb{R}^n)}$. Since $Mu \in L^\infty(\mathbb{R}^n)$, by Theorem 2.21 (c) we conclude that $Mu \in W^{1,\infty}(\mathbb{R}^n)$ and

$$\begin{aligned}\|Mu\|_{W^{1,\infty}(\mathbb{R}^n)} &= \|Mu\|_{L^\infty(\mathbb{R}^n)} + \|\nabla Mu\|_{L^\infty(\mathbb{R}^n)} \\ &\le \|u\|_{L^\infty(\mathbb{R}^n)} + \|\nabla u\|_{L^\infty(\mathbb{R}^n)} = \|u\|_{W^{1,\infty}(\mathbb{R}^n)}.\end{aligned}$$ □

The proof above shows that if $u \in \operatorname{Lip}(\mathbb{R}^n)$ and $Mu(x_0) < \infty$ for some $x_0 \in \mathbb{R}^n$, then $Mu \in \operatorname{Lip}(\mathbb{R}^n)$ with the same Lipschitz constant as u, see Remark 1.18. Theorem 2.9 implies that then Mu is differentiable almost everywhere.

In general, the question of differentiability for maximal functions is more delicate and the maximal function of a positive differentiable function is not necessarily differentiable. Nevertheless, certain weak differentiability properties are preserved.

THEOREM 4.21. *Let $1 < p < \infty$ and assume that $u \in W^{1,p}(\mathbb{R}^n)$. Then $Mu \in W^{1,p}(\mathbb{R}^n)$ and there exists a constant $C = C(n,p)$ such that*

$$\|Mu\|_{W^{1,p}(\mathbb{R}^n)} \le C\|u\|_{W^{1,p}(\mathbb{R}^n)}. \tag{4.15}$$

PROOF. The proof is based on the characterization of $W^{1,p}(\mathbb{R}^n)$ by difference quotients, see Theorem 2.45. By the maximal function theorem with $1 < p < \infty$, see Theorem 1.15 (c), we have $Mu \in L^p(\mathbb{R}^n)$. Using Lemma 1.5, Theorem 1.15 (c), and Theorem 2.45 (a), we obtain

$$\begin{aligned}\|(\tau_{he_j} Mu) - Mu\|_{L^p(\mathbb{R}^n)} &= \|M(\tau_{he_j} u) - Mu\|_{L^p(\mathbb{R}^n)} \le \|M(\tau_{he_j} u - u)\|_{L^p(\mathbb{R}^n)} \\ &\le C(n,p)\|\tau_{he_j} u - u\|_{L^p(\mathbb{R}^n)} \le C(n,p)\|\nabla u\|_{L^p(\mathbb{R}^n)}|h|\end{aligned}$$

for every $h \in \mathbb{R} \setminus \{0\}$ and $j = 1, \dots, n$. Theorem 2.45 (c) implies that $Mu \in W^{1,p}(\mathbb{R}^n)$ with

$$\|\nabla Mu\|_{L^p(\mathbb{R}^n)} \le C(n,p)\|\nabla u\|_{L^p(\mathbb{R}^n)}.$$

Thus, again by Theorem 1.15 (c),

$$\begin{aligned}\|Mu\|_{W^{1,p}(\mathbb{R}^n)} &\le \|Mu\|_{L^p(\mathbb{R}^n)} + \|\nabla Mu\|_{L^p(\mathbb{R}^n)}\\ &\le C(n,p)\big(\|u\|_{L^p(\mathbb{R}^n)} + \|\nabla u\|_{L^p(\mathbb{R}^n)}\big) \le C(n,p)\|u\|_{W^{1,p}(\mathbb{R}^n)}.\end{aligned}$$ □

A more careful analysis gives a pointwise estimate for the gradient.

THEOREM 4.22. *Let $1 < p < \infty$ and assume that $u \in W^{1,p}(\mathbb{R}^n)$. Then $Mu \in W^{1,p}(\mathbb{R}^n)$ and*

$$|\nabla Mu(x)| \le M|\nabla u|(x) \tag{4.16}$$

for almost every $x \in \mathbb{R}^n$.

PROOF. For $r > 0$ and $x \in \mathbb{R}^n$, we define

$$\chi_r(x) = \frac{\chi_{B(0,r)}(x)}{|B(0,r)|}, \tag{4.17}$$

where $\chi_{B(0,r)}$ is the characteristic function of $B(0,r)$. Then

$$\fint_{B(x,r)} |u(y)|\,dy = (|u| * \chi_r)(x),$$

for every $x \in \mathbb{R}^n$. For almost every $x \in \mathbb{R}^n$ we have

$$\nabla(|u| * \chi_r)(x) = (\nabla|u| * \chi_r)(x),$$

where the convolution on the right-hand side is taken componentwise. Young's convolution inequality (2.13) and Theorem 2.25 (a) give $|u| * \chi_r \in W^{1,p}(\mathbb{R}^n)$.

Let $(r_j)_{j\in\mathbb{N}}$ be an enumeration of positive rational numbers. Then

$$Mu(x) = \sup_{j\in\mathbb{N}}(|u| * \chi_{r_j})(x)$$

for every $x \in \mathbb{R}^n$. We define functions $v_k\colon \mathbb{R}^n \to \mathbb{R}$, with $k \in \mathbb{N}$, by

$$v_k(x) = \max_{1\le j\le k}(|u| * \chi_{r_j})(x).$$

By applying Theorem 2.25 (c) inductively, we find that $(v_k)_{k\in\mathbb{N}}$ is a pointwise increasing sequence of functions in $W^{1,p}(\mathbb{R}^n)$ with

$$\lim_{k\to\infty} v_k(x) = Mu(x)$$

for every $x \in \mathbb{R}^n$. Moreover, for every $k \in \mathbb{N}$, we have

$$|\nabla v_k(x)| \le \max_{1\le j\le k}\big|\nabla(|u| * \chi_{r_j})(x)\big| = \max_{1\le j\le k}\big|(\nabla|u| * \chi_{r_j})(x)\big| \le M|\nabla u|(x)$$

for almost every $x \in \mathbb{R}^n$.

Theorem 1.15 (c) gives

$$\begin{aligned}\sup_{k\in\mathbb{N}}\|v_k\|_{W^{1,p}(\mathbb{R}^n)} &\le \|Mu\|_{L^p(\mathbb{R}^n)} + \big\|M|\nabla u|\big\|_{L^p(\mathbb{R}^n)}\\ &\le C(n,p)\big(\|u\|_{L^p(\mathbb{R}^n)} + \|\nabla u\|_{L^p(\mathbb{R}^n)}\big) < \infty,\end{aligned}$$

and thus the sequence $(v_k)_{k\in\mathbb{N}}$ is bounded in $W^{1,p}(\mathbb{R}^n)$. Since $v_k \to Mu$ pointwise, as $k \to \infty$, Theorem 2.40 (a) implies that $Mu \in W^{1,p}(\mathbb{R}^n)$, $v_k \to Mu$ weakly in

$L^p(\mathbb{R}^n)$, and $\nabla v_k \to \nabla Mu$ weakly in $L^p(\mathbb{R}^n;\mathbb{R}^n)$ as $k \to \infty$. Theorem 2.35, with $f_k = \nabla v_k$ and $g_k = M|\nabla u|$ for every $k \in \mathbb{N}$, implies (4.16). □

REMARK 4.23. Theorem 4.22 gives an alternative proof for Theorem 4.21. To see this, let $1 < p < \infty$ and assume that $u \in W^{1,p}(\mathbb{R}^n)$. By Theorem 1.15 (c) and (4.16), we have

$$\begin{aligned}\|Mu\|_{W^{1,p}(\mathbb{R}^n)} &\le \|Mu\|_{L^p(\mathbb{R}^n)} + \|\nabla Mu\|_{L^p(\mathbb{R}^n)} \\ &\le C(n,p)\big(\|u\|_{L^p(\mathbb{R}^n)} + \big\|M|\nabla u|\big\|_{L^p(\mathbb{R}^n)}\big) \\ &\le C(n,p)\big(\|u\|_{L^p(\mathbb{R}^n)} + \|\nabla u\|_{L^p(\mathbb{R}^n)}\big) \le C(n,p)\|u\|_{W^{1,p}(\mathbb{R}^n)}.\end{aligned}$$

4.5. Maximal function with respect to an open set

We study boundedness of the maximal operator with respect to an open set, see Definition 1.30, on Sobolev spaces. The arguments in the previous section do not apply directly, mainly because this maximal operator does not commute with translations. Results in this section hold for every nontrivial open set in $\mathbb{R}^n$.

We assume throughout this section that $\Omega \subsetneq \mathbb{R}^n$ is an open set, and thus $d(x,\partial\Omega)$ is finite and positive for every $x \in \Omega$. For $u \in L^1_{\mathrm{loc}}(\Omega)$, we define integral average functions $u_t\colon \Omega \to [0,\infty)$, $0 < t < 1$, by

$$u_t(x) = \fint_{B(x,t\delta(x))} |u(y)|\,dy,$$

where $\delta(x) = d(x,\partial\Omega) \in (0,\infty)$ for every $x \in \Omega \subsetneq \mathbb{R}^n$. Then

$$M_\Omega u(x) = \sup_{0<t<1} u_t(x)$$

for every $x \in \Omega$.

LEMMA 4.24. *Let $1 < p < \infty$, let $\Omega \subsetneq \mathbb{R}^n$ be an open set, and assume that $u \in W^{1,p}(\Omega)$. For every $0 < t < 1$, we have $u_t \in W^{1,p}(\Omega)$ and*

$$|\nabla u_t(x)| \le 2M_\Omega|\nabla u|(x), \tag{4.18}$$

for almost every $x \in \Omega$.

PROOF. Assume first that $u \in C^\infty(\Omega)$ is nonnegative and let $0 < t < 1$. By Theorem 2.9, the Lipschitz function δ is differentiable almost everywhere in Ω. Moreover, $|\nabla\delta(x)| \le 1$ for almost every $x \in \Omega$. By the Leibniz rule

$$\frac{\partial u_t}{\partial x_i}(x) = \frac{\partial}{\partial x_i}\left(\frac{1}{\omega_n(t\delta(x))^n}\right)\int_{B(x,t\delta(x))} u(y)\,dy + \frac{1}{\omega_n(t\delta(x))^n}\frac{\partial}{\partial x_i}\int_{B(x,t\delta(x))} u(y)\,dy,$$

for almost every $x \in \Omega$, and by the chain rule

$$\begin{aligned}\frac{\partial}{\partial x_i}\int_{B(x,t\delta(x))} u(y)\,dy &= \int_{B(x,t\delta(x))}\frac{\partial u}{\partial x_i}(y)\,dy \\ &\quad + t\frac{\partial\delta}{\partial x_i}(x)\int_{\partial B(x,t\delta(x))} u(y)\,d\sigma(y),\end{aligned} \tag{4.19}$$

for almost every $x \in \Omega$. Here we also used the fact that

$$\frac{\partial}{\partial r}\int_{B(x,r)} u(y)\,dy = \int_{\partial B(x,r)} u(y)\,d\sigma(y),$$

which follows by differentiating the integral over spheres. By collecting the terms, we obtain

$$\begin{aligned}\nabla u_t(x) &= n\frac{\nabla\delta(x)}{\delta(x)}\bigg(\fint_{\partial B(x,t\delta(x))} u(y)\,d\sigma(y) - \fint_{B(x,t\delta(x))} u(y)\,dy\bigg)\\ &\quad + \fint_{B(x,t\delta(x))} \nabla u(y)\,dy\end{aligned} \tag{4.20}$$

for almost every $x \in \Omega$. The last integral in (4.20) is taken componentwise.

In order to estimate the difference of the two integrals in (4.20), we take into account a cancellation effect. Let $B(x,R) \subset \Omega$ be a ball such that $x \in \Omega$ and $0 < R < d(x,\partial\Omega)$. Green's first identity states that

$$\int_{\partial B(x,R)} u(y)\frac{\partial v}{\partial \nu}(y)\,d\sigma(y) = \int_{B(x,R)} \big(u(y)\Delta v(y) + \nabla u(y)\cdot\nabla v(y)\big)\,dy,$$

for every $v \in C^\infty(\Omega)$, where $\nu(y) = \frac{y-x}{R}$ is the unit exterior normal of $B(x,R)$ at $y \in \partial B(x,R)$. By choosing $v(y) = \frac{1}{2}|y-x|^2$ for every $y \in \Omega$, we have $\nabla v(y) = y - x$ and $\Delta v(y) = n$ for every $y \in B(x,R)$, and $\frac{\partial v}{\partial \nu}(y) = R$ for every $y \in \partial B(x,R)$. This implies

$$\fint_{\partial B(x,R)} u(y)\,d\sigma(y) - \fint_{B(x,R)} u(y)\,dy = \frac{1}{n}\fint_{B(x,R)} \nabla u(y)\cdot(y-x)\,dy, \tag{4.21}$$

and consequently

$$\begin{aligned}\bigg|\fint_{\partial B(x,R)} u(y)\,d\sigma(y) - \fint_{B(x,R)} u(y)\,dy\bigg| &\le \frac{1}{n}\fint_{B(x,R)} |\nabla u(y)||y-x|\,dy\\ \le \frac{R}{n}\fint_{B(x,R)} |\nabla u(y)|\,dy &\le \frac{R}{n}M_\Omega|\nabla u|(x).\end{aligned} \tag{4.22}$$

Using (4.20) and (4.22) with $R = t\delta(x)$, we obtain

$$\begin{aligned}|\nabla u_t(x)| &\le n\frac{|\nabla\delta(x)|}{\delta(x)}\frac{t\delta(x)}{n}M_\Omega|\nabla u|(x) + \bigg|\fint_{B(x,t\delta(x))} \nabla u(y)\,dy\bigg|\\ &\le tM_\Omega|\nabla u|(x) + \fint_{B(x,t\delta(x))} |\nabla u(y)|\,dy\\ &\le (t+1)M_\Omega|\nabla u|(x) \le 2M_\Omega|\nabla u|(x)\end{aligned}$$

for almost every $x \in \Omega$. This proves (4.18) for nonnegative smooth functions.

The case $u \in W^{1,p}(\Omega)$, for $1 < p < \infty$, follows from an approximation argument. By Theorem 2.25 we have $|u| \in W^{1,p}(\Omega)$ and $\big|\nabla|u|(x)\big| = |\nabla u(x)|$ for almost every $x \in \Omega$. Thus we may assume that u is nonnegative. By Remark 2.22 (b), there are nonnegative functions $u_j \in W^{1,p}(\Omega) \cap C^\infty(\Omega)$, with $j \in \mathbb{N}$, such that $u_j \to u$ in $W^{1,p}(\Omega)$ as $j \to \infty$. Since $\chi_{B(x,t\delta(x))} \in L^{\frac{p}{p-1}}(\Omega)$, we obtain from the reverse triangle inequality and Hölder's inequality that

$$u_t(x) = \lim_{j\to\infty} (u_j)_t(x)$$

for every $x \in \Omega$. It is clear that

$$|(u_j)_t(x)| = \fint_{B(x,t\delta(x))} |u_j(y)|\,dy \le M_\Omega u_j(x)$$

for every $x \in \Omega$ and every $j \in \mathbb{N}$. Since (4.18) holds for nonnegative smooth functions, we have

$$|\nabla (u_j)_t(x)| \le 2M_\Omega|\nabla u_j|(x) \tag{4.23}$$

for almost every $x \in \Omega$ and every $j \in \mathbb{N}$. The inequalities above and Theorem 1.15 imply

$$\begin{aligned}\|(u_j)_t\|_{W^{1,p}(\Omega)} &\le \|(u_j)_t\|_{L^p(\Omega)} + \|\nabla(u_j)_t\|_{L^p(\Omega)}\\ &\le \big\|M_\Omega u_j\big\|_{L^p(\Omega)} + 2\big\|M_\Omega|\nabla u_j|\big\|_{L^p(\Omega)}\\ &\le \big\|M(u_j\chi_\Omega)\big\|_{L^p(\mathbb{R}^n)} + 2\big\|M(|\nabla u_j|\chi_\Omega)\big\|_{L^p(\mathbb{R}^n)}\\ &\le C(n,p)\big(\|u_j\|_{L^p(\Omega)} + \|\nabla u_j\|_{L^p(\Omega)}\big) \le C(n,p)\|u_j\|_{W^{1,p}(\Omega)}\end{aligned}$$

for every $j \in \mathbb{N}$.

Thus $((u_j)_t)_{j\in\mathbb{N}}$ is a bounded sequence in $W^{1,p}(\Omega)$. Since $(u_j)_t \to u_t$ pointwise, as $j \to \infty$, Theorem 2.40 implies that $u_t \in W^{1,p}(\Omega)$ and $\nabla(u_j)_t \to \nabla u_t$ weakly in $L^p(\Omega;\mathbb{R}^n)$ as $j \to \infty$. To prove (4.18) we want to proceed to the limit in (4.23) as $j \to \infty$. Arguing as in Remark 1.16, we obtain

$$\begin{aligned}\big\|M_\Omega|\nabla u_j| - M_\Omega|\nabla u|\big\|_{L^p(\Omega)} &\le \big\|M_\Omega(|\nabla u_j| - |\nabla u|)\big\|_{L^p(\Omega)}\\ &\le C(n,p)\big\||\nabla u_j| - |\nabla u|\big\|_{L^p(\Omega)}.\end{aligned}$$

Hence $M_\Omega|\nabla u_j| \to M_\Omega|\nabla u|$ weakly in $L^p(\Omega)$ as $j \to \infty$, and (4.18) follows from Theorem 2.35 and (4.23). □

Next we obtain the boundedness of the maximal operator M_Ω in the Sobolev space $W^{1,p}(\Omega)$ when $1 < p < \infty$.

THEOREM 4.25. *Let $1 < p < \infty$, let $\Omega \subsetneq \mathbb{R}^n$ be an open set, and assume that $u \in W^{1,p}(\Omega)$. Then $M_\Omega u \in W^{1,p}(\Omega)$ and*

$$|\nabla M_\Omega u(x)| \le 2M_\Omega|\nabla u|(x),$$

for almost every $x \in \Omega$.

PROOF. The proof is a straightforward application of Lemma 4.24. Let $(t_j)_{j\in\mathbb{N}}$ be an enumeration of $\mathbb{Q} \cap (0,1)$ and let $u_j = u_{t_j}$. By Lemma 4.24, $u_j \in W^{1,p}(\Omega)$ for every $j \in \mathbb{N}$. Define $v_k \colon \Omega \to [0,\infty]$, with $k \in \mathbb{N}$, as

$$v_k(x) = \max_{1\le j\le k} u_j(x).$$

Then $(v_k)_{k\in\mathbb{N}}$ is a pointwise increasing sequence of functions in $W^{1,p}(\Omega)$ converging to $M_\Omega u$ pointwise in Ω. By Theorem 2.25 (c) and inequality (4.18), we have

$$|\nabla v_k(x)| = \Big|\nabla \max_{1\le j\le k} u_j(x)\Big| \le \max_{1\le j\le k}|\nabla u_j(x)| \le 2M_\Omega|\nabla u|(x) \tag{4.24}$$

for almost every $x \in \Omega$ and every $j \in \mathbb{N}$. Using Theorem 1.15 (c), we obtain

$$\begin{aligned}\|v_k\|_{W^{1,p}(\Omega)} &\le \|v_k\|_{L^p(\Omega)} + \|\nabla v_k\|_{L^p(\Omega)} \le \|M_\Omega u\|_{L^p(\Omega)} + 2\big\|M_\Omega|\nabla u|\big\|_{L^p(\Omega)}\\ &\le \big\|M(u\chi_\Omega)\big\|_{L^p(\mathbb{R}^n)} + 2\big\|M(|\nabla u|\chi_\Omega)\big\|_{L^p(\mathbb{R}^n)}\\ &\le C(n,p)\|u\|_{W^{1,p}(\Omega)}.\end{aligned}$$

Thus $(v_k)_{k\in\mathbb{N}}$ is a bounded sequence in $W^{1,p}(\Omega)$ and $v_k \to M_\Omega u$ pointwise in Ω as $k \to \infty$. Theorem 2.40 implies that $M_\Omega u \in W^{1,p}(\Omega)$, $v_k \to M_\Omega u$ weakly in $L^p(\Omega)$, and $\nabla v_k \to \nabla M_\Omega u$ weakly in $L^p(\Omega;\mathbb{R}^n)$ as $k \to \infty$. Moreover, by the estimate in (4.24) and Theorem 2.35,

$$|\nabla M_\Omega u(x)| \le 2M_\Omega|\nabla u|(x)$$

for almost every $x \in \Omega$. This proves the claim. □

REMARK 4.26. Let $1 < p < \infty$ and let $\Omega \subsetneq \mathbb{R}^n$ be an open set. If $u \in W^{1,p}(\Omega)$, then by Theorem 4.25 and the maximal function theorem we have

$$\begin{aligned}\|M_\Omega u\|_{W^{1,p}(\Omega)} &\le \|M_\Omega u\|_{L^p(\Omega)} + \|\nabla M_\Omega u\|_{L^p(\Omega)}\\ &\le \|M_\Omega u\|_{L^p(\Omega)} + 2\big\|M_\Omega|\nabla u|\big\|_{L^p(\Omega)}\\ &\le C(n,p)\|u\|_{W^{1,p}(\Omega)}.\end{aligned}$$

Thus $M_\Omega\colon W^{1,p}(\Omega) \to W^{1,p}(\Omega)$ is a bounded operator for every $1 < p < \infty$.

We have shown above that the operator M_Ω preserves the Sobolev spaces $W^{1,p}(\Omega)$ when $1 < p < \infty$. Next we show that this maximal operator also preserves boundary values in the Sobolev sense.

THEOREM 4.27. *Let $1 < p < \infty$, let $\Omega \subsetneq \mathbb{R}^n$ be an open set, and assume that $u \in W^{1,p}(\Omega)$. Then $|u| - M_\Omega u \in W_0^{1,p}(\Omega)$.*

PROOF. By Theorem 2.25 we have $|u| \in W^{1,p}(\Omega)$ and $\big|\nabla|u|(x)\big| = |\nabla u(x)|$ for almost every $x \in \Omega$. Let t with $0 < t < 1$, and let $x \in \Omega$ be a Lebesgue point of $|u|$. Using a telescoping argument, as in the proof of Lemma 4.13, and the Poincaré inequality in Theorem 3.14, we obtain

$$\begin{aligned}\big||u(x)| - u_t(x)\big| &= \left||u(x)| - \fint_{B(x,td(x,\partial\Omega))} |u(y)|\,dy\right|\\ &\le C(n)td(x,\partial\Omega)M_\Omega\big|\nabla|u|\big|(x)\\ &\le C(n)d(x,\partial\Omega)M_\Omega|\nabla u|(x).\end{aligned}$$

There exists a sequence $(t_j)_{j\in\mathbb{N}}$, with $0 < t_j < 1$ for every $j \in \mathbb{N}$, such that

$$M_\Omega u(x) = \lim_{j\to\infty} u_{t_j}(x),$$

and thus

$$\big||u(x)| - M_\Omega u(x)\big| = \lim_{j\to\infty}\big||u(x)| - u_{t_j}(x)\big| \le C(n)d(x,\partial\Omega)M_\Omega|\nabla u|(x).$$

By Theorem 1.15 (c), and the fact that almost every point in Ω is a Lebesgue point of $|u|$, we obtain

$$\begin{aligned}\int_\Omega \frac{\big||u(x)| - M_\Omega u(x)\big|^p}{d(x,\partial\Omega)^p}\,dx &\le C(n,p)\int_\Omega \big(M_\Omega|\nabla u|(x)\big)^p\,dx\\ &\le C(n,p)\int_\Omega |\nabla u(x)|^p\,dx < \infty.\end{aligned}$$

This shows that

$$\frac{|u(x)| - M_\Omega u(x)}{d(x,\partial\Omega)} \in L^p(\Omega).$$

By Theorem 4.25 we have $|u| - M_\Omega u \in W^{1,p}(\Omega)$, and Lemma 4.10 implies $|u| - M_\Omega u \in W_0^{1,p}(\Omega)$. □

COROLLARY 4.28. *Let $1 < p < \infty$, let $\Omega \subsetneq \mathbb{R}^n$ be an open set, and assume that $u \in W_0^{1,p}(\Omega)$. Then also $M_\Omega u \in W_0^{1,p}(\Omega)$.*

PROOF. Since $u \in W_0^{1,p}(\Omega)$, we have $|u| \in W_0^{1,p}(\Omega)$, see Theorem 2.30. Theorem 4.27 implies

$$M_\Omega u = |u| - (|u| - M_\Omega u) \in W_0^{1,p}(\Omega). \qquad \square$$

4.6. Fractional maximal operator on Sobolev spaces

The fractional maximal operator in Definition 1.38 has similar smoothing properties as the Riesz potentials. We begin by proving a boundedness result for the fractional maximal operator on Sobolev spaces.

THEOREM 4.29. *Let $1 < p < \infty$, $0 < \alpha < \frac{n}{p}$ and $p^* = \frac{np}{n-\alpha p}$. Assume that $u \in W^{1,p}(\mathbb{R}^n)$. Then $M_\alpha u \in W^{1,p^*}(\mathbb{R}^n)$ and there exists a constant $C = C(n,p,\alpha)$ such that*

$$\|M_\alpha u\|_{W^{1,p^*}(\mathbb{R}^n)} \le C\|u\|_{W^{1,p}(\mathbb{R}^n)}.$$

PROOF. The argument is similar to the proof of Theorem 4.21. By arguing as in the proof of Theorem 1.5, we see that the fractional maximal operator commutes with translations and is sublinear. Since $M_\alpha\colon L^p(\mathbb{R}^n) \to L^{p^*}(\mathbb{R}^n)$ is bounded by Theorem 1.42, we can apply Theorem 2.45 (a) to estimate

$$\begin{aligned}
\big\|(\tau_{he_j} M_\alpha u) - M_\alpha u\big\|_{L^{p^*}(\mathbb{R}^n)} &= \big\|M_\alpha(\tau_{he_j} u) - M_\alpha u\big\|_{L^{p^*}(\mathbb{R}^n)} \\
&\le \big\|M_\alpha(\tau_{he_j} u - u)\big\|_{L^{p^*}(\mathbb{R}^n)} \\
&\le C(n,p,\alpha)\|\tau_{he_j} u - u\|_{L^p(\mathbb{R}^n)} \\
&\le C(n,p,\alpha)\|\nabla u\|_{L^p(\mathbb{R}^n)}|h|
\end{aligned}$$

for every $h \in \mathbb{R} \setminus \{0\}$ and $j = 1, \ldots, n$. Theorem 2.45 (c) implies $M_\alpha u \in W^{1,p^*}(\mathbb{R}^n)$ with

$$\|\nabla M_\alpha u\|_{L^{p^*}(\mathbb{R}^n)} \le C(n,p,\alpha)\|\nabla u\|_{L^p(\mathbb{R}^n)}.$$

This, together with Theorem 1.42, gives

$$\begin{aligned}
\|M_\alpha u\|_{W^{1,p^*}(\mathbb{R}^n)} &\le \|M_\alpha u\|_{L^{p^*}(\mathbb{R}^n)} + \|\nabla M_\alpha u\|_{L^{p^*}(\mathbb{R}^n)} \\
&\le C(n,p,\alpha)\big(\|u\|_{L^p(\mathbb{R}^n)} + \|\nabla u\|_{L^p(\mathbb{R}^n)}\big) \\
&\le C(n,p,\alpha)\|u\|_{W^{1,p}(\mathbb{R}^n)}. \qquad \square
\end{aligned}$$

REMARK 4.30. Let $1 < p < \infty$ and assume that $u \in W^{1,p}(\mathbb{R}^n)$. A more careful analysis, as in the proof of Theorem 4.22, shows that

$$|\nabla M_\alpha u(x)| \le M_\alpha |\nabla u|(x)$$

for almost every $x \in \mathbb{R}^n$.

Next we show that the fractional maximal operator has smoothing properties.

THEOREM 4.31. *Let $1 < p < n$ and $1 \le \alpha < \frac{n}{p}$, and assume that $f \in L^p(\mathbb{R}^n)$.*

(a) *The weak gradient $\nabla M_\alpha f$ exists almost everywhere and there exists a constant $C = C(n,\alpha)$ such that*

$$|\nabla M_\alpha f(x)| \le C M_{\alpha-1} f(x)$$

for almost every $x \in \mathbb{R}^n$.

(b) *We have* $M_\alpha f \in L^{p^*}(\mathbb{R}^n)$ *and* $\nabla M_\alpha f \in L^q(\mathbb{R}^n)$ *with* $p^* = \frac{np}{n-\alpha p}$ *and* $q = \frac{np}{n-(\alpha-1)p}$. *Moreover, there exists a constant* $C = C(n, p, \alpha)$ *such that*

$$\|M_\alpha f\|_{L^{p^*}(\mathbb{R}^n)} \leq C\|f\|_{L^p(\mathbb{R}^n)}$$

and

$$\|\nabla M_\alpha f\|_{L^q(\mathbb{R}^n)} \leq C\|f\|_{L^p(\mathbb{R}^n)}.$$

PROOF. First we assume that $f \in C_0^\infty(\mathbb{R}^n)$ is not identically zero. We show that $M_\alpha f$ is Lipschitz continuous in every compact set. Let $K \subset \mathbb{R}^n$ be compact set. Since f is bounded, integrable and not identically zero, there are constants r_0 and R_0 with $0 < r_0 < R_0$, only depending on α, f and K, such that

$$M_\alpha f(x) = \sup_{r_0 \leq r \leq R_0} r^\alpha \fint_{B(x,r)} |f(y)|\,dy \tag{4.25}$$

for every $x \in K$. Here it is important to note that $0 < 1 \leq \alpha < \frac{n}{p} < n$, and $r^\alpha \to 0$ as $r \to 0$ and $r^{\alpha-n} \to 0$ as $r \to \infty$. For $r > 0$ we define, as in (4.17),

$$\chi_r(x) = \frac{\chi_{B(0,r)}(x)}{|B(0,r)|}$$

for every $x \in \mathbb{R}^n$. Then $r^\alpha(\chi_r * |f|) \in W^{1,\infty}(\mathbb{R}^n)$ is continuous in $\mathbb{R}^n$, and by Hölder's inequality

$$\begin{aligned}
\big\|\nabla(r^\alpha(\chi_r * |f|))\big\|_{L^\infty(\mathbb{R}^n)} &= \big\|r^\alpha(\chi_r * \nabla|f|)\big\|_{L^\infty(\mathbb{R}^n)} \\
&\leq \big\|r^\alpha \chi_r\big\|_{L^{p'}(\mathbb{R}^n)} \big\|\nabla|f|\big\|_{L^p(\mathbb{R}^n)} \\
&\leq r^\alpha |B(0,r)|^{\frac{1}{p'}-1} \big\|\nabla|f|\big\|_{L^p(\mathbb{R}^n)},
\end{aligned}$$

where $p' = \frac{p}{p-1}$. Theorem 2.21 implies that $r^\alpha \chi_r * |f|$ is Lipschitz continuous in $\mathbb{R}^n$ with a constant that is uniformly bounded over $r_0 \leq r \leq R_0$. The Lipschitz continuity for $M_\alpha f$ in K follows from this since

$$M_\alpha f(x) = \sup_{r_0 \leq r \leq R_0} r^\alpha(\chi_r * |f|)(x)$$

for every $x \in K$ and the Lipschitz constant of the supremum does not exceed the supremum of the individual Lipschitz constants because $M_\alpha f$ is finite everywhere.

In order to prove (a) for $f \in C_0^\infty(\mathbb{R}^n)$, we let $x \in \mathbb{R}^n$. Write $K = \overline{B}(x, 2)$, let r_0 and R_0 be as in (4.25), corresponding to the compact set K, and define

$$D^+ M_\alpha f(x) = \limsup_{h \to 0} \frac{M_\alpha f(x+h) - M_\alpha f(x)}{|h|}.$$

Since $M_\alpha f$ is Lipschitz continuous in K, we have $-\infty < D^+ M_\alpha f(x) < \infty$. Thus there exist $h_k \in B(0,2) \setminus \{0\}$, for $k \in \mathbb{N}$, such that $h_k \to 0$ as $k \to \infty$ and

$$\frac{M_\alpha f(x+h_k) - M_\alpha f(x)}{|h_k|} \geq D^+ M_\alpha f(x) - \frac{1}{k}.$$

Consequently, for a fixed $k \in \mathbb{N}$, we have

$$|h_k| D^+ M_\alpha f(x) \leq M_\alpha f(x+h_k) - M_\alpha f(x) + \frac{|h_k|}{k}.$$

For every $j \in \mathbb{N}$, choose $r_j > 0$ such that

$$r_j^\alpha \fint_{B(x+h_k,r_j)} |f(y)|\,dy \geq M_\alpha f(x+h_k) - \frac{1}{j}.$$

Since $x + h_k \in K$, we may assume that $r_0 \leq r_j \leq R_0$ for every $j \in \mathbb{N}$.

Let $j \in \mathbb{N}$ to be chosen later. Since $B(x+h_k, r_j) \subset B(x, r_j + |h_k|)$, we have

$$\begin{aligned} M_\alpha f(x) &\geq (r_j + |h_k|)^\alpha \fint_{B(x,r_j+|h_k|)} |f(y)|\,dy \\ &\geq \omega_n^{-1}(r_j+|h_k|)^{\alpha-n} \int_{B(x+h_k,r_j)} |f(y)|\,dy. \end{aligned}$$

From the estimates above we obtain

$$|h_k| D^+ M_\alpha f(x) \leq \omega_n^{-1}\left(r_j^{\alpha-n} - (r_j+|h_k|)^{\alpha-n}\right) \int_{B(x+h_k,r_j)} |f(y)|\,dy + \frac{1}{j} + \frac{|h_k|}{k}.$$

By the mean value theorem there exists $\rho_{j,k}$, with $r_j < \rho_{j,k} < r_j + |h_k|$, such that

$$r_j^{\alpha-n} - (r_j+|h_k|)^{\alpha-n} \leq (n-\alpha)\rho_{j,k}^{\alpha-n-1}|h_k|,$$

and hence

$$\begin{aligned} |h_k| D^+ M_\alpha f(x) &\leq C(n,\alpha)\rho_{j,k}^{\alpha-n-1}|h_k| \int_{B(x+h_k,r_j)} |f(y)|\,dy + \frac{1}{j} + \frac{|h_k|}{k} \\ &\leq C(n,\alpha) r_j^{\alpha-n-1}|h_k| \int_{B(x,r_j+|h_k|)} |f(y)|\,dy + \frac{1}{j} + \frac{|h_k|}{k} \\ &\leq C(n,\alpha)\left(\frac{r_j}{r_j+|h_k|}\right)^{\alpha-n-1} |h_k| M_{\alpha-1} f(x) + \frac{1}{j} + \frac{|h_k|}{k}. \end{aligned}$$

We now choose $j \in \mathbb{N}$ to be so large that $\frac{1}{j} \leq \frac{|h_k|}{k}$. Then

$$\begin{aligned} D^+ M_\alpha f(x) &\leq C(n,\alpha)\left(\frac{r_j}{r_j+|h_k|}\right)^{\alpha-n-1} M_{\alpha-1} f(x) + \frac{2}{k} \\ &\leq C(n,\alpha)\left(\frac{r_0}{r_0+|h_k|}\right)^{\alpha-n-1} M_{\alpha-1} f(x) + \frac{2}{k}, \end{aligned}$$

and by letting $k \to \infty$ we conclude that

$$D^+ M_\alpha f(x) \leq C(n,\alpha) M_{\alpha-1} f(x) \tag{4.26}$$

for every $x \in \mathbb{R}^n$. Clearly, this estimate also holds if f is identically zero.

Then assume that $f \in L^p(\mathbb{R}^n)$, and let $f_j \in C_0^\infty(\mathbb{R}^n)$, $j \in \mathbb{N}$, be such that $f_j \to f$ in $L^p(\mathbb{R}^n)$ as $j \to \infty$. Let $j \in \mathbb{N}$. Since $M_\alpha f_j$ is Lipschitz continuous in compact subsets of $\mathbb{R}^n$, it is differentiable almost everywhere in $\mathbb{R}^n$ by Theorem 2.9. If $x \in \mathbb{R}^n$ is a point of differentiability, then inequality (4.26) implies

$$\left|\frac{\partial M_\alpha f_j}{\partial x_i}(x)\right| \leq D^+ M_\alpha f_j(x) \leq C(n,\alpha) M_{\alpha-1} f_j(x)$$

for every $i = 1, \ldots, n$. Thus

$$|\nabla M_\alpha f_j(x)| \leq C(n,\alpha) M_{\alpha-1} f_j(x) \tag{4.27}$$

for almost every $x \in \mathbb{R}^n$. Theorem 1.42, or the maximal function theorem if $\alpha = 1$, implies

$$\|\nabla M_\alpha f_j\|_{L^q(\mathbb{R}^n;\mathbb{R}^n)} \leq C(n,\alpha)\|M_{\alpha-1} f_j\|_{L^q(\mathbb{R}^n)} \leq C(n,p,\alpha)\|f_j\|_{L^p(\mathbb{R}^n)},$$

and so $(\nabla M_\alpha f_j)_{j\in\mathbb{N}}$ is a bounded sequence in $L^q(\mathbb{R}^n;\mathbb{R}^n)$. Using Theorem 1.42 we obtain

$$\begin{aligned}\|M_\alpha f_j - M_\alpha f\|_{L^{p^*}(\mathbb{R}^n)} &\leq \|M_\alpha (f_j - f)\|_{L^{p^*}(\mathbb{R}^n)} \\ &\leq C(n,p,\alpha)\|f_j - f\|_{L^p(\mathbb{R}^n)} \xrightarrow{j\to\infty} 0,\end{aligned}$$

and therefore $M_\alpha f_j \to M_\alpha f$ in $L^{p^*}(\mathbb{R}^n)$ as $j \to \infty$. A similar argument shows that $M_{\alpha-1} f_j \to M_{\alpha-1} f$ in $L^q(\mathbb{R}^n)$ as $j \to \infty$. By Theorem 2.36, there exists a subsequence of $(\nabla M_\alpha f_j)_{j\in\mathbb{N}}$, which converges to some $v \in L^q(\mathbb{R}^n;\mathbb{R}^n)$. Since $M_\alpha f_j \to M_\alpha f$ in $L^{p^*}(\mathbb{R}^n)$, it follows from a straightforward adaptation of the proof of Theorem 2.39 that $\nabla M_\alpha f = v \in L^q(\mathbb{R}^n;\mathbb{R}^n)$ and, by passing to a subsequence, that $\nabla M_\alpha f_j \to \nabla M_\alpha f$ weakly in $L^q(\mathbb{R}^n;\mathbb{R}^n)$ as $j \to \infty$. By Theorem 2.35 we also see that the estimate in (4.27) is preserved up to the weak limit and hence

$$|\nabla M_\alpha f(x)| \leq C(n,\alpha) M_{\alpha-1} f(x)$$

for almost every $x \in \mathbb{R}^n$. This concludes the proof of part (a).

Part (b) follows immediately from part (a) and Theorem 1.42, or the maximal function theorem if $\alpha = 1$, since

$$\|M_\alpha f\|_{L^{p^*}(\mathbb{R}^n)} \leq C(n,p,\alpha)\|f\|_{L^p(\mathbb{R}^n)}$$

and

$$\|\nabla M_\alpha f\|_{L^q(\mathbb{R}^n)} \leq C(n,\alpha)\|M_{\alpha-1} f\|_{L^q(\mathbb{R}^n)} \leq C(n,p,\alpha)\|f\|_{L^p(\mathbb{R}^n)}. \qquad \square$$

4.7. Notes

The estimate in Theorem 4.1 has been studied by Calderón in [**69**] and Calderón and Scott [**70**]. Such pointwise inequalities are also discussed in Bennett and Sharpley [**28**, Section V.5] and DeVore and Sharpley [**100**]. Definition 4.5 on metric measure spaces is from Hajłasz [**165**], see also Varopoulos [**391**] and Vodop′yanov [**392**] for special cases. For Theorem 4.6 we refer to Bennett and Sharpley [**28**, Theorem V.5.6, Corollary V.5.9], Calderón [**69**] and Hajłasz [**165**]. See also Hajłasz [**164**], Hajłasz and Martio [**175**] and Vodop′yanov [**392**]. The equivalence fails for $p = 1$, see Hajłasz [**164**]. For $p = 1$ there is a characterization for the Hardy–Sobolev space in Koskela and Saksman [**245**]. See also Miyachi [**328**]. A counterpart of Theorem 4.6 also holds for open subsets of $\mathbb{R}^n$ with a sufficiently regular boundary, but not in general. However, we have $M^{1,p}(\Omega) \subset W^{1,p}(\Omega)$ for every open set $\Omega \subset \mathbb{R}^n$ and $1 \leq p < \infty$, see [**164**, Proposition 1] and [**175**, Lemma 6].

For Lemma 4.13, we refer to Calderón and Scott [**70**], DeVore and Sharpley [**100**, Theorem 2.5] and Macías and Segovia [**296**]. See also Bojarski [**48**] and Hajłasz and Kinnunen [**170**, Lemma 3.6]. The equivalence of (a) and (c) in Theorem 4.16 is by Calderón [**69**], see also DeVore and Sharpley [**100**, Theorem 6.2]. For a general version of Theorem 4.16, see Hajłasz and Kinnunen [**170**, Theorem 3.4]. Franchi, Hajłasz and Koskela [**132**] showed that the equivalence of (a) and (c) in Theorem 4.16 also holds for $p = 1$. For a version of Theorem 4.16 in the range $1 \leq p < \infty$ we refer to Hajłasz [**169**, Theorem 2.1]. See also Franchi, Lu and Wheeden [**133**–**135**], Hajłasz [**168**], Hajłasz and Koskela [**171**], Heinonen and Koskela [**188**] and Koskela and MacManus [**244**].

The approximation property in Theorem 4.9 has been studied by Calderón and Zygmund [**71**]. Liu [**284**] showed the approximation in the Sobolev norm. See also Michael and Ziemer [**326**]. For extensions of Theorem 4.9, see Bojarski and Hajłasz [**50**], Bojarski, Hajłasz and Strzelecki [**51**], Hajłasz [**165**], Hajłasz and Kinnunen [**170**, Theorem 5.3] and Malý [**299**]. See also Bennett and Sharpley [**28**, Theorem V.5.10], Malý and Ziemer [**303**, Theorem 1.69], Maz′ya [**321**] and Ziemer [**407**]. The Lipschitz truncation method in Theorem 4.9 will be applied in Section 12.3. See also the notes at the end of Chapter 12. The argument in the proof of Lemma 4.10 is from Zhong [**405**, Theorem 1.9]. See also Kinnunen and Martio [**235**, Theorem 3.13] and Kilpeläinen, Kinnunen and Martio [**219**, Theorem 5.1].

Theorem 4.22 is from Kinnunen [**224**]. A result of Luiro [**287**] shows that $M\colon W^{1,p}(\mathbb{R}^n) \to W^{1,p}(\mathbb{R}^n)$ is a continuous operator for $1 < p < \infty$. Kurka [**251**] and Tanaka [**378**] study boundedness of maximal operators for $u \in W^{1,1}(\mathbb{R})$. For extensions of Tanaka's result to functions of bounded variation in the one-dimensional case we refer to Aldaz [**10**], Aldaz and Pérez Lázaro [**12**, **13**], Bernal [**29**] and Liu, Chen and Wu [**283**]. For extensions of Theorem 4.22, see Aldaz, Colzani and Pérez Lázaro [**11**], Buckley [**62**], Carneiro and Moreira [**77**], Carneiro and Sveiter [**78**], Korry [**239**, **240**], Hajłasz and Liu [**173**], Hajłasz and Malý [**174**] and Pérez, Picon, Saari and Sousa [**346**].

Theorem 4.25 has been proved by Kinnunen and Lindqvist [**232**]. Continuity of the maximal operator with respect to an open set is discussed by Luiro [**288**–**290**]. For extensions see Aalto and Kinnunen [**1**], Hajłasz and Liu [**172**], Hajłasz and Onninen [**176**] and Luiro and Vähäkangas [**291**, **292**]. Theorem 4.31 is proved by Kinnunen and Saksman [**237**]. For a corresponding result on an open set, we refer to Heikkinen, Kinnunen, Korvenpää and Tuominen [**183**]. See also Beltran, Ramos and Saari [**26**], Carneiro, Finder and Sousa [**74**], Carneiro and Madrid [**75**], Carneiro, Madrid and Pierce [**76**] and Heikkinen and Tuominen [**184**].

CHAPTER 5

Capacities and Fine Properties of Sobolev Functions

Capacity is an appropriate outer measure in the study of pointwise behavior of Sobolev functions. We begin with the definition, basic properties and estimates for Sobolev capacity. Section 5.3 studies quasicontinuity of Sobolev functions. A capacitary version of the Lebesgue differentiation theorem for Sobolev functions is proved in Section 5.4, and Section 5.5 discusses pointwise characterizations of Sobolev spaces with zero boundary values. At the end of the chapter we consider a capacity relative to an open set, which is applied, for example, in the capacity density condition in Definition 6.17 and in Maz′ya's inequality in Theorem 5.47.

5.1. Sobolev capacity

The Sobolev capacity is induced by the Sobolev norm.

DEFINITION 5.1. Let $1 < p < \infty$. The Sobolev p-capacity of a set $E \subset \mathbb{R}^n$ is defined by

$$\operatorname{Cap}_p(E) = \inf_{u \in \mathcal{A}_p(E)} \|u\|^p_{W^{1,p}(\mathbb{R}^n)} = \inf_{u \in \mathcal{A}_p(E)} \big(\|u\|^p_{L^p(\mathbb{R}^n)} + \|\nabla u\|^p_{L^p(\mathbb{R}^n)}\big),$$

where $\mathcal{A}_p(E)$ consists of $u \in W^{1,p}(\mathbb{R}^n)$ with $u \geq 1$ almost everywhere in a neighborhood of E. Functions in $\mathcal{A}_p(E)$ are called p-admissible functions for E. If $\mathcal{A}_p(E) = \emptyset$, we set $\operatorname{Cap}_p(E) = \infty$.

It is not enough to require $u \geq 1$ on E instead of $u \geq 1$ almost everywhere in an open set containing E in the definition of $\mathcal{A}_p(E)$. The reason for this is that Sobolev functions are defined almost everywhere, and we are often interested in capacities of sets of Lebesgue measure zero. If $|E| = 0$ and we only require $u \geq 1$ on E, we could choose $u \in W^{1,p}(\mathbb{R}^n)$ to be zero in $\mathbb{R}^n \setminus E$, and thus all sets of measure zero would also be of capacity zero.

We may restrict the class of admissible functions in the definition of capacity.

LEMMA 5.2. *Let $1 < p < \infty$. For $E \subset \mathbb{R}^n$, we have*

$$\operatorname{Cap}_p(E) = \inf_{u \in \mathcal{A}'_p(E)} \|u\|^p_{W^{1,p}(\mathbb{R}^n)},$$

where $\mathcal{A}'_p(E)$ consists of $u \in W^{1,p}(\mathbb{R}^n)$ with $0 \leq u \leq 1$ almost everywhere in $\mathbb{R}^n$ and $u = 1$ in a neighborhood of E.

PROOF. Since $\mathcal{A}'_p(E) \subset \mathcal{A}_p(E)$, we have

$$\operatorname{Cap}_p(E) \leq \inf_{u \in \mathcal{A}'_p(E)} \|u\|^p_{W^{1,p}(\mathbb{R}^n)}.$$

For the converse inequality, we may assume that $\mathcal{A}_p(E) \neq \emptyset$. Let $\varepsilon > 0$ and let $u_\varepsilon \in \mathcal{A}_p(E)$ be such that

$$\|u_\varepsilon\|^p_{W^{1,p}(\mathbb{R}^n)} \le \operatorname{Cap}_p(E) + \varepsilon.$$

By Theorem 2.25, we have $v = \min\{|u_\varepsilon|, 1\} \in \mathcal{A}'_p(E)$, $|v| \le |u_\varepsilon|$ and $|\nabla v| \le |\nabla u_\varepsilon|$ almost everywhere in $\mathbb{R}^n$. This implies

$$\inf_{u\in\mathcal{A}'_p(E)} \|u\|^p_{W^{1,p}(\mathbb{R}^n)} \le \|v\|^p_{W^{1,p}(\mathbb{R}^n)} \le \|u_\varepsilon\|^p_{W^{1,p}(\mathbb{R}^n)} \le \operatorname{Cap}_p(E) + \varepsilon,$$

and by letting $\varepsilon \to 0$ we obtain

$$\inf_{u\in\mathcal{A}'_p(E)} \|u\|^p_{W^{1,p}(\mathbb{R}^n)} \le \operatorname{Cap}_p(E). \qquad \square$$

The Sobolev capacity is an outer measure.

THEOREM 5.3. *Let $1 < p < \infty$.*

(a) $\operatorname{Cap}_p(\emptyset) = 0$.
(b) *For $E_1 \subset E_2 \subset \mathbb{R}^n$, we have $\operatorname{Cap}_p(E_1) \le \operatorname{Cap}_p(E_2)$.*
(c) *For $E_i \subset \mathbb{R}^n$, $i \in \mathbb{N}$, we have*

$$\operatorname{Cap}_p\left(\bigcup_{i=1}^{\infty} E_i\right) \le \sum_{i=1}^{\infty} \operatorname{Cap}_p(E_i).$$

PROOF. Clearly $\operatorname{Cap}_p(\emptyset) = 0$. If $E_1 \subset E_2 \subset \mathbb{R}^n$, then $\mathcal{A}_p(E_2) \subset \mathcal{A}_p(E_1)$ and thus $\operatorname{Cap}_p(E_1) \le \operatorname{Cap}_p(E_2)$. This proves assertion (b).

In order to prove the countable subadditivity in (c), we may assume that $\sum_{i=1}^{\infty} \operatorname{Cap}_p(E_i) < \infty$ since otherwise there is nothing to prove. Let $\varepsilon > 0$. For every $i \in \mathbb{N}$, let $u_i \in \mathcal{A}'_p(E_i) \neq \emptyset$ with

$$\|u_i\|^p_{W^{1,p}(\mathbb{R}^n)} \le \operatorname{Cap}_p(E_i) + 2^{-i}\varepsilon. \tag{5.1}$$

We claim that $v = \sup_{i\in\mathbb{N}} u_i$ is p-admissible for $\bigcup_{i=1}^{\infty} E_i$.

First we show that $v \in W^{1,p}(\mathbb{R}^n)$. For $k \in \mathbb{N}$, define $v_k(x) = \max_{1\le i\le k} u_i(x)$ for every $x \in \mathbb{R}^n$. The sequence $(v_k)_{k\in\mathbb{N}}$ is increasing and $v_k \to v$ pointwise almost everywhere as $k \to \infty$. Moreover,

$$0 \le v_k(x) = \max_{1\le i\le k} u_i(x) \le \sup_{i\in\mathbb{N}} u_i(x) = v(x),$$

and, by Theorem 2.25,

$$|\nabla v_k(x)| \le \max_{1\le i\le k} |\nabla u_i(x)| \le \sup_{i\in\mathbb{N}} |\nabla u_i(x)|,$$

for almost every $x \in \mathbb{R}^n$ and every $k \in \mathbb{N}$. By (5.1), we obtain

$$
\begin{aligned}
\|v_k\|^p_{W^{1,p}(\mathbb{R}^n)} &= \int_{\mathbb{R}^n} |v_k(x)|^p\,dx + \int_{\mathbb{R}^n} |\nabla v_k(x)|^p\,dx \\
&\le \int_{\mathbb{R}^n} \sup_{i\in\mathbb{N}} |u_i(x)|^p\,dx + \int_{\mathbb{R}^n} \sup_{i\in\mathbb{N}} |\nabla u_i(x)|^p\,dx \\
&\le \int_{\mathbb{R}^n} \sum_{i=1}^{\infty} |u_i(x)|^p\,dx + \int_{\mathbb{R}^n} \sum_{i=1}^{\infty} |\nabla u_i(x)|^p\,dx \\
&= \sum_{i=1}^{\infty} \Big(\int_{\mathbb{R}^n} |u_i(x)|^p\,dx + \int_{\mathbb{R}^n} |\nabla u_i(x)|^p\,dx \Big) \\
&\le \sum_{i=1}^{\infty} (\operatorname{Cap}_p(E_i) + 2^{-i}\varepsilon) = \sum_{i=1}^{\infty} \operatorname{Cap}_p(E_i) + \varepsilon < \infty,
\end{aligned}
\tag{5.2}
$$

for every $k \in \mathbb{N}$. This shows that $(v_k)_{k\in\mathbb{N}}$ is a a bounded sequence in $W^{1,p}(\mathbb{R}^n)$. Theorem 2.40 implies that $v \in W^{1,p}(\mathbb{R}^n)$, $v_k \to v$ weakly in $L^p(\mathbb{R}^n)$, and $\nabla v_k \to \nabla v$ weakly in $L^p(\mathbb{R}^n;\mathbb{R}^n)$ as $k \to \infty$.

Let $i \in \mathbb{N}$. Since $u_i \in \mathcal{A}'_p(E_i)$, there exists an open set $G_i \supset E_i$ such that $u_i(x) = 1$ for almost every $x \in G_i$ and $0 \le u_i(x) \le 1$ for almost every $x \in \mathbb{R}^n$. It follows that $v(x) = \sup_{i\in\mathbb{N}} u_i(x) = 1$ for almost every $x \in \bigcup_{i=1}^{\infty} G_i$. Since $\bigcup_{i=1}^{\infty} G_i$ is an open set that contains $\bigcup_{i=1}^{\infty} E_i$, we conclude that v is p-admissible for $\bigcup_{i=1}^{\infty} E_i$. Thus

$$
\begin{aligned}
\operatorname{Cap}_p\Big(\bigcup_{i=1}^{\infty} E_i\Big) &\le \|v\|^p_{W^{1,p}(\mathbb{R}^n)} = \|v\|^p_{L^p(\mathbb{R}^n)} + \|\nabla v\|^p_{L^p(\mathbb{R}^n)} \\
&\le \liminf_{k\to\infty} \|v_k\|^p_{L^p(\mathbb{R}^n)} + \liminf_{k\to\infty} \|\nabla v_k\|^p_{L^p(\mathbb{R}^n)} \\
&\le \liminf_{k\to\infty} \|v_k\|^p_{W^{1,p}(\mathbb{R}^n)} \le \sum_{i=1}^{\infty} \operatorname{Cap}_p(E_i) + \varepsilon,
\end{aligned}
$$

where we applied the lower semicontinuity of L^p norms with respect to weak limits, see inequality (2.18), and the estimate in (5.2). Property (c) follows by letting $\varepsilon \to 0$. □

The Sobolev capacity is outer regular in the sense that capacity of an arbitrary set can be approximated by capacity of a larger open set.

LEMMA 5.4. *Let $1 < p < \infty$. For every $E \subset \mathbb{R}^n$, we have*

$$\operatorname{Cap}_p(E) = \inf\{\operatorname{Cap}_p(G) : E \subset G,\ G \text{ is open}\}.$$

PROOF. Theorem 5.3 (b) implies

$$\operatorname{Cap}_p(E) \le \inf\{\operatorname{Cap}_p(G) : E \subset G,\ G \text{ is open}\}.$$

In order to prove the converse inequality, we may assume that $\mathcal{A}_p(E) \neq \emptyset$. Let $\varepsilon > 0$ and let $u \in \mathcal{A}_p(E)$ with

$$\|u\|^p_{W^{1,p}(\mathbb{R}^n)} \le \operatorname{Cap}_p(E) + \varepsilon.$$

Since $u \in \mathcal{A}_p(E)$, there exists an open set $G_\varepsilon \supset E$ such that $u(x) \ge 1$ for almost every $x \in G_\varepsilon$. Thus

$$\inf\{\operatorname{Cap}_p(G) : E \subset G,\ G \text{ is open}\} \le \operatorname{Cap}_p(G_\varepsilon) \le \|u\|^p_{W^{1,p}(\mathbb{R}^n)} \le \operatorname{Cap}_p(E) + \varepsilon,$$

and the claim follows by letting $\varepsilon \to 0$. □

5.2. Estimates for capacity

We collect basic estimates for Sobolev capacity in this section. More refined estimates are considered in Section 5.7. We are mainly interested in sets of zero capacity since they are negligible in the theory of Sobolev spaces. Sets of capacity zero are of measure zero and thus capacity is a finer outer measure than Lebesgue outer measure.

LEMMA 5.5. *Let $1 < p < \infty$. For every $E \subset \mathbb{R}^n$, we have $|E| \le \operatorname{Cap}_p(E)$.*

PROOF. If $\operatorname{Cap}_p(E) = \infty$, there is nothing to prove, and thus we may assume that $\operatorname{Cap}_p(E) < \infty$. Let $\varepsilon > 0$ and let $u \in \mathcal{A}_p(E)$ with

$$\|u\|^p_{W^{1,p}(\mathbb{R}^n)} \le \operatorname{Cap}_p(E) + \varepsilon.$$

There exists an open set G such that $E \subset G$ and $u(x) \ge 1$ for almost every $x \in G$. This implies

$$|E| \le |G| \le \int_G |u(x)|^p \, dx \le \|u\|^p_{L^p(\mathbb{R}^n)} \le \|u\|^p_{W^{1,p}(\mathbb{R}^n)} \le \operatorname{Cap}_p(E) + \varepsilon,$$

and the claim follows by letting $\varepsilon \to 0$. □

Lemma 5.5 shows that $\operatorname{Cap}_p(B(x,r)) > 0$ for every $x \in \mathbb{R}^n$ and $r > 0$. This implies that capacity is nontrivial in the sense that every nonempty open set has positive capacity. On the other hand, every bounded set has finite Sobolev capacity.

LEMMA 5.6. *Let $1 < p < \infty$ and let $B(x,r) \subset \mathbb{R}^n$ be a ball. There exists a constant $C = C(n)$ such that*

$$\operatorname{Cap}_p(B(x,r)) \le \begin{cases} Cr^{n-p}, & \text{if } 0 < r \le 1, \\ Cr^n, & \text{if } 1 \le r < \infty. \end{cases}$$

PROOF. Let

$$u(y) = \max\left\{0, 1 - \frac{1}{r} d(y, B(x,r))\right\} = \begin{cases} 1, & y \in B(x,r), \\ 2 - \frac{|y-x|}{r}, & y \in B(x,2r) \setminus B(x,r), \\ 0, & y \in \mathbb{R}^n \setminus B(x,2r). \end{cases}$$

Then $u \in \operatorname{Lip}_0(B(x,2r))$ with $0 \le u(y) \le \chi_{B(x,2r)}(y)$ for every $y \in \mathbb{R}^n$ and u is $\frac{1}{r}$-Lipschitz. This implies $u \in \mathcal{A}_p(B(x,r))$. Moreover, $|\nabla u(y)| \le \frac{1}{r}$ for almost every $y \in \mathbb{R}^n$. Hence we obtain, for $0 < r \le 1$,

$$\begin{aligned} \operatorname{Cap}_p(B(x,r)) &\le \int_{B(x,2r)} |u(y)|^p \, dy + \int_{B(x,2r)} |\nabla u(y)|^p \, dy \le (1 + r^{-p})|B(x,2r)| \\ &\le (r^{-p} + r^{-p})|B(x,2r)| = 2r^{-p}|B(x,2r)| = C(n) r^{n-p}. \end{aligned}$$

For $r \ge 1$ we apply above the estimate $1 + r^{-p} \le 2$, and the claim follows. □

For $1 < p < n$, there is a corresponding estimate for the capacity of a ball from below.

LEMMA 5.7. *Let $1 < p < n$ and let $B(x,r) \subset \mathbb{R}^n$ be a ball. There exists a constant $C = C(n,p)$ such that*

$$\operatorname{Cap}_p(B(x,r)) \ge \begin{cases} Cr^{n-p}, & \text{if } 0 < r \le 1, \\ Cr^n, & \text{if } 1 \le r < \infty. \end{cases}$$

PROOF. In the case $1 \le r < \infty$ the claim follows from Lemma 5.5 since

$$\operatorname{Cap}_p(B(x,r)) \ge |B(x,r)| = C(n)r^n.$$

Assume then that $r > 0$ and let $u \in \mathcal{A}_p(B(x,r))$. Since $u(y) \ge 1$ for almost every $y \in B(x,r)$, by Theorem 3.6 we obtain

$$1 \le \left(\fint_{B(x,r)} |u(y)|^{\frac{np}{n-p}} \, dy \right)^{\frac{n-p}{np}} \le |B(x,r)|^{\frac{p-n}{np}} \left(\int_{\mathbb{R}^n} |u(y)|^{\frac{np}{n-p}} \, dy \right)^{\frac{n-p}{np}}$$
$$\le C(n,p) r^{\frac{p-n}{p}} \left(\int_{\mathbb{R}^n} |\nabla u(y)|^p \, dy \right)^{\frac{1}{p}}.$$

This implies

$$C(n,p)r^{n-p} \le \|\nabla u\|^p_{L^p(\mathbb{R}^n)} \le \|u\|^p_{W^{1,p}(\mathbb{R}^n)},$$

and consequently $\operatorname{Cap}_p(B(x,r)) \ge C(n,p)r^{n-p}$ for every $r > 0$. □

REMARK 5.8. Let $x \in \mathbb{R}^n$. For $1 < p < n$ and $0 < r \le 1$, Lemma 5.6 implies

$$\operatorname{Cap}_p(\{x\}) \le \operatorname{Cap}_p(B(x,r)) \le C(n)r^{n-p}.$$

On the other hand, for $p = n$ and $0 < r \le \frac{1}{2}$, we apply the admissible function

$$u(y) = \begin{cases} \left(\log \frac{1}{r}\right)^{-1} \log \frac{1}{|y-x|}, & y \in B(x,1) \setminus B(x,r), \\ 1, & y \in B(x,r), \\ 0, & y \in \mathbb{R}^n \setminus B(x,1), \end{cases}$$

and obtain

$$\operatorname{Cap}_n(\{x\}) \le \operatorname{Cap}_n(B(x,r)) \le C(n)\Big(\log \frac{1}{r}\Big)^{1-n}.$$

By letting $r \to 0$, we conclude that $\operatorname{Cap}_p(\{x\}) = 0$ for $1 < p \le n$. Countable subadditivity of capacity implies that all countable sets $E \subset \mathbb{R}^n$ have zero capacity for $1 < p \le n$.

All points have positive capacity for $p > n$, and thus there are no nontrivial sets of capacity zero when $p > n$. In practice this means that capacity is useful only for $1 < p \le n$.

LEMMA 5.9. *Let $n < p < \infty$. For every $x \in \mathbb{R}^n$, we have* $\operatorname{Cap}_p(\{x\}) > 0$.

PROOF. Let $z \in \mathbb{R}^n$. By Lemma 5.6 and monotonicity,

$$\operatorname{Cap}_p(\{z\}) \le \operatorname{Cap}_p(B(z,1)) \le C(n) < \infty.$$

Let $u \in \mathcal{A}_p(\{z\})$. There exists $0 < r \le 1$ such that $u(x) \ge 1$ for almost every $x \in B(z,r)$. By redefining u on a set of measure zero, if necessary, we may assume that $u(x) \ge 1$ for every $x \in B(z,r)$. Let

$$\psi(x) - \max\{0, 1 - d\big(x, B(z,1)\big)\}$$

for every $x \in \mathbb{R}^n$. Then $\psi \in \operatorname{Lip}_0(B(z,2))$, $0 \le \psi \le 1$, $\psi(x) = 1$ for every $x \in B(z,r)$, and $|\nabla \psi(x)| \le 1$ for almost every $x \in B(z,2)$. Let $v(x) = \psi(x)u(x)$ for every $x \in \mathbb{R}^n$. By (2.17), we have $v \in W^{1,p}(\mathbb{R}^n)$. By Morrey's inequality, see Theorem 3.23, there exists a set $E \subset \mathbb{R}^n$ such that $|E| = 0$ and

$$|v(x) - v(y)| \le C(n,p)\|\nabla v\|_{L^p(\mathbb{R}^n)}|x-y|^{1-\frac{n}{p}}$$

for every $x, y \in \mathbb{R}^n \setminus E$. Choose $x \in B(z,r)$ and $y \in B(z,4) \setminus B(z,2)$ such that $x, y \notin E$. Then $v(x) \geq 1$ and $v(y) = 0$, and thus

$$\int_{B(z,2)} |\nabla v(x)|^p \, dx = \|\nabla v\|^p_{L^p(\mathbb{R}^n)} \geq C(n,p)|x-y|^{n-p}|v(x)-v(y)|^p$$
$$\geq C(n,p)5^{n-p} > 0.$$

On the other hand, we have

$$\begin{aligned}\int_{B(z,2)} |\nabla v(x)|^p \, dx &\leq \int_{B(z,2)} \left(|\nabla \psi(x) u(x)| + |\psi(x)\nabla u(x)|\right)^p dx \\ &\leq 2^p \left(\int_{B(z,2)} |\nabla \psi(x)|^p |u(x)|^p \, dx + \int_{B(z,2)} |\psi(x)|^p |\nabla u(x)|^p \, dx \right) \\ &\leq 2^p \left(\int_{B(z,2)} |u(x)|^p \, dx + \int_{B(z,2)} |\nabla u(x)|^p \, dx \right) \\ &\leq 2^p \|u\|^p_{W^{1,p}(\mathbb{R}^n)}.\end{aligned}$$

This shows that there exists a constant $C = C(n,p)$ such that $\|u\|^p_{W^{1,p}(\mathbb{R}^n)} \geq C > 0$ for every $u \in \mathcal{A}_p(\{z\})$, and consequently $\operatorname{Cap}_p(\{z\}) \geq C > 0$. □

5.3. Quasicontinuity and fine properties of capacity

We discuss properties of Sobolev functions outside a set of small or vanishing capacity.

DEFINITION 5.10. Let $1 < p < \infty$ and $G \subset \mathbb{R}^n$. We say that a property holds p-quasieverywhere in G, or for p-quasievery $x \in G$, if there is a set $E \subset \mathbb{R}^n$ with $\operatorname{Cap}_p(E) = 0$ such that the property holds for every $x \in G \setminus E$. If $G = \mathbb{R}^n$, we say that the property holds p-quasieverywhere.

We consider a Sobolev space version of the result which asserts that for every Cauchy sequence in $L^p(\mathbb{R}^n)$ there is a subsequence that converges pointwise almost everywhere. The part concerning the uniform convergence is a Sobolev space version of Egorov's theorem.

THEOREM 5.11. *Let $1 < p < \infty$. Assume that $u_i \in W^{1,p}(\mathbb{R}^n) \cap C(\mathbb{R}^n)$, for every $i \in \mathbb{N}$, and that $(u_i)_{i\in\mathbb{N}}$ is a Cauchy sequence in $W^{1,p}(\mathbb{R}^n)$. There is a subsequence of $(u_i)_{i\in\mathbb{N}}$ that converges pointwise p-quasieverywhere in $\mathbb{R}^n$. Moreover, the convergence is uniform outside a set of arbitrarily small Sobolev p-capacity.*

PROOF. By induction we obtain a subsequence, which we still denote by $(u_i)_{i\in\mathbb{N}}$, such that

$$\sum_{i=1}^{\infty} 2^{ip} \|u_i - u_{i+1}\|^p_{W^{1,p}(\mathbb{R}^n)} < \infty. \tag{5.3}$$

Let $E_i = \{x \in \mathbb{R}^n : |u_i(x) - u_{i+1}(x)| > 2^{-i}\}$ for every $i \in \mathbb{N}$. Since $u_i \in C(\mathbb{R}^n)$ for every $i \in \mathbb{N}$, we have $2^i|u_i - u_{i+1}| \in \mathcal{A}_p(E_i)$ and thus

$$\operatorname{Cap}_p(E_i) \leq 2^{ip}\|u_i - u_{i+1}\|^p_{W^{1,p}(\mathbb{R}^n)}.$$

Let $j \in \mathbb{N}$ and $F_j = \bigcup_{i=j}^{\infty} E_i$. Using Theorem 5.3 (c), we obtain

$$\operatorname{Cap}_p(F_j) \leq \sum_{i=j}^{\infty} \operatorname{Cap}_p(E_i) \leq \sum_{i=j}^{\infty} 2^{ip}\|u_i - u_{i+1}\|^p_{W^{1,p}(\mathbb{R}^n)}.$$

By monotonicity and (5.3), we have

$$\operatorname{Cap}_p\left(\bigcap_{j=1}^{\infty} F_j\right) \le \liminf_{j\to\infty} \operatorname{Cap}_p(F_j) \le \liminf_{j\to\infty} \sum_{i=j}^{\infty} 2^{ip}\|u_i - u_{i+1}\|^p_{W^{1,p}(\mathbb{R}^n)} = 0,$$

since the tail of a convergent series converges to zero.

Observe that $(u_i)_{i\in\mathbb{N}}$ converges pointwise in $\mathbb{R}^n \setminus \bigcap_{j=1}^{\infty} F_j$. Moreover,

$$|u_m(x) - u_k(x)| \le \sum_{i=m}^{k-1} |u_i(x) - u_{i+1}(x)| \le \sum_{i=m}^{k-1} 2^{-i} \le 2^{1-m}$$

for every $x \in \mathbb{R}^n \setminus F_j$ and every $k > m > j$. Cauchy's criterion for uniform convergence in $\mathbb{R}^n\setminus F_j$ is satisfied, and thus $(u_i)_{i\in\mathbb{N}}$ convergences uniformly in $\mathbb{R}^n\setminus F_j$ for each $j \in \mathbb{N}$. Since $\lim_{j\to\infty} \operatorname{Cap}_p(F_j) = 0$, we conclude that the convergence is uniform outside a set of arbitrarily small Sobolev p-capacity. □

DEFINITION 5.12. Let $1 < p < \infty$. A function $u\colon \mathbb{R}^n \to [-\infty,\infty]$ is p-quasicontinuous in $\mathbb{R}^n$ if for every $\varepsilon > 0$ there is a set $E \subset \mathbb{R}^n$ such that $\operatorname{Cap}_p(E) < \varepsilon$ and the restriction of u to $\mathbb{R}^n \setminus E$, denoted by $u|_{\mathbb{R}^n\setminus E}$, is a continuous real-valued function.

By outer regularity of capacity, see Lemma 5.4, we may assume that E is open and Lebesgue measurable in the definition above.

REMARK 5.13. Let $1 < p < \infty$ and assume that u is p-quasicontinuous in $\mathbb{R}^n$. If we redefine or modify u on a set of zero Sobolev p-capacity, the function remains p-quasicontinuous. Moreover, by definition

$$\operatorname{Cap}_p\big(\{x \in \mathbb{R}^n : |u(x)| = \infty\}\big) = 0.$$

If v is p-quasicontinuous in $\mathbb{R}^n$ and $\varphi\colon \mathbb{R} \to \mathbb{R}$ is continuous, then it is possible to redefine u and v on a set of p-capacity zero in such a way that uv, $u+v$, and $\varphi \circ u$ are well-defined p-quasicontinuous functions in $\mathbb{R}^n$.

Every Sobolev function has a quasicontinuous representative.

THEOREM 5.14. *Let $1 < p < \infty$ and assume that $u \in W^{1,p}(\mathbb{R}^n)$. There exists a p-quasicontinuous $v \in W^{1,p}(\mathbb{R}^n)$ such that $u = v$ almost everywhere in $\mathbb{R}^n$.*

PROOF. By Lemma 2.20, there are functions $u_i \in W^{1,p}(\mathbb{R}^n) \cap C(\mathbb{R}^n)$, $i \in \mathbb{N}$, such that $u_i \to u$ in $W^{1,p}(\mathbb{R}^n)$ and almost everywhere in $\mathbb{R}^n$ as $i \to \infty$. From Theorem 5.11 we obtain a subsequence of $(u_i)_{i\in\mathbb{N}}$ that converges p-quasieverywhere to a function $v\colon \mathbb{R}^n \to [-\infty,\infty]$, and the convergence is uniform outside a set of arbitrarily small Sobolev p-capacity. In particular $u = v$ almost everywhere in $\mathbb{R}^n$, by Lemma 5.5, and so also $v \in W^{1,p}(\mathbb{R}^n)$. Uniform convergence implies continuity of the limit function, and thus v is a continuous real-valued function outside a set of arbitrarily small Sobolev p-capacity. This implies that v is p-quasicontinuous. □

Theorem 5.16 below shows that the quasicontinuous representative given by Theorem 5.14 is unique. In the proof of Theorem 5.16 we apply the following useful observation.

LEMMA 5.15. *Let $1 < p < \infty$. For every open set $G \subset \mathbb{R}^n$ and every $E \subset \mathbb{R}^n$ with $|E| = 0$, we have $\operatorname{Cap}_p(G) = \operatorname{Cap}_p(G \setminus E)$.*

PROOF. By monotonicity $\operatorname{Cap}_p(G) \geq \operatorname{Cap}_p(G \setminus E)$. To prove the reverse inequality, we may assume that $\mathcal{A}_p(G \setminus E) \neq \emptyset$. Let $\varepsilon > 0$ and let $u \in \mathcal{A}_p(G \setminus E)$ with

$$\|u\|^p_{W^{1,p}(\mathbb{R}^n)} \leq \operatorname{Cap}_p(G \setminus E) + \varepsilon.$$

There is an open set $U \subset \mathbb{R}^n$ such that $G \setminus E \subset U$ and $u(x) \geq 1$ for almost every $x \in U$. Then $U \cup G$ is open and $G \subset U \cup G$. Since $|E| = 0$ and $u \geq 1$ almost everywhere in $U \cup (G \setminus E)$, we have $u \geq 1$ almost everywhere in $U \cup G$. This implies $u \in \mathcal{A}_p(U \cup G)$ and

$$\operatorname{Cap}_p(G) \leq \operatorname{Cap}_p(U \cup G) \leq \|u\|^p_{W^{1,p}(\mathbb{R}^n)} \leq \operatorname{Cap}_p(G \setminus E) + \varepsilon.$$

The claim follows by letting $\varepsilon \to 0$. □

THEOREM 5.16. *Let $1 < p < \infty$. Assume that u and v are p-quasicontinuous functions in $\mathbb{R}^n$ and $u = v$ almost everywhere. Then $u = v$ holds p-quasieverywhere in $\mathbb{R}^n$.*

PROOF. Let $\varepsilon > 0$ and let $V \subset \mathbb{R}^n$ be an open set such that $\operatorname{Cap}_p(V) < \varepsilon$ and the restrictions of u and v to $\mathbb{R}^n \setminus V$ are continuous. The set

$$\{x \in \mathbb{R}^n \setminus V : u(x) \neq v(x)\}$$

is open in the relative topology of $\mathbb{R}^n \setminus V$, that is, there exists an open set $U \subset \mathbb{R}^n$ with

$$U \setminus V = U \cap (\mathbb{R}^n \setminus V) = \{x \in \mathbb{R}^n \setminus V : u(x) \neq v(x)\}.$$

Moreover,

$$\{x \in \mathbb{R}^n : u(x) \neq v(x)\} \subset V \cup \{x \in \mathbb{R}^n \setminus V : u(x) \neq v(x)\} = U \cup V,$$

and, by assumption, we have

$$|U \setminus V| = |\{x \in \mathbb{R}^n \setminus V : u(x) \neq v(x)\}| = 0.$$

Lemma 5.15, with $G = U \cup V$ and $E = U \setminus V$, implies

$$\operatorname{Cap}_p(\{x \in \mathbb{R}^n : u(x) \neq v(x)\}) \leq \operatorname{Cap}_p(U \cup V) = \operatorname{Cap}_p(V) < \varepsilon.$$

The claim follows by letting $\varepsilon \to 0$. □

REMARK 5.17. Let $1 < p < \infty$. If u and v are p-quasicontinuous in $\mathbb{R}^n$ and $u \geq v$ almost everywhere in $\mathbb{R}^n$, then $\max\{v - u, 0\} = 0$ almost everywhere in $\mathbb{R}^n$ and $\max\{v - u, 0\}$ is p-quasicontinuous; see also Remark 5.13. Theorem 5.16 implies that $\max\{v - u, 0\} = 0$ p-quasieverywhere in $\mathbb{R}^n$, and consequently $u \geq v$ p-quasieverywhere in $\mathbb{R}^n$.

The reasoning above can be generalized to open subsets of $\mathbb{R}^n$.

LEMMA 5.18. *Let $1 < p < \infty$ and let $G \subset \mathbb{R}^n$ be an open set. Assume that u and v are p-quasicontinuous functions in $\mathbb{R}^n$ and $u \geq v$ almost everywhere in G. Then $u \geq v$ holds p-quasieverywhere in G.*

PROOF. If $G = \mathbb{R}^n$, the claim follows from Remark 5.17, and thus we may assume that $G \subsetneq \mathbb{R}^n$. Let

$$G_k = \left\{x \in G : d(x, \mathbb{R}^n \setminus G) > \frac{1}{k}\right\}$$

for $k \in \mathbb{N}$, and define

$$u_k(x) = u(x) \min\{1, kd(x, \mathbb{R}^n \setminus G)\}$$

and

$$v_k(x) = v(x)\min\{1, kd(x, \mathbb{R}^n \setminus G)\}$$

for every $k \in \mathbb{N}$ and $x \in \mathbb{R}^n$. Then $G = \bigcup_{k=1}^{\infty} G_k$ and $u_k(x) = u(x)$ and $v_k(x) = v(x)$ for every $k \in \mathbb{N}$ and $x \in G_k$. Let $k \in \mathbb{N}$. Since the functions u_k and v_k are p-quasicontinuous in $\mathbb{R}^n$ and $u_k \geq v_k$ almost everywhere in $\mathbb{R}^n$, by Theorem 5.16 $u_k \geq v_k$ holds p-quasieverywhere in $\mathbb{R}^n$. This implies $\operatorname{Cap}_p(E_k) = 0$ for every $k \in \mathbb{N}$, where $E_k = \{x \in \mathbb{R}^n : u_k(x) < v_k(x)\}$. It follows that

$$\operatorname{Cap}_p\big(\{x \in G : u(x) < v(x)\}\big) \leq \operatorname{Cap}_p\left(\bigcup_{k=1}^{\infty} E_k\right) \leq \sum_{k=1}^{\infty} \operatorname{Cap}_p(E_k) = 0,$$

and hence $u \geq v$ p-quasieverywhere in G, as claimed. □

Sobolev capacity can be characterized in terms of quasicontinuous functions.

Definition 5.19. Let $1 < p < \infty$. For $E \subset \mathbb{R}^n$, we define

$$\operatorname{Cap}_p^*(E) = \inf_{u \in \mathcal{A}_p^*(E)} \|u\|^p_{W^{1,p}(\mathbb{R}^n)},$$

where $\mathcal{A}_p^*(E)$ consists of $u \in W^{1,p}(\mathbb{R}^n)$ such that u is p-quasicontinuous and $u \geq 1$ p-quasieverywhere in E. Here we use the convention $\operatorname{Cap}_p^*(E) = \infty$ if $\mathcal{A}_p^*(E) = \emptyset$.

Theorem 5.20. *Let $1 < p < \infty$. For $E \subset \mathbb{R}^n$ we have $\operatorname{Cap}_p^*(E) = \operatorname{Cap}_p(E)$.*

Proof. For inequality $\operatorname{Cap}_p^*(E) \leq \operatorname{Cap}_p(E)$ we may assume that $\operatorname{Cap}_p(E) < \infty$. Let $u \in \mathcal{A}_p(E)$. Then $u \in W^{1,p}(\mathbb{R}^n)$ and $u \geq 1$ almost everywhere in a neighborhood G of E. Let v be the p-quasicontinuous representative of u, given by Theorem 5.14. Since $u = v$ almost everywhere, we have $v \geq 1$ almost everywhere in G. Lemma 5.18 implies that $v \geq 1$ p-quasieverywhere in G, and hence $v \geq 1$ p-quasieverywhere in E. Thus $v \in \mathcal{A}_p^*(E)$ and it follows that $\operatorname{Cap}_p^*(E) \leq \operatorname{Cap}_p(E)$.

For the reverse inequality, we may assume that $\operatorname{Cap}_p^*(E) < \infty$. Let $v \in \mathcal{A}_p^*(E)$. By considering the p-quasicontinuous function $\min\{|v|, 1\}$ and arguing as in the proof of Lemma 5.2, we may assume that $0 \leq v \leq 1$ everywhere and $v = 1$ p-quasieverywhere in E. Let $0 < \varepsilon < 1$ and let V be an open set such that $\operatorname{Cap}_p(V) < \varepsilon^p$, $v = 1$ in $E \setminus V$, and the restriction $v|_{\mathbb{R}^n \setminus V}$ is continuous. By relative topology, there exists an open set $U \subset \mathbb{R}^n$ satisfying

$$\{x \in \mathbb{R}^n \setminus V : v(x) > 1 - \varepsilon\} = U \cap (\mathbb{R}^n \setminus V) = U \setminus V.$$

Since $v = 1$ in $E \setminus V$, we have $E \setminus V \subset U \setminus V$.

Let $w \in \mathcal{A}_p'(V)$ with $\|w\|_{W^{1,p}(\mathbb{R}^n)} < \varepsilon$, and let

$$u(x) = \frac{v(x)}{1 - \varepsilon} + w(x)$$

for every $x \in \mathbb{R}^n$. Then $u \geq 1$ almost everywhere in $(U \setminus V) \cup V = U \cup V$, which is an open neighborhood of E. This implies $u \in \mathcal{A}_p(E)$ and

$$\begin{aligned}\operatorname{Cap}_p(E)^{\frac{1}{p}} &\leq \|u\|_{W^{1,p}(\mathbb{R}^n)} \leq \frac{1}{1-\varepsilon}\|v\|_{W^{1,p}(\mathbb{R}^n)} + \|w\|_{W^{1,p}(\mathbb{R}^n)} \\ &\leq \frac{1}{1-\varepsilon}\|v\|_{W^{1,p}(\mathbb{R}^n)} + \varepsilon.\end{aligned}$$

Since $\varepsilon > 0$ and $v \in \mathcal{A}_p^*(E)$ were arbitrary, and v is independent of ε, we obtain $\operatorname{Cap}_p(E) \leq \operatorname{Cap}_p^*(E)$. □

5.4. Lebesgue points of Sobolev functions

The Lebesgue differentiation theorem, see Theorem 1.21, states that a locally integrable function has Lebesgue points almost everywhere. In this section we consider a capacitary version of this result for Sobolev functions, see Theorem 5.23. In the proof we apply the following capacitary weak type estimate for Sobolev functions.

THEOREM 5.21. *Let $1 < p < \infty$ and assume that $u \in W^{1,p}(\mathbb{R}^n)$. There exists a constant $C = C(n,p)$ such that*

$$\operatorname{Cap}_p\big(\{x \in \mathbb{R}^n : Mu(x) > t\}\big) \leq \frac{C}{t^p} \|u\|^p_{W^{1,p}(\mathbb{R}^n)}$$

for every $t > 0$.

PROOF. Let $t > 0$ and let $E_t = \{x \in \mathbb{R}^n : Mu(x) > t\}$. The set E_t is open by Lemma 1.3. Theorem 4.21 implies $Mu \in W^{1,p}(\mathbb{R}^n)$, and thus

$$\frac{Mu}{t} \in \mathcal{A}_p(E_t).$$

Since the maximal operator M is bounded on $W^{1,p}(\mathbb{R}^n)$, see inequality (4.15), we conclude that

$$\operatorname{Cap}_p(E_t) \leq \frac{1}{t^p} \|Mu\|^p_{W^{1,p}(\mathbb{R}^n)} \leq \frac{C(n,p)}{t^p} \|u\|^p_{W^{1,p}(\mathbb{R}^n)}. \qquad \square$$

Also the following measure theoretic result will be needed in the proof of Theorem 5.23. Roughly speaking, this lemma asserts that the set where an L^p function blows up rapidly is of capacity zero.

LEMMA 5.22. *Let $1 < p < \infty$. Assume that $f \in L^p(\mathbb{R}^n)$ and define*

$$E = \left\{ x \in \mathbb{R}^n : \limsup_{r \to 0} r^p \fint_{B(x,r)} |f(y)|^p \, dy > 0 \right\}.$$

Then $\operatorname{Cap}_p(E) = 0$.

PROOF. Let

$$E_\varepsilon = \left\{ x \in \mathbb{R}^n : \limsup_{r \to 0} r^p \fint_{B(x,r)} |f(y)|^p \, dy > \varepsilon \right\},$$

for $\varepsilon > 0$ We show that $\operatorname{Cap}_p(E_\varepsilon) = 0$ for every $\varepsilon > 0$, and the claim follows by subadditivity of capacity. Let $\varepsilon > 0$ and $0 < \delta < \frac{1}{5}$. For every $x \in E_\varepsilon$ there exists r_x, with $0 < r_x \leq \delta$, such that

$$r_x^p \fint_{B(x,r_x)} |f(y)|^p \, dy > \varepsilon.$$

Then $\mathcal{B} = \{B(x,r_x) : x \in E_\varepsilon\}$ is a cover of the set E_ε. By Lemma 1.13, there exists a countable collection of pairwise disjoint balls $B(x_i,r_i) \in \mathcal{B}$, $i \in \mathbb{N}$, such that $E_\varepsilon \subset \bigcup_{i=1}^\infty B(x_i, 5r_i)$. Using countable subadditivity of capacity, the estimate in Lemma 5.6, the choice of the radii r_i, and the pairwise disjointness of the balls

$B(x_i, r_i)$, we obtain

$$\begin{aligned}\operatorname{Cap}_p(E_\varepsilon) &\le \sum_{i=1}^{\infty} \operatorname{Cap}_p(B(x_i, 5r_i)) \le C(n) \sum_{i=1}^{\infty} r_i^{n-p} \\ &\le \frac{C(n)}{\varepsilon} \sum_{i=1}^{\infty} \int_{B(x_i,r_i)} |f(y)|^p \, dy = \frac{C(n)}{\varepsilon} \int_{\bigcup_{i=1}^{\infty} B(x_i,r_i)} |f(y)|^p \, dy.\end{aligned}$$

The pairwise disjointness of the balls $B(x_i, r_i)$, $i \in \mathbb{N}$, also gives

$$\begin{aligned}\left| \bigcup_{i=1}^{\infty} B(x_i, r_i) \right| &= \sum_{i=1}^{\infty} |B(x_i, r_i)| \le \sum_{i=1}^{\infty} \frac{r_i^p}{\varepsilon} \int_{B(x_i,r_i)} |f(y)|^p \, dy \\ &\le \frac{\delta^p}{\varepsilon} \int_{\mathbb{R}^n} |f(y)|^p \, dy \xrightarrow{\delta \to 0} 0.\end{aligned}$$

By absolute continuity of integral, we obtain

$$\operatorname{Cap}_p(E_\varepsilon) \le \frac{C(n)}{\varepsilon} \int_{\bigcup_{i=1}^{\infty} B(x_i,r_i)} |f(y)|^p \, dy \xrightarrow{\delta \to 0} 0.$$

This implies $\operatorname{Cap}_p(E_\varepsilon) = 0$ for every $\varepsilon > 0$, and the claim follows. □

We are ready to state and prove a refined version of the Lebesgue differentiation theorem for Sobolev functions. This result gives an expression for the quasicontinuous representative in Theorem 5.14.

THEOREM 5.23. *Let $1 < p < \infty$ and assume that $u \in W^{1,p}(\mathbb{R}^n)$. There exists a set $E \subset \mathbb{R}^n$ with $\operatorname{Cap}_p(E) = 0$ such that the limit*

$$\lim_{r \to 0} \fint_{B(x,r)} u(y) \, dy = u^*(x) \tag{5.4}$$

exists for every $x \in \mathbb{R}^n \setminus E$ and any extension of $u^* \colon \mathbb{R}^n \setminus E \to \mathbb{R}$ to entire $\mathbb{R}^n$ is a p-quasicontinuous representative of u. Moreover, there exists a set $E' \supset E$ such that $\operatorname{Cap}_p(E') = 0$ and

$$\lim_{r \to 0} \fint_{B(x,r)} |u(y) - u^*(x)| \, dy = 0 \tag{5.5}$$

for every $x \in \mathbb{R}^n \setminus E'$.

PROOF. By Lemma 2.20, there exist functions $u_i \in C(\mathbb{R}^n) \cap W^{1,p}(\mathbb{R}^n)$ such that

$$\|u - u_i\|_{W^{1,p}(\mathbb{R}^n)}^p \le 2^{-i(p+1)}$$

for every $i \in \mathbb{N}$. Let $F_i = \{x \in \mathbb{R}^n : M(u - u_i)(x) > 2^{-i}\}$ for $i \in \mathbb{N}$. Theorem 5.21 gives

$$\operatorname{Cap}_p(F_i) \le C(n,p) 2^{ip} \|u - u_i\|_{W^{1,p}(\mathbb{R}^n)}^p \le C(n,p) 2^{-i}, \tag{5.6}$$

for every $i \in \mathbb{N}$. Let $i \in \mathbb{N}$ and note that

$$\begin{aligned}|u_i(x) - u_{B(x,r)}| &\le \fint_{B(x,r)} |u_i(x) - u(y)| \, dy \\ &\le \fint_{B(x,r)} |u_i(x) - u_i(y)| \, dy + \fint_{B(x,r)} |u_i(y) - u(y)| \, dy,\end{aligned}$$

for every $x \in \mathbb{R}^n$. Since u_i is continuous, we have

$$\limsup_{r\to 0} \fint_{B(x,r)} |u_i(x) - u_i(y)|\,dy = 0,$$

and thus

$$\begin{aligned}\limsup_{r\to 0} |u_i(x) - u_{B(x,r)}| &\le \limsup_{r\to 0} \fint_{B(x,r)} |u_i(y) - u(y)|\,dy \\ &\le M(u_i - u)(x) \le 2^{-i}\end{aligned} \tag{5.7}$$

for every $x \in \mathbb{R}^n \setminus F_i$.

Let $k \in \mathbb{N}$ and define $E_k = \bigcup_{m=k}^{\infty} F_m$. By countable subadditivity of capacity and inequality (5.6), we obtain

$$\operatorname{Cap}_p(E_k) \le \sum_{m=k}^{\infty} \operatorname{Cap}_p(F_m) \le C(n,p) \sum_{m=k}^{\infty} 2^{-m} = C(n,p)2^{1-k}. \tag{5.8}$$

If $i > j \ge k$ and $x \in \mathbb{R}^n \setminus E_k$, then (5.7) implies

$$\begin{aligned}|u_i(x) - u_j(x)| &\le \limsup_{r\to 0} |u_i(x) - u_{B(x,r)}| + \limsup_{r\to 0} |u_{B(x,r)} - u_j(x)| \\ &\le 2^{-i} + 2^{-j} \le 2^{1-j}.\end{aligned}$$

Hence $(u_i)_{i\in\mathbb{N}}$ satisfies the Cauchy criterion for uniform convergence in $\mathbb{R}^n \setminus E_k$. Since the functions u_i are continuous, the sequence $(u_i)_{i\in\mathbb{N}}$ converges uniformly in $\mathbb{R}^n \setminus E_k$ to a continuous function $v_k\colon \mathbb{R}^n \setminus E_k \to \mathbb{R}$. Furthermore, if $x \in \mathbb{R}^n \setminus E_k$ and $i \ge k$, then

$$\begin{aligned}\limsup_{r\to 0} |v_k(x) - u_{B(x,r)}| &\le |v_k(x) - u_i(x)| + \limsup_{r\to 0} |u_i(x) - u_{B(x,r)}| \\ &\le |v_k(x) - u_i(x)| + 2^{-i}.\end{aligned}$$

The right-hand side of the previous inequality tends to zero as $i \to \infty$. Thus

$$\limsup_{r\to 0} |v_k(x) - u_{B(x,r)}| = 0,$$

and consequently

$$v_k(x) = \lim_{r\to 0} \fint_{B(x,r)} u(y)\,dy = u^*(x)$$

for every $x \in \mathbb{R}^n \setminus E_k$.

Let $E = \bigcap_{k=1}^{\infty} E_k$. Then the limit in (5.4) exists for every $x \in \mathbb{R}^n \setminus E$, and by monotonicity of capacity and (5.8), we have

$$\operatorname{Cap}_p(E) \le \liminf_{k\to\infty} \operatorname{Cap}_p(E_k) \le C(n,p) \liminf_{k\to\infty} 2^{1-k} = 0.$$

Let $\varepsilon > 0$ and choose $k \in \mathbb{N}$ to be so large that $\operatorname{Cap}_p(E_k) < \frac{\varepsilon}{2}$. Since $u^*|_{\mathbb{R}^n \setminus E_k} = v_k$ is continuous, any extension of $u^*\colon \mathbb{R}^n \setminus E \to \mathbb{R}$ to entire $\mathbb{R}^n$ is p-quasicontinuous in $\mathbb{R}^n$. Lemma 5.5 implies $|E| = 0$ and by Theorem 1.21 we have $u = u^*$ almost everywhere in $\mathbb{R}^n \setminus E$. Thus any extension of $u^*\colon \mathbb{R}^n \setminus E$ is a p-quasicontinuous representative of u.

Finally, let

$$F = \left\{ x \in \mathbb{R}^n : \limsup_{r\to 0} r^p \fint_{B(x,r)} |\nabla u(y)|^p\,dy > 0 \right\}.$$

Lemma 5.22 shows that $\operatorname{Cap}_p(F) = 0$. The p-Poincaré inequality, see Theorem 3.14, implies

$$\begin{aligned}\limsup_{r\to 0}\biggl(\fint_{B(x,r)} |u(y) - u_{B(x,r)}|^p\,dy\biggr)^{\frac{1}{p}} \\ \le C(n,p)\limsup_{r\to 0} r\biggl(\fint_{B(x,r)} |\nabla u(y)|^p\,dy\biggr)^{\frac{1}{p}} = 0\end{aligned} \tag{5.9}$$

for every $x \in \mathbb{R}^n \setminus F$. Choose $E' = E \cup F$. Then $\operatorname{Cap}_p(E') \le \operatorname{Cap}_p(E) + \operatorname{Cap}_p(F) = 0$. Since $E \subset E'$, the limit in (5.4) exists for every $x \in \mathbb{R}^n \setminus E' \subset \mathbb{R}^n \setminus E$. By (5.9), we have

$$\begin{aligned}\limsup_{r\to 0}\fint_{B(x,r)} |u(y) - u^*(x)|\,dy &\le \limsup_{r\to 0}\biggl(\fint_{B(x,r)} |u(y) - u^*(x)|^p\,dy\biggr)^{\frac{1}{p}} \\ &\le \limsup_{r\to 0}\biggl(\fint_{B(x,r)} |u(y) - u_{B(x,r)}|^p\,dy\biggr)^{\frac{1}{p}} + \limsup_{r\to 0}|u_{B(x,r)} - u^*(x)| = 0,\end{aligned}$$

for every $x \in \mathbb{R}^n \setminus E' \subset \mathbb{R}^n \setminus F$, and thus (5.5) holds for every $x \in \mathbb{R}^n \setminus E'$. □

REMARK 5.24. Let $n < p < \infty$. Lemma 5.9 implies that $\operatorname{Cap}_p(\{x\}) > 0$ for every $x \in \mathbb{R}^n$, and Theorem 5.23 holds for every $u \in W^{1,p}(\mathbb{R}^n)$ with $E = \emptyset$. Consequently, every function $u \in W^{1,p}(\mathbb{R}^n)$ has a continuous representative and every point $x \in \mathbb{R}^n$ is a Lebesgue point of u. Recall that this fact follows more directly from Morrey's inequality, see Theorem 3.23, which is also applied in the proof of Lemma 5.9.

REMARK 5.25. Let $1 < p < \infty$ and assume that $u \in W^{1,p}(\mathbb{R}^n)$. Let $1 \le q \le \frac{np}{n-p}$ for $1 < p < n$, and $1 \le q < \infty$ for $n \le p < \infty$. By applying the (q,p)-Poincaré inequality, see Theorem 3.14, instead of the p-Poincaré inequality in (5.9), we conclude that there exists a set $E \subset \mathbb{R}^n$ such that $\operatorname{Cap}_p(E) = 0$ and

$$\lim_{r\to 0}\biggl(\fint_{B(x,r)} |u(y) - u^*(x)|^q\,dy\biggr)^{\frac{1}{q}} = 0$$

for every $x \in \mathbb{R}^n \setminus E$.

5.5. Sobolev spaces with zero boundary values revisited

We return to Sobolev spaces with zero boundary values, see Definition 2.27. Applying pointwise properties of Sobolev functions, we obtain characterizations for these spaces. The first result is a version of Theorem 5.14.

THEOREM 5.26. *Let $1 < p < \infty$, let $\Omega \subset \mathbb{R}^n$ be an open set, and assume that $u \in W_0^{1,p}(\Omega)$. There exists a p-quasicontinuous function $v \in W^{1,p}(\mathbb{R}^n)$ such that $v = u$ almost everywhere in Ω and $v = 0$ holds p-quasieverywhere in $\mathbb{R}^n \setminus \Omega$.*

PROOF. Since $u \in W_0^{1,p}(\Omega)$, there exist $u_i \in C_0^\infty(\Omega)$, $i \in \mathbb{N}$, such that $u_i \to u$ in $W^{1,p}(\Omega)$ as $i \to \infty$. By taking a subsequence, if necessary, we may assume that $u_i \to u$ pointwise almost everywhere in Ω, as $i \to \infty$. Since $(u_i)_{i\in\mathbb{N}}$ is a Cauchy sequence in $W^{1,p}(\mathbb{R}^n)$, by Theorem 5.11 it has a subsequence that converges pointwise p-quasieverywhere in $\mathbb{R}^n$ to a function v. By Lemma 5.5 and Theorem 2.40, we have $v \in W^{1,p}(\mathbb{R}^n)$ and $v = u$ almost everywhere in Ω. Moreover, the convergence is uniform outside a set of arbitrarily small Sobolev p-capacity and, as in Theorem 5.14, the limit function v is p-quasicontinuous. Observe that $v = 0$ p-quasieverywhere in $\mathbb{R}^n \setminus \Omega$ since $u_i = 0$ in $\mathbb{R}^n \setminus \Omega$ for every $i \in \mathbb{N}$. □

Next we consider a converse of Theorem 5.26.

THEOREM 5.27. *Let $1 < p < \infty$ and let $\Omega \subset \mathbb{R}^n$ be an open set. Assume that $u \in W^{1,p}(\mathbb{R}^n)$ is p-quasicontinuous and $u = 0$ holds p-quasieverywhere in $\mathbb{R}^n \setminus \Omega$. Then $u \in W_0^{1,p}(\Omega)$.*

PROOF. Since $W^{1,p}(\mathbb{R}^n) = W_0^{1,p}(\mathbb{R}^n)$, see Remark 2.22, we may assume that $\Omega \subsetneq \mathbb{R}^n$. By Remark 2.28, see also Lemma 4.10, and the fact that $W_0^{1,p}(\Omega)$ is a closed subspace of $W^{1,p}(\Omega)$, it suffices to show that u can be approximated in $W^{1,p}(\mathbb{R}^n)$ by functions having a compact support in Ω. If we can construct such a sequence for $u_+ = \max\{u, 0\}$, then we can do it for $u_- = -\min\{u, 0\}$ as well, and we obtain the result for $u = u_+ - u_-$. Thus we may assume that $u \geq 0$ everywhere in $\mathbb{R}^n$. Moreover, we may also assume that u has a compact support in $\mathbb{R}^n$, and, by considering the truncations $\min\{u, t\}$, for $t > 0$, that u is bounded.

Let $\delta > 0$ and let $G' \subset \mathbb{R}^n$ be an open set such that $\operatorname{Cap}_p(G') < \delta$ and the restriction of u to $\mathbb{R}^n \setminus G'$ is continuous. Define $E = \{x \in \mathbb{R}^n \setminus \Omega : u(x) \neq 0\}$. By assumption $\operatorname{Cap}_p(E) = 0$. Since

$$\operatorname{Cap}_p(G' \cup E) \leq \operatorname{Cap}_p(G') + \operatorname{Cap}_p(E) < \delta,$$

there exists a function $v = v_\delta \in \mathcal{A}'_p(G' \cup E)$ with $0 \leq v \leq 1$ almost everywhere, $v = 1$ almost everywhere in a neighborhood of $G' \cup E$ and $\|v\|^p_{W^{1,p}(\mathbb{R}^n)} < \delta$, see Lemma 5.2. By modifying v on a set of measure zero, if necessary, we may assume that $v = 1$ in an open set G containing $G' \cup E$.

Let $0 < \varepsilon < 1$ and define

$$u_\varepsilon(x) = \max\{u(x) - \varepsilon, 0\},$$

for every $x \in \mathbb{R}^n$. Let $x \in (\mathbb{R}^n \setminus \Omega) \setminus G$. Since $u(x) = 0$ and the restriction of u to $\mathbb{R}^n \setminus G$ is continuous, there exists $r_x > 0$ such that $u_\varepsilon = 0$ in $B(x, r_x) \setminus G$. Thus $(1 - v)u_\varepsilon = 0$ in $B(x, r_x)$ for every $x \in (\mathbb{R}^n \setminus \Omega) \setminus G$, and so $(1 - v)u_\varepsilon$ is zero in the neighborhood

$$G \cup \bigcup_{x \in (\mathbb{R}^n \setminus \Omega) \setminus G} B(x, r_x)$$

of $\mathbb{R}^n \setminus \Omega$. Since u, and therefore also $(1 - v)u_\varepsilon$, are compactly supported in $\mathbb{R}^n$, we conclude that $(1 - v)u_\varepsilon$ is compactly supported in Ω for every $0 < \varepsilon < 1$. Moreover, since both $(1 - v)$ and u are bounded, we have $(1 - v)u_\varepsilon \in W^{1,p}(\mathbb{R}^n)$ and $(1 - v)u_\varepsilon \in W_0^{1,p}(\Omega)$ by Lemma 4.10. To conclude that $u \in W_0^{1,p}(\Omega)$, it suffices to show that the functions $(1 - v)u_\varepsilon$ converge to u in $W^{1,p}(\mathbb{R}^n)$.

Observe that

$$\|u - (1 - v)u_\varepsilon\|_{W^{1,p}(\mathbb{R}^n)} \leq \|u - u_\varepsilon\|_{W^{1,p}(\mathbb{R}^n)} + \|vu_\varepsilon\|_{W^{1,p}(\mathbb{R}^n)}. \tag{5.10}$$

Hence it is enough to prove that each of the terms on the right hand side can be made arbitrarily small by choosing δ and ε to be small enough; recall here that $v = v_\delta$ depends on δ. Since

$$u_\varepsilon = \begin{cases} u - \varepsilon & \text{in } \{x \in \mathbb{R}^n : u(x) \geq \varepsilon\}, \\ 0 & \text{in } \{x \in \mathbb{R}^n : u(x) \leq \varepsilon\}, \end{cases}$$

by Theorem 2.25 (c) we have

$$\nabla u_\varepsilon = \begin{cases} \nabla u, & \text{a.e. in } \{x \in \mathbb{R}^n : u(x) \geq \varepsilon\}, \\ 0, & \text{a.e. in } \{x \in \mathbb{R}^n : u(x) \leq \varepsilon\}. \end{cases}$$

Thus

$$\|\nabla u - \nabla u_\varepsilon\|_{L^p(\mathbb{R}^n)} \leq \|\chi_{\{0\leq u\leq\varepsilon\}}\nabla u\|_{L^p(\mathbb{R}^n)}.$$

By the dominated convergence theorem and Theorem 2.25 (b), we obtain

$$\begin{aligned}\lim_{\varepsilon\to 0}\|\chi_{\{0\leq u\leq\varepsilon\}}\nabla u\|_{L^p(\mathbb{R}^n)} &= \left(\lim_{\varepsilon\to 0}\int_{\mathbb{R}^n}\chi_{\{0\leq u\leq\varepsilon\}}(x)|\nabla u(x)|^p\,dx\right)^{\frac{1}{p}} \\ &= \left(\int_{\mathbb{R}^n}\lim_{\varepsilon\to 0}\chi_{\{0\leq u\leq\varepsilon\}}(x)|\nabla u(x)|^p\,dx\right)^{\frac{1}{p}} \\ &= \left(\int_{\mathbb{R}^n}\chi_{\{u=0\}}(x)|\nabla u(x)|^p\,dx\right)^{\frac{1}{p}} = 0.\end{aligned}$$

Here $\chi_{\{0\leq u\leq\varepsilon\}}|\nabla u|^p \leq |\nabla u|^p$ for every $0 < \varepsilon < 1$, and so $|\nabla u|^p \in L^1(\mathbb{R}^n)$ may be used as an integrable majorant. Since $0 \leq u - u_\varepsilon \leq \varepsilon$ and $\operatorname{supp}(u - u_\varepsilon) \subset \operatorname{supp} u$, we have

$$\begin{aligned}\|u - u_\varepsilon\|_{W^{1,p}(\mathbb{R}^n)} &\leq \|u - u_\varepsilon\|_{L^p(\mathbb{R}^n)} + \|\nabla u - \nabla u_\varepsilon\|_{L^p(\mathbb{R}^n)} \\ &\leq \varepsilon\|\chi_{\operatorname{supp} u}\|_{L^p(\mathbb{R}^n)} + \|\chi_{\{0\leq u\leq\varepsilon\}}\nabla u\|_{L^p(\mathbb{R}^n)},\end{aligned}$$

and it follows that

$$\lim_{\varepsilon\to 0}\|u - u_\varepsilon\|_{W^{1,p}(\mathbb{R}^n)} = 0. \tag{5.11}$$

Then we consider the term $\|vu_\varepsilon\|_{W^{1,p}(\mathbb{R}^n)}$ in (5.10). Since

$$|u_\varepsilon(x)| \leq |u(x)| \leq \|u\|_{L^\infty(\mathbb{R}^n)}$$

for almost every $x \in \mathbb{R}^n$, we have

$$\|vu_\varepsilon\|_{L^p(\mathbb{R}^n)} \leq \|u\|_{L^\infty(\mathbb{R}^n)}\|v\|_{L^p(\mathbb{R}^n)} \leq \|u\|_{L^\infty(\mathbb{R}^n)}\|v\|_{W^{1,p}(\mathbb{R}^n)}.$$

On the other hand, $|\nabla u_\varepsilon| \leq |\nabla u|$ almost everywhere, and using (2.17) we obtain

$$\begin{aligned}\|\nabla(vu_\varepsilon)\|_{L^p(\mathbb{R}^n)} &\leq \|u_\varepsilon\nabla v\|_{L^p(\mathbb{R}^n)} + \|v\nabla u_\varepsilon\|_{L^p(\mathbb{R}^n)} \\ &\leq \|u\nabla v\|_{L^p(\mathbb{R}^n)} + \|v\nabla u_\varepsilon\|_{L^p(\mathbb{R}^n)} \\ &\leq \|u\|_{L^\infty(\mathbb{R}^n)}\|\nabla v\|_{L^p(\mathbb{R}^n)} + \|v\nabla u\|_{L^p(\mathbb{R}^n)} \\ &\leq \|u\|_{L^\infty(\mathbb{R}^n)}\|v\|_{W^{1,p}(\mathbb{R}^n)} + \|v\nabla u\|_{L^p(\mathbb{R}^n)}.\end{aligned}$$

Recalling that $\|v\|^p_{W^{1,p}(\mathbb{R}^n)} < \delta$, we arrive at the estimate

$$\begin{aligned}\|vu_\varepsilon\|_{W^{1,p}(\mathbb{R}^n)} &\leq \|vu_\varepsilon\|_{L^p(\mathbb{R}^n)} + \|\nabla(vu_\varepsilon)\|_{L^p(\mathbb{R}^n)} \\ &\leq 2\|u\|_{L^\infty(\mathbb{R}^n)}\|v\|_{W^{1,p}(\mathbb{R}^n)} + \|v\nabla u\|_{L^p(\mathbb{R}^n)} \\ &\leq 2\delta^{\frac{1}{p}}\|u\|_{L^\infty(\mathbb{R}^n)} + \|v\nabla u\|_{L^p(\mathbb{R}^n)}.\end{aligned} \tag{5.12}$$

Let $\delta_i > 0$, $i \in \mathbb{N}$, be such that $\delta_i \to 0$ as $i \to \infty$, and let $v_i = v_{\delta_i}$ be the corresponding functions as above, with $\|v_i\|^p_{W^{1,p}(\mathbb{R}^n)} < \delta_i$. Then $v_i \to 0$ in $L^p(\mathbb{R}^n)$.

By passing to a subsequence, we may also assume that $v_i \to 0$ almost everywhere in $\mathbb{R}^n$ as $i \to \infty$. By the dominated convergence theorem, we have

$$\begin{aligned}\limsup_{i\to\infty} \|v_i \nabla u\|_{L^p(\mathbb{R}^n)} &\le \left(\lim_{i\to\infty} \int_{\mathbb{R}^n} |v_i(x)|^p |\nabla u(x)|^p \, dx\right)^{\frac{1}{p}} \\ &= \left(\int_{\mathbb{R}^n} \lim_{i\to\infty} |v_i(x)|^p |\nabla u(x)|^p \, dx\right)^{\frac{1}{p}} = 0.\end{aligned}$$

Here $|v_i|^p |\nabla u|^p \le |\nabla u|^p$, and so $|\nabla u|^p \in L^1(\mathbb{R}^n)$ may be used as an integrable majorant. Hence we obtain from (5.12), with with $v = v_i$, that

$$\limsup_{i\to\infty} \|v_i u_\varepsilon\|_{W^{1,p}(\mathbb{R}^n)} \le \limsup_{i\to\infty} \Big(2\delta_i^{\frac{1}{p}} \|u\|_{L^\infty(\mathbb{R}^n)} + \|v_i \nabla u\|_{L^p(\mathbb{R}^n)}\Big) = 0. \tag{5.13}$$

By combining (5.10) and (5.11), with $v = v_i$, and (5.13), we conclude that

$$\|u - (1 - v_i) u_\varepsilon\|_{W^{1,p}(\mathbb{R}^n)} \longrightarrow 0$$

as $\varepsilon \to 0$ and $i \to \infty$, and the claim $u \in W_0^{1,p}(\Omega)$ follows. □

REMARK 5.28. Let $1 < p < \infty$ and let $\Omega \subset \mathbb{R}^n$ be an open set. If $u \in W^{1,p}(\mathbb{R}^n)$ is continuous and zero everywhere in $\mathbb{R}^n \setminus \Omega$, then $u \in W_0^{1,p}(\Omega)$ by Theorem 5.27.

By combining Theorem 5.26 and Theorem 5.27, we obtain a Havin and Bagby type characterization of Sobolev spaces with zero boundary values on an arbitrary open set.

COROLLARY 5.29. *Let $1 < p < \infty$, let $\Omega \subset \mathbb{R}^n$ be an open set, and assume that $u \in W^{1,p}(\Omega)$. Then $u \in W_0^{1,p}(\Omega)$ if and only if there exists a p-quasicontinuous function $v \in W^{1,p}(\mathbb{R}^n)$ such that $v = u$ almost everywhere in Ω and $v = 0$ holds p-quasieverywhere in $\mathbb{R}^n \setminus \Omega$.*

It is also possible to characterize Sobolev spaces with zero boundary values in terms of Lebesgue points of Sobolev functions.

THEOREM 5.30. *Let $1 < p < \infty$, let $\Omega \subset \mathbb{R}^n$ be an open set, and assume that $u \in W^{1,p}(\Omega)$. Then $u \in W_0^{1,p}(\Omega)$ if and only if there exists $v \in W^{1,p}(\mathbb{R}^n)$ such that $u = v$ almost everywhere in Ω and*

$$\lim_{r\to 0} \fint_{B(x,r)} v(y) \, dy = 0$$

for p-quasievery $x \in \mathbb{R}^n \setminus \Omega$.

PROOF. Assume first that $u \in W_0^{1,p}(\Omega)$. By applying Theorem 5.26 to u, we obtain a p-quasicontinuous function $v \in W^{1,p}(\mathbb{R}^n)$ satisfying $v = u$ almost everywhere in Ω and $v = 0$ p-quasieverywhere in $\mathbb{R}^n \setminus \Omega$. On the other hand, Theorem 5.23 shows that the limit

$$\lim_{r\to 0} \fint_{B(x,r)} v(y) \, dy = v^*(x)$$

exists for p-quasievery $x \in \mathbb{R}^n$ and that the zero extension $v^* \colon \mathbb{R}^n \to \mathbb{R}$ is a p-quasicontinuous representative of v. Since $v^* = v$ almost everywhere in $\mathbb{R}^n$ and both v^* and v are p-quasicontinuous, Theorem 5.16 implies that $v^* = v$ holds p-quasieverywhere in $\mathbb{R}^n$. From this we conclude that

$$\lim_{r\to 0} \fint_{B(x,r)} v(y)\,dy = v^*(x) = v(x) = 0$$

for p-quasievery $x \in \mathbb{R}^n \setminus \Omega$, as desired.

Conversely, let $u \in W^{1,p}(\Omega)$ and $v \in W^{1,p}(\mathbb{R}^n)$ be such that $u = v$ almost everywhere in Ω and

$$\lim_{r\to 0} \fint_{B(x,r)} v(y)\,dy = 0 \tag{5.14}$$

for p-quasievery $x \in \mathbb{R}^n \setminus \Omega$. As above, by Theorem 5.23 the limit

$$v^*(x) = \lim_{r\to 0} \fint_{B(x,r)} v(y)\,dy \tag{5.15}$$

exists for p-quasievery $x \in \mathbb{R}^n$ and the zero extension $v^* : \mathbb{R}^n \to \mathbb{R}$ is a p-quasicontinuous representative of v. Hence $v^* \in W^{1,p}(\mathbb{R}^n)$. From (5.14) and (5.15) we obtain $v^*(x) = 0$ for p-quasievery $x \in \mathbb{R}^n \setminus \Omega$. Since $v^* = v = u$ almost everywhere in Ω, Corollary 5.29 implies that $u \in W_0^{1,p}(\Omega)$. □

EXAMPLE 5.31. Let $\Omega = B(0,1) \setminus \{0\}$ and consider $u\colon \Omega \to \mathbb{R}$, $u(x) = 1 - |x|$, for every $x \in \Omega$. Then $u \in W_0^{1,p}(\Omega)$ for $1 < p \le n$ and $u \notin W_0^{1,p}(\Omega)$ for $p > n$. This is a consequence of Remark 5.8, Lemma 5.9 and Theorem 5.30. Hence, a function that belongs to a Sobolev space with zero boundary values does not have to be zero at every Lebesgue point in the complement.

5.6. Variational capacity

Variational capacity is defined relative to an open subset of $\mathbb{R}^n$ with admissible functions having zero boundary values. Moreover, only the integral of the gradient is taken into account instead of the full Sobolev norm. This notion of capacity will be useful, for example, in connection with the capacity density condition in Definition 6.17.

DEFINITION 5.32. Let $1 < p < \infty$ and let $\Omega \subset \mathbb{R}^n$ be an open set. The variational p-capacity of a subset $E \Subset \Omega$ is

$$\operatorname{cap}_p(E,\Omega) = \inf \int_\Omega |\nabla u(x)|^p\,dx,$$

where the infimum is taken over all $u \in W_0^{1,p}(\Omega)$ with $u \ge 1$ almost everywhere in a neighborhood of E.

Let $\Omega \subset \mathbb{R}^n$ be an open set and let $E \Subset \Omega$. Since $d(\overline{E}, \partial\Omega) > 0$, there exists a neighborhood G of E with $G \Subset \Omega$. Then $u\colon \mathbb{R}^n \to \mathbb{R}$,

$$u(x) = \max\left\{0, 1 - \frac{d(x,G)}{d(G,\Omega^c)}\right\},$$

satisfies $u = 1$ in G and $u \in \operatorname{Lip}_0(\Omega) \subset W_0^{1,p}(\Omega)$, see Lemma 2.29. It follows that u is an admissible test function for $E \Subset \Omega$ and thus $\operatorname{cap}_p(E,\Omega) < \infty$. Variational capacity can be defined for all $E \subset \Omega$, for instance, by using the characterization in Theorem 5.46 below, but then $\operatorname{cap}_p(E,\Omega) = \infty$ may occur.

REMARK 5.33. As in Lemma 5.2, we may assume $0 \le u \le 1$ in Definition 5.32. This follows by considering the function $v = \min\{|u|, 1\}$ since $|\nabla v| \le |\nabla u|$ almost everywhere and $v \in W_0^{1,p}(\Omega)$ by Corollary 5.29.

Variational capacity has similar basic properties as the Sobolev capacity.

THEOREM 5.34. *Let $1 < p < \infty$ and let $\Omega \subset \mathbb{R}^n$ be an open set.*

(a) $\operatorname{cap}_p(\emptyset, \Omega) = 0$.

(b) *For $E_1 \subset E_2 \Subset \Omega$, we have* $\operatorname{cap}_p(E_1, \Omega) \le \operatorname{cap}_p(E_2, \Omega)$.

(c) *For $E \Subset \Omega \subset \Omega' \subset \mathbb{R}^n$, where Ω' is an open set, we have*

$$\operatorname{cap}_p(E, \Omega') \le \operatorname{cap}_p(E, \Omega).$$

(d) *For unbounded Ω and $E \Subset \Omega$, we have*

$$\operatorname{cap}_p(E, \Omega) = \lim_{j\to\infty} \operatorname{cap}_p(E, \Omega \cap B(0, j)).$$

(e) *For bounded Ω and $\bigcup_{i=1}^{\infty} E_i \Subset \Omega$, we have*

$$\operatorname{cap}_p\left(\bigcup_{i=1}^{\infty} E_i, \Omega\right) \le \sum_{i=1}^{\infty} \operatorname{cap}_p(E_i, \Omega).$$

PROOF. Properties (a) and (b) are immediate from the definition. Property (c) follows easily from Theorem 2.31.

For the proof of (d), let $u \in W_0^{1,p}(\Omega)$ with $u \ge 1$ almost everywhere in a neighborhood of E. Let $\varphi \in C_0^\infty(B(0,1))$ be such that $\varphi = 1$ in $B(0, \frac{1}{2})$ and let $v_j(x) = u(x)\varphi(\frac{x}{j})$ for every $x \in \Omega$. Then $v_j \in W_0^{1,p}(\Omega \cap B(0,j))$ and $v_j \ge 1$ almost everywhere in a neighborhood of E for j large enough. Observe also that

$$\nabla v_j(x) = \nabla u(x)\varphi(\tfrac{x}{j}) + \tfrac{1}{j} u(x) \nabla\varphi(\tfrac{x}{j})$$

for almost every $x \in \Omega$. By the dominated convergence theorem

$$\limsup_{j\to\infty} \operatorname{cap}_p(E, \Omega \cap B(0,j)) \le \limsup_{j\to\infty} \int_{\Omega\cap B(0,j)} |\nabla v_j(x)|^p \, dx = \int_\Omega |\nabla u(x)|^p \, dx.$$

Inequality

$$\limsup_{j\to\infty} \operatorname{cap}_p(E, \Omega \cap B(0,j)) \le \operatorname{cap}_p(E, \Omega)$$

follows by taking infimum over all functions u as above. The reverse inequality

$$\operatorname{cap}_p(E, \Omega) \le \liminf_{j\to\infty} \operatorname{cap}_p(E, \Omega \cap B(0,j))$$

is a consequence of (c). This proves (d).

The proof of the subadditivity in (e) is similar to the proof of the corresponding property for the Sobolev capacity in Theorem 5.3 (c). We may assume that $\sum_{i=1}^{\infty} \operatorname{cap}_p(E_i, \Omega) < \infty$. Let $\varepsilon > 0$ and let $u_i \in W_0^{1,p}(\Omega)$ be an admissible test function for E_i, with $0 \le u_i \le 1$ almost everywhere in Ω and

$$\int_\Omega |\nabla u_i(x)|^p \, dx \le \operatorname{cap}_p(E_i, \Omega) + 2^{-i}\varepsilon \tag{5.16}$$

for every $i \in \mathbb{N}$. Since $u_i \in W_0^{1,p}(\Omega)$ and Ω is bounded, we have by Corollary 3.7,

$$\int_\Omega |u_i(x)|^p \, dx \le C(n, p, \Omega) \int_\Omega |\nabla u_i(x)|^p \, dx$$

for every $i \in \mathbb{N}$. Define $v = \sup_{i \in \mathbb{N}} u_i$. It can be shown, as in the proof of Theorem 5.3, that $v \in W^{1,p}(\Omega)$. Moreover, it follows from Theorem 2.30 and Theorem 2.40 (b) that $v \in W_0^{1,p}(\Omega)$. There is a neighborhood G of $\bigcup_{i=1}^\infty E_i$ such that $v = 1$ almost everywhere in G, and thus v is admissible for $\bigcup_{i=1}^\infty E_i$. This implies

$$\operatorname{cap}_p\Bigl(\bigcup_{i=1}^\infty E_i, \Omega\Bigr) \le \int_\Omega |\nabla v(x)|^p\,dx \le \sum_{i=1}^\infty \int_\Omega |\nabla u_i(x)|^p\,dx \le \sum_{i=1}^\infty \operatorname{cap}_p(E_i, \Omega) + \varepsilon,$$

and we obtain property (e) for bounded Ω by letting $\varepsilon \to 0$. □

We have the following measure estimates for variational capacity.

LEMMA 5.35. *Let $1 < p < \infty$, let $B(x,r) \subset \mathbb{R}^n$ be a ball, and assume that $E \subset \overline{B}(x,r)$. There exist constants $C_1 = C(n,p)$ and $C_2 = C(n)$ such that*

$$C_1 \frac{|E|}{r^p} \le \operatorname{cap}_p(E, B(x,2r)) \le C_2 \frac{|B(x,r)|}{r^p}.$$

In particular,

$$C(n,p) r^{n-p} \le \operatorname{cap}_p\bigl(\overline{B}(x,r), B(x,2r)\bigr) \le C(n) r^{n-p}.$$

PROOF. First we prove the upper bound. Let $r < \rho < 2r$ and define

$$u(y) = \max\Bigl\{0, 1 - \frac{1}{2r - \rho} d(y, B(x,\rho))\Bigr\}$$

for every $y \in \mathbb{R}^n$. Then $u \in \operatorname{Lip}_0(B(x,2r)) \subset W_0^{1,p}(B(x,2r))$ and $u = 1$ in the open neighborhood $G = B(x,\rho)$ of E. Thus u is an admissible test function for $\operatorname{cap}_p(E, B(x,2r))$. Since $|\nabla u(y)| \le \frac{1}{2r-\rho}$ for almost every $y \in B(x,2r)$, we conclude that

$$\operatorname{cap}_p(E, B(x,2r)) \le \int_{B(x,2r)} |\nabla u(y)|^p\,dy \le \frac{|B(x,2r)|}{(2r-\rho)^p} = C(n) \frac{|B(x,r)|}{(2r-\rho)^p}.$$

The upper bound follows by letting $\rho \to r$ from above.

For the lower bound we apply the Poincaré inequality in Theorem 3.17. Let $u \in W_0^{1,p}(B(x,2r))$ be such that $u \ge 1$ almost everywhere in a neighborhood $G \subset B(x,2r)$ of E. By (3.11) with $q = p$, we have

$$|E| \le |G| \le \int_G |u(y)|^p\,dy \le \int_{B(x,2r)} |u(y)|^p\,dy \le C(n,p) r^p \int_{B(x,2r)} |\nabla u(y)|^p\,dy.$$

The lower bound follows by taking infimum over all admissible test functions. The final claim holds by choosing $E = \overline{B}(x,r)$. □

We review capacity estimates for concentric balls. For detailed formulas, see also Heinonen, Kilpeläinen and Martio [**187**, Example 2.12]

LEMMA 5.36. *Let $x \in \mathbb{R}^n$ and assume that $0 < r \le \frac{R}{2}$.*

(a) *For $1 < p < n$, there exists a constant $C = C(n,p)$ such that*

$$C^{-1} r^{n-p} \le \operatorname{cap}_p\bigl(\overline{B}(x,r), B(x,R)\bigr) \le C r^{n-p}.$$

(b) *For $n < p < \infty$, there exists a constant $C = C(n,p)$ such that*

$$C^{-1} R^{n-p} \le \operatorname{cap}_p\bigl(\overline{B}(x,r), B(x,R)\bigr) \le C R^{n-p}.$$

(c) *There exists a constant* $C = C(n)$ *such that*

$$C^{-1}\Big(\log\frac{R}{r}\Big)^{1-n} \le \operatorname{cap}_n\big(\overline{B}(x,r), B(x,R)\big) \le C\Big(\log\frac{R}{r}\Big)^{1-n}.$$

PROOF. First assume that $1 < p < n$. The upper bound in (a) follows from Theorem 5.34 (c) and Lemma 5.35 since

$$\operatorname{cap}_p\big(\overline{B}(x,r), B(x,R)\big) \le \operatorname{cap}_p\big(\overline{B}(x,r), B(x,2r)\big) \le C(n)r^{n-p}.$$

Here we applied the assumption $2r \le R$.

To prove the lower bound in (a), let $u \in W_0^{1,p}(B(x,R))$ be such that $u \ge 1$ almost everywhere in an open set G satisfying $\overline{B}(x,r) \subset G$. Then $G \subset B(x,R)$, and we obtain from the Sobolev inequality, see Remark 3.8, that

$$\begin{aligned} r^{\frac{n-p}{p}} &= C(n,p)|B(x,r)|^{\frac{n-p}{np}} \le C(n,p)|G|^{\frac{n-p}{np}} \\ &\le C(n,p)\Big(\int_{B(x,R)} |u(y)|^{\frac{np}{n-p}}\,dy\Big)^{\frac{n-p}{np}} \le C(n,p)\Big(\int_{B(x,R)} |\nabla u(y)|^p\,dy\Big)^{\frac{1}{p}}. \end{aligned}$$

The claim follows by taking infimum over all functions u as above.

Then we consider the case $n < p < \infty$. The upper bound in (b) follows from Theorem 5.34 (b) and Lemma 5.35 since

$$\operatorname{cap}_p\big(\overline{B}(x,r), B(x,R)\big) \le \operatorname{cap}_p\big(\overline{B}(x,\tfrac{R}{2}), B(x,R)\big) \le C(n)R^{n-p}.$$

For the lower bound we use Morrey's inequality. Let $u \in W_0^{1,p}(B(x,R))$ be an admissible test function for $\operatorname{cap}_p\big(\overline{B}(x,r), B(x,R)\big)$ and extend u as zero so that $u \in W^{1,p}(\mathbb{R}^n)$, see Theorem 2.31. By Theorem 3.23 we may assume that u is continuous, and thus $u = 1$ in $\overline{B}(x,r)$ and $u = 0$ in $\mathbb{R}^n \setminus B(x,R)$. Let $z \in \partial B(x,R)$. Then $u(x) = 1$ and $u(z) = 0$, and by Corollary 3.21 and continuity of u we have

$$\begin{aligned} 1 = |u(x) - u(z)| &\le C(n,p)2R\Big(\fint_{B(x,2R)} |\nabla u(y)|^p\,dy\Big)^{\frac{1}{p}} \\ &\le C(n,p)R^{1-\frac{n}{p}}\Big(\int_{B(x,R)} |\nabla u(y)|^p\,dy\Big)^{\frac{1}{p}}. \end{aligned}$$

Hence

$$R^{n-p} \le C(n,p)\int_{B(x,R)} |\nabla u(y)|^p\,dy,$$

and the lower bound in (b) follows.

The upper bound in (c) follows by testing capacity with the Lipschitz function

$$u(y) = \begin{cases} 1, & \text{if } y \in B(x,r), \\ \big(\log\frac{R}{r}\big)^{-1}\log\frac{R}{|y-x|}, & \text{if } y \in B(x,R)\setminus B(x,r), \\ 0, & \text{if } y \notin B(x,R), \end{cases}$$

compare to Remark 5.8. We omit the details concerning the lower bound in (c), but this can be obtained with a direct computation, as in Heinonen, Kilpeläinen and Martio [**187**, Example 2.12], or using a telescoping argument, as in Björn, Björn and Lehrbäck [**32**, Proposition 7.1]. The latter approach also applies in more general metric spaces. □

REMARK 5.37. Let $1 < p < n$. Assume that $\Omega \subset \mathbb{R}^n$ is an open set and $B(x, 2r) \subset \Omega$. By replacing $B(x, R)$ in the proof of Lemma 5.36 (a) by Ω, we obtain

$$C^{-1} r^{n-p} \leq \operatorname{cap}_p(\overline{B}(x,r), \Omega) \leq C r^{n-p},$$

where $C = C(n,p)$.

Next we make comparisons between the Sobolev p-capacity and the variational p-capacity.

THEOREM 5.38. *Let $1 < p < \infty$ and assume that $E \subset \overline{B}(x,r) \subset \mathbb{R}^n$. There exists a constant $C = C(n,p)$ such that*

$$\frac{1}{1 + Cr^p} \operatorname{Cap}_p(E) \leq \operatorname{cap}_p(E, B(x, 2r)) \leq 2^p \Big(1 + \frac{1}{r^p}\Big) \operatorname{Cap}_p(E). \tag{5.17}$$

PROOF. Let $u \in W_0^{1,p}(B(x,2r))$ be admissible for $\operatorname{cap}_p(E, B(x,2r))$. We extend u as zero outside $B(x,2r)$. Then $u \in \mathcal{A}_p(E)$, where $\mathcal{A}_p(E)$ is as in Definition 5.1. Using the p-Poincaré inequality for Sobolev functions with zero boundary values, see Theorem 3.17, we obtain

$$\begin{aligned} \operatorname{Cap}_p(E) \leq \|u\|^p_{W^{1,p}(\mathbb{R}^n)} &= \int_{B(x,2r)} |u(y)|^p \, dy + \int_{B(x,2r)} |\nabla u(y)|^p \, dy \\ &\leq (C(n,p) r^p + 1) \int_{B(x,2r)} |\nabla u(y)|^p \, dy. \end{aligned}$$

The first inequality in (5.17) follows by taking infimum over all such u.

To obtain the second inequality in (5.17), we let $u \in \mathcal{A}_p(E)$ and $r < \rho < 2r$, and consider

$$\psi(y) = \max\Big\{0, 1 - \frac{1}{2r - \rho} d\big(y, B(x, 2^{-1}(r + \rho))\big)\Big\},$$

for every $y \in \mathbb{R}^n$. Then $0 \leq \psi(y) \leq 1$ and $|\nabla \psi(y)| \leq \frac{1}{2r-\rho}$ for almost every $y \in B(x,2r)$, and $\psi u \in W_0^{1,p}(B(x,2r))$ is admissible for $\operatorname{cap}_p(E, B(x,2r))$. By Leibniz's rule (2.17),

$$\begin{aligned} \operatorname{cap}_p(E, B(x,2r)) &\leq \int_{B(x,2r)} |\nabla(\psi u)(y)|^p \, dy \\ &\leq \int_{B(x,2r)} \big(|\nabla \psi(y)||u(y)| + \psi(y)|\nabla u(y)|\big)^p \, dy \\ &\leq 2^p \Big(\int_{B(x,2r)} (2r - \rho)^{-p} |u(y)|^p \, dy + \int_{B(x,2r)} |\nabla u(y)|^p \, dy \Big) \\ &\leq 2^p \left((2r - \rho)^{-p} + 1 \right) \|u\|^p_{W^{1,p}(\mathbb{R}^n)}. \end{aligned}$$

Taking $\rho \to r$ and then infimum over $u \in \mathcal{A}_p(E)$ concludes the proof. □

5.7. Capacity and Hausdorff content

There are close connections between capacities and Hausdorff contents, see Definition 1.43. We begin with an upper bound. In general, upper bounds for capacity are easier to obtain than lower bounds.

THEOREM 5.39. *Let $1 < p \leq n$ and assume that $E \subset \overline{B}(x,r) \subset \mathbb{R}^n$. There exists a constant $C = C(n)$ such that*

$$\operatorname{cap}_p(E, B(x,2r)) \leq C \mathcal{H}^{n-p}_\infty(E).$$

PROOF. Let $\{B(x_i, r_i) : i \in \mathbb{N}\}$ be a cover of E. Without loss of generality, we may assume that $B(x_i, r_i) \cap E \neq \emptyset$ for every $i \in \mathbb{N}$. If $r_j \geq \frac{r}{4}$ for some $j \in \mathbb{N}$, we obtain from Lemma 5.35 that

$$\operatorname{cap}_p(E, B(x, 2r)) \leq C(n)r^{n-p} \leq 4^n C(n) r_j^{n-p} \leq 4^n C(n) \sum_{i=1}^{\infty} r_i^{n-p}.$$

On the other hand, if $0 < r_i < \frac{r}{4}$ for every $i \in \mathbb{N}$, then $B(x_i, 2r_i) \subset B(x, 2r)$ for every $i \in \mathbb{N}$, and Theorem 5.34 (c) and Lemma 5.35 imply

$$\operatorname{cap}_p\big(B(x_i, r_i), B(x, 2r)\big) \leq \operatorname{cap}_p\big(B(x_i, r_i), B(x_i, 2r_i)\big) \leq C(n) r_i^{n-p}$$

for every $i \in \mathbb{N}$. Observe also that $\bigcup_{i=1}^{\infty} B(x_i, r_i) \subset B(x, \frac{3}{2}r) \Subset B(x, 2r)$. By monotonicity and countable subadditivity, we have

$$\begin{aligned}\operatorname{cap}_p(E, B(x, 2r)) &\leq \operatorname{cap}_p\left(\bigcup_{i=1}^{\infty} B(x_i, r_i), B(x, 2r)\right) \\ &\leq \sum_{i=1}^{\infty} \operatorname{cap}_p\big(B(x_i, r_i), B(x, 2r)\big) \leq C(n) \sum_{i=1}^{\infty} r_i^{n-p}.\end{aligned}$$

The claim follows by taking infimum over all such covers. □

The reverse inequality does not hold in Theorem 5.39, see Remark 7.23. Theorem 5.41 gives a slightly weaker estimate to this direction. Before that, we prove a comparison result for the Hausdorff contents with different dimensions.

LEMMA 5.40. *Let $0 \leq \lambda_1 < \lambda_2$ and assume that $E \subset \overline{B}(x, r) \subset \mathbb{R}^n$. Then*

$$\mathcal{H}_\infty^{\lambda_2}(E) \leq r^{\lambda_2 - \lambda_1} \mathcal{H}_\infty^{\lambda_1}(E). \tag{5.18}$$

PROOF. Since $E \subset B(x, r + \varepsilon)$, for any $\varepsilon > 0$, we have $\mathcal{H}_\infty^{\lambda_2}(E) \leq (r + \varepsilon)^{\lambda_2}$. By letting $\varepsilon \to 0$ we obtain $\mathcal{H}_\infty^{\lambda_2}(E) \leq r^{\lambda_2}$. Let $\{B(x_i, r_i : i \in \mathbb{N})\}$ be a cover of E. Assume first that $r_j > r$ for some $j \in \mathbb{N}$. Then

$$\mathcal{H}_\infty^{\lambda_2}(E) \leq r^{\lambda_2} = r^{\lambda_2 - \lambda_1} r^{\lambda_1} \leq r^{\lambda_2 - \lambda_1} r_j^{\lambda_1} \leq r^{\lambda_2 - \lambda_1} \sum_{i=1}^{\infty} r_i^{\lambda_1}.$$

On the other hand, if $r_i \leq r$ for every $i \in \mathbb{N}$, then $r_i^{\lambda_2} \leq r^{\lambda_2 - \lambda_1} r_i^{\lambda_1}$ for every $i \in \mathbb{N}$. Hence,

$$\mathcal{H}_\infty^{\lambda_2}(E) \leq \sum_{i=1}^{\infty} r_i^{\lambda_2} \leq r^{\lambda_2 - \lambda_1} \sum_{i=1}^{\infty} r_i^{\lambda_1}.$$

The claim follows by taking infimum over all such covers of E in both cases. □

Let $E \subset \overline{B}(x, r)$ and $1 < p \leq n$. By Lemma 5.40 we have

$$\mathcal{H}_\infty^{\lambda}(E) r^{n-p-\lambda} \leq \mathcal{H}_\infty^{n-p}(E)$$

for $\lambda_2 = \lambda > n - p = \lambda_1 \geq 0$. A combination of Theorem 5.39 and the following Theorem 5.41 implies

$$C(n, p, \lambda) \mathcal{H}_\infty^{\lambda}(E) r^{n-p-\lambda} \leq \operatorname{cap}_p(E, B(x, 2r)) \leq C(n) \mathcal{H}_\infty^{n-p}(E).$$

THEOREM 5.41. *Let $1 < p \leq n$ and $\lambda > n - p$, and assume that $E \subset \overline{B}(x, r) \subset \mathbb{R}^n$. There exists a constant $C = C(n, p, \lambda)$ such that*

$$\mathcal{H}_\infty^{\lambda}(E) r^{n-p-\lambda} \leq C \operatorname{cap}_p(E, B(x, 2r)).$$

PROOF. Let $u \in W_0^{1,p}(B(x,2r))$ be admissible for $\operatorname{cap}_p(E, B(x,2r))$. Extending u as zero outside $B(x,2r)$, we may assume that $u \in W^{1,p}(\mathbb{R}^n)$, see Theorem 2.31. We may also assume that $0 \le u \le 1$, $u = 0$ in $\mathbb{R}^n \setminus B(x,2r)$, and that there exists a neighborhood G of E such that $u(y) = 1$ for every $y \in G$.

Let $z \in E$ and let $r_k = 2^{2-k} r$ for $k \in \mathbb{N}_0$. Since $B(x,2r) \subset B(z,4r)$, $u = 0$ in $B(z,4r) \setminus B(x,2r)$, and $0 \le u \le 1$, we have

$$|u_{B(z,4r)}| = \frac{1}{|B(z,4r)|} \int_{B(x,2r)} u(y)\, dy \le \frac{|B(x,2r)|}{|B(z,4r)|} = \frac{(2r)^n}{(4r)^n} = 2^{-n} < 1.$$

Since $u = 1$ in G, there exists $j \in \mathbb{N}$ with $u_{B(z,r_j)} = 1$, and thus

$$|u_{B(z,r_j)} - u_{B(z,4r)}| = 1 - 2^{-n} > 0.$$

Using this lower bound and a telescoping argument, as in the proof of Theorem 3.19 for the balls $B(z, r_k)$, we obtain

$$1 \le C(n)|u_{B(z,r_j)} - u_{B(z,4r)}| \le C(n,p) \sum_{k=0}^{\infty} r_k \left(\fint_{B(z,r_k)} |\nabla u(y)|^p \, dy \right)^{\frac{1}{p}}. \tag{5.19}$$

On the other hand, for $\delta > 0$, we have $\sum_{k=0}^{\infty} 2^{-k\delta} = C(\delta)$, and (5.19) gives

$$\sum_{k=0}^{\infty} 2^{-k\delta} \le C(n,p,\delta) \sum_{k=0}^{\infty} r_k \left(\fint_{B(z,r_k)} |\nabla u(y)|^p \, dy \right)^{\frac{1}{p}}.$$

From this we conclude that there exist an index $k_z \in \mathbb{N}_0$ such that

$$r_{k_z} \left(\fint_{B(z,r_{k_z})} |\nabla u(y)|^p \, dy \right)^{\frac{1}{p}} \ge C(n,p,\delta) 2^{-k_z \delta} = C(n,p,\delta) r^{-\delta} (r_{k_z})^{\delta}, \tag{5.20}$$

where $C(n,p,\delta) > 0$. We choose $\delta = \frac{\lambda - n + p}{p} > 0$ and write $r_z = r_{k_z}$ for the corresponding radius in (5.20).

By raising both sides of (5.20) to power p, we obtain for every $z \in E$ a ball $B(z, r_z)$ with

$$r_z^{\lambda} \le C(n,p,\lambda) r^{\lambda-n+p} \int_{B(z,r_z)} |\nabla u(y)|^p \, dy. \tag{5.21}$$

Observe that the radii r_z with $z \in E$ are uniformly bounded. We apply Lemma 1.13 for the balls $B(z,r_z)$, $z \in E$, and obtain points $z_i \in E$, $i \in \mathbb{N}$, such that the balls $B(z_i, r_{z_i})$ are pairwise disjoint and $E \subset \bigcup_{i=1}^{\infty} B(z_i, 5r_{z_i})$. From (5.21), pairwise disjointness of the balls $B(z_i, r_{z_i})$, and the fact that $\nabla u = 0$ almost everywhere in $\mathbb{R}^n \setminus B(x,2r)$, we conclude that

$$\begin{aligned}
\mathcal{H}^{\lambda}_{\infty}(E) &\le \sum_{i=1}^{\infty} (5r_{z_i})^{\lambda} = 5^{\lambda} \sum_{i=1}^{\infty} r_{z_i}^{\lambda} \\
&\le C(n,p,\lambda) \sum_{i=1}^{\infty} r^{\lambda-n+p} \int_{B(z_i,r_{z_i})} |\nabla u(y)|^p \, dy \\
&\le C(n,p,\lambda) r^{\lambda-n+p} \int_{B(x,2r)} |\nabla u(y)|^p \, dy.
\end{aligned}$$

The claim follows by taking infimum over all admissible test functions. □

The comparison results above give relations between zero Sobolev capacity and the Hausdorff dimension, see Definition 1.44.

THEOREM 5.42. *Let $1 < p < n$ and let $E \subset \mathbb{R}^n$.*

(a) *If $\operatorname{Cap}_p(E) = 0$, then $\dim_{\mathrm{H}}(E) \le n - p$.*
(b) *If $\dim_{\mathrm{H}}(E) < n - p$, then $\operatorname{Cap}_p(E) = 0$.*

PROOF. First assume that $\operatorname{Cap}_p(E) = 0$. Let $\lambda > n - p$ and $r > 0$. Then $\operatorname{Cap}_p(E \cap B(0,r)) = 0$, and Theorem 5.41 and Theorem 5.38 imply

$$\begin{aligned}\mathcal{H}^\lambda_\infty(E \cap B(0,r))r^{n-p-\lambda} &\le C(n,p,\lambda)\operatorname{cap}_p(E \cap B(0,r), B(0,2r))\\ &\le C(n,p,\lambda)\Big(1 + \frac{1}{r^p}\Big)\operatorname{Cap}_p(E \cap B(0,r)) = 0.\end{aligned}$$

Since this holds for every $r > 0$, by countable subadditivity we conclude that

$$\mathcal{H}^\lambda_\infty(E) \le \sum_{j=1}^{\infty} \mathcal{H}^\lambda_\infty(E \cap B(0,j)) = 0$$

for every $\lambda > n - p$. Thus $\dim_{\mathrm{H}}(E) \le n - p$, proving part (a).

Then assume that $\dim_{\mathrm{H}}(E) < n - p$ and let $r > 0$. Since

$$\mathcal{H}^{n-p}_\infty(E \cap B(0,r)) \le \mathcal{H}^{n-p}_\infty(E) = 0,$$

we obtain from Theorem 5.38 and Theorem 5.39 that

$$\begin{aligned}\frac{1}{1 + Cr^p}\operatorname{Cap}_p(E \cap B(0,r)) &\le \operatorname{cap}_p(E \cap B(0,r), B(0,2r))\\ &\le C(n,p)\mathcal{H}^{n-p}_\infty(E \cap B(0,r)) = 0.\end{aligned}$$

Countable subadditivity implies that $\operatorname{Cap}_p(E) = 0$. □

REMARK 5.43. The claim in Theorem 5.42 (b) can be improved to the following: if $\mathcal{H}^{n-p}(E) < \infty$, then $\operatorname{Cap}_p(E) = 0$. We refer to Evans and Gariepy [**122**, Section 4.7.2] for the proof.

5.8. Lipschitz test functions for variational capacity

For compact sets, variational capacity can be characterized in terms of Lipschitz functions with zero boundary values.

THEOREM 5.44. *Let $1 < p < \infty$, let $\Omega \subset \mathbb{R}^n$ be an open set, and assume that $E \subset \Omega$ is compact. Then*

$$\operatorname{cap}_p(E, \Omega) = \inf \int_\Omega |\nabla u(x)|^p \, dx, \tag{5.22}$$

where the infimum is taken over all $u \in \operatorname{Lip}_0(\Omega)$ with $u = 1$ in E.

PROOF. Assume first that $u \in \operatorname{Lip}_0(\Omega)$ satisfies $u(x) = 1$ for every $x \in E$. Let $0 < \varepsilon < 1$ and let $v(x) = \frac{1}{1-\varepsilon}u(x)$, for every $x \in \Omega$. Then the set $G = \{x \in \Omega : u(x) > 1 - \varepsilon\}$ is a neighborhood of E and the function $v \in \operatorname{Lip}_0(\Omega)$ is admissible for the capacity $\operatorname{cap}_p(E, \Omega)$, by Theorem 2.29. Thus

$$\operatorname{cap}_p(E, \Omega) \le \int_\Omega |\nabla v(x)|^p \, dx = \Big(\frac{1}{1 - \varepsilon}\Big)^p \int_\Omega |\nabla u(x)|^p \, dx.$$

By letting $\varepsilon \to 0$ and taking infimum over all u as above, we conclude that $\operatorname{cap}_p(E, \Omega)$ is at most the infimum in (5.22).

For the proof of the reverse inequality, we may assume by Theorem 5.34 (d) that Ω is bounded. Let $0<\varepsilon'<1$ and let $u\in W_0^{1,p}(\Omega)$ be such that $u=1$ almost everywhere in a neighborhood $G'\subset\Omega$ of E and

$$\int_\Omega |\nabla u(x)|^p\,dx \le \operatorname{cap}_p(E,\Omega)+\varepsilon'.$$

Let $v\in W^{1,p}(\mathbb{R}^n)$ be a p-quasicontinuous function such that $v=u$ almost everywhere in Ω and $v=0$ p-quasieverywhere in Ω^c, see Corollary 5.29. Since $v=u$ almost everywhere in Ω, we have $v=1$ almost everywhere in G'. Lemma 5.18 implies that $v=1$ p-quasieverywhere in G'. By modifying v on a set of Sobolev p-capacity zero, if necessary, we may assume that $v=1$ in the open set G' and $v=0$ in Ω^c.

Let $0<\varepsilon<1$ and let $G\Subset G'$ be an open set with $E\subset G$. Choose an open set $V\subset G^c$ such that $\operatorname{Cap}_p(V)<\varepsilon^p$ and the restriction $v|_{\mathbb{R}^n\setminus V}$ is continuous; here we need the fact that $v=1$ in the open set $G'\Supset G$. Let $w\in\mathcal{A}_p'(V)$ be such that $\|w\|_{W^{1,p}(\mathbb{R}^n)}<\varepsilon$. By modifying w on a set of measure zero, we may assume that $w=1$ in V. By quasicontinuity and relative topology, there exists an open set $U_1\subset\mathbb{R}^n$ such that

$$\Omega^c\setminus V\subset\{x\in\mathbb{R}^n\setminus V: v(x)<\varepsilon\}=U_1\cap(\mathbb{R}^n\setminus V)=U_1\setminus V.$$

This implies $\Omega^c\subset U_1\cup V$. Since Ω is bounded, $(U_1\cup V)^c$ is a compact subset of Ω. Thus there exists an open set U_2 with $\Omega^c\subset U_2\subset U_1\cup V$ and $d\big(U_2,(U_1\cup V)^c\big)>0$. Let G_1 and G_2 be open sets satisfying $E\subset G_2\Subset G_1\Subset G$. This is possible since E is compact. Observe that $d(V,G_1)\ge d(G^c,G_1)=\delta>0$ since $V\subset G^c$. Here it is important that δ is independent of ε. Let

$$\psi(x)=\max\left\{0,1-\frac{1}{\delta}d(x,V)\right\},$$

for every $x\in\mathbb{R}^n$. Then $\psi\in\operatorname{Lip}(\mathbb{R}^n)$, $0\le\psi\le 1$, $\psi=1$ in V, $\psi=0$ in G_1, and $|\nabla\psi|\le\frac{1}{\delta}$ for almost every $x\in\mathbb{R}^n$.

Let

$$\varphi(x)=\min\left\{\frac{(v(x)-\varepsilon)_+}{1-\varepsilon},1-\psi(x)w(x)\right\}$$

for every $x\in\mathbb{R}^n$. Then $\varphi\in W^{1,p}(\mathbb{R}^n)$, $\varphi=1$ in G_1, and $\varphi=0$ in $U_1\cup V$. By Theorem 2.25 (c) and Leibniz's rule (2.17) we have

$$\begin{aligned}\|\nabla\varphi\|_{L^p(\mathbb{R}^n)}&\le\frac{1}{1-\varepsilon}\|\nabla v\|_{L^p(\mathbb{R}^n)}+\|\nabla w\|_{L^p(\mathbb{R}^n)}+\frac{1}{\delta}\|w\|_{L^p(\mathbb{R}^n)}\\&\le\frac{1}{1-\varepsilon}\|\nabla u\|_{L^p(\Omega)}+\|w\|_{W^{1,p}(\mathbb{R}^n)}+\frac{1}{\delta}\|w\|_{W^{1,p}(\mathbb{R}^n)}\\&\le\frac{1}{1-\varepsilon}(\operatorname{cap}_p(E,\Omega)+\varepsilon')^{\frac{1}{p}}+\left(1+\frac{1}{\delta}\right)\|w\|_{W^{1,p}(\mathbb{R}^n)}.\end{aligned}\tag{5.23}$$

Let $r_1=d(U_2,(U_1\cup V)^c)>0$ and $r_2=d(G_2,G_1^c)>0$. If $z\in U_2$ and $0<r<r_1$, then

$$\fint_{B(z,r)}|\nabla\varphi(x)|\,dx=0.$$

By Hölder's inequality, we obtain

$$
\begin{aligned}
M|\nabla\varphi|(z) &= \sup_{r\geq r_1} \fint_{B(z,r)} |\nabla\varphi(x)|\,dx \leq \sup_{r\geq r_1}\left(\fint_{B(z,r)} |\nabla\varphi(x)|^p\,dx\right)^{\frac{1}{p}} \\
&\leq \frac{1}{|B(z,r_1)|^{\frac{1}{p}}}\|\nabla\varphi\|_{L^p(\mathbb{R}^n)} = C_1 < \infty
\end{aligned}
\tag{5.24}
$$

for every $z \in U_2$. Similarly, we have

$$
M|\nabla\varphi|(z) \leq \frac{1}{|B(z,r_2)|^{\frac{1}{p}}}\|\nabla\varphi\|_{L^p(\mathbb{R}^n)} = C_2 < \infty
$$

for every $z \in G_2$.

For $t > 0$, let

$$
F_t = \{x \in \mathbb{R}^n \setminus A : M|\nabla\varphi|(x) \leq t \text{ and } |\varphi(x)| \leq t\},
$$

as in the proof of Theorem 4.9, where $A \subset \mathbb{R}^n$ is an exceptional set with $|A| = 0$ from Theorem 4.2. Since $0 \leq \varphi(z) \leq 1$ and $M|\nabla\varphi|(z) \leq \max\{C_1, C_2\}$ for every $z \in U_2 \cup G_2$, we have $(U_2 \cup G_2) \setminus A \subset F_t$ for large t. By the proof of Theorem 4.9, we obtain $t > 0$ and a Lipschitz function φ_t such that $\|\varphi - \varphi_t\|_{W^{1,p}(\mathbb{R}^n)} < \varepsilon$ and $\varphi_t = \varphi$ in $F_t \supset (U_2 \cup G_2) \setminus A$. Since $U_2 \cup G_2$ is open and φ_t is continuous, we conclude that $\varphi_t = 0$ in U_2 and $\varphi_t = 1$ in G_2. Therefore $\varphi_t \in \operatorname{Lip}_0(\Omega)$ and $\varphi_t = 1$ in E, and hence φ_t is an admissible function for (5.22). By (5.23) and the choice of w, we finally obtain

$$
\begin{aligned}
&\left(\inf\left\{\int_\Omega |\nabla u(x)|^p\,dx : u \in \operatorname{Lip}_0(\Omega),\ u = 1 \text{ in } E\right\}\right)^{\frac{1}{p}} \leq \|\nabla\varphi_t\|_{L^p(\mathbb{R}^n)} \\
&\quad\leq \|\nabla\varphi\|_{L^p(\mathbb{R}^n)} + \|\varphi - \varphi_t\|_{W^{1,p}(\mathbb{R}^n)} \\
&\quad\leq \frac{1}{1-\varepsilon}(\operatorname{cap}_p(E,\Omega) + \varepsilon')^{\frac{1}{p}} + \left(1 + \frac{1}{\delta}\right)\varepsilon + \varepsilon.
\end{aligned}
$$

Since δ is independent of ε, the claim follows by letting $\varepsilon \to 0$ and then $\varepsilon' \to 0$. □

5.9. Maz′ya's inequality

Variational capacity can be characterized using quasicontinuous functions; compare to Definition 5.12 and Corollary 5.29.

Definition 5.45. Let $1 < p < \infty$ and let $\Omega \subset \mathbb{R}^n$ be an open set. A function $u \in W_0^{1,p}(\Omega)$ is p-quasicontinuous if the zero extension $v \in W^{1,p}(\mathbb{R}^n)$ of u to entire $\mathbb{R}^n$ is p-quasicontinuous in $\mathbb{R}^n$.

We have a version of Theorem 5.20 for variational capacity. This characterization is applied in Maz′ya's inequality, see Theorem 5.47, which is a capacitary version of the Poincaré inequality.

Theorem 5.46. *Let $1 < p < \infty$, let $\Omega \subset \mathbb{R}^n$ be an open set, and assume that $E \Subset \Omega$. Then*

$$
\operatorname{cap}_p(E,\Omega) = \inf \int_\Omega |\nabla u(x)|^p\,dx,
\tag{5.25}
$$

where the infimum is taken over all p-quasicontinuous functions $u \in W_0^{1,p}(\Omega)$ with $u \geq 1$ p-quasieverywhere in E.

PROOF. Assume first that $u \in W_0^{1,p}(\Omega)$ is admissible for $\operatorname{cap}_p(E,\Omega)$, see Definition 5.32. Let $v \in W^{1,p}(\mathbb{R}^n)$ be the p-quasicontinuous representative of u given by Corollary 5.29. Then $v = u$ almost everywhere in Ω and $v = 0$ holds p-quasieverywhere in $\mathbb{R}^n \setminus \Omega$. Since $u = v$ almost everywhere in Ω and $u \geq 1$ almost everywhere in a neighborhood $G \subset \Omega$ of E, also $v \geq 1$ almost everywhere in G. Lemma 5.18 implies that $v \geq 1$ holds p-quasieverywhere in G, and hence $v \geq 1$ p-quasieverywhere in E. Thus $v|_\Omega \in W_0^{1,p}(\Omega)$ is admissible for (5.25). Since $\nabla u(x) = \nabla v(x) = \nabla v|_\Omega(x)$ for almost every $x \in \Omega$, $\operatorname{cap}_p(E,\Omega)$ is at least the infimum in (5.25).

For the converse inequality, assume that $v \in W_0^{1,p}(\Omega)$ is p-quasicontinuous and $v \geq 1$ p-quasieverywhere in E. We extend v as zero to Ω^c. Then $v \in W^{1,p}(\mathbb{R}^n)$ is p-quasicontinuous in $\mathbb{R}^n$ and, by truncation, we may assume that $0 \leq v(x) \leq 1$ for every $x \in \mathbb{R}^n$. Fix $0 < \varepsilon < 1$ and choose an open set V such that $\operatorname{Cap}_p(V) < \varepsilon^p$, $v = 1$ in $E \setminus V$, and the restriction $v|_{\mathbb{R}^n \setminus V}$ is continuous. By relative topology, there exists an open set $U \subset \mathbb{R}^n$ satisfying

$$\{x \in \mathbb{R}^n \setminus V : v(x) > 1 - \varepsilon\} = U \cap (\mathbb{R}^n \setminus V) = U \setminus V.$$

Then $E \setminus V \subset U \setminus V$ since $v = 1$ on $E \setminus V$.

As $E \Subset \Omega$, we have $d(\Omega^c, E) = \delta > 0$. Define

$$\psi(x) = \max\Big\{0, 1 - \frac{2}{\delta} d(x,E)\Big\},$$

for every $x \in \mathbb{R}^n$. Then $\psi \in \operatorname{Lip}_c(\Omega)$, $0 \leq \psi \leq 1$, $\psi = 1$ in E, and $|\nabla \psi| \leq \frac{2}{\delta}$ for almost every $x \in \Omega$. Let $G = \{x \in \Omega : \psi(x) > \frac{1}{2}\}$. Let $w \in \mathcal{A}_p'(V)$ with $\|w\|_{W^{1,p}(\mathbb{R}^n)} < \varepsilon$ and define

$$u(x) = \frac{v(x)}{1-\varepsilon} + 2\psi(x) w(x)$$

for every $x \in \mathbb{R}^n$. Then $u \in W_0^{1,p}(\Omega)$ and $u \geq 1$ almost everywhere in

$$((U \setminus V) \cup V) \cap G = (U \cup V) \cap G,$$

which is a neighborhood of E. Hence u is an admissible test function for $\operatorname{cap}_p(E,\Omega)$. By Leibniz's rule (2.17) and the triangle inequality we have

$$|\nabla u(x)| \leq \frac{|\nabla v(x)|}{1-\varepsilon} + 2|\nabla w(x)| + 2w(x)|\nabla \psi(x)|,$$

for almost every $x \in \mathbb{R}^n$, and thus

$$\begin{aligned}
\operatorname{cap}_p(E,\Omega)^{\frac{1}{p}} &\leq \frac{1}{1-\varepsilon}\|\nabla v\|_{L^p(\Omega)} + 2\|\nabla w\|_{L^p(\Omega)} + \frac{4}{\delta}\|w\|_{L^p(\Omega)} \\
&\leq \frac{1}{1-\varepsilon}\|\nabla v\|_{L^p(\Omega)} + 4\Big(1 + \frac{1}{\delta}\Big)\|w\|_{W^{1,p}(\mathbb{R}^n)} \\
&\leq \frac{1}{1-\varepsilon}\Big(\int_\Omega |\nabla v(x)|^p\,dx\Big)^{\frac{1}{p}} + 4\Big(1 + \frac{1}{\delta}\Big)\varepsilon.
\end{aligned}$$

The claim follows by letting $\varepsilon \to 0$ and taking infimum over all functions v as above. □

The following Maz′ya's inequality for quasicontinuous representatives provides an important link between capacities and Poincaré inequalities. Note that the claim is not true for an arbitrary function $u \in W^{1,p}_{\mathrm{loc}}(\mathbb{R}^n)$.

THEOREM 5.47. *Let $1 < p < \infty$, let $B(z,r) \subset \mathbb{R}^n$ be a ball, and assume that $u \in W^{1,p}_{\mathrm{loc}}(\mathbb{R}^n)$ is a p-quasicontinuous function. Let $1 \le q \le \frac{np}{n-p}$ for $1 < p < n$, and $1 \le q < \infty$ for $n \le p < \infty$. There exists a constant $C = C(n,p,q)$ such that*

$$\left(\fint_{B(z,r)} |u(x)|^q \, dx\right)^{\frac{p}{q}} \le \frac{C}{\operatorname{cap}_p(\{u=0\} \cap \overline{B}(z,\frac{r}{2}), B(z,r))} \int_{B(z,r)} |\nabla u(x)|^p \, dx.$$

Here $\{u = 0\} = \{y \in \mathbb{R}^n : u(y) = 0\}$.

PROOF. By considering $|u|$ instead of u, we may assume that $u \ge 0$ in $B(z,r)$. Write $B = B(z,r)$ and

$$u_{B,q} = \left(\fint_B u(x)^q \, dx\right)^{\frac{1}{q}} = |B|^{-\frac{1}{q}} \|u\|_{L^q(B)}.$$

By the (q,p)-Poincaré inequality in Theorem 3.14 and Hölder's inequality we have

$$\begin{aligned} u_{B,q} &\le \left(\fint_B |u(x) - u_B|^q \, dx\right)^{\frac{1}{q}} + u_B \\ &\le C(n,p,q) r \left(\fint_B |\nabla u(x)|^p \, dx\right)^{\frac{1}{p}} + \left(\fint_B u(x)^p \, dx\right)^{\frac{1}{p}} < \infty. \end{aligned}$$

If $u_{B,q} = 0$ the claim is true, and thus we may assume that $u_{B,q} > 0$. Let

$$\psi(x) = \max\Big\{0, 1 - \frac{3}{r} d\big(x, B(z,\tfrac{r}{2})\big)\Big\},$$

for every $x \in \mathbb{R}^n$. Then $\psi \in \operatorname{Lip}_c(B(z,r))$, $0 \le \psi \le 1$, $\psi = 1$ in $\overline{B}(z,\frac{r}{2})$, and $|\nabla \psi| \le \frac{3}{r}$ for almost every $x \in B(z,r)$. Let

$$v(x) = \psi(x)\left(1 - \frac{u(x)}{u_{B,q}}\right),$$

for every $x \in \mathbb{R}^n$. Then $v \in W^{1,p}_0(B(z,r))$ is p-quasicontinuous and $v = 1$ in $\{u = 0\} \cap \overline{B}(z,\frac{r}{2})$. The Leibniz rule implies

$$|\nabla v(x)| \le \frac{3}{r} \frac{|u_{B,q} - u(x)|}{u_{B,q}} + \frac{|\nabla u(x)|}{u_{B,q}}$$

for almost every $x \in B(z,r)$ and thus, by Theorem 5.46,

$$\begin{aligned} \operatorname{cap}_p(\{u=0\} \cap \overline{B}(z,\tfrac{r}{2}), B(z,r)) &\le \int_B |\nabla v(x)|^p \, dx \\ &\le \frac{C(p)}{r^p (u_{B,q})^p} \int_B |u(x) - u_{B,q}|^p \, dx + \frac{C(p)}{(u_{B,q})^p} \int_B |\nabla u(x)|^p \, dx. \end{aligned} \tag{5.26}$$

We use Minkowski's inequality and the p-Poincaré inequality, see Theorem 3.14, to estimate the first term on the right-hand side of (5.26), and obtain

$$\begin{aligned} \left(\fint_B |u(x) - u_{B,q}|^p \, dx\right)^{\frac{1}{p}} &\le \left(\fint_B |u(x) - u_B|^p \, dx\right)^{\frac{1}{p}} + |u_{B,q} - u_B| \\ &\le C(n,p) r \left(\fint_B |\nabla u(x)|^p \, dx\right)^{\frac{1}{p}} + |B|^{-\frac{1}{q}} \big| \|u\|_{L^q(B)} - \|u_B\|_{L^q(B)} \big|. \end{aligned} \tag{5.27}$$

By the triangle inequality and the (q,p)-Poincaré inequality (3.8), we have

$$\begin{aligned} |B|^{-\frac{1}{q}}\big|\|u\|_{L^q(B)} - \|u_B\|_{L^q(B)}\big| &\le |B|^{-\frac{1}{q}}\|u-u_B\|_{L^q(B)} \\ &= \Big(\fint_B |u(x)-u_B|^q\,dx\Big)^{\frac{1}{q}} \\ &\le C(n,p,q)r\Big(\fint_B |\nabla u(x)|^p\,dx\Big)^{\frac{1}{p}}. \end{aligned}$$

Together with (5.27) this gives

$$\Big(\fint_B |u(x)-u_{B,q}|^p\,dx\Big)^{\frac{1}{p}} \le C(n,p,q)r\Big(\fint_B |\nabla u(x)|^p\,dx\Big)^{\frac{1}{p}},$$

and thus

$$\int_B |u(x)-u_{B,q}|^p\,dx \le C(n,p,q)r^p\int_B |\nabla u(x)|^p\,dx.$$

Substituting this to (5.26) and recalling that $B=B(z,r)$, we arrive at

$$\operatorname{cap}_p\big(\{u=0\}\cap \overline{B}(z,\tfrac{r}{2}),B(z,r)\big) \le \frac{C(n,p,q)}{(u_{B,q})^p}\int_{B(z,r)} |\nabla u(x)|^p\,dx.$$

The claim follows by reorganizing the terms. □

REMARK 5.48. If $n\le p<\infty$ and $1\le q\le 2p$, then the constant C in Theorem 5.47 can be chosen to be independent of q. This follows immediately from Hölder's inequality on the left-hand side.

5.10. Notes

For the capacity theory, we refer to Adams and Hedberg [**4**], Björn and Björn [**31**, Chapter 6], Evans and Gariepy [**122**, Section 4.7], Diening, Harjulehto, Hästö and Růžička [**104**, Chapter 10], Heinonen, Kilpeläinen and Martio [**187**, Chapter 2], Malý and Ziemer [**303**, Section 2.1], Maz′ya [**318, 321**], Ponce [**350**], Reshetnyak [**354**], Turesson [**388**] and Ziemer [**407**]. See also Kinnunen and Martio [**234**]. The proof of Theorem 5.16 is from Kilpeläinen [**218**], see also Kilpeläinen, Kinnunen and Martio [**219**]. Theorem 5.23 is discussed in Adams and Hedberg [**4**, Section 6.2], Evans and Gariepy [**122**, Section 4.8], Federer and Ziemer [**126**], Malý and Ziemer [**303**, 2.1.8] and Ziemer [**407**, Chapter 3]. See also Kinnunen and Latvala [**229**].

We considered capacity for $p>1$. The case $p=1$ has been studied in Federer and Ziemer [**126**], Hakkarainen and Kinnunen [**177**], Turesson [**388**] and Ziemer [**407**, Section 5.12]. Fleming proved in [**130**] that in this case capacity has the same zero sets as the Hausdorff measure of codimension one. Moreover, capacity is equivalent by two sided estimates to the Hausdorff content of codimension one, see Kinnunen, Korte, Shanmugalingam and Tuominen [**227**], Malý [**300, 301**], Malý, Swanson and Ziemer [**302**], Mäkäläinen [**298**], Turesson [**388**, Theorem 3.5.5] and Ziemer [**407**, Lemma 5.12.3].

Sobolev spaces with zero boundary values are discussed in Adams and Hedberg [**4**, Chapters 10–11], Diening, Harjulehto, Hästö and Růžička [**104**, Section 11.3], Heinonen, Kilpeläinen and Martio [**187**, Chapter 1] and Maz′ya [**318, 321**]. See also Kilpeläinen, Kinnunen and Martio [**219**]. For Corollary 5.29, we refer to Adams and Hedberg [**4**, Theorem 9.1.3]. See also Diening, Harjulehto, Hästö and

Růžička [**104**, Section 11.2], Heinonen, Kilpeläinen and Martio [**187**, Theorem 4.5] and Turesson [**388**, Section 4.4].

Theorem 5.41 is based on Heinonen and Koskela [**188**, Theorem 5.9]. For the connections between capacities and Hausdorff dimension, see also Adams and Hedberg [**4**, Section 5.1], Aikawa and Essén [**8**, Section I.6], Evans and Gariepy [**122**, Section 4.7.2], Heinonen, Kilpeläinen and Martio [**187**, Chapter 2], Maz′ya and Havin [**322**] and Reshetnyak [**354**, Section 3.4]. The proof of Theorem 5.44 is inspired by Björn and Björn [**31**, Theorem 6.19 (x) and Lemma 5.4]. Theorem 5.47 can be found in Maz′ya [**318**, Chapter 10] and [**321**, Chapter 14]. See also Adams and Hedberg [**4**, Section 8.2], Kilpeläinen and Koskela [**220**] and Mikkonen [**327**, Lemma 8.11]. For a general version, see Björn and Björn [**31**, Theorem 6.21].

CHAPTER 6

Hardy's Inequalities

This chapter studies Hardy's inequalities for functions with zero boundary values. A pointwise Hardy inequality captures local phenomena better than an integral Hardy inequality and, by the Hardy–Littlewood–Wiener maximal function theorem, it implies the integral Hardy inequality. Hardy's inequalities hold in every open set for $p > n$, but the corresponding result is not true for $1 < p \leq n$. Several sufficient and necessary conditions for Hardy's inequalities are obtained in Chapter 6, Chapter 7 and Chapter 10.

6.1. Introduction to Hardy's inequalities

We begin with an integral version of Hardy's inequality.

DEFINITION 6.1. Let $1 \leq p < \infty$ and let $\Omega \subsetneq \mathbb{R}^n$ be an open set. We say that the p-Hardy inequality holds in Ω, with a constant C, if

$$\int_\Omega \frac{|u(x)|^p}{d(x,\partial\Omega)^p}\,dx \leq C\int_\Omega |\nabla u(x)|^p\,dx, \tag{6.1}$$

for every $u \in \mathrm{Lip}_0(\Omega)$.

It is essential that the constant in (6.1) is independent of u. This implies that we may consider (6.1) not only for $u \in \mathrm{Lip}_0(\Omega)$ but also for some other classes of functions with zero boundary values.

LEMMA 6.2. *Let $1 \leq p < \infty$ and $0 < C < \infty$, and assume that $\Omega \subsetneq \mathbb{R}^n$ is an open set. The following conditions are equivalent.*

(a) *The p-Hardy inequality holds in Ω with the constant C.*
(b) *Inequality (6.1) holds for every $u \in \mathrm{Lip}_c(\Omega)$ with the constant C.*
(c) *Inequality (6.1) holds for every $u \in W_0^{1,p}(\Omega)$ with the constant C.*

PROOF. Since $\mathrm{Lip}_c(\Omega) \subset W_0^{1,p}(\Omega)$, the implication from (c) to (b) is clear.
Next we show that (b) implies (a). Let $u \in \mathrm{Lip}_0(\Omega)$ and define

$$u_j(x) = \max\{0, |u(x)| - \tfrac{1}{j}\} \tag{6.2}$$

for every $x \in \Omega$ and $j \in \mathbb{N}$. The sequence $(u_j(x))_{j\in\mathbb{N}}$ converges to $|u(x)|$ for every $x \in \Omega$. Moreover, $u_j \in \mathrm{Lip}_c(\Omega)$ and $|\nabla u_j(x)| \leq |\nabla u(x)|$ for almost every $x \in \Omega$ and every $j \in \mathbb{N}$. By Fatou's lemma and (6.1) for u_j, with $j \in \mathbb{N}$, we obtain

$$\begin{aligned}\int_\Omega \frac{|u(x)|^p}{d(x,\partial\Omega)^p}\,dx &\leq \liminf_{j\to\infty}\int_\Omega \frac{|u_j(x)|^p}{d(x,\partial\Omega)^p}\,dx\\ &\leq C\liminf_{j\to\infty}\int_\Omega |\nabla u_j(x)|^p\,dx \leq C\int_\Omega |\nabla u(x)|^p\,dx.\end{aligned} \tag{6.3}$$

To complete the proof, we show that (a) implies (c). To this end, let $u \in W_0^{1,p}(\Omega)$ and consider a sequence of functions $u_j \in C_0^\infty(\Omega) \subset \operatorname{Lip}_0(\Omega)$, with $j \in \mathbb{N}$, such that $u_j \to u$ in $W^{1,p}(\Omega)$ as $j \to \infty$. By passing to a subsequence, we may assume that $u_j(x) \to u(x)$ for almost every $x \in \Omega$. An application of Fatou's lemma, as in (6.3), gives (6.1) for u with the constant C. □

REMARK 6.3. The proof of Lemma 6.2 shows that the p-Hardy inequality holds in Ω if and only if (6.1) holds for every $u \in C_0^\infty(\Omega)$.

Next we consider a pointwise variant of Hardy's inequality, in terms of the restricted Hardy–Littlewood maximal operator from Definition 1.27.

DEFINITION 6.4. Let $1 \le p < \infty$ and let $\Omega \subsetneq \mathbb{R}^n$ be an open set. We say that the pointwise p-Hardy inequality holds in Ω, with a constant C, if

$$|u(x)| \le C d(x, \partial\Omega)\big(M_{2d(x,\partial\Omega)}|\nabla u|^p(x)\big)^{\frac{1}{p}}, \tag{6.4}$$

for every $x \in \Omega$ and every $u \in \operatorname{Lip}_0(\Omega)$.

As in (6.1) it is essential that the constant in (6.4) is independent of u.

REMARK 6.5. Let $1 \le q < p < \infty$ and assume that the pointwise q-Hardy inequality holds in an open set $\Omega \subsetneq \mathbb{R}^n$. Hölder's inequality implies that also the pointwise p-Hardy inequality holds in Ω.

REMARK 6.6. Let $1 \le q < p < \infty$ and let $\Omega \subsetneq \mathbb{R}^n$ be an open set. It follows from the maximal function theorem that if the pointwise q-Hardy inequality holds in Ω, then also the p-Hardy inequality holds in Ω. To see this, let $u \in \operatorname{Lip}_0(\Omega)$ and assume that (6.4) holds for u with a constant C_1 and the exponent q. Since $\frac{p}{q} > 1$, Theorem 1.15 (c) for the restricted maximal function implies

$$\begin{aligned}\int_\Omega \frac{|u(x)|^p}{d(x,\partial\Omega)^p}\,dx &\le C_1^p \int_\Omega \big(M_{2d(x,\partial\Omega)}|\nabla u|^q(x)\big)^{\frac{p}{q}}\,dx \\ &\le C(n,p,q,C_1)\int_\Omega \big(|\nabla u(x)|^q\big)^{\frac{p}{q}}\,dx \\ &= C(n,p,q,C_1)\int_\Omega |\nabla u(x)|^p\,dx.\end{aligned} \tag{6.5}$$

In (6.5) we used the fact that u vanishes in $\Omega^c = \mathbb{R}^n \setminus \Omega$ and thus $\nabla u = 0$ almost everywhere in Ω^c.

The reasoning in (6.5) does not apply for $q = p$ since the maximal operator is not bounded on $L^1(\mathbb{R}^n)$. However, it is true that the pointwise p-Hardy inequality implies the p-Hardy inequality, but the proof is more complicated. We discuss this question in Section 6.5, see Corollary 6.26.

The pointwise p-Hardy inequality for $p > n$ is an easy consequence of Morrey's inequality, see Theorem 3.19.

THEOREM 6.7. *Let $n < p < \infty$ and let $\Omega \subsetneq \mathbb{R}^n$ be an open set. The pointwise p-Hardy inequality and the p-Hardy inequality hold in Ω with a constant $C = C(n,p)$.*

PROOF. Let $u \in \operatorname{Lip}_0(\Omega)$ and $x \in \Omega$, and choose $z \in \partial\Omega$ with $|x-z| = d(x,\partial\Omega)$. Since $u(z) = 0$, Theorem 3.19 implies

$$
\begin{aligned}
|u(x)| = |u(x) - u(z)| &\le C(n,p) 2d(x,\partial\Omega)\left(\fint_{B(x,2d(x,\partial\Omega))} |\nabla u(y)|^p \, dy\right)^{\frac{1}{p}} \\
&\le C(n,p) d(x,\partial\Omega)\big(M_{2d(x,\partial\Omega)}|\nabla u|^p(x)\big)^{\frac{1}{p}}.
\end{aligned}
$$

This shows that the pointwise p-Hardy inequality holds in Ω with a constant $C = C(n,p)$. In particular, the pointwise q-Hardy inequality holds in Ω for $q = \frac{n+p}{2} > n$ with a constant $C = C(n,q) = C(n,p)$. Since $q < p$, also the p-Hardy inequality holds in Ω by Remark 6.6. $\square$

Pointwise Hardy inequalities also hold for Sobolev functions with zero boundary values.

LEMMA 6.8. *Let $1 \le p < \infty$. Assume that the pointwise p-Hardy inequality holds in an open set $\Omega \subsetneq \mathbb{R}^n$ with a constant C and let $u \in W_0^{1,p}(\Omega)$. Then inequality (6.4) holds for the zero extension of u at almost every $x \in \Omega$ with the same constant C.*

PROOF. By considering the zero extension of u to Ω^c as in Theorem 2.31, we may assume that $u \in W^{1,p}(\mathbb{R}^n)$. Since $\operatorname{Lip}_c(\Omega)$ is dense in $W_0^{1,p}(\Omega)$, there exist functions $u_j \in \operatorname{Lip}_c(\Omega)$ with $j \in \mathbb{N}$ such that $u_j \to u$ in $W^{1,p}(\mathbb{R}^n)$ as $j \to \infty$. By Minkowski's inequality

$$
\begin{aligned}
\big(M_{2d(x,\partial\Omega)}|\nabla u_j|^p(x)\big)^{\frac{1}{p}} &= \big(M_{2d(x,\partial\Omega)}|\nabla u + (\nabla u_j - \nabla u)|^p(x)\big)^{\frac{1}{p}} \\
&\le \big(M_{2d(x,\partial\Omega)}|\nabla u|^p(x)\big)^{\frac{1}{p}} + \big(M_{2d(x,\partial\Omega)}|\nabla u_j - \nabla u|^p(x)\big)^{\frac{1}{p}}
\end{aligned}
$$

for every $x \in \Omega$. Theorem 1.15 (b) implies

$$
\big|\{x \in \mathbb{R}^n : M_{2d(x,\partial\Omega)}|\nabla u_j - \nabla u|^p(x) > \varepsilon^p\}\big| \le \frac{C(n)}{\varepsilon^p} \int_{\mathbb{R}^n} |\nabla u_j(x) - \nabla u(x)|^p \, dx
$$

for every $\varepsilon > 0$. Thus

$$
\big(M_{2d(x,\partial\Omega)}|\nabla u_j - \nabla u|^p(x)\big)^{\frac{1}{p}} \to 0
$$

in measure as $j \to \infty$. By passing twice to a subsequence, if necessary, we obtain a set $E \subset \Omega$ with $|E| = 0$ such that $u_j(x) \to u(x) < \infty$ as $j \to \infty$ and

$$
\lim_{j\to\infty} \big(M_{2d(x,\partial\Omega)}|\nabla u_j - \nabla u|^p(x)\big)^{\frac{1}{p}} = 0
$$

for every $x \in \Omega \setminus E$. Since $\operatorname{Lip}_c(\Omega) \subset \operatorname{Lip}_0(\Omega)$, the pointwise p-Hardy inequality and the estimates above give

$$
\begin{aligned}
|u(x)| = \lim_{j\to\infty} |u_j(x)| &\le C \limsup_{j\to\infty} d(x,\partial\Omega)\big(M_{2d(x,\partial\Omega)}|\nabla u_j|^p(x)\big)^{\frac{1}{p}} \\
&\le C d(x,\partial\Omega)\big(M_{2d(x,\partial\Omega)}|\nabla u|^p(x)\big)^{\frac{1}{p}}
\end{aligned}
$$

for every $x \in \Omega \setminus E$. $\square$

REMARK 6.9. Consider the Rayleigh quotient

$$
\lambda = \lambda(p,\Omega) = \inf \frac{\int_\Omega |\nabla u(x)|^p \, dx}{\int_\Omega \frac{|u(x)|^p}{d(x,\partial\Omega)^p} \, dx}, \tag{6.6}
$$

where the infimum is taken over all $u \in W_0^{1,p}(\Omega)$ such that u is not zero almost everywhere. This variational problem is related to the p-Hardy inequality since $\lambda > 0$ if and only if the p-Hardy inequality holds in Ω. Note that $u \in W_0^{1,p}(\Omega)$ is a minimizer of (6.6) if and only if u is a weak solution to the nonlinear eigenvalue problem

$$\operatorname{div}(|\nabla u(x)|^{p-2}\nabla u(x)) + \lambda \frac{|u(x)|^{p-2}u(x)}{d(x,\partial\Omega)^p} = 0. \tag{6.7}$$

The p-Laplace operator $\operatorname{div}(|\nabla u(x)|^{p-2}\nabla u(x))$ will be discussed in Chapter 11 and Chapter 12. This approach to Hardy inequalities is considered, for example, by Barbatis, Filippas and Tertikas [**23–25**], Brezis, Marcus, Mizel, Pinchover and Shafrir [**57, 58, 305, 306**], Davies [**96**], Matskewich and Sobolevskii [**314**], Pinchover and Tintarev [**347–349**] and Tidblom [**380, 381**].

REMARK 6.10. The origins of Hardy's inequalities are in the one-dimensional case. Let $1 < p < \infty$, $\beta \neq p-1$ and $u \in \operatorname{Lip}_0(0,\infty)$. Then

$$\int_0^\infty \frac{|u(x)|^p}{x^p}\, x^\beta \, dx \le \frac{p^p}{|p-1-\beta|^p} \int_0^\infty |u'(x)|^p \, x^\beta \, dx, \tag{6.8}$$

where the constant on the right-hand side is the best possible. The proof of this weighted one-dimensional Hardy inequality is based on integration by parts and Hölder's inequality. We refer to Hardy, Littlewood and Pólya [**180**, § 330] and Kufner [**248**, Theorem 5.2] for more details.

The inequality in (6.8) implies the p-Hardy inequality for $\Omega = \mathbb{R}^n \setminus \{0\}$; this will be applied in Example 6.13 below. Let $1 < p < \infty$ with $p \neq n$, let $u \in \operatorname{Lip}_0(\mathbb{R}^n \setminus \{0\})$, and write $x \in \mathbb{R}^n$ as $x = rz$, where $0 < r = |x| < \infty$ and $z \in \partial B(0,1)$. By (3.1), Fubini's theorem, and (6.8) with $\beta = n-1 \neq p-1$, we obtain

$$\begin{aligned}
\int_{\mathbb{R}^n} \frac{|u(x)|^p}{|x|^p}\, dx &= \int_0^\infty \int_{\partial B(0,r)} \frac{|u(x)|^p}{r^p} \, d\sigma(x)\, dr \\
&= \int_{\partial B(0,1)} \int_0^\infty \frac{|u(rz)|^p}{r^p} r^{n-1} \, dr \, d\sigma(z) \\
&\le C \int_{\partial B(0,1)} \int_0^\infty \left|\tfrac{\partial}{\partial r} u(rz)\right|^p r^{n-1} \, dr \, d\sigma(z) \\
&\le C \int_{\mathbb{R}^n} |\nabla u(x)|^p \, dx,
\end{aligned} \tag{6.9}$$

where $C = C(n,p)$. In the final step we used again Fubini's theorem, (3.1), and the fact that $\left|\frac{\partial}{\partial r} u(rz)\right| \le |\nabla u(rz)| = |\nabla u(x)|$ for every $x = rz \in \mathbb{R}^n$.

6.2. Measure density and Hardy's inequality

In contrast with the case $p > n$ in Theorem 6.7, for $1 \le p \le n$ there exist open sets where the p-Hardy inequality, or the pointwise p-Hardy inequality, does not hold, see Example 6.13. The following geometric property is a sufficient condition for the Hardy inequality.

DEFINITION 6.11. A set $E \subset \mathbb{R}^n$ satisfies the measure density condition if there exists a constant C, with $0 < C \le 1$, such that

$$|E \cap B(x,r)| \ge C|B(x,r)| \tag{6.10}$$

for every $x \in E$ and $r > 0$.

THEOREM 6.12. *Let $1 < p < \infty$ and assume that $\Omega \subsetneq \mathbb{R}^n$ is an open set. If Ω^c satisfies the measure density condition, with a constant C_1, then the pointwise p-Hardy inequality and the p-Hardy inequality hold in Ω with a constant $C = C(n,p,C_1)$.*

PROOF. Let $u \in \operatorname{Lip}_0(\Omega)$ and $x \in \Omega$. Choose a point $z \in \partial\Omega$ with $|x-z| = d(x,\partial\Omega)$ and let $R = d(x,\partial\Omega)$. Then

$$|u(x)| \le |u(x) - u_{B(x,R)}| + |u_{B(x,R)} - u_{B(z,R)}| + |u_{B(z,R)}|. \tag{6.11}$$

For the first term on the right-hand side we apply Theorem 3.4, inequality (4.1) and Remark 1.29, as in (4.2), and obtain

$$|u(x) - u_{B(x,R)}| \le C(n)R\big(M_{2R}|\nabla u|^p(x)\big)^{\frac{1}{p}}. \tag{6.12}$$

For the second term on the right-hand side of (6.11) we use the Poincaré inequality in Theorem 3.14 and the inclusion $B(x,R) \cup B(z,R) \subset B(x,2R)$. This gives

$$\begin{aligned}|u_{B(x,R)} - u_{B(z,R)}| &\le |u_{B(x,R)} - u_{B(x,2R)}| + |u_{B(z,R)} - u_{B(x,2R)}| \\ &\le C(n) \fint_{B(x,2R)} |u(y) - u_{B(x,2R)}|\,dy \\ &\le C(n,p)R\Big(\fint_{B(x,2R)} |\nabla u(y)|^p\,dy\Big)^{\frac{1}{p}} \\ &\le C(n,p)R\big(M_{2R}|\nabla u|^p(x)\big)^{\frac{1}{p}}.\end{aligned}$$

The measure density assumption is applied to estimate the last term on the right-hand side of (6.11). Here we employ the same idea as in the proof of Theorem 3.17, see also Remark 3.18. Let $B = B(z,R)$. By Minkowski's inequality, the p-Poincaré inequality, and Hölder's inequality, we obtain

$$\begin{aligned}\Big(\fint_B |u(y)|^p\,dy\Big)^{\frac{1}{p}} &\le \Big(\fint_B |u(y) - u_B|^p\,dy\Big)^{\frac{1}{p}} + |u_B| \\ &\le C(n,p)R\Big(\fint_B |\nabla u(y)|^p\,dy\Big)^{\frac{1}{p}} + \frac{1}{|B|}\int_{B\setminus\Omega^c} |u(y)|\,dy \\ &\le C(n,p)R\Big(\fint_B |\nabla u(y)|^p\,dy\Big)^{\frac{1}{p}} + \Big(\frac{|B\setminus\Omega^c|}{|B|}\Big)^{1-\frac{1}{p}}\Big(\fint_B |u(y)|^p\,dy\Big)^{\frac{1}{p}}.\end{aligned} \tag{6.13}$$

The measure density condition, with $0 < C_1 \le 1$, implies

$$|B\setminus\Omega^c| = |B| - |\Omega^c \cap B| \le (1-C_1)|B|,$$

and thus

$$\Big(\frac{|B\setminus\Omega^c|}{|B|}\Big)^{1-\frac{1}{p}} \le (1-C_1)^{1-\frac{1}{p}} < 1.$$

The last term in (6.13) can be absorbed to the left-hand side. Since $B = B(z,R) \subset B(x,2R)$, we obtain

$$\begin{aligned}|u_B| &\le \Big(\fint_B |u(y)|^p\,dy\Big)^{\frac{1}{p}} \le C(n,p,C_1)R\Big(\fint_B |\nabla u(y)|^p\,dy\Big)^{\frac{1}{p}} \\ &\le C(n,p,C_1)R\Big(\fint_{B(x,2R)} |\nabla u(y)|^p\,dy\Big)^{\frac{1}{p}} \le C(n,p,C_1)R\big(M_{2R}|\nabla u|^p(x)\big)^{\frac{1}{p}}.\end{aligned}$$

Combining the estimates above for the right-hand side of (6.11) we conclude that

$$|u(x)| \leq C(n,p,C_1)d(x,\partial\Omega)\big(M_{2d(x,\partial\Omega)}|\nabla u|^p(x)\big)^{\frac{1}{p}}.$$

This proves that the pointwise p-Hardy inequality holds in Ω with a constant $C = C(n,p,C_1)$. By the argument above, the pointwise q-Hardy inequality holds in Ω for $q = \frac{p+1}{2} \in (1,p)$ with a constant $C = C(n,q,C_1) = C(n,p,C_1)$. Since $q < p$, also the p-Hardy inequality holds in Ω by Remark 6.6. $\square$

In Section 6.4 we obtain more sensitive sufficient conditions for Hardy's inequalities based on a capacity density condition, which is a refined version of the measure density condition. Next example shows that there exist open sets for which Hardy's inequality does not hold.

EXAMPLE 6.13. Let $\Omega = B(0,1) \setminus \{0\} \subset \mathbb{R}^n$. We show that the p-Hardy inequality holds in Ω when $1 < p < n$ or $p > n$, but the n-Hardy inequality does not hold in Ω. For $p > n$, the p-Hardy inequality follows immediately from Theorem 6.7. Thus we may assume that $1 < p < n$. Let $u \in \operatorname{Lip}_0(\Omega)$. Since $B(0,1)^c$ satisfies the measure density condition with $C = \frac{1}{2}$ and $u \in \operatorname{Lip}_0(B(0,1))$, the p-Hardy inequality with respect to $B(0,1)$ holds for u with a constant $C = C(n,p)$ by Theorem 6.12. On the other hand, Remark 6.10 implies that the p-Hardy inequality (6.9) holds for $u \in \operatorname{Lip}_0(\Omega) \subset \operatorname{Lip}_0(\mathbb{R}^n \setminus \{0\})$. Since $d(x,\partial\Omega) = \min\{|x|, 1-|x|\}$ for every $x \in \Omega$, where $1-|x| = d(x,\partial B(0,1))$, and $|\nabla u(x)| = 0$ for almost every $x \in \Omega^c$, we conclude that

$$\begin{aligned}\int_\Omega \frac{|u(x)|^p}{d(x,\partial\Omega)^p}\,dx &\leq \int_{B(0,1)} \frac{|u(x)|^p}{(1-|x|)^p}\,dx + \int_{\mathbb{R}^n} \frac{|u(x)|^p}{|x|^p}\,dx \\ &\leq C(n,p)\Big(\int_{B(0,1)} |\nabla u(x)|^p\,dx + \int_{\mathbb{R}^n} |\nabla u(x)|^p\,dx\Big) \\ &\leq C(n,p)\int_\Omega |\nabla u(x)|^p\,dx.\end{aligned}$$

Next we show that the n-Hardy inequality does not hold in Ω. Let $A_i = B(0,2^{-i}) \setminus B(0,2^{-i-1})$ with $i \in \mathbb{N}$. For $j \in \mathbb{N}$, consider $u_j \in \operatorname{Lip}_0(\Omega)$ defined as

$$u_j(x) = \begin{cases} 1 - 2^{j+1}d(x,A_j), & |x| \leq 2^{-j-1} \\ 1, & 2^{-j-1} < |x| < \frac{1}{2}, \\ \max\{0, 1-2d(x,A_1)\}, & |x| \geq \frac{1}{2}. \end{cases}$$

Then

$$\int_\Omega \frac{|u_j(x)|^n}{d(x,\partial\Omega)^n}\,dx \geq \sum_{i=1}^{j}\int_{A_i} \frac{1}{d(x,\partial\Omega)^n}\,dx \geq C(n)\sum_{i=1}^{j} 2^{-in}2^{in} = C(n)j \xrightarrow{j\to\infty} \infty,$$

while the left-hand side of the n-Hardy inequality remains bounded since

$$\begin{aligned}\int_\Omega |\nabla u_j(x)|^n\,dx &\leq \int_\Omega 2^n\,dx + \int_{B(0,2^{-j-1})} (2^{j+1})^n\,dx \\ &\leq C(n) + C(n)2^{-(j+1)n}2^{(j+1)n} = C(n) < \infty\end{aligned}$$

for every $j \in \mathbb{N}$. Thus the n-Hardy inequality does not hold in Ω.

Example 6.13 illustrates general features related to Hardy's inequality that will be discussed more in the following sections and chapters. For instance, the p-Hardy inequality holds in Ω if the complement Ω^c is sufficiently large or sufficiently small, or an appropriate combination of large and small parts. Hardy's inequality in these three cases will be considered in Theorem 7.17, Corollary 10.36 and Theorem 10.40, respectively. On the other hand, if Ω^c contains an isolated $(n-p)$-dimensional part, then the p-Hardy inequality does not hold in Ω. This occurs in Example 6.13 with the 0-dimensional part $\{0\} \subset \Omega^c$. A precise necessary condition is given in Theorem 10.46, and Example 10.49 generalizes the idea in Example 6.13.

6.3. Self-improvement of Hardy's inequality

If the p-Hardy inequality holds in an open set $\Omega \subsetneq \mathbb{R}^n$ for some $1<p<\infty$, then it is not obvious whether the q-Hardy inequality holds in Ω for some $q \neq p$. In this respect there is a difference between integral and pointwise Hardy inequalities since the pointwise p-Hardy inequality implies pointwise q-Hardy inequalities for every $p<q<\infty$, see Remark 6.5. On the other hand, Example 6.13 shows that the p-Hardy inequality does not necessarily imply the q-Hardy inequality for every $p<q<\infty$, nor for every $1 \le q < p$.

In this section we show that if the p-Hardy inequality holds in Ω, for some $1<p<\infty$, then there exists $\varepsilon>0$ such that also the q-Hardy inequality holds in Ω for $p-\varepsilon<q<p+\varepsilon$, see Theorem 6.16. The proofs of the cases $q>p$ and $q<p$ are rather different and we consider them separately.

LEMMA 6.14. *Let $1<p<\infty$ and assume that the p-Hardy inequality holds in an open set $\Omega \subsetneq \mathbb{R}^n$ with a constant C_1. There exists $\varepsilon=\varepsilon(p,C_1)>0$ such that the q-Hardy inequality holds in Ω whenever $p<q<p+\varepsilon$, with a constant $C=C(p,C_1)$.*

PROOF. By Lemma 6.2 (b), we may assume that $u \in \operatorname{Lip}_c(\Omega)$. Let $\varepsilon>0$ and define

$$v(x)=|u(x)|^{1+\frac{\varepsilon}{p}} d(x,\partial\Omega)^{-\frac{\varepsilon}{p}}$$

for every $x \in \Omega$. Then $v \in \operatorname{Lip}_c(\Omega)$, and by the Leibniz rule (2.4) and the chain rule (2.5) we obtain

$$|\nabla v(x)| \le \Big(1+\frac{\varepsilon}{p}\Big)|u(x)|^{\frac{\varepsilon}{p}}|\nabla u(x)|\,d(x,\partial\Omega)^{-\frac{\varepsilon}{p}}+\frac{\varepsilon}{p}|u(x)|^{1+\frac{\varepsilon}{p}} d(x,\partial\Omega)^{-\frac{\varepsilon}{p}-1}$$

for almost every $x \in \Omega$. Here we also used the fact that $|\nabla d(x,\Omega)| \le 1$ for almost every $x \in \Omega$. The p-Hardy inequality for v and the estimate above for $|\nabla v(x)|$ give

$$\begin{aligned}\int_\Omega\Big(\frac{|u(x)|}{d(x,\partial\Omega)}\Big)^{p+\varepsilon}dx &= \int_\Omega\Big(\frac{|v(x)|}{d(x,\partial\Omega)}\Big)^{p}dx \le C_1\int_\Omega|\nabla v(x)|^p\,dx\\ &\le 2^pC_1\Big(1+\frac{\varepsilon}{p}\Big)^p\int_\Omega|u(x)|^{\varepsilon}|\nabla u(x)|^p d(x,\partial\Omega)^{-\varepsilon}\,dx\\ &\quad+2^pC_1\Big(\frac{\varepsilon}{p}\Big)^p\int_\Omega|u(x)|^{p+\varepsilon}d(x,\partial\Omega)^{-p-\varepsilon}\,dx.\end{aligned} \tag{6.14}$$

By choosing $\varepsilon<\frac{p}{2}C_1^{-\frac{1}{p}}$ we have $2^pC_1(\frac{\varepsilon}{p})^p<1$, and we can absorb the last term in (6.14) to the left-hand side. Note that this term is finite for $u \in \operatorname{Lip}_c(\Omega)$. Let

$\alpha = 1 + \frac{\varepsilon}{p}$. Then $\alpha p = p + \varepsilon = \frac{\varepsilon\alpha}{\alpha - 1}$, and Hölder's inequality gives

$$
\begin{aligned}
&\Big(1 - 2^p C_1 \Big(\frac{\varepsilon}{p}\Big)^p\Big) \int_\Omega \Big(\frac{|u(x)|}{d(x,\partial\Omega)}\Big)^{p+\varepsilon} dx \\
&\quad \le 2^p C_1 \alpha^p \int_\Omega |\nabla u(x)|^p |u(x)|^\varepsilon d(x,\partial\Omega)^{-\varepsilon}\, dx \\
&\quad \le 2^p C_1 \alpha^p \Big(\int_\Omega |\nabla u(x)|^{\alpha p}\, dx\Big)^{\frac{1}{\alpha}} \Big(\int_\Omega |u(x)|^{\frac{\varepsilon\alpha}{\alpha-1}} d(x,\partial\Omega)^{\frac{-\varepsilon\alpha}{\alpha-1}}\, dx\Big)^{\frac{\alpha-1}{\alpha}} \\
&\quad = 2^p C_1 \alpha^p \Big(\int_\Omega |\nabla u(x)|^{p+\varepsilon}\, dx\Big)^{\frac{1}{\alpha}} \Big(\int_\Omega \Big(\frac{|u(x)|}{d(x,\partial\Omega)}\Big)^{p+\varepsilon} dx\Big)^{1-\frac{1}{\alpha}}.
\end{aligned} \tag{6.15}
$$

This implies

$$
\int_\Omega \Big(\frac{|u(x)|}{d(x,\partial\Omega)}\Big)^{p+\varepsilon} dx \le \Big(\frac{2^p C_1 \alpha^p}{1 - 2^p C_1 (\frac{\varepsilon}{p})^p}\Big)^\alpha \int_\Omega |\nabla u(x)|^{p+\varepsilon}\, dx.
$$

We conclude that if $0 < \varepsilon < \frac{p}{2} C_1^{-\frac{1}{p}}$, then the $(p+\varepsilon)$-Hardy inequality holds in Ω with the constant

$$
C(p,\varepsilon,C_1) = \Big(\frac{2^p C_1 (1 + \frac{\varepsilon}{p})^p}{1 - 2^p C_1 (\frac{\varepsilon}{p})^p}\Big)^{1+\frac{\varepsilon}{p}}. \tag{6.16}
$$

From this expression we see that for $0 < \varepsilon < \frac{p}{4} C_1^{-\frac{1}{p}}$ the constant in the $(p+\varepsilon)$-Hardy inequality can be chosen to be independent of ε. □

Next we consider self-improvement for $q < p$. The proof is based on Lipschitz truncation and maximal function techniques, similar to the proof of Theorem 4.11, and integration over distribution sets. This approach is applied for self-improvement properties several times, see the proofs of Theorem 7.32, Theorem 8.11, Lemma 8.24, Theorem 8.43 and Theorem 12.23.

LEMMA 6.15. *Let $1 < p < \infty$ and assume that the p-Hardy inequality holds in an open set $\Omega \subsetneq \mathbb{R}^n$ with a constant C_1. There exists $\varepsilon = \varepsilon(n,p,C_1) > 0$ such that the q-Hardy inequality holds in Ω whenever $p - \varepsilon < q < p$, with a constant $C = C(n,p,C_1)$.*

PROOF. Let $u \in \operatorname{Lip}_0(\Omega)$. We may assume that u is not identically zero. For $t > 0$, define $F_t = G_t \cap H_t$, where

$$
G_t = \{x \in \Omega : M|\nabla u|(x) \le t\} \quad \text{and} \quad H_t = \{x \in \Omega : |u(x)| \le t d(x,\partial\Omega)\}.
$$

The pointwise estimate in Theorem 4.1 implies

$$
|u(x) - u(y)| \le C(n)|x-y|\big(M|\nabla u|(x) + M|\nabla u|(y)\big) \le C(n)t|x-y|
$$

for every $x, y \in F_t$. For $x \in F_t$ and $y \in \Omega^c$, we have

$$
|u(x) - u(y)| = |u(x)| \le t d(x,\partial\Omega) \le t|x-y|.
$$

Since u vanishes in Ω^c, we conclude from the estimates above that the restriction $u|_{F_t\cup\Omega^c}$ is a $C(n)t$-Lipschitz function. We use Theorem 2.7 and extend $u|_{F_t\cup\Omega^c}$ to a $C(n)t$-Lipschitz function $\widetilde{u}$ in $\mathbb{R}^n$. Then $\widetilde{u} = u = 0$ in Ω^c. Moreover, the support of $\widetilde{u}$ is bounded. To see this, fix $x_0 \in \operatorname{supp} u$. There exists $R = R(t,u) > 0$ such that if $x \in \Omega \setminus B(x_0,R)$, then $u(x) = 0$ and $x \in F_t$. Here the inclusion $x \in G_t$ is a consequence of the fact that $|\nabla u| = |\nabla u|\chi_{\operatorname{supp} u} \in L^1(\mathbb{R}^n)$. Thus $\widetilde{u} = u = 0$ outside $B(x_0, R)$, and this proves the boundedness of $\operatorname{supp} \widetilde{u}$. It follows that $\widetilde{u} \in \operatorname{Lip}_0(\Omega)$.

Since the p-Hardy inequality holds for $\widetilde{u}$ with the constant C_1 and

$$|\nabla \widetilde{u}(x)| \leq |\nabla u(x)|\chi_{F_t}(x) + C(n)t\chi_{\Omega\setminus F_t}(x)$$

for almost every $x \in \Omega$, we obtain

$$\begin{aligned}\int_{F_t} \left(\frac{|u(x)|}{d(x,\partial\Omega)}\right)^p dx &= \int_{F_t} \left(\frac{|\widetilde{u}(x)|}{d(x,\partial\Omega)}\right)^p dx \\ &\leq \int_{\Omega} \left(\frac{|\widetilde{u}(x)|}{d(x,\partial\Omega)}\right)^p dx \leq C_1 \int_{\Omega} |\nabla \widetilde{u}(x)|^p \, dx \\ &\leq C_1 \int_{F_t} |\nabla u(x)|^p \, dx + C_1 C(n,p) \int_{\Omega\setminus F_t} t^p \, dx.\end{aligned}$$

Thus

$$\begin{aligned}\int_{H_t} \left(\frac{|u(x)|}{d(x,\partial\Omega)}\right)^p dx &\leq C_1 \int_{F_t} |\nabla u(x)|^p \, dx + C_1 C(n,p) \int_{\Omega\setminus F_t} t^p \, dx \\ &\quad + \int_{H_t\setminus F_t} \left(\frac{|u(x)|}{d(x,\partial\Omega)}\right)^p dx \\ &\leq C_1 \int_{G_t} |\nabla u(x)|^p \, dx + C_1 C(n,p) \int_{\Omega\setminus H_t} t^p \, dx \\ &\quad + (1 + C_1 C(n,p)) \int_{\Omega\setminus G_t} t^p \, dx,\end{aligned} \tag{6.17}$$

where we have used the definition of H_t and the inclusions $\Omega \setminus F_t \subset (\Omega\setminus H_t) \cup (\Omega\setminus G_t)$ and $H_t \setminus F_t \subset \Omega \setminus G_t$.

Next we multiply (6.17) by $t^{-\varepsilon-1}$, where $0 < \varepsilon < p-1$, and then integrate with respect to t over $(0,\infty)$. After changing the order of integration on the left-hand side, this leads to

$$\begin{aligned}&\frac{1}{\varepsilon} \int_{\Omega} \left(\frac{|u(x)|}{d(x,\partial\Omega)}\right)^{p-\varepsilon} dx \\ &\quad \leq C_1 \int_0^\infty t^{-\varepsilon-1} \int_{G_t} |\nabla u(x)|^p \, dx \, dt + C_1 C(n,p) \int_0^\infty t^{p-\varepsilon-1} |\Omega \setminus H_t| \, dt \\ &\quad + (1 + C_1 C(n,p)) \int_0^\infty t^{p-\varepsilon-1} |\Omega \setminus G_t| \, dt.\end{aligned} \tag{6.18}$$

Since $|\nabla u(x)| \leq M|\nabla u|(x)$ for almost every $x \in \Omega$, see Corollary 1.23, the first integral on the right-hand side of (6.18) can be estimated as

$$\begin{aligned}\int_0^\infty t^{-\varepsilon-1} \int_{G_t} |\nabla u(x)|^p \, dx \, dt &\leq \int_0^\infty t^{-\varepsilon-1} \int_{\{x\in\Omega : |\nabla u(x)| < t\}} |\nabla u(x)|^p \, dx \, dt \\ &= \frac{1}{\varepsilon} \int_{\Omega} |\nabla u(x)|^{p-\varepsilon} \, dx.\end{aligned}$$

For the second integral on the right-hand side of (6.18), we have

$$\int_0^\infty t^{p-\varepsilon-1} |\Omega \setminus H_t| \, dt = \frac{1}{p-\varepsilon} \int_{\Omega} \left(\frac{|u(x)|}{d(x,\partial\Omega)}\right)^{p-\varepsilon} dx.$$

Finally, we estimate the last integral on the right-hand side of (6.18) by using the definition of G_t and Theorem 1.15 (c) with $p-\varepsilon>1$, and obtain

$$\int_0^\infty t^{p-\varepsilon-1}|\Omega\setminus G_t|\,dt=\frac{1}{p-\varepsilon}\int_\Omega\big(M|\nabla u|(x)\big)^{p-\varepsilon}\,dx\le\frac{C_2}{p-\varepsilon}\int_\Omega|\nabla u(x)|^{p-\varepsilon}\,dx,$$

where $C_2=C_2(n,p,\varepsilon)=5^n2^{p-\varepsilon}\frac{p-\varepsilon}{p-\varepsilon-1}$. In the final step, we also used the fact that u vanishes in $\Omega^c=\mathbb{R}^n\setminus\Omega$ and thus $\nabla u=0$ almost everywhere in Ω^c.

By choosing $0<\varepsilon<p-1$ so small that $(1+C_1C(n,p))\frac{\varepsilon}{p-\varepsilon}\le\frac12$, it follows from (6.18) and the previous estimates that

$$\begin{aligned}\int_\Omega\left(\frac{|u(x)|}{d(x,\partial\Omega)}\right)^{p-\varepsilon}dx&\le2\Big(C_1+(1+C_1C(n,p))C_2\frac{\varepsilon}{p-\varepsilon}\Big)\int_\Omega|\nabla u(x)|^{p-\varepsilon}\,dx\\&\le(2C_1+C_2)\int_\Omega|\nabla u(x)|^{p-\varepsilon}\,dx.\end{aligned}$$

If we in addition require that $\varepsilon\le\frac{p-1}{2}$, we may assume that the constant C_2 above only depends on p and n. Thus there exists $\varepsilon_0=\varepsilon_0(n,p,C_1)>0$ such that if $0<\varepsilon\le\varepsilon_0$, then the $(p-\varepsilon)$-Hardy inequality holds in Ω with the constant $C=C(n,p,C_1)=2C_1+C_2$. □

Combination of the previous lemmas gives two-sided improvement for Hardy's inequality.

THEOREM 6.16. *Let $1<p<\infty$ and assume that the p-Hardy inequality holds in an open set $\Omega\subsetneq\mathbb{R}^n$ with a constant C_1. Then there exists $\varepsilon=\varepsilon(n,p,C_1)>0$ such that the q-Hardy inequality holds in Ω whenever $p-\varepsilon<q<p+\varepsilon$, with a constant $C=C(n,p,C_1)$.*

PROOF. Let $\varepsilon_0=\varepsilon_0(n,p,C_1)>0$ be as in Lemma 6.15 and choose

$$\varepsilon=\min\left\{\frac{p}{4}C_1^{-\frac1p},\varepsilon_0\right\}>0.$$

It follows from Lemma 6.14 and Lemma 6.15 that the q-Hardy inequality holds in Ω whenever $p-\varepsilon<q<p+\varepsilon$, with a constant $C=C(n,p,C_1)$. □

We emphasize that the constant in the q-Hardy inequality in Theorem 6.16 is independent of q.

6.4. Capacity density and pointwise Hardy inequalities

We characterize validity of the pointwise Hardy inequality with a capacity density condition for the complement of the open set. This is a generalization of the measure density condition in Definition 6.11.

DEFINITION 6.17. Let $1\le p<\infty$. A closed set $E\subset\mathbb{R}^n$ satisfies the p-capacity density condition if there exists a constant $C>0$ such that

$$\operatorname{cap}_p\big(E\cap\overline{B}(x,r),B(x,2r)\big)\ge C\operatorname{cap}_p\big(\overline{B}(x,r),B(x,2r)\big)\tag{6.19}$$

for every $x\in E$ and $r>0$.

From Lemma 5.35 we see that (6.19) is equivalent to the requirement that

$$\operatorname{cap}_p\big(E\cap\overline{B}(x,r),B(x,2r)\big)\ge C_1r^{n-p}$$

for every $x\in E$ and $r>0$. We may also assume that $0<C\le1$ in (6.19).

EXAMPLE 6.18. If $1 \le p < \infty$ and a closed set $E \subset \mathbb{R}^n$ satisfies the measure density condition (6.10) with a constant C_1, then E satisfies the p-capacity density condition, with a constant $C = C(n,p,C_1)$. This follows from the lower bound in Lemma 5.35 since

$$\begin{aligned}\operatorname{cap}_p\big(E \cap \overline{B}(x,r), B(x,2r)\big) &\ge C(n,p)|E \cap B(x,r)|r^{-p}\\ &\ge C_1 C(n,p)|B(x,r)|r^{-p} = C(n,p,C_1)r^{n-p}\end{aligned}$$

for every $x \in E$ and $r > 0$.

LEMMA 6.19. *Let $n < p < \infty$ and let $E \subset \mathbb{R}^n$ be a nonempty closed set. Then E satisfies the p-capacity density condition, with a constant $C = C(n,p)$.*

PROOF. Let $x \in E$ and $r > 0$. Let $u \in \operatorname{Lip}_0(B(x,2r))$ with $u = 1$ in $E \cap \overline{B}(x,r)$, and let $z \in \partial B(x,2r)$. By Morrey's inequality, see Theorem 3.19, we have

$$1 = |u(x)| = |u(x) - u(z)| \le C(n,p) r \left(\fint_{B(x,2r)} |\nabla u(y)|^p \, dy \right)^{\frac{1}{p}},$$

and thus

$$\int_{B(x,2r)} |\nabla u(y)|^p \, dy \ge C(n,p)|B(x,2r)|r^{-p} = C(n,p)r^{n-p}.$$

By taking infimum over all functions u as above, Theorem 5.44 gives

$$\operatorname{cap}_p\big(E \cap \overline{B}(x,r), B(x,2r)\big) \ge C(n,p)r^{n-p}. \qquad \square$$

In Chapter 7, we will encounter more examples of sets that satisfy the p-capacity density condition, see Theorem 7.9 and Example 7.10.

REMARK 6.20. If $1 \le p < q < \infty$ and a closed set $E \subset \mathbb{R}^n$ satisfies the p-capacity density condition, with a constant C_1, then E satisfies the q-capacity density condition, with a constant $C = C(n,p,q,C_1)$. To see this, let $x \in E$ and $r > 0$, and let $u \in \operatorname{Lip}_0(B(x,2r))$ with $u = 1$ in $E \cap \overline{B}(x,r)$. By Theorem 5.44, we can test $\operatorname{cap}_p\big(E \cap \overline{B}(x,r), B(x,2r)\big)$ with u. Thus the p-capacity density condition and Hölder's inequality imply

$$\begin{aligned} r^{n-p} &\le C \operatorname{cap}_p\big(E \cap \overline{B}(x,r), B(x,2r)\big) \le C \int_{B(x,2r)} |\nabla u(y)|^p \, dy \\ &\le C|B(x,2r)|^{1-\frac{p}{q}} \left(\int_{B(x,2r)} |\nabla u(y)|^q \, dy \right)^{\frac{p}{q}} \\ &= C r^{n - \frac{np}{q}} \left(\int_{B(x,2r)} |\nabla u(y)|^q \, dy \right)^{\frac{p}{q}}, \end{aligned}$$

where $C = C(n,p,q,C_1)$. A straightforward simplification gives

$$r^{n-q} \le C(n,p,q,C_1) \int_{B(x,2r)} |\nabla u(y)|^q \, dy.$$

The claim follows from Theorem 5.44 by taking infimum over all functions u as above.

REMARK 6.21. The capacity density condition has several applications in the theory of partial differential equation. For example, the p-capacity density condition for Ω^c is stronger than the Wiener criterion

$$\int_0^1 \left(\frac{\operatorname{cap}_p(\Omega^c \cap \overline{B}(y,r), B(y,2r))}{\operatorname{cap}_p(\overline{B}(y,r), B(y,2r))} \right)^{\frac{1}{p-1}} \frac{dr}{r} = \infty,$$

which characterizes regular boundary points $y \in \partial\Omega$ for the Dirichlet problem for the p-Laplace equation. See Remark 11.43 and Remark 11.44 for more details, and Chapter 12 for other applications.

The p-capacity density condition implies the following boundary Poincaré inequality. Notice that if the complement of an open set satisfies the measure density condition in (6.10), then Theorem 6.22 follows from Remark 3.18, but under the more general capacity density condition we apply Maz′ya's inequality from Theorem 5.47. By approximation, Theorem 6.22 holds for every $u \in W_0^{1,p}(\Omega)$.

THEOREM 6.22. *Let $1 < p < \infty$ and assume that $\Omega \subsetneq \mathbb{R}^n$ is an open set such that the complement Ω^c satisfies the p-capacity density condition with a constant C_1. Let $x \in \Omega^c$ and $r > 0$. Let $1 \le q \le \frac{np}{n-p}$ for $1 \le p < n$, and $1 \le q < \infty$ for $n \le p < \infty$. There exists a constant $C = C(n,p,q,C_1)$ such that*

$$\left(\fint_{B(x,r)} |u(y)|^q \, dy \right)^{\frac{1}{q}} \le Cr \left(\fint_{B(x,r)} |\nabla u(y)|^p \, dy \right)^{\frac{1}{p}}$$

for every $u \in \operatorname{Lip}_0(\Omega)$.

PROOF. Let $u \in \operatorname{Lip}_0(\Omega)$. Then $\Omega^c \subset \{u = 0\} = \{y \in \mathbb{R}^n : u(y) = 0\}$. Since Ω^c satisfies the p-capacity density condition, we obtain

$$\begin{aligned} \operatorname{cap}_p(\{u=0\} \cap \overline{B}(x,\tfrac{r}{2}), B(x,r)) &\ge \operatorname{cap}_p(\Omega^c \cap \overline{B}(x,\tfrac{r}{2}), B(x,r)) \\ &\ge C(n,p,C_1) r^{n-p} = C(n,p,C_1)|B(x,r)|r^{-p}. \end{aligned}$$

Together with Theorem 5.47, this gives

$$\begin{aligned} \left(\fint_{B(x,r)} |u(y)|^q \, dy \right)^{\frac{p}{q}} &\le \frac{C(n,p,q)}{\operatorname{cap}_p(\{u=0\} \cap \overline{B}(x,\frac{r}{2}), B(x,r))} \int_{B(x,r)} |\nabla u(y)|^p \, dy \\ &\le C(n,p,q,C_1) r^p \fint_{B(x,r)} |\nabla u(y)|^p \, dy. \end{aligned}$$

□

The pointwise p-Hardy inequality holds in an open set if and only if the complement satisfies the p-capacity density condition. In addition, validity of the pointwise p-Hardy inequality in Ω is equivalent to the boundary Poincaré inequality in Theorem 6.22 and to a mean value version of the pointwise Hardy inequality. Compare also to Theorem 6.12, where the pointwise Hardy inequality was obtained from the measure density condition instead of the capacity density condition.

THEOREM 6.23. *Let $1 < p < \infty$ and assume that $\Omega \subsetneq \mathbb{R}^n$ is an open set. The following conditions are equivalent.*

(a) *Ω^c satisfies the p-capacity density condition.*

(b) *For every $x \in \Omega^c$ and $r > 0$, and every $u \in \operatorname{Lip}_0(\Omega)$,*

$$\fint_{B(x,r)} |u(y)|^p \, dy \le Cr^p \fint_{B(x,r)} |\nabla u(y)|^p \, dy. \tag{6.20}$$

(c) *For every $x \in \Omega$ and every $u \in \operatorname{Lip}_0(\Omega)$,*

$$|u_{B(x,d(x,\partial\Omega))}|^p \le C d(x,\partial\Omega)^p \fint_{B(x,2d(x,\partial\Omega))} |\nabla u(y)|^p \, dy.$$

(d) *The pointwise p-Hardy inequality holds in Ω.*

Moreover, the constants in each of the conditions only depend on each other, n and p.

PROOF. The implication from (a) to (b) follows from Theorem 6.22 with $q = p$.

Next we show that (b) implies (c). Assume that (6.20) holds with a constant C_1. Let $u \in \operatorname{Lip}_0(\Omega)$, $x \in \Omega$ and $R = d(x,\partial\Omega)$. Choose $z \in \partial\Omega$ with $|x-z| = R$. Then

$$|u_{B(x,R)}| \le |u_{B(x,R)} - u_{B(z,R)}| + |u_{B(z,R)}|.$$

By the $(1,p)$-Poincaré inequality, see Theorem 3.14, and the inclusion $B(x,R) \cup B(z,R) \subset B(x,2R)$, we have

$$\begin{aligned}
|u_{B(x,R)} - u_{B(z,R)}| &\le |u_{B(x,R)} - u_{B(x,2R)}| + |u_{B(x,2R)} - u_{B(z,R)}| \\
&\le C(n) \fint_{B(x,2R)} |u(y) - u_{B(x,2R)}| \, dy \\
&\le C(n,p) R \Big(\fint_{B(x,2R)} |\nabla u(y)|^p \, dy \Big)^{\frac{1}{p}}.
\end{aligned}$$

On the other hand, Hölder's inequality and assumption (b) imply

$$\begin{aligned}
|u_{B(z,R)}| &\le \Big(\fint_{B(z,R)} |u(y)|^p \, dy \Big)^{\frac{1}{p}} \le C_1^{\frac{1}{p}} R \Big(\fint_{B(z,R)} |\nabla u(y)|^p \, dy \Big)^{\frac{1}{p}} \\
&\le C(n,p,C_1) R \Big(\fint_{B(x,2R)} |\nabla u(y)|^p \, dy \Big)^{\frac{1}{p}}.
\end{aligned}$$

Condition (c) follows by combining the estimates above.

Then we show that (c) implies (d). Assume that (c) holds with a constant C_1. Let $u \in \operatorname{Lip}_0(\Omega)$, $x \in \Omega$ and $R = d(x,\partial\Omega)$. By the estimate in (6.12) and assumption (c), we obtain

$$\begin{aligned}
|u(x)| &\le |u(x) - u_{B(x,R)}| + |u_{B(x,R)}| \le C(n) R \big(M_{2R} |\nabla u|^p(x) \big)^{\frac{1}{p}} + |u_{B(x,R)}| \\
&\le C(n) R \big(M_{2R} |\nabla u|^p(x) \big)^{\frac{1}{p}} + C_1^{\frac{1}{p}} R \Big(\fint_{B(x,2R)} |\nabla u(y)|^p \, dy \Big)^{\frac{1}{p}} \\
&\le C(n) R \big(M_{2R} |\nabla u|^p(x) \big)^{\frac{1}{p}} + C_1^{\frac{1}{p}} R \big(M_{2R} |\nabla u|^p(x) \big)^{\frac{1}{p}} \\
&\le C(n,p,C_1) d(x,\partial\Omega) \big(M_{2d(x,\partial\Omega)} |\nabla u|^p(x) \big)^{\frac{1}{p}}.
\end{aligned}$$

Finally, we establish the implication from (d) to (a). Assume that the pointwise p-Hardy inequality (6.4) holds in Ω with a constant C_1. Let $x \in \Omega^c$ and $R > 0$. By Theorem 5.44 and Lemma 5.35, it suffices to find a constant $C = C(n,p,C_1)$ such that

$$R^{n-p} \le C \int_{B(x,2R)} |\nabla v(y)|^p \, dy \tag{6.21}$$

whenever $v \in \operatorname{Lip}_0(B(x,2R))$ and $v = 1$ in $\Omega^c \cap \overline{B}(x,R)$. Moreover, by truncation we may assume that $0 \le v \le 1$, compare to Remark 5.33.

Let $\gamma = \frac{1}{6}$. If

$$\fint_{B(x,R)} v(y)\,dy > \frac{\gamma^n}{4},$$

then the Sobolev–Poincaré inequality (3.11) with $q = 1$, see Theorem 3.17, gives

$$\begin{aligned} 1 < 4\gamma^{-n} \fint_{B(x,R)} v(y)\,dy &\le C(n) \fint_{B(x,2R)} |v(y)|\,dy \\ &\le C(n,p)R\Big(\fint_{B(x,2R)} |\nabla v(y)|^p\,dy\Big)^{\frac{1}{p}}, \end{aligned}$$

and (6.21) follows.

We may hence assume that

$$\fint_{B(x,R)} v(y)\,dy \le \frac{\gamma^n}{4}.$$

Let $F = \{y \in B(x,\gamma R) : v(y) < \frac{1}{2}\}$. Since $v = 1$ in $\Omega^c \cap \overline{B}(x,R)$, we have $F \subset \Omega$. By definition $v \ge \frac{1}{2}$ in $B(x,\gamma R) \setminus F$, and thus

$$|B(x,\gamma R) \setminus F| \le 2 \int_{B(x,\gamma R)\setminus F} v(y)\,dy \le 2 \int_{B(x,R)} v(y)\,dy \le \frac{\gamma^n}{2}|B(x,R)|.$$

This implies

$$\begin{aligned} |F| &= |B(x,\gamma R)| - |B(x,\gamma R) \setminus F| \\ &\ge \gamma^n |B(x,R)| - \frac{\gamma^n}{2}|B(x,R)| = \frac{\gamma^n}{2}|B(x,R)|. \end{aligned} \tag{6.22}$$

Let

$$\psi(y) = \max\Big\{0, 1 - \tfrac{2}{R} d\big(y, B(x, \tfrac{R}{2})\big)\Big\}$$

for every $y \in \mathbb{R}^n$. Then $\psi \in \operatorname{Lip}_0(B(x,R))$ and $\psi = 1$ in $\overline{B}(x, \frac{R}{2})$. Define

$$u(x) = \psi(x)(1 - v(x))$$

for every $y \in \mathbb{R}^n$. Then $u \in \operatorname{Lip}_0(\Omega)$ and $u = 1 - v$ in $\overline{B}(x, \frac{R}{2})$. In particular $u = 0$ in $\overline{B}(x, \frac{R}{2}) \cap \Omega^c$ and $|\nabla u| = |\nabla v|$ almost everywhere in $B(x, \frac{R}{2})$.

For every $z \in F$ there exists a radius $0 < r_z \le 2d(z, \partial\Omega)$ such that

$$M_{2d(z,\partial\Omega)}|\nabla u|^p(z) \le 2 \fint_{B(z,r_z)} |\nabla u(y)|^p\,dy.$$

By Lemma 1.13, there exist pairwise disjoint balls $B(z_i, r_i)$, where $z_i \in F$ and $r_i = r_{z_i}$ are as above, such that $F \subset \bigcup_{i=1}^\infty B(z_i, 5r_i)$. It follows from (6.22) that

$$|B(x,R)| \le \frac{2}{\gamma^n}|F| \le C(n) \sum_{i=1}^\infty |B(z_i, r_i)|. \tag{6.23}$$

Let $i \in \mathbb{N}$. Since $z_i \in F \cap B(x,\gamma R) \subset \Omega \cap B(x,\gamma R)$ and $x \notin \Omega$, we have $d(z_i, \partial\Omega) < \gamma R$. If $y \in B(z_i, r_i)$, then

$$|x - y| \le |x - z_i| + |z_i - y| \le \gamma R + 2d(z_i, \partial\Omega) < \gamma R(1+2) = \frac{R}{2},$$

and thus $B(z_i, r_i) \subset B(x, \frac{R}{2})$. This implies that $|\nabla u| = |\nabla v|$ almost everywhere in $B(z_i, r_i)$. Since $z_i \in F$, we have $u(z_i) = 1 - v(z_i) > \frac{1}{2}$. The pointwise p-Hardy inequality and the choice of the radius r_i imply

$$\begin{aligned}\left(\frac{1}{2}\right)^p &\le |u(z_i)|^p \le C_1^p d(z_i, \partial\Omega)^p M_{2d(z_i,\partial\Omega)} |\nabla u|^p(z_i)\\ &\le C_1^p \gamma^p R^p \fint_{B(z_i,r_i)} |\nabla u(y)|^p \, dy,\end{aligned}$$

and consequently

$$|B(z_i, r_i)| \le C(p, C_1) R^p \int_{B(z_i,r_i)} |\nabla v(y)|^p \, dy$$

for every $i \in \mathbb{N}$. Substituting these estimates into (6.23) gives

$$\begin{aligned}|B(x, R)| &\le C(n, p, C_1) R^p \sum_{i=1}^{\infty} \int_{B(z_i,r_i)} |\nabla v(y)|^p \, dy\\ &\le C(n, p, C_1) R^p \int_{B(x,2R)} |\nabla v(y)|^p \, dy,\end{aligned}$$

where we also used the fact that the balls $B(z_i, r_i) \subset B(x, 2R)$ are pairwise disjoint. This shows that (6.21) holds, and the proof is complete. □

Theorem 6.23 settles the question of the validity of pointwise p-Hardy inequalities, for $1 < p < \infty$. By Remark 6.6, this also implies sufficient conditions for q-Hardy inequalities for $p < q$. However, Theorem 6.23 does not characterize the validity of the p-Hardy inequality (6.1). For instance, let $\Omega = B(0,1) \setminus \{0\}$ as in Example 6.13. Then the complement Ω^c satisfies the p-capacity density condition only when $p > n$. Indeed, if $1 < p \le n$ and $0 < r < \frac{1}{2}$, then by monotonicity and the upper bounds in parts (a) and (c) of Lemma 5.36, we obtain

$$\begin{aligned}\operatorname{cap}_p\big(\Omega^c \cap \overline{B}(0, \tfrac{1}{2}), B(0,1)\big) &= \operatorname{cap}_p\big(\{0\}, B(0,1)\big)\\ &\le \operatorname{cap}_p\big(\overline{B}(0,r), B(0,1)\big) \xrightarrow{r\to 0} 0.\end{aligned}$$

Since $\operatorname{cap}_p\big(\overline{B}(0, \frac{1}{2}), B(0,1)\big) > 0$, it follows that Ω^c does not satisfy the p-capacity density condition when $1 < p \le n$. Hence we conclude from Theorem 6.23 that the pointwise p-Hardy inequality holds in Ω if and only if $p > n$. On the other hand, by Example 6.13 the p-Hardy inequality holds in Ω also when $1 < p < n$. In particular, this shows that the pointwise and integral versions of the p-Hardy inequality are not equivalent.

6.5. Wannebo's approach

If $1 \le q < p$ and the pointwise q-Hardy inequality holds in $\Omega \subset \mathbb{R}^n$, then the p-Hardy inequality holds in Ω, see Remark 6.6. In this section we give a direct proof for the fact that the p-capacity density condition for the complement implies the p-Hardy inequality with the same $p > 1$. By the equivalence in Theorem 6.23, this also shows that the pointwise p-Hardy inequality implies the p-Hardy inequality, see Corollary 6.26. We apply a quantitative version of a weighted Hardy inequality under the assumption that the complement satisfies the capacity density condition.

LEMMA 6.24. *Let $1 < p < \infty$ and $\beta < 0$. Assume that $\Omega \subsetneq \mathbb{R}^n$ is an open set and that Ω^c satisfies the p-capacity density condition with a constant C_1. There exists a constant $C = C(n,p,\beta,C_1)$ such that*

$$\int_\Omega \left(\frac{|u(x)|}{d(x,\partial\Omega)}\right)^p d(x,\partial\Omega)^\beta\,dx \le C\int_\Omega |\nabla u(x)|^p d(x,\partial\Omega)^\beta\,dx \tag{6.24}$$

for every $u \in \operatorname{Lip}_c(\Omega)$. Moreover, for $-\frac{1}{2} < \beta < 0$, we have $C \le |\beta|^{-1}C(n,p,C_1)$.

PROOF. For $j \in \mathbb{Z}$, let $A_j = \{x \in \Omega : 2^{-j-1} \le d(x,\partial\Omega) < 2^{-j}\}$ and

$$\Omega_j = \bigcup_{k=j}^\infty A_k = \{x \in \Omega : d(x,\partial\Omega) < 2^{-j}\}.$$

Let $j \in \mathbb{Z}$. By Lemma 1.13, there exists a collection $\{B(x_i,2^{-j}) : i \in \mathbb{N},\ x_i \in A_j\}$ such that $A_j \subset \bigcup_{i=1}^\infty B(x_i,2^{-j})$ and the balls $B(x_i,\frac{1}{5}2^{-j})$, $i \in \mathbb{N}$, are pairwise disjoint. For each $B(x_i,2^{-j})$ there exists $B(y_i,2^{-j+1})$, with $y_i \in \partial\Omega$, such that $B(x_i,2^{-j}) \subset B(y_i,2^{-j+1})$ and $B(y_i,2^{-j+1}) \cap \Omega \subset \Omega_{j-1}$. Since the measures of $B(x_i,\frac{1}{5}2^{-j})$ and $B(y_i,2^{-j+1})$ are comparable and the balls $B(x_i,\frac{1}{5}2^{-j})$ are pairwise disjoint, the balls $B(y_i,2^{-j+1})$, $i \in \mathbb{N}$, have bounded overlap, that is,

$$\sum_{i=1}^\infty \chi_{B(y_i,2^{-j+1})}(x) \le C(n) < \infty,$$

for every $x \in \mathbb{R}^n$.

Let $u \in \operatorname{Lip}_c(\Omega)$. Since Ω^c satisfies the p-capacity density condition, it follows from the case $q = p$ of the boundary Poincaré inequality in Theorem 6.22 that

$$\begin{aligned}\int_{B(x_i,2^{-j})} |u(x)|^p\,dx &\le \int_{B(y_i,2^{-j+1})} |u(x)|^p\,dx\\ &\le C(n,p,C_1)2^{-jp}\int_{B(y_i,2^{-j+1})} |\nabla u(x)|^p\,dx.\end{aligned}$$

By summing up these inequalities over $i \in \mathbb{N}$ and using the bounded overlap of the corresponding balls and the fact that $\nabla u = 0$ in Ω^c, we obtain

$$\begin{aligned}\int_{A_j} |u(x)|^p\,dx &\le \sum_{i=1}^\infty \int_{B(x_i,2^{-j})} |u(x)|^p\,dx\\ &\le C(n,p,C_1)2^{-jp}\sum_{i=1}^\infty \int_{B(y_i,2^{-j+1})} |\nabla u(x)|^p\,dx\\ &\le C(n,p,C_1)2^{-jp}\int_{\Omega_{j-1}} |\nabla u(x)|^p\,dx\\ &= C(n,p,C_1)2^{-jp}\sum_{k=j-1}^\infty \int_{A_k} |\nabla u(x)|^p\,dx,\end{aligned}$$

for every $j \in \mathbb{Z}$. This implies

$$\begin{aligned}
\int_\Omega |u(x)|^p d(x,\partial\Omega)^{\beta-p}\,dx &\le \sum_{j=-\infty}^{\infty} 2^{p-\beta}2^{jp-j\beta}\int_{A_j}|u(x)|^p\,dx \\
&\le C(n,p,C_1)\sum_{j=-\infty}^{\infty} 2^{p-\beta}2^{-j\beta}\sum_{k=j-1}^{\infty}\int_{A_k}|\nabla u(x)|^p\,dx \\
&= C(n,p,C_1)2^p\sum_{k=-\infty}^{\infty}\sum_{j=-\infty}^{k+1} 2^{-(j+1)\beta}\int_{A_k}|\nabla u(x)|^p\,dx \\
&\le C(n,p,C_1)2^p\sum_{k=-\infty}^{\infty}\frac{2^{-(k+2)\beta}}{1-2^\beta}\int_{A_k}|\nabla u(x)|^p\,dx \\
&\le C(n,p,C_1)\frac{2^{-2\beta}}{1-2^\beta}\sum_{k=-\infty}^{\infty}\int_{A_k}|\nabla u(x)|^p d(x,\partial\Omega)^\beta\,dx \\
&= C(n,p,C_1)\frac{2^{-2\beta}}{1-2^\beta}\int_\Omega|\nabla u(x)|^p d(x,\partial\Omega)^\beta\,dx.
\end{aligned}$$

Thus (6.24) holds with the constant $C = C(n,p,C_1)\frac{2^{-2\beta}}{1-2^\beta}$.

Finally, if $-\frac{1}{2} < \beta < 0$, then $2^{-2\beta} \le 2$ and

$$1 - 2^\beta = \log 2\int_\beta^0 e^{t\log 2}\,dt \ge \log 2\cdot e^{-\frac{\log 2}{2}}|\beta|.$$

Therefore

$$C = C(n,p,C_1)\frac{2^{-2\beta}}{1-2^\beta} \le C(n,p,C_1)e^{\frac{\log 2}{2}}\frac{2}{|\beta|\log 2},$$

proving the claim concerning the constant for these values of β. □

THEOREM 6.25. *Let $1 < p < \infty$. Assume that $\Omega \subsetneq \mathbb{R}^n$ is an open set and that Ω^c satisfies the p-capacity density condition, with a constant C_1. Then the p-Hardy inequality holds in Ω with a constant $C(n,p,C_1)$.*

PROOF. By Lemma 6.2 (b) we may assume that $u \in \operatorname{Lip}_c(\Omega)$. Let $-\frac{1}{2} < \beta < 0$ and

$$v(x) = u(x)d(x,\partial\Omega)^{-\frac{\beta}{p}},$$

for every $x \in \mathbb{R}^n$. Then $v \in \operatorname{Lip}_c(\Omega)$ and by the Leibniz rule (2.4) and the chain rule (2.5) we have

$$|\nabla v(x)| \le |\nabla u(x)|d(x,\partial\Omega)^{-\frac{\beta}{p}} + \frac{|\beta|}{p}|u(x)|d(x,\partial\Omega)^{-\frac{\beta}{p}-1}$$

for almost every $x \in \Omega$. By Lemma 6.24, the weighted Hardy inequality (6.24) holds for $v \in \operatorname{Lip}_c(\Omega)$, and we obtain

$$\begin{aligned}
\int_\Omega \frac{|u(x)|^p}{d(x,\partial\Omega)^p}\,dx &= \int_\Omega \frac{|v(x)|^p}{d(x,\partial\Omega)^p}d(x,\partial\Omega)^\beta\,dx \\
&\le \frac{C(n,p,C_1)}{|\beta|}\int_\Omega|\nabla v(x)|^p d(x,\partial\Omega)^\beta\,dx \qquad (6.25)\\
&\le \frac{C(n,p,C_1)}{|\beta|}\int_\Omega|\nabla u(x)|^p\,dx + \frac{C(n,p,C_1)}{|\beta|}\frac{|\beta|^p}{p^p}\int_\Omega\frac{|u(x)|^p}{d(x,\partial\Omega)^p}\,dx.
\end{aligned}$$

We choose $-\frac{1}{2} < \beta < 0$ to be so close to zero that

$$\frac{C(n,p,C_1)}{|\beta|}\frac{|\beta|^p}{p^p} = C(n,p,C_1)p^{-p}|\beta|^{p-1} \le \frac{1}{2}.$$

Since $u \in \operatorname{Lip}_c(\Omega)$, the last integral on the right-hand side of (6.25) is finite. Thus the last term in (6.25) can be absorbed to the left-hand side, and we arrive at

$$\int_\Omega \frac{|u(x)|^p}{d(x,\partial\Omega)^p}\,dx \le C(n,p,C_1)\int_\Omega |\nabla u(x)|^p\,dx. \qquad \square$$

COROLLARY 6.26. *Let $1 < p < \infty$ and assume that the pointwise p-Hardy inequality holds in an open set $\Omega \subsetneq \mathbb{R}^n$. Then the p-Hardy inequality holds in Ω.*

PROOF. By Theorem 6.23, Ω^c satisfies the p-capacity density condition. Thus Theorem 6.25 implies that the p-Hardy inequality holds in Ω. $\square$

6.6. Stability of Sobolev spaces with zero boundary values

There is a close connection between Hardy's inequalities and Sobolev spaces with zero boundary values. By Lemma 4.10, finiteness of the right-hand side of Hardy's inequality implies that a function has zero boundary values in the Sobolev sense. This condition is necessary and sufficient if the complement satisfies the capacity density condition.

THEOREM 6.27. *Let $1 < p < \infty$ and let $\Omega \subsetneq \mathbb{R}^n$ be an open set. Assume that $u \in W^{1,p}(\Omega)$ and that the complement Ω^c satisfies the p-capacity density condition. Then $u \in W_0^{1,p}(\Omega)$ if and only if*

$$\int_\Omega \frac{|u(x)|^p}{d(x,\partial\Omega)^p}\,dx < \infty. \tag{6.26}$$

PROOF. The sufficiency of the condition in (6.26) was proved in Lemma 4.10. To prove the necessity, assume that $u \in W_0^{1,p}(\Omega)$. Theorem 6.25 implies that the p-Hardy inequality holds in Ω, with a constant C, and by Lemma 6.2 (c) we have

$$\int_\Omega \frac{|u(x)|^p}{d(x,\partial\Omega)^p}\,dx \le C\int_\Omega |\nabla u(x)|^p\,dx < \infty. \qquad \square$$

The main purpose of this section is to find when the Sobolev spaces with zero boundary values satisfy the stability property

$$W^{1,p}(\Omega) \cap \bigcap_{1<q<p} W_0^{1,q}(\Omega) = W_0^{1,p}(\Omega), \tag{6.27}$$

where $\Omega \subset \mathbb{R}^n$ is an open set. If Ω is bounded, then $W_0^{1,p}(\Omega) \subset W_0^{1,q}(\Omega)$ for every $1 \le q < p$ by Hölder's inequality. The converse is not necessarily true since there are bounded open sets Ω for which (6.27) does not hold.

EXAMPLE 6.28. Let $1 < p < n$. By Adams and Hedberg [**4**, Theorem 5.4.1] there exists a compact set $K \subset B(0,1)$ such that $\operatorname{Cap}_p(K) > 0$ and $\dim_{\mathrm{H}}(K) = n-p$. For instance, we can apply the construction in [**4**] with gauge function

$$h(r) = r^{n-p}\Big(\log\frac{1}{r}\Big)^{-1}.$$

By the construction, the generalized Hausdorff measure $\mathcal{H}^h(K)$ is zero. From this it follows that also $\mathcal{H}^{n-q}(K)=0$ for every $q<p$, and hence $\dim_{\mathrm{H}}(K)\le n-p$. On the other hand, since $\mathrm{Cap}_p(K)>0$, by Theorem 5.42 (b) we have $\dim_{\mathrm{H}}(K)\ge n-p$.

We show that (6.27) does not hold for $\Omega=B(0,2)\setminus K$. Let $1<q<p$. Since $\dim_{\mathrm{H}}(K)=n-p<n-q$, Theorem 5.42 (b) implies $\mathrm{Cap}_q(K)=0$. Let

$$u(x)=\max\{0,1-d(x,B(0,1))\}$$

for every $x\in\mathbb{R}^n$. Then $u\in \mathrm{Lip}_c(\mathbb{R}^n)\subset W^{1,q}(\mathbb{R}^n)$ and $u=0$ q-quasieverywhere in Ω^c since $\mathrm{Cap}_q(K)=0$ and $u=0$ everywhere in $\mathbb{R}^n\setminus B(0,2)$. Corollary 5.29 implies that $u\in W_0^{1,q}(\Omega)$ for every $1<q<p$. As $u\in W^{1,p}(\Omega)$, u belongs to the left-hand side of (6.27).

Assume, for a contradiction, that $u\in W_0^{1,p}(\Omega)$. By Theorem 5.30 there exists $v\in W^{1,p}(\mathbb{R}^n)$ such that $v=u$ almost everywhere in Ω and

$$\lim_{r\to 0}\fint_{B(x,r)} v(y)\,dy=0$$

for p-quasievery $x\in\Omega^c$. On the other hand, since $|K|=0$ and $v=u$ almost everywhere in $B(0,1)\setminus K\subset\Omega$, by the definition of u we have

$$\lim_{r\to 0}\fint_{B(x,r)} v(y)\,dy=\lim_{r\to 0}\fint_{B(x,r)} u(y)\,dy=1$$

for every $x\in K\subset\Omega^c$. As $\mathrm{Cap}_p(K)>0$, we arrive at a contradiction with the claim that this limit is zero for p-quasievery $x\in\Omega^c$. Hence $u\notin W_0^{1,p}(\Omega)$.

If $p>n$, then (6.27) holds for all bounded open sets $\Omega\subset\mathbb{R}^n$.

THEOREM 6.29. *Let $n<p<\infty$ and let $\Omega\subset\mathbb{R}^n$ be an open set. Assume that $u\in W^{1,p}(\Omega)$ and that $u\in W_0^{1,q}(\Omega)$ for every $1<q<p$. Then $u\in W_0^{1,p}(\Omega)$.*

PROOF. Let $n<q<p$. Since $u\in W_0^{1,q}(\Omega)$, the zero extension u_0 of u belongs to $W^{1,q}(\mathbb{R}^n)$ by Theorem 2.31. Furthermore, the weak gradient ∇u_0 equals almost everywhere in $\mathbb{R}^n$ to the zero extension of the weak gradient ∇u. Thus

$$\|u_0\|_{W^{1,p}(\mathbb{R}^n)}=\|u\|_{W^{1,p}(\Omega)}<\infty,$$

since $u\in W^{1,p}(\Omega)$. This shows that $u_0\in W^{1,p}(\mathbb{R}^n)$.

Theorem 3.23 implies that $u_0\in W^{1,p}(\mathbb{R}^n)$ has a continuous representative u^*, and by continuity

$$u^*(x)=\lim_{r\to 0}\fint_{B(x,r)} u_0(y)\,dy$$

for every $x\in\mathbb{R}^n$. Since all points in $\mathbb{R}^n$ have positive q-capacity by Lemma 5.9, we obtain from Theorem 5.26 a continuous function $v\in W^{1,q}(\mathbb{R}^n)$ satisfying $v=u$ almost everywhere in Ω and $v=0$ everywhere in $\mathbb{R}^n\setminus\Omega$. Then $u^*=u_0=v$ almost everywhere in $\mathbb{R}^n$, and by continuity $u^*=v$ everywhere in $\mathbb{R}^n$. Thus $u^*\in W^{1,p}(\mathbb{R}^n)$ is continuous, $u^*=u$ almost everywhere in Ω, and $u^*=0$ in $\mathbb{R}^n\setminus\Omega$. The claim $u\in W_0^{1,p}(\Omega)$ follows from Corollary 5.29. □

As Example 6.28 indicates, in the case $1<p<n$ some additional conditions must be imposed on Ω in order to obtain (6.27).

DEFINITION 6.30. Let $1 < p < \infty$. The pointwise upper p-capacity density of a set $E \subset \mathbb{R}^n$ at a point $x \in \mathbb{R}^n$ is

$$\overline{\operatorname{dens}}_p(E, x) = \limsup_{r\to 0} \frac{\operatorname{cap}_p\big(E \cap \overline{B}(x,r), B(x,2r)\big)}{r^{n-p}}.$$

A closed set $E \subset \mathbb{R}^n$ satisfies the pointwise p-capacity density condition, with exponents $P \subset (1,p)$, if for p-quasievery $x \in E$ there exists $q \in P$ such that $\overline{\operatorname{dens}}_q(E,x) > 0$.

The pointwise capacity density condition for the complement implies a stability result for Sobolev functions with zero boundary values.

THEOREM 6.31. *Let $1 < p < \infty$ and let $\Omega \subset \mathbb{R}^n$ be an open set. Assume that Ω^c satisfies the pointwise p-capacity density condition, with exponents $P \subset (1,p)$, and that $u \in W^{1,p}(\Omega)$ and $u \in W_0^{1,q}(\Omega)$ for every $q \in P$. Then $u \in W_0^{1,p}(\Omega)$.*

PROOF. Let $q \in P$. As in the proof of Theorem 6.29, the zero extension u_0 of u belongs to $W^{1,q}(\mathbb{R}^n) \cap W^{1,p}(\mathbb{R}^n)$. Define $u^* \colon \mathbb{R}^n \to \mathbb{R}$ by

$$u^*(x) = \begin{cases} \lim_{r\to 0} \fint_{B(x,r)} u_0(y)\,dy, & \text{if } x \in \operatorname{Leb}(u_0), \\ 0, & \text{if } x \notin \operatorname{Leb}(u_0). \end{cases}$$

Theorem 5.23 implies that u^* is both a p-quasicontinuous and a q-quasicontinuous representative of $u_0 \in W^{1,p}(\mathbb{R}^n) \cap W^{1,q}(\mathbb{R}^n)$. Thus $u^* = 0$ holds q-quasieverywhere in Ω^c for every $q \in P$, by Theorem 5.26 and Theorem 5.16.

By the assumptions, p-quasievery point $x \in \Omega^c$ satisfies $\overline{\operatorname{dens}}_q(\Omega^c, x) > 0$ for some $q \in P$. Fix such $x \in \Omega^c$ and a corresponding $q \in P$, and choose $\delta = \frac{1}{2}\overline{\operatorname{dens}}_q(\Omega^c, x) > 0$. Then there is a sequence $(r_i)_{i\in\mathbb{N}}$ with $\lim_{i\to\infty} r_i = 0$ and

$$\delta r_i^{n-q} \le \operatorname{cap}_q\big(\Omega^c \cap \overline{B}(x,r_i), B(x,2r_i)\big) \tag{6.28}$$

for every $i \in \mathbb{N}$. The q-quasicontinuity of u^* and Theorem 5.47 give

$$\begin{aligned} &\fint_{B(x,2r_i)} |u^*(y)|\,dy \\ &\quad\le \left(\frac{C(n,q)}{\operatorname{cap}_q\big(\{u^*=0\} \cap \overline{B}(x,r_i), B(x,2r_i)\big)} \int_{B(x,2r_i)} |\nabla u^*(y)|^q\,dy \right)^{\frac{1}{q}}. \end{aligned}$$

Since $u^* = 0$ q-quasieverywhere in Ω^c, using the monotonicity and subadditivity of the variational capacity with Theorem 5.38, we obtain from (6.28)

$$\operatorname{cap}_q\big(\{u^*=0\} \cap \overline{B}(x,r_i), B(x,2r_i)\big) \ge \operatorname{cap}_q\big(\Omega^c \cap \overline{B}(x,r_i), B(x,2r_i)\big) \ge \delta r_i^{n-q}$$

for every $i \in \mathbb{N}$. By Hölder's inequality

$$\begin{aligned} \fint_{B(x,2r_i)} |u^*(y)|\,dy &\le C\left(r^q \fint_{B(x,2r_i)} |\nabla u^*(y)|^q\,dy \right)^{\frac{1}{q}} \\ &\le C\left(r^p \fint_{B(x,2r_i)} |\nabla u^*(y)|^p\,dy \right)^{\frac{1}{p}}, \end{aligned}$$

where $C = C(n,q,\delta)$. As $u^* = u_0$ and $\nabla u^* = \nabla u_0$ almost everywhere in $\mathbb{R}^n$, we have

$$\liminf_{r\to 0} \fint_{B(x,r)} |u_0(y)|\,dy \le C\left(\limsup_{r\to 0} r^p \fint_{B(x,r)} |\nabla u_0(y)|^p\,dy \right)^{\frac{1}{p}} \tag{6.29}$$

for p-quasievery $x \in \Omega^c$, where the constant $C = C(n, q, \delta)$ may depend on x via q and δ.

By Lemma 5.22, the right-hand side of (6.29) is zero p-quasieverywhere in $\mathbb{R}^n$. Thus

$$|u^*(x)| = \lim_{r\to 0} \left| \fint_{B(x,r)} u_0(y)\, dy \right| \le \liminf_{r\to 0} \fint_{B(x,r)} |u_0(y)|\, dy = 0$$

for p-quasievery $x \in \Omega^c$, by Theorem 5.23, and hence $u^* = 0$ p-quasieverywhere in Ω^c. Since $u^* \in W^{1,p}(\mathbb{R}^n)$ is p-quasicontinuous and $u^* = u$ almost everywhere in Ω, we conclude from Corollary 5.29 that $u \in W_0^{1,p}(\Omega)$. $\square$

As a consequence of Theorem 6.31 we obtain the following characterization of $W_0^{1,p}(\Omega)$ under the pointwise p-capacity density condition.

THEOREM 6.32. *Let $1 < p < \infty$ and let $\Omega \subset \mathbb{R}^n$ be a bounded open set. Assume that $u \in W^{1,p}(\Omega)$ and that Ω^c satisfies the pointwise p-capacity density condition, with exponents $P \subset (1, p)$. Then $u \in W_0^{1,p}(\Omega)$ if and only if $u \in W_0^{1,q}(\Omega)$ for every $q \in P$.*

PROOF. If $u \in W_0^{1,p}(\Omega)$, then $u \in W_0^{1,q}(\Omega)$ for every $q \in P$ by Hölder's inequality and the fact that $W_0^{1,p}(\Omega)$ is the completion of $C_0^\infty(\Omega)$ in $W^{1,p}(\Omega)$. The assumption that Ω is bounded is needed here. The reverse implication follows directly from Theorem 6.31. $\square$

REMARK 6.33. Let $1 < p < \infty$ and assume that $\Omega \subset \mathbb{R}^n$ is a bounded open set and that Ω^c satisfies the p-capacity density condition. In Chapter 7 we will show that then there is $1 < q < p$ such that the q-capacity density condition holds for Ω^c, see Theorem 7.21. Hence Ω^c satisfies the pointwise p-capacity density condition with $P = \{q\} \subset (1, p)$. Thus we obtain from Theorem 6.27 and Theorem 6.32 that under these assumptions the following conditions are equivalent for a function $u \in W^{1,p}(\Omega)$:

(a) $u \in W_0^{1,p}(\Omega)$,
(b) $u \in W_0^{1,q}(\Omega)$ for every $1 < q < p$,
(c) $\int_\Omega \frac{|u(x)|^p}{d(x,\partial\Omega)^p}\, dx < \infty$.

6.7. Notes

Several sufficient Lipschitz and Hölder type boundary conditions for Hardy's inequality have been given by Balinsky, Evans and Lewis [**22**], Edmunds and Evans [**115, 116**], Kufner [**248**], Nečas [**337**] and Opic and Kufner [**340**]. Capacitary characterizations of Hardy's inequality are given in Maz′ya [**318, 321**], see also Section 10.7. For details on the origins and history of one-dimensional Hardy's inequalities, we refer to Hardy, Littlewood and Pólya [**180**, Chapter IX] and Kufner, Maligranda and Persson [**249**]. General theory of one-dimensional Hardy-type inequalities can be found in Kufner and Persson [**250**].

The measure density condition plays a role in harmonic analysis, function spaces and the regularity theory up to the boundary for partial differential equations, see Brudnyi and Brudnyi [**59, 60**], Giaquinta [**147, 148**], Giusti [**153**], Heinonen, Kilpeläinen and Martio [**187**], Jonsson and Wallin [**215**] and Shvartsman [**361**]. Theorem 6.25 for $p = 2$ is by Ancona [**14**] and for $p \ge 1$ by Lewis [**271**] and Wannebo [**393**]. See also Mikkonen [**327**]. The proof of Theorem 6.25 in Section 6.5

is based on Wannebo [**393**], see also Buckley and Koskela [**64**]. Hardy's inequalities for functions vanishing only on a part of the boundary are considered by Egert, Haller-Dintelmann and Rehberg [**118**].

The pointwise Hardy inequality has been studied by Hajłasz [**166**] and Kinnunen and Martio [**235**]. Weighted pointwise Hardy inequalities are considered in Koskela and Lehrbäck [**243**]. Self-improvement of the pointwise Hardy inequality is proved by Eriksson-Bique and Vähäkangas [**119**]. Theorem 6.23 has been studied by Kinnunen and Korte [**226**], Korte, Lehrbäck and Tuominen [**241**] and Lehrbäck [**256**]. More characterizations in the borderline case $p = n$ are discussed in Theorem 7.20. Section 6.3 is from Koskela and Zhong [**246**], see also Lehrbäck [**257**] for weighted results. Section 6.6 is based on Hedberg and Kilpeläinen [**182**].

Many arguments related to Hardy's inequality are based on general principles which apply even on metric measure spaces, see Björn, MacManus and Shanmugalingam [**36**], Kilpeläinen, Kinnunen and Martio [**219**], Korte, Lehrbäck and Tuominen [**241**], Korte and Shanmugalingam [**242**], Koskela and Zhong [**246**] and Lehrbäck [**260, 261**]. In [**95**] Danielli, Garofalo and Phuc studied Hardy's inequalities in Carnot–Carathéodory spaces. We refer to Buckley and Koskela [**64**] and Cianci [**86**] for results in Orlicz–Sobolev spaces. For other function spaces, see Edmunds, Hurri-Syrjänen and Vähäkangas [**117**], Ihnatsyeva, Lehrbäck, Tuominen and Vähäkangas [**201**] and Ihnatsyeva and Vähäkangas [**203, 204**].

CHAPTER 7

Density Conditions

Geometric characterizations of the capacity density condition in Definition 6.17 are given in terms of Hausdorff content, Ahlfors–David regular sets and the lower dimension in Section 7.1, Section 7.2 and Section 7.3, respectively. The case $p > n$ is not of interest since by Lemma 6.19 every nonempty closed set satisfies the p-capacity density condition. On the other hand, the borderline case $p = n$ is special for density conditions and also for Hardy's inequalities. Several analytic, metric and geometric characterizations for the n-capacity density condition and the n-Hardy inequality are given in Section 7.4. These include a self-improvement property for the n-capacity density condition. Section 7.5 discusses a corresponding self-improvement property of the capacity density condition for $1 < p < n$. The argument is completed in Section 7.6, Section 7.7 and Section 7.8.

7.1. Hausdorff content density

A density condition for the Hausdorff content is analogous to the measure and capacity density conditions, see Definition 6.11 and Definition 6.17. Recall the definitions of the Hausdorff content and the Hausdorff measure from Definition 1.43.

DEFINITION 7.1. Let $0 \le \lambda \le n$. A set $E \subset \mathbb{R}^n$ satisfies the λ-Hausdorff content density condition if there exists a constant $C > 0$ such that

$$\mathcal{H}^{\lambda}_{\infty}(E \cap \overline{B}(x,r)) \ge C r^{\lambda} \tag{7.1}$$

for every $x \in E$ and $r > 0$.

For $\lambda = n$, the n-dimensional Hausdorff content is equivalent to the n-dimensional Lebesgue outer measure, and hence n-Hausdorff content density condition is equivalent to the measure density condition. Observe that if a set $\emptyset \neq E \subset \mathbb{R}^n$ satisfies the λ-Hausdorff content density condition with $\lambda > 0$, then E is unbounded.

REMARK 7.2. Let $0 \le \lambda_1 < \lambda_2$ and assume that $E \subset \mathbb{R}^n$ satisfies the λ_2-Hausdorff content density condition, with a constant C_1. Let $x \in E$ and $r > 0$. Lemma 5.40 implies

$$C_1 r^{\lambda_2} \le \mathcal{H}^{\lambda_2}_{\infty}(E \cap \overline{B}(x,r)) \le r^{\lambda_2-\lambda_1}\mathcal{H}^{\lambda_1}_{\infty}(E \cap \overline{B}(x,r)), \tag{7.2}$$

and thus $\mathcal{H}^{\lambda_1}_{\infty}(E \cap \overline{B}(x,r)) \ge C_1 r^{\lambda_1}$. This shows that E satisfies the λ_1-Hausdorff content density condition, with the same constant C_1.

THEOREM 7.3. *Let $1 < p < \infty$ and $0 \le \lambda \le n$, with $\lambda > n - p$, and assume that a closed set $E \subset \mathbb{R}^n$ satisfies the λ-Hausdorff content density condition, with a constant C_1. Then E satisfies the p-capacity density condition, with a constant $C = C(n, p, \lambda, C_1)$.*

PROOF. For $p > n$, the claim holds by Lemma 6.19. Hence we may assume $1 < p \leq n$. Theorem 5.41 and (7.1) imply

$$\begin{aligned}\operatorname{cap}_p\big(E \cap \overline{B}(x,r), B(x,2r)\big) &\geq C(n,p,\lambda)\mathcal{H}^\lambda_\infty\big(E \cap \overline{B}(x,r)\big)r^{n-p-\lambda}\\ &\geq C(n,p,\lambda,C_1)r^{n-p}\end{aligned}$$

for every $x \in E$ and $r > 0$. □

REMARK 7.4. Theorem 7.3 has also a converse, see Theorem 7.22, but this is based on the self-improvement property of the capacity density condition discussed in Section 7.5. However, we observe that if $1 < p \leq n$ and a closed set $E \subset \mathbb{R}^n$ satisfies the p-capacity density condition, with a constant C_1, then by Theorem 5.39 we have

$$r^{n-p} \leq C(n,p,C_1)\operatorname{cap}_p\big(E \cap \overline{B}(x,r), B(x,2r)\big) \leq C(n,p,C_1)\mathcal{H}^{n-p}_\infty\big(E \cap \overline{B}(x,r)\big)$$

for every $x \in E$ and $r > 0$. Hence E satisfies the λ-Hausdorff content density condition with $\lambda = n - p$.

The Hausdorff content density condition gives a sufficient condition for Hardy's inequality without applying Theorem 6.25.

COROLLARY 7.5. *Let $1 < p < \infty$ and $0 \leq \lambda \leq n$, with $\lambda > n - p$. Assume that $\Omega \subsetneq \mathbb{R}^n$ is an open set and that Ω^c satisfies the λ-Hausdorff content density condition. Then the pointwise p-Hardy inequality and the p-Hardy inequality hold in Ω.*

PROOF. Theorem 7.3 implies that Ω^c satisfies the q-capacity density condition with $p > q > \max\{1, n - \lambda\}$. The pointwise q-Hardy inequality holds in Ω by Theorem 6.23 and the p-Hardy inequality follows as in Remark 6.6. □

7.2. Ahlfors–David regular sets

Ahlfors–David regularity implies the capacity density condition, see Theorem 7.9. Examples of Ahlfors–David regular sets include subspaces and compact submanifolds of $\mathbb{R}^n$ and many self-similar fractals.

DEFINITION 7.6. Let $0 < \lambda \leq n$. A closed set $E \subset \mathbb{R}^n$ is (Ahlfors–David) λ-regular, or a λ-set, if there exists a constant $C \geq 1$ such that

$$C^{-1}r^\lambda \leq \mathcal{H}^\lambda\big(E \cap \overline{B}(x,r)\big) \leq Cr^\lambda \tag{7.3}$$

for every $x \in E$ and $0 < r \leq \operatorname{diam}(E)$.

It is easy to see that if E is λ-regular, then $\dim_{\mathrm{H}}(E) = \lambda$. Cantor type constructions show that there exist λ-regular sets for every λ with $0 < \lambda \leq n$. We refer to Falconer [**125**] and Mattila [**315**] for more details concerning fractal geometry and self-similar sets. The next remark is related to the definitions of the Hausdorff content and the Hausdorff measure, see Definition 1.43.

REMARK 7.7. In Definition 1.43 it is sometimes convenient to impose the additional restriction that the covering balls are centered in the set. This leads to quantities which are comparable to $\mathcal{H}^\lambda_\infty$ and $\mathcal{H}^\lambda$ with two-sided bounds. Let $0 < \delta \leq \infty$ and let $\{B(x_i, R_i) : i \in \mathbb{N}\}$ be a cover of $E \subset \mathbb{R}^n$, with $0 < R_i \leq \frac{\delta}{2}$ for every $i \in \mathbb{N}$. We may assume that for every $i \in \mathbb{N}$ there exists $z_i \in E \cap B(x_i, R_i)$ since otherwise

$B(x_i, R_i)$ is redundant in the cover. Then $B(x_i, R_i) \subset B(z_i, 2R_i)$, and the balls $B(z_i, 2R_i)$, $i \in \mathbb{N}$, cover the set E. This implies

$$\inf\left\{\sum_{i=1}^{\infty} r_i^\lambda : E \subset \bigcup_{i=1}^{\infty} B(y_i, r_i),\ y_i \in E,\ 0 < r_i \leq \delta\right\} \leq \sum_{i=1}^{\infty}(2R_i)^\lambda = 2^\lambda \sum_{i=1}^{\infty} R_i^\lambda.$$

By taking infimum over all such covers of E, we obtain

$$\mathcal{H}^\lambda_\delta(E) \leq \inf\left\{\sum_{i=1}^{\infty} r_i^\lambda : E \subset \bigcup_{i=1}^{\infty} B(y_i, r_i),\ y_i \in E,\ 0 < r_i \leq \delta\right\} \leq C(\lambda)\mathcal{H}^\lambda_{\frac{\delta}{2}}(E).$$

For $\delta = \infty$ we have two-sided bounds by $\mathcal{H}^\lambda_\infty(E)$, and at the limit $\delta \to 0$ we obtain two-sided bounds by $\mathcal{H}^\lambda(E)$.

By the following lemma, unbounded λ-regular sets satisfy the λ-Hausdorff content density condition.

LEMMA 7.8. *Let $0 < \lambda \leq n$ and assume that a closed set $E \subset \mathbb{R}^n$ is λ-regular with a constant C_1. There exists a constant $C = C(\lambda, C_1)$ such that*

$$\mathcal{H}^\lambda_\infty\big(E \cap \overline{B}(x, r)\big) \geq Cr^\lambda$$

for every $x \in E$ and $0 < r \leq \operatorname{diam}(E)$.

PROOF. Fix $x \in E$ and $0 < r \leq \operatorname{diam}(E)$. Let $\{B(x_i, r_i) : i \in \mathbb{N}\}$ be a cover of $E \cap \overline{B}(x, r)$, with $x_i \in E \cap \overline{B}(x, r) \subset E$ and $r_i > 0$ for every $i \in \mathbb{N}$. If $r_j > \operatorname{diam}(E)$ for some $j \in \mathbb{N}$, then

$$\sum_{i=1}^{\infty} r_i^\lambda \geq r_j^\lambda > \operatorname{diam}(E)^\lambda \geq r^\lambda.$$

On the other hand, if $0 < r_i \leq \operatorname{diam}(E)$ for every $i \in \mathbb{N}$, we apply (7.3) and obtain

$$r^\lambda \leq C_1 \mathcal{H}^\lambda\big(E \cap \overline{B}(x, r)\big) \leq C_1 \sum_{i=1}^{\infty} \mathcal{H}^\lambda\big(E \cap \overline{B}(x_i, r_i)\big) \leq C_1^2 \sum_{i=1}^{\infty} r_i^\lambda.$$

By taking infimum over all such covers, and recalling that by Remark 7.7 it suffices to consider balls centered at $E \cap \overline{B}(x, r)$, we conclude that

$$r^\lambda \leq C(\lambda, C_1)\mathcal{H}^\lambda_\infty\big(E \cap \overline{B}(x, r)\big).$$

□

THEOREM 7.9. *Let $1 < p < \infty$ and $0 < \lambda \leq n$, with $\lambda > n - p$, and assume that $E \subset \mathbb{R}^n$ is a closed and unbounded λ-regular set. Then E satisfies the p-capacity density condition.*

PROOF. Since E is unbounded, Lemma 7.8 implies that the λ-Hausdorff content density condition in (7.1) holds for every $x \in E$ and $r > 0$. The claim follows from Theorem 7.3. □

EXAMPLE 7.10. Assume that $\Omega_0 \subset \mathbb{R}^n$ is a bounded open set whose complement Ω_0^c satisfies the measure density condition. Let $0 < \lambda \leq n$, let $K \subset \Omega_0$ be a compact λ-regular set, with a constant C_1, and define $\Omega = \Omega_0 \setminus K$. We claim that Ω^c satisfies the λ-Hausdorff content density condition in (7.1). For $x \in \Omega_0^c$, this follows from Remark 7.2 since the measure density condition is equivalent to the n-Hausdorff content density condition and $\lambda \leq n$. Assume then that $x \in K$. For $0 < r \leq 2\operatorname{diam}(\Omega_0)$ the λ-Hausdorff content density condition, with a constant depending on C_1, λ, $\operatorname{diam}(K)$ and $\operatorname{diam}(\Omega_0)$, follows from the λ-regularity of K

and Theorem 7.8. On the other hand, if $r > 2\operatorname{diam}(\Omega_0)$, then the measure density condition of Ω_0^c implies the λ-Hausdorff content density condition.

By Theorem 7.3, the complement Ω^c satisfies the p-capacity density condition for every $p > \max\{1, n-\lambda\}$. Consequently, the pointwise p-Hardy inequality holds in Ω for all such p.

7.3. Lower dimension and capacity density

The λ-Hausdorff content density condition holds for a set $E \subset \mathbb{R}^n$ if and only if there is a constant $C > 0$ such that if $\{B(x_i, r_i) : i \in \mathbb{N}\}$ is a cover of $E \cap \overline{B}(x, R)$, where $x \in E$ and $R > 0$, then

$$\sum_{i=1}^{\infty} r_i^\lambda \geq CR^\lambda. \tag{7.4}$$

Next we consider this condition for finite covers of $E \cap \overline{B}(x, R)$ with balls that have a fixed radius $0 < r_i = r \leq R$. For such covers $\{B(x_i, r) : i = 1, \dots, N\}$, where $N \in \mathbb{N}$, the condition corresponding to (7.4) is

$$\sum_{i=1}^{N} r^\lambda \geq CR^\lambda, \quad \text{or equivalently, } N \geq C\Big(\frac{R}{r}\Big)^\lambda.$$

This leads to the definition of the lower dimension. It turns out that this concept of dimension can be used to characterize the Hausdorff content density condition and thus also the capacity density condition, see Theorem 7.14 and Theorem 7.22. See also Section 10.3 for the dual concept, the Assouad dimension.

DEFINITION 7.11. The lower dimension $\dim_{\mathrm{L}}(E)$ of a set $E \subset \mathbb{R}^n$ is the supremum of the numbers $\lambda \geq 0$ for which there exists a constant $C > 0$ such that

$$N\big(E \cap \overline{B}(x, R), r\big) \geq C\Big(\frac{R}{r}\Big)^\lambda \tag{7.5}$$

for every $x \in E$ and $0 < r < R \leq \operatorname{diam}(E)$. Here $N(\cdot, r)$ denotes the minimal number of open balls of radius r that are needed to cover the set.

In particular, (7.5) holds whenever $0 \leq \lambda < \dim_{\mathrm{L}}(E)$, and possibly also for $\lambda = \dim_{\mathrm{L}}(E)$. Observe that $0 \leq \dim_{\mathrm{L}}(E) \leq n$ for every $E \subset \mathbb{R}^n$. In the case $E = \{x_0\}$, $x_0 \in \mathbb{R}^n$, we remove the requirement $R \leq \operatorname{diam}(E)$ from the definition above and thus $\dim_{\mathrm{L}}(\{x_0\}) = 0$. It should be noted that unlike many other notions of dimension, the lower dimension is not monotone. For instance, if $E = \{0\} \cup [1, 2] \subset \mathbb{R}$, then $\dim_{\mathrm{L}}(E) = 0$ due to the isolated point 0, but for the subset $[1, 2] \subset E$ we have $\dim_{\mathrm{L}}([1, 2]) = 1$. The latter fact is a special case of the following theorem.

THEOREM 7.12. *Let $0 < \lambda \leq n$ and assume that a closed set $E \subset \mathbb{R}^n$ is λ-regular. Then* $\dim_{\mathrm{L}}(E) = \lambda$.

PROOF. Fix $x \in E$ and $0 < r < R \leq \operatorname{diam}(E)$, and consider balls $B(x_i, r)$, $i = 1, \dots, N$, which cover $E \cap \overline{B}(x, R)$. By Lemma 7.8 and the definition of the Hausdorff content, we obtain

$$R^\lambda \leq C\mathcal{H}_\infty^\lambda(E \cap \overline{B}(x, R)) \leq C \sum_{i=1}^{N} r^\lambda = CNr^\lambda.$$

Thus $N \geq C\big(\frac{R}{r}\big)^\lambda$, which implies $\dim_{\mathrm{L}}(E) \geq \lambda$.

On the other hand, by Lemma 1.13 and boundedness, there exist $N_0 \in \mathbb{N}$ and points $x_i \in E \cap \overline{B}(x,R)$, $i = 1, 2, \dots, N_0$, such that $E \cap \overline{B}(x,R) \subset \bigcup_{i=1}^{N_0} B(x_i, r)$ and the balls $B(x_i, \frac{r}{5})$, $i = 1, \dots, N_0$, are pairwise disjoint. If $2R \le \operatorname{diam}(E)$, then by the λ-regularity we have

$$R^\lambda \ge C\mathcal{H}^\lambda(E \cap \overline{B}(x,2R)) \ge C\sum_{i=1}^{N_0} \mathcal{H}^\lambda(E \cap B(x_i, \tfrac{r}{5})) \ge C\sum_{i=1}^{N_0} r^\lambda = CN_0 r^\lambda.$$

The same estimate also holds if $2R > \operatorname{diam}(E)$ since in this case

$$\mathcal{H}^\lambda(E \cap \overline{B}(x,2R)) \le \mathcal{H}^\lambda(E \cap \overline{B}(x, \operatorname{diam}(E))) \le C \operatorname{diam}(E)^\lambda \le CR^\lambda.$$

Thus $N_0 \le C\big(\frac{R}{r}\big)^\lambda$. Here the constant C is independent of x, r and R. It follows that the lower bound in (7.5) cannot hold for any exponent larger than λ. Hence $\dim_{\mathrm{L}}(E) = \lambda$, as desired. □

The following lemma gives a connection between the lower dimension and the Hausdorff content density condition. Roughly speaking, the lemma asserts that the estimate in (7.6), for covers using balls of fixed radii r, implies a corresponding estimate (7.7) for covers where balls of all radii are allowed. However, we obtain the latter only for dimensional parameters $0 \le \lambda < \lambda_0$, and so the endpoint λ_0 is lost.

LEMMA 7.13. *Let $E \subset \mathbb{R}^n$ be a closed set. Assume that there exist $0 < \lambda_0 \le n$ and a constant $C_1 > 0$ such that*

$$N(E \cap \overline{B}(x,R), r) \ge C_1\left(\frac{R}{r}\right)^{\lambda_0} \tag{7.6}$$

for every $x \in E$ and $0 < r < R \le \operatorname{diam}(E)$. Then, for every $0 \le \lambda < \lambda_0$, there exists a constant $C = C(n, \lambda_0, \lambda, C_1) > 0$ such that

$$\mathcal{H}^\lambda_\infty(E \cap \overline{B}(x,R)) \ge CR^\lambda \tag{7.7}$$

for every $x \in E$ and $0 < R \le \operatorname{diam}(E)$.

PROOF. Let $0 \le \lambda < \lambda_0$, $x \in E$ and $0 < R \le \operatorname{diam}(E)$. We may assume $C_1 < 1$. Choose $K \in \mathbb{N}$ to be so large that

$$\lambda < \frac{\lambda_0 \log K}{\log K - \log C_1 + \log 11^{\lambda_0}}.$$

The purpose of K is to give the number of the pairwise disjoint balls that we use in each step of the following Cantor type construction, and the number

$$m = 11\left(\frac{K}{C_1}\right)^{\frac{1}{\lambda_0}} > 11$$

gives the ratio of the radii used in the consecutive steps of the construction. Notice that

$$\lambda < \frac{\log K}{\log m} < \lambda_0.$$

Let $\overline{B}_0 = \overline{B}(x,R)$. In the first step of the construction we take $r_1 = Rm^{-1} < \frac{R}{11}$. Using the assumption (7.6) and Lemma 1.13 with radius $\frac{11}{5}r_1$, we find closed balls $\overline{B}_i = \overline{B}(z_i, r_1)$, $i = 1, 2, \ldots, N_0$, with $z_i \in E \cap \overline{B}_0$, such that the balls $\overline{B}(z_i, 2r_1) \subset \overline{B}(x, 2R)$, $i = 1, \ldots, N_0$, are pairwise disjoint,

$$E \cap \overline{B}_0 \subset \bigcup_{i=1}^{N_0} B(z_i, 11r_1),$$

and $N_0(11r_1)^{\lambda_0} \geq C_1 R^{\lambda_0}$. By the choices of r_1 and m, we have $N_0 \geq K$. We proceed the construction with the balls $\overline{B}_i$, $i = 1, \ldots, K$, and omit the rest of them.

In the second step we take $r_2 = Rm^{-2} < \frac{r_1}{11}$. We find balls $\overline{B}_{i_1 i_2} = \overline{B}(z_{i_1 i_2}, r_2)$, $i_1 = 1, 2, \ldots, K$ and $i_2 = 1, 2, \ldots, N_{i_1}$, with $z_{i_1 i_2} \in E \cap \overline{B}_{i_1}$, such that, for every $i_1 = 1, 2, \ldots, K$, the balls $\overline{B}(z_{i_1 i_2}, 2r_2) \subset \overline{B}(z_{i_1}, 2r_1)$, $i_2 = 1 \ldots, N_{i_1}$, are pairwise disjoint,

$$E \cap \overline{B}_{i_1} \subset \bigcup_{i_2=1}^{N_{i_1}} B(z_{i_1 i_2}, 11r_2),$$

and, by (7.6), we have $N_{i_1}(11r_2)^{\lambda_0} \geq C_1 r_1^{\lambda_0}$. Again, $N_{i_1} \geq K$ for every $i_1 = 1, \ldots, K$. We continue with the balls $\overline{B}_{i_1 i_2} = \overline{B}(z_{i_1 i_2}, r_2)$, $i_1, i_2 = 1, 2, \ldots, K$. Since $\overline{B}(z_{i_1}, 2r_1) \cap \overline{B}(z_{j_1}, 2r_1) = \emptyset$, if $i_1 \neq j_1$, and $\overline{B}(z_{i_1 i_2}, 2r_2) \subset \overline{B}(z_{i_1}, 2r_1)$, for every $i_1, i_2 = 1, \ldots, K$, we conclude that balls $\overline{B}(z_{i_1 i_2}, 2r_2)$, $i_1, i_2 = 1, \ldots, K$, are pairwise disjoint.

Continuing recursively, in the kth step of the construction we obtain a collection of closed balls $\overline{B}_{i_1 i_2 \ldots i_k}$, where $i_j = 1, \ldots, K$ for $j \in \{1, \ldots, k-1\}$ and $i_k = 1, \ldots, N_{i_1 i_2 \ldots i_{k-1}}$, with center points $z_{i_1 i_2 \ldots i_k} \in E \cap \overline{B}_{i_1 i_2 \ldots i_{k-1}}$ and radius $r_k = Rm^{-k}$, satisfying the following properties: $\overline{B}(z_{i_1 i_2 \ldots i_k}, 2r_k) \subset \overline{B}(z_{i_1 i_2 \ldots i_{k-1}}, 2r_{k-1})$ are pairwise disjoint,

$$E \cap \overline{B}_{i_1 i_2 \ldots i_{k-1}} \subset \bigcup_{i_k=1}^{N_{i_1 i_2 \cdots i_{k-1}}} B(z_{i_1 i_2 \ldots i_k}, 11r_k),$$

and, by (7.6), we have $N_{i_1 i_2 \ldots i_{k-1}}(11r_k)^{\lambda_0} \geq C_1 r_{k-1}^{\lambda_0}$. Since $r_{k-1}^{\lambda_0} = m^{\lambda_0} r_k^{\lambda_0}$, the definition of m implies that $N_{i_1 i_2 \ldots i_{k-1}} \geq K$ for every $i_1 i_2 \ldots i_{k-1}$, where $i_j = 1, \ldots, K$ for $j \in \{1, \ldots, k-1\}$. The construction continues with the balls $\overline{B}_{i_1 i_2 \ldots i_k}$, where $i_1, i_2, \ldots, i_k = 1, 2, \ldots, K$. Observe that the corresponding balls $\overline{B}(z_{i_1 i_2 \ldots i_k}, 2r_k)$ are pairwise disjoint.

We define a Cantor-type set

$$F = \bigcap_{k=1}^{\infty} \bigcup_{i_1, \ldots, i_k=1}^{K} \overline{B}(z_{i_1 i_2 \ldots i_k}, 2r_k).$$

Observe that $F \subset E \cap \overline{B}(x, 2R)$. Indeed, if $z \in F$, then $z \in \overline{B}(x, 2R)$, and since $z_{i_1 i_2 \ldots i_k} \in E$ and $r_k \to 0$ as $k \to \infty$, we have $z \in \overline{E} = E$.

Let μ be the equally distributed probability measure supported on F. That is, μ is constructed by distributing a unit mass to the set F via the balls $\overline{B}(z_{i_1 i_2 \ldots i_k}, 2r_k)$ in such a way that in each step all the balls $\overline{B}(z_{i_1 i_2 \ldots i_k}, 2r_k)$ obtain the same amount of mass. Then

$$\mu(E \cap \overline{B}(x, 2R)) = \mu(F) = 1, \tag{7.8}$$

and more generally

$$\mu\big(E \cap \overline{B}(z_{i_1 i_2 \dots i_k}, 2r_k)\big) = K^{-k}, \tag{7.9}$$

for every $1 \le i_1, i_2, \dots, i_k \le K$. The existence of such a measure μ can be proved using the Carathéodory construction with respect to the balls $\overline{B}(z_{i_1 i_2 \dots i_k}, 2r_k)$. We refer to Falconer [**125**, pp. 13–14] and Mattila [**315**, pp. 54–55] for further details.

Let $y \in \mathbb{R}^n$ and $0 < r < R$, and choose $k \in \mathbb{N}$ with $Rm^{-k} \le r < Rm^{-k+1}$. By the pairwise disjointness of the balls $\overline{B}(z_{i_1 i_2 \dots i_k}, 2r_k)$, there exists a constant $C_2 = C_2(n, m)$ such that $\overline{B}(y, r)$ intersects at most C_2 of the balls $\overline{B}(z_{i_1 i_2 \dots i_k}, 2r_k)$ from the kth step of the construction. By (7.9) and the choice of k, we obtain

$$\mu(B(y,r)) \le \mu\big(F \cap \overline{B}(y,r)\big) \le C_2 K^{-k} = C_2 m^{-k\frac{\log K}{\log m}} \le C_2 \Big(\frac{r}{R}\Big)^{\lambda}, \tag{7.10}$$

where we also used the fact that $\lambda < \frac{\log K}{\log m}$.

Let $\{B(x_i, r_i) : i \in \mathbb{N}\}$ be a cover of $E \cap \overline{B}(x, 2R)$. First assume that $r_i < R$ for every $i \in \mathbb{N}$. Applying (7.8) and (7.10), we obtain

$$1 = \mu\big(E \cap \overline{B}(x, 2R)\big) \le \sum_{i=1}^{\infty} \mu(B(x_i, r_i)) \le \sum_{i=1}^{\infty} C(n, \lambda_0, \lambda, C_1) \Big(\frac{r_i}{R}\Big)^{\lambda}.$$

Thus

$$(2R)^{\lambda} \le C(n, \lambda_0, \lambda, C_1) \sum_{i=1}^{\infty} r_i^{\lambda}. \tag{7.11}$$

Observe that (7.11) also holds in the case that $r_j \ge R$ for some $j \in \mathbb{N}$. The claim (7.7) for $2R$ follows by taking infimum over all such covers. □

Lemma 7.13 has several consequences. In the following theorem the Hausdorff content density condition is obtained for all radii up to the diameter of the set.

THEOREM 7.14. *Let $E \subset \mathbb{R}^n$ be a closed set and assume that $0 \le \lambda < \dim_{\mathrm{L}}(E)$. Then there exists a constant $C = C(n, \lambda, E)$ such that*

$$\mathcal{H}^{\lambda}_{\infty}\big(E \cap \overline{B}(x, R)\big) \ge CR^{\lambda}$$

for every $x \in E$ and $0 < R \le \mathrm{diam}(E)$. Moreover, $\dim_{\mathrm{L}}(E)$ is the supremum of the exponents $\lambda \ge 0$ for which such a constant C exists.

PROOF. Let $0 \le \lambda < \lambda_0 < \dim_{\mathrm{L}}(E)$. The definition of the lower dimension implies that (7.6) holds with a constant C_1 for every $x \in E$ and $0 < r < R \le \mathrm{diam}(E)$. Thus we obtain from Lemma 7.13 that

$$\mathcal{H}^{\lambda}_{\infty}\big(E \cap \overline{B}(x, R)\big) \ge C(n, \lambda, \lambda_0, C_1) R^{\lambda}$$

for every $x \in E$ and $0 < R \le \mathrm{diam}(E)$. By choosing $\lambda_0 = \frac{1}{2}(\dim_{\mathrm{L}}(E) + \lambda)$, the constant above only depends on n, λ and E.

Assume then that

$$\mathcal{H}^{\lambda}_{\infty}\big(E \cap \overline{B}(x, R)\big) \ge CR^{\lambda}$$

for every $x \in E$ and $0 < R \le \mathrm{diam}(E)$. Let $x \in E$ and $0 < r < R \le \mathrm{diam}(E)$, and let $\{B(x_i, r) : i = 1, \dots, N\}$ be a cover of $E \cap \overline{B}(x, R)$. Then

$$R^{\lambda} \le C\mathcal{H}^{\lambda}_{\infty}\big(E \cap \overline{B}(x, R)\big) \le C \sum_{i=1}^{N} r^{\lambda} = CNr^{\lambda},$$

and thus $N \geq C\left(\frac{R}{r}\right)^{\lambda}$. Since this holds for all such covers, we have

$$N\big(E \cap \overline{B}(x,R), r\big) \geq C\left(\frac{R}{r}\right)^{\lambda}.$$

This implies that $\dim_{\mathrm{L}}(E) \geq \lambda$, and the proof is complete. □

Theorem 7.14 gives a comparison between Hausdorff and lower dimensions of closed sets.

COROLLARY 7.15. *Assume that $E \subset \mathbb{R}^n$ is a closed set and let $x \in E$ and $r > 0$. Then $\dim_{\mathrm{L}}(E) \leq \dim_{\mathrm{H}}\big(E \cap \overline{B}(x,r)\big)$.*

PROOF. We may clearly assume that $\dim_{\mathrm{L}}(E) > 0$ and $0 < r \leq \operatorname{diam}(E)$. Let $0 \leq \lambda < \dim_{\mathrm{L}}(E)$. By Theorem 7.14, we have $\mathcal{H}^{\lambda}_{\infty}\big(E \cap \overline{B}(x,r)\big) > 0$. Hence $\lambda \leq \dim_{\mathrm{H}}(E \cap \overline{B}(x,r))$, and the claim follows. □

REMARK 7.16. The assumption that E is closed is necessary in Corollary 7.15. It is easy to see that $\dim_{\mathrm{L}}(\overline{E}) = \dim_{\mathrm{L}}(E)$ for all $E \subset \mathbb{R}^n$, and hence for instance

$$\dim_{\mathrm{L}}(\mathbb{Q}^n) = \dim_{\mathrm{L}}(\mathbb{R}^n) = n \not\leq 0 = \dim_{\mathrm{H}}\big(\mathbb{Q}^n \cap B(x,r)\big)$$

for every $x \in \mathbb{Q}^n$ and $r > 0$.

By combining Theorem 7.14 and Corollary 7.5, we obtain a sufficient condition for Hardy inequalities in terms of the lower dimension of the complement of the set. See also Theorem 7.22 for a converse implication and Corollary 10.36 for a dual pair of this statement for Hardy inequalities in terms of the Assouad dimension of the complement of the set.

THEOREM 7.17. *Let $1 < p < \infty$. Assume that $\Omega \subset \mathbb{R}^n$ is an open set such that Ω^c is unbounded and $\dim_{\mathrm{L}}(\Omega^c) > n-p$. Then the pointwise p-Hardy inequality and the p-Hardy inequality hold in Ω.*

PROOF. Let $\lambda \geq 0$ be such that $\dim_{\mathrm{L}}(\Omega^c) > \lambda > n-p$. By Theorem 7.14 and the unboundedness of Ω^c, the complement Ω^c satisfies the λ-Hausdorff content density condition (7.1). The claim follows from Corollary 7.5. □

7.4. Density conditions and Hardy's inequality in the borderline case

This section discusses density conditions and Hardy's inequalities in the borderline case $p = n$. We begin with the following metric definition.

DEFINITION 7.18. A set $E \subset \mathbb{R}^n$ is uniformly perfect if E contains more than one point and there exists a constant C, with $0 < C < 1$, such that

$$E \cap \big(\overline{B}(x,R) \setminus B(x,CR)\big) \neq \emptyset$$

for every $x \in E$ and $0 < R \leq \operatorname{diam}(E)$.

Uniform perfectness can be characterized in terms of the lower dimension.

LEMMA 7.19. *A set $E \subset \mathbb{R}^n$ is uniformly perfect if and only if $\dim_{\mathrm{L}}(E) > 0$.*

PROOF. Assume first that $E \subset \mathbb{R}^n$ is uniformly perfect, with a constant $0 < C < 1$. We show that there exists $\lambda > 0$ such that (7.5) holds for every $x \in E$ and $0 < r < R \leq \operatorname{diam}(E)$. Fix $x_0 \in E$ and $0 < r < R \leq \operatorname{diam}(E)$. By uniform perfectness, there exists $x_1 \in E \cap \overline{B}\big(x_0, \frac{R}{2}\big)$ such that $|x_1 - x_0| \geq C\frac{R}{2}$, and thus at least two disjoint balls $B_0, B_1 \subset B(x_0,R)$ of radius $8^{-1}CR$ are needed to cover $E \cap \overline{B}(x_0,R)$.

By applying a similar reasoning to the balls $\overline{B}(x_0, 8^{-1}CR)$ and $\overline{B}(x_1, 8^{-1}CR)$, which do not intersect, we obtain that at least four pairwise disjoint balls of radius $(8^{-1}C)^2R$ are needed to cover $E \cap \overline{B}(x_0,R)$. Continuing recursively we conclude that, for every $N \in \mathbb{N}_0$, at least 2^N balls of radius $(8^{-1}C)^N R$ are needed to cover $E \cap \overline{B}(x_0,R)$.

Let $N \in \mathbb{N}_0$ be such that

$$(8^{-1}C)^{N+1} \leq \frac{r}{R} < (8^{-1}C)^N,$$

and let

$$\lambda = -\frac{\log 2}{\log(8^{-1}C)} > 0.$$

We have

$$\begin{aligned} N\big(E \cap \overline{B}(x_0,R), r\big) &\geq N\big(E \cap \overline{B}(x_0,R), (8^{-1}C)^N R\big) \\ &\geq 2^N = \frac{1}{2}2^{N+1} = \frac{1}{2}\big((8^{-1}C)^{N+1}\big)^{-\lambda} \geq \frac{1}{2}\left(\frac{R}{r}\right)^{\lambda}, \end{aligned}$$

and thus $\dim_{\mathrm{L}}(E) \geq \lambda > 0$.

Assume then that $\dim_{\mathrm{L}}(E) > 0$ and let $0 < \lambda < \dim_{\mathrm{L}}(E)$. There exists $0 < C_1 < 1$ such that for every $x \in E$ and $0 < r < R \leq \operatorname{diam}(E)$ we have

$$N\big(E \cap \overline{B}(x,R), r\big) \geq C_1 \left(\frac{R}{r}\right)^{\lambda}.$$

Choose $C = \left(\frac{C_1}{2}\right)^{\frac{1}{\lambda}}$ and $r = CR$. Then $N\big(E \cap \overline{B}(x,R), CR\big) \geq 2$, and thus $B(x,CR)$ does not cover $E \cap \overline{B}(x,R)$. We conclude that $E \cap \big(\overline{B}(x,R) \setminus B(x,CR)\big) \neq \emptyset$, and this proves that E is uniformly perfect. $\square$

The next theorem contains several equivalent statements, which are specific to the borderline case $p = n$. The proof below avoids the use of the improvement of Hardy inequalities (Lemma 6.15) and the fact that the p-capacity density condition implies the p-Hardy inequality (Theorem 6.25). However, the crux of the proof is the fact that uniform perfectness, through positivity of the lower dimension, implies the Hausdorff content density condition.

THEOREM 7.20. *Assume that $\Omega \subsetneq \mathbb{R}^n$ is an open set. The following conditions are equivalent.*

(a) *The n-Hardy inequality holds in Ω.*
(b) *Ω^c satisfies the n-capacity density condition.*
(c) *Ω^c is uniformly perfect and unbounded.*
(d) *$\dim_{\mathrm{L}}(\Omega^c) > 0$ and Ω^c is unbounded.*
(e) *There exists $\lambda > 0$ such that Ω^c satisfies the λ-Hausdorff content density condition.*
(f) *Ω^c satisfies the p-capacity density condition for some $1 < p < n$.*
(g) *The pointwise p-Hardy inequality holds in Ω for some $1 < p < n$.*
(h) *The pointwise n-Hardy inequality holds in Ω.*

Moreover, the constants in each of the conditions only depend on each other, n and p.

PROOF. First we show that (a) implies (c). Let $y \in \Omega^c$ and $R > 0$. If

$$\Omega^c \cap \big(\overline{B}(y,R) \setminus B(y, \tfrac{1}{8}R)\big) \neq \emptyset,$$

then the uniform perfectness condition holds for y and R with the constant $\frac{1}{8}$ and the diameter of Ω^c is at least $\frac{1}{8}R$. Hence it suffices to consider the cases where $\Omega^c \cap \big(\overline{B}(y,R) \setminus B(y,\frac{1}{8}R)\big) = \emptyset$, which implies that $\Omega^c \cap \overline{B}(y,R) \subset B(y, C_1 R)$ for some $0 < C_1 \leq \frac{1}{8}$. We show that there is a uniform lower bound for such $C_1 > 0$, and this proves the claim.

Let $u \colon \mathbb{R}^n \to \mathbb{R}$,

$$u(x) = \begin{cases} 0, & \text{if } |x-y| \leq C_1 R \text{ or } |x-y| \geq R, \\ \frac{1}{C_1 R}|x-y| - 1, & \text{if } C_1 R < |x-y| < 2C_1 R, \\ 1, & \text{if } 2C_1 R \leq |x-y| \leq \frac{R}{2}, \\ 2 - \frac{2}{R}|x-y|, & \text{if } \frac{R}{2} < |x-y| < R. \end{cases}$$

Observe that u is supported in the annulus $\overline{B}(y,R) \setminus B(y, C_1 R) \subset \Omega$ and $u \in \operatorname{Lip}_0(\Omega)$. Let $A = B(y, \frac{R}{2}) \setminus B(y, 2C_1 R)$. For every $x \in A$, we have $u(x) = 1$ and $d(x, \partial\Omega) \leq |x-y|$, and thus

$$\begin{aligned} \int_\Omega \frac{|u(x)|^n}{d(x,\partial\Omega)^n}\,dx &\geq \int_A d(x,\partial\Omega)^{-n}\,dx \geq \int_A |x-y|^{-n}\,dx \\ &= C(n)\Big(\log\frac{R}{2} - \log(2C_1 R)\Big) = -C(n)\log(4C_1). \end{aligned}$$

The n-Hardy inequality implies

$$\begin{aligned} -C(n)\log(4C_1) &\leq \int_\Omega \frac{|u(x)|^n}{d(x,\partial\Omega)^n}\,dx \leq C_2 \int_\Omega |\nabla u(x)|^n\,dx \\ &\leq C_2\bigg(|B(y,2C_1R)|(C_1R)^{-n} + |B(y,R)|\Big(\frac{R}{2}\Big)^{-n}\bigg) \leq C(n,C_2), \end{aligned}$$

where C_2 is the constant in the assumed n-Hardy inequality. Hence

$$-\log(4C_1) \leq \frac{C(n,C_2)}{C(n)}, \quad \text{and so} \quad C_1 \geq \frac{1}{4}e^{-\frac{C(n,C_2)}{C(n)}}.$$

This shows that Ω^c is uniformly perfect and unbounded.

Then we show that (b) implies (c). Let $y \in \Omega^c$ and $R > 0$, and assume that $\Omega^c \cap \overline{B}(y,R) \subset B(y, C_1 R)$ for some $0 < C_1 < 1$. The n-capacity density condition of Ω^c, monotonicity of the capacity, and Lemma 5.36 (c) give

$$\begin{aligned} C(n,C_2) &\leq \operatorname{cap}_n\big(\Omega^c \cap \overline{B}(y,R), B(y,2R)\big) \\ &\leq \operatorname{cap}_n\big(\overline{B}(y,C_1R), B(y,2R)\big) \leq C(n)\Big(\log\frac{2}{C_1}\Big)^{1-n}, \end{aligned}$$

where C_2 is the constant in the n-capacity density condition. Hence

$$\log\frac{2}{C_1} \leq \bigg(\frac{C(n,C_2)}{C(n)}\bigg)^{\frac{1}{1-n}}, \quad \text{and so} \quad C_1 \geq 2e^{-\left(\frac{C(n,C_2)}{C(n)}\right)^{\frac{1}{1-n}}} > 0.$$

This shows that Ω^c is uniformly perfect and also unbounded.

The assertion (c) implies (d) is the necessity part of Lemma 7.19.

The assertion (d) implies (e) follows from Theorem 7.14 for every $0 < \lambda < \dim_{\mathrm{L}}(\Omega^c)$. In particular, since Ω^c is assumed to be unbounded, the density condition holds for every $0 < R < \infty$.

The assertion (e) implies (f) follows from Theorem 7.3.

The assertion (f) implies (g) follows from the implication from (a) to (d) in Theorem 6.23.

The assertion (g) implies (h) is a special case of Remark 6.5.

The assertion (h) implies (b) is the implication from (d) to (a) in Theorem 6.23, for $p = n$.

The assertion (g) implies (a) follows from Remark 6.6. □

7.5. Self-improvement of the capacity density condition

The implication from (b) to (f) in Theorem 7.20 gives a relatively elementary proof for the self-improvement of the capacity density condition for $p = n$. Our next goal is to show that a corresponding improvement holds for $1 < p < n$.

THEOREM 7.21. *Let $1 < p < \infty$ and assume that a closed set $E \subset \mathbb{R}^n$ satisfies the p-capacity density condition, with a constant C_0. Then there exists $1 < q < p$ such that E satisfies the q-capacity density condition. Moreover, the exponent q and the constant in the q-capacity density condition depend on n, p and C_0 in an explicit and quantitative fashion.*

The proof of Theorem 7.21, which will be completed in Section 7.6, Section 7.7 and Section 7.8, is based on quantitative estimates and absorption arguments, where it is crucial to track the dependencies of constants. Before proceeding to this somewhat lengthy proof, we record the following consequence of Theorem 7.21. Compare to Theorem 7.20 for $p = n$.

THEOREM 7.22. *Let $1 < p \leq n$ and assume that $E \subset \mathbb{R}^n$ is a closed set. The following conditions are equivalent.*

(a) *E satisfies the p-capacity density condition.*
(b) *There exists $\lambda > n-p$ such that E satisfies the λ-Hausdorff content density condition.*
(c) *$\dim_{\mathrm{L}}(E) > n - p$ and E is unbounded.*
(d) *The pointwise p-Hardy inequality holds in the open set $\Omega = E^c$.*

Moreover, the constants in each of the conditions only depend on each other, n and p.

PROOF. First we show that (a) implies (b). By Theorem 7.21, there exists $1 < q < p$ such that E satisfies the q-capacity density condition, with a constant C_1. Let $\lambda = n - q > n - p \geq 0$. As in Remark 7.4, we obtain

$$\begin{aligned} r^\lambda = r^{n-q} &\leq C(n,q,C_1)\,\mathrm{cap}_q\big(E \cap \overline{B}(x,r), B(x,2r)\big) \\ &\leq C(n,q,C_1)\mathcal{H}^\lambda_\infty\big(E \cap \overline{B}(x,r)\big) \end{aligned}$$

for every $x \in E$ and $r > 0$, and this proves the claim.

The assertion (b) implies (c) follows from Theorem 7.14. Note that if E is nonempty and bounded, then the λ-Hausdorff content density condition fails for large $r > 0$.

The assertion (c) implies (d) is Theorem 7.17.

The assertion (d) implies (a) is the implication from (d) to (a) in Theorem 6.23. □

REMARK 7.23. Theorem 7.22 implies that $\operatorname{cap}_p(E, B(x,2r))$ and $\mathcal{H}^{n-p}_\infty(E)$ of a set $E \subset \overline{B}(x,r)$ are not equivalent for $1 < p < n$. To see this, let $F \subset \mathbb{R}^n$ be $(n-p)$-regular and unbounded, and assume for contradiction that there exists a constant C_1 such that

$$\mathcal{H}^{n-p}_\infty\big(F \cap \overline{B}(x,r)\big) \le C_1 \operatorname{cap}_p\big(F \cap \overline{B}(x,r), B(x,2r)\big)$$

for every $x \in F$ and $r > 0$. Since $\mathcal{H}^{n-p}_\infty\big(F \cap \overline{B}(x,r)\big) \ge C_2 r^{n-p}$ by Lemma 7.8, we conclude that F satisfies the p-capacity density condition. By Theorem 7.22 there exists $\lambda > n-p$ such that F satisfies the λ-Hausdorff content density condition. Consequently $\dim_{\mathrm{H}}(F) \ge \lambda$, which contradicts the fact that $\dim_{\mathrm{H}}(F) = n - p < \lambda$ by the $(n-p)$-regularity.

To obtain the improvement of the p-capacity density condition in Theorem 7.21, it will be sufficient to show that the p-capacity density condition of E implies (6.20) for every $z \in E$ and $r > 0$, but with an exponent $q < p$ instead of p. This is the content of the next theorem. The proof of Theorem 7.21 is a combination of Theorem 6.23 and Theorem 7.24.

THEOREM 7.24. *Let $1 < p < \infty$ and assume that $E \subset \mathbb{R}^n$ is a closed set which satisfies the p-capacity density condition with a constant C_0. There exist $\varepsilon = \varepsilon(n,p,C_0)$, with $0 < \varepsilon < p-1$, and a constant $C = C(n,p,C_0)$ such that*

$$\int_{B(z,r)} |u(x)|^{p-\varepsilon}\,dx \le C r^{p-\varepsilon} \int_{B(z,r)} |\nabla u(x)|^{p-\varepsilon}\,dx$$

for every $u \in \operatorname{Lip}_0(E^c)$, $z \in E$ and $r > 0$.

PROOF. Let $z \in E$, $r > 0$ and $u \in \operatorname{Lip}_0(E^c)$. Let $F \subset E \cap \overline{B}(z, \frac{r}{12})$ be any closed set with $z \in F$. Since

$$\begin{aligned}\int_{B(z,r)} \frac{|u(x)|^{p-\varepsilon}}{r^{p-\varepsilon}}\,dx &= \int_{B(z,r)\cap F} \frac{|u(x)|^{p-\varepsilon}}{r^{p-\varepsilon}}\,dx + \int_{B(z,r)\setminus F} \frac{|u(x)|^{p-\varepsilon}}{r^{p-\varepsilon}}\,dx\\ &\le \int_{B(z,r)\setminus F} \frac{|u(x)|^{p-\varepsilon}}{d(x,F)^{p-\varepsilon}}\,dx,\end{aligned}$$

it suffices to find ε and C, and a closed set F as above, such that

$$\int_{B(z,r)\setminus F} \frac{|u(x)|^{p-\varepsilon}}{d(x,F)^{p-\varepsilon}}\,dx \le C \int_{B(z,r)} |\nabla u(x)|^{p-\varepsilon}\,dx. \tag{7.12}$$

We prove this improved local Hardy inequality in Theorem 7.32 below, and the claim follows. □

7.6. Truncation and absorption

In this section we establish some technical tools that will be used in the proofs of the local Hardy inequalities in Sections 7.7 and 7.8. The following truncation procedure provides us with the closed set F that is needed in the proof of Theorem 7.24. Here the ball $\overline{B}(z,r)$ corresponds to the ball $\overline{B}(z, \frac{r}{12})$ in the proof of Theorem 7.24.

LEMMA 7.25. *Assume that $E \subset \mathbb{R}^n$ is a closed set and let $z \in E$ and $r > 0$. Let $F_0 = E \cap \overline{B}(z, \frac{r}{2})$ and define recursively*

$$F_j = \bigcup_{x \in F_{j-1}} E \cap \overline{B}(x, 2^{-j-1}r)$$

for every $j \in \mathbb{N}$. Let $F = \overline{\bigcup_{j \in \mathbb{N}_0} F_j}$. Then $z \in F \subset E \cap \overline{B}(z, r)$, and $F_{j-1} \subset F_j \subset F$ for every $j \in \mathbb{N}$.

PROOF. The claim $z \in F$ is clear since $z \in F_0 \subset F$, and the claim $F \subset E$ follows since E is closed and $\bigcup_{j \in \mathbb{N}_0} F_j \subset E$ by definition. To verify $F \subset \overline{B}(z, r)$, let $x \in F_j$. If $j = 0$, then $x \in \overline{B}(z, r)$. If $j \in \mathbb{N}$, then by induction we find a sequence $x_j, \ldots, x_0$ such that $x_j = x$, $x_k \in F_k$ for every $k = 0, \ldots, j$, and $x_k \in E \cap \overline{B}(x_{k-1}, 2^{-k-1}r)$ for every $k = 1, \ldots, j$. It follows that

$$|x - z| \leq \sum_{k=1}^{j} |x_k - x_{k-1}| + |x_0 - z| \leq \sum_{k=1}^{j} 2^{-k-1}r + 2^{-1}r < r,$$

and thus $x \in \overline{B}(z, r)$. We have shown that $F_j \subset \overline{B}(z, r)$ for every $j \in \mathbb{N}_0$, and hence $F \subset \overline{B}(z, r)$. To prove $F_{j-1} \subset F_j \subset F$, let $j \in \mathbb{N}$ and $x \in F_{j-1}$. Since $x \in E$, we conclude that $x \in E \cap \overline{B}(x, 2^{-j-1}r) \subset F_j$. □

Next we show that the truncated set F in Lemma 7.25 has the property that there always exist certain balls whose intersection with F contain big pieces of the original set E.

LEMMA 7.26. *Assume that E, $B(z, r)$ and F are as in Lemma 7.25. Let $m \in \mathbb{N}_0$ and $x \in \mathbb{R}^n$ be such that $d(x, F) < 2^{-m+1}r$, and write $r_m = 2^{-m-1}r$. There exists $y_{x,m} \in E$ such that $\overline{B}(y_{x,m}, r_m) \subset B(x, 8r_m)$ and*

$$E \cap \overline{B}(y_{x,m}, \tfrac{1}{2}r_m) = F \cap \overline{B}(y_{x,m}, \tfrac{1}{2}r_m). \tag{7.13}$$

PROOF. In this proof we will apply Lemma 7.25 without further notice. Since $d(x, F) < 2^{-m+1}r$, there exists $y \in \bigcup_{j \in \mathbb{N}_0} F_j \subset E$ satisfying $|y - x| < 2^{-m+1}r$. Let $j \in \mathbb{N}_0$ be such that $y \in F_j$. We consider two cases.

First, assume that $j > m \geq 0$. By induction, there are points $y_k \in F_k$ with $k = m, \ldots, j$ such that $y_j = y$ and $y_k \in E \cap \overline{B}(y_{k-1}, 2^{-k-1}r)$ for every $k = m+1, \ldots, j$. It follows that

$$|y_m - y| = |y_j - y_m| \leq \sum_{k=m+1}^{j} |y_k - y_{k-1}| \leq \sum_{k=m+1}^{j} 2^{-k-1}r < 2^{-m-1}r.$$

Let $y_{x,m} = y_m \in F_m \subset E$. If $y' \in \overline{B}(y_m, 2^{-m-1}r)$, then

$$\begin{aligned} |y' - x| &\leq |y' - y_m| + |y_m - y| + |y - x| \\ &< 2^{-m-1}r + 2^{-m-1}r + 2^{-m+1}r < 2^{-m+2}r, \end{aligned}$$

and thus $\overline{B}(y_m, 2^{-m-1}r) \subset B(x, 2^{-m+2}r)$. Moreover, since $y_m \in F_m$, we have

$$E \cap \overline{B}(y_m, 2^{-m-2}r) \subset \bigcup_{y' \in F_m} E \cap \overline{B}(y', 2^{-m-2}r) = F_{m+1} \subset F.$$

On the other hand $F \subset E$, and (7.13) follows.

Then consider the case $m \geq j \geq 0$. Let $y_{x,m} = y \in E$. We have

$$|y' - x| \leq |y' - y| + |y - x| < 2^{-m-1}r + 2^{-m+1}r < 2^{-m+2}r$$

for every $y' \in \overline{B}(y, 2^{-m-1}r)$. This implies $\overline{B}(y, 2^{-m-1}r) \subset B(x, 2^{-m+2}r)$. Since $y \in F_j \subset F_m$, we can now repeat the final part of the argument above, with y_m replaced by y, and (7.13) follows also in this case. $\square$

The main reason for truncating the set E is to obtain the following absorption result that is needed twice during the subsequent arguments. Dependencies of the constants below are delicate. Note that inequality (7.15) is useful only if $C_3 > 0$. This can be ensured by requiring that C_1 is sufficiently small, depending on n, p, σ and ς, but not depending on q.

LEMMA 7.27. *Let $1 < p < \infty$ and $1 \le q \le p$, and assume that E, $B(z,r)$ and F are as in Lemma 7.25. Moreover, let $\sigma \ge 1$ and $\varsigma \ge 2$, and assume that $u \in \operatorname{Lip}_0(F^c)$ satisfies the inequality*

$$
\begin{aligned}
&\int_{B(z,\varsigma r)\setminus F} \frac{|u(x)|^q}{d(x,F)^q}\,dx \\
&\qquad \le C_1 \int_{B(z,\sigma\varsigma r)\setminus F} \frac{|u(x)|^q}{d(x,F)^q}\,dx + C_2 \int_{B(z,\sigma\varsigma r)} |\nabla u(x)|^q\,dx
\end{aligned}
\tag{7.14}
$$

with some constants C_1, C_2. Then

$$
C_3 \int_{B(z,\sigma\varsigma r)\setminus F} \frac{|u(x)|^q}{d(x,F)^q}\,dx \le C_4 \int_{B(z,\sigma\varsigma r)} |\nabla u(x)|^q\,dx,
\tag{7.15}
$$

where $C_3 = 1 - C_1(1 + C(n,p,\sigma,\varsigma))$ and $C_4 = (1 + C_2)C(n,q,\sigma,\varsigma)$.

PROOF. Since $F \subset \overline{B}(z,r)$ and $\varsigma \ge 2$, we have

$$
\begin{aligned}
&\int_{(B(z,\sigma\varsigma r)\setminus F)\setminus B(z,\varsigma r)} \frac{|u(x)|^q}{d(x,F)^q}\,dx \le r^{-q} \int_{B(z,\sigma\varsigma r)} |u(x)|^q\,dx \\
&\le 3^q r^{-q} \Big(\int_{B(z,\sigma\varsigma r)} \big|u(x) - u_{B(z,\sigma\varsigma r)}\big|^q\,dx \\
&\qquad + |B(z,\sigma\varsigma r)| \big|u_{B(z,\sigma\varsigma r)} - u_{B(z,\varsigma r)}\big|^q + |B(z,\sigma\varsigma r)| \big|u_{B(z,\varsigma r)}\big|^q \Big).
\end{aligned}
$$

Theorem 3.14 implies

$$
\begin{aligned}
&3^q r^{-q} \Big(\int_{B(z,\sigma\varsigma r)} \big|u(x) - u_{B(z,\sigma\varsigma r)}\big|^q\,dx + |B(z,\sigma\varsigma r)| \big|u_{B(z,\sigma\varsigma r)} - u_{B(z,\varsigma r)}\big|^q \Big) \\
&\le C(n,\sigma,q) r^{-q} \int_{B(z,\sigma\varsigma r)} \big|u(x) - u_{B(z,\sigma\varsigma r)}\big|^q\,dx \\
&\le C(n,q,\sigma,\varsigma) \int_{B(z,\sigma\varsigma r)} |\nabla u(x)|^q\,dx.
\end{aligned}
$$

On the other hand, since $q \le p$ and $u = 0$ in F, by (7.14) we obtain

$$\begin{aligned}3^q r^{-q}|B(z,\sigma\varsigma r)|\big|u_{B(z,\varsigma r)}\big|^q &\le 3^q C(n,\sigma) r^{-q}\int_{B(z,\varsigma r)\setminus F}|u(x)|^q\,dx\\ &\le 3^q\varsigma^q C(n,\sigma)\int_{B(z,\varsigma r)\setminus F}\frac{|u(x)|^q}{d(x,F)^q}\,dx\\ &\le 3^p\varsigma^p C(n,\sigma)C_1\int_{B(z,\sigma\varsigma r)\setminus F}\frac{|u(x)|^q}{d(x,F)^q}\,dx\\ &\quad+3^q\varsigma^q C(n,\sigma)C_2\int_{B(z,\sigma\varsigma r)}|\nabla u(x)|^q\,dx.\end{aligned}$$

Combining the estimates above, we find that

$$\begin{aligned}&\int_{B(z,\sigma\varsigma r)\setminus F}\frac{|u(x)|^q}{d(x,F)^q}\,dx\\ &\quad=\int_{B(z,\varsigma r)\setminus F}\frac{|u(x)|^q}{d(x,F)^q}\,dx+\int_{(B(z,\sigma\varsigma r)\setminus F)\setminus B(z,\varsigma r)}\frac{|u(x)|^q}{d(x,F)^q}\,dx\\ &\quad\le C_1(1+C(n,p,\sigma,\varsigma))\int_{B(z,\sigma\varsigma r)\setminus F}\frac{|u(x)|^q}{d(x,F)^q}\,dx+C_4\int_{B(z,\sigma\varsigma r)}|\nabla u(x)|^q\,dx,\end{aligned}$$

where $C_4 = (1 + C_2)C(n, q, \sigma, \varsigma)$. Let $0 \le L < \infty$ be a Lipschitz constant of $u \in \operatorname{Lip}_0(F^c)$. Then

$$\sup_{x\in F^c}|u(x)|^q d(x,F)^{-q}\le L^q,$$

and therefore the first integral on the right-hand side is finite. Absorption of this integral to the left-hand side completes the proof. □

7.7. Local Hardy inequality

In this section we prove a local Hardy inequality with respect to the truncated set F, see Theorem 7.31. The strategy is to adapt Wannebo's argument in Section 6.5, together with covering and absorption arguments, to the present context.

For the rest of this section, we let $1 < p < \infty$ and assume that a closed set $E \subset \mathbb{R}^n$ satisfies the p-capacity density condition with a constant C_0 and that F is as in Lemma 7.25, with fixed $z \in E$ and $r > 0$. We begin with Lemma 7.28 and Lemma 7.29, which provide information concerning the individual balls in the following collections $\mathcal{B}_m$.

Let $A_m = \{x \in B(z, 8r) : 2^{-m}r \le d(x, F) < 2^{-m+1}r\}$ and

$$\Omega_m=\bigcup_{k=m}^{\infty}A_k=\{x\in B(z,8r):0<d(x,F)<2^{-m+1}r\}$$

for $m \in \mathbb{Z}$. By Lemma 1.13, for every A_m there exists a cover

$$\mathcal{B}_m=\{B(x_i,2^{-m+2}r):x_i\in A_m,\ i=1,\dots,N_m\},$$

such that the balls $B(x_i, \frac{1}{5}2^{-m+2}r)$, $i = 1,\dots,N_m$, are pairwise disjoint. The balls in $\mathcal{B}_m$ have uniformly bounded overlap, that is,

$$\sum_{i=1}^{N_m}\chi_{B(x_i,2^{-m+2}r)}(x)\le C(n), \tag{7.16}$$

for every $x \in \mathbb{R}^n$ and $m \in \mathbb{N}_0$.

LEMMA 7.28. *Let $m \in \mathbb{N}_0$ and assume that $B(x_i, 2^{-m+2}r) \in \mathcal{B}_m$. Then*

$$B(x_i, 2^{-m+2}r) \setminus F \subset \Omega_{m-2}. \tag{7.17}$$

PROOF. Let $x \in B(x_i, 2^{-m+2}r) \setminus F$. Then $d(x, F) > 0$, and

$$d(x, F) \leq |x - x_i| + d(x_i, F) < 2^{-m+2}r + 2^{-m+1}r < 2^{-(m-2)+1}r.$$

Since $m \in \mathbb{N}_0$, we have

$$|x - z| \leq d(x, \overline{B}(z, r)) + r \leq d(x, F) + r < 2^{-m+2}r + 2^{-m+1}r + r < 8r,$$

and thus $x \in B(z, 8r)$. Estimates above imply that $x \in \Omega_{m-2}$. □

The following lemma is a substitute for the boundary Poincaré inequality, see Theorem 6.22, to the present setting. Note that this is the only place in the proof of Theorem 7.21 where the p-capacity density condition, with the constant C_0, is applied.

LEMMA 7.29. *Let $m \in \mathbb{N}_0$ and assume that $B(x_i, 2^{-m+2}r) \in \mathcal{B}_m$ and $v \in \operatorname{Lip}_0(F^c)$. Then*

$$\int_{B(x_i, 2^{-m+2}r)} |v(x)|^p \, dx \leq C 2^{-mp} r^p \int_{B(x_i, 2^{-m+2}r)} |\nabla v(x)|^p \, dx, \tag{7.18}$$

with $C = C(n, p, C_0)$.

PROOF. Let $B_i = B(x_i, 2^{-m+2}r)$. Lemma 7.26, applied to m and x_i, gives a closed ball $\overline{B}(y, r_m) \subset B_i$, with $y \in E$ and $r_m = 2^{-m-1}r$. Note that

$$\fint_{B_i} |v(x)|^p \, dx \leq C(p) \Big(\fint_{B_i} |v(x) - v_{B_i}|^p \, dx + \big|v_{B_i} - v_{\overline{B}(y, r_m)}\big|^p + \big|v_{\overline{B}(y, r_m)}\big|^p \Big). \tag{7.19}$$

By Hölder's inequality,

$$\big|v_{B_i} - v_{\overline{B}(y, r_m)}\big|^p \leq \Big(\fint_{\overline{B}(y, r_m)} |v(x) - v_{B_i}| \, dx \Big)^p \leq C(n) \fint_{B_i} |v(x) - v_{B_i}|^p \, dx,$$

and by Theorem 3.14,

$$\begin{aligned} &C(p) \Big(\fint_{B_i} |v(x) - v_{B_i}|^p \, dx + \big|v_{B_i} - v_{\overline{B}(y, r_m)}\big|^p \Big) \\ &\qquad \leq C(n, p) 2^{-mp} r^p \fint_{B_i} |\nabla v(x)|^p \, dx. \end{aligned} \tag{7.20}$$

To estimate the last term in (7.19), we recall that by assumption $v = 0$ in F and, by Lemma 7.26, we have $E \cap \overline{B}(y, \frac{1}{2} r_m) = F \cap \overline{B}(y, \frac{1}{2} r_m)$. By the p-capacity density condition and the fact that $|B_i| = C(n)(2^{-m+2}r)^n$, we obtain

$$\begin{aligned} \operatorname{cap}_p\big(\{v = 0\} \cap \overline{B}(y, \tfrac{1}{2} r_m), B(y, r_m)\big) &\geq \operatorname{cap}_p\big(F \cap \overline{B}(y, \tfrac{1}{2} r_m), B(y, r_m)\big) \\ &= \operatorname{cap}_p\big(E \cap \overline{B}(y, \tfrac{1}{2} r_m), B(y, r_m)\big) \\ &\geq C(n, p, C_0)(2^{-m-2}r)^{n-p} \\ &= C(n, p, C_0)|B_i| 2^{mp} r^{-p}. \end{aligned}$$

Theorem 5.47 and the inclusions $B(y,r_m)\subset\overline{B}(y,r_m)\subset B_i$ imply

$$\begin{aligned}\left|v_{\overline{B}(y,r_m)}\right|^p&\le\fint_{\overline{B}(y,r_m)}|v(x)|^p\,dx\\&\le\frac{C(n,p)}{\operatorname{cap}_p\big(\{v=0\}\cap\overline{B}(y,\frac12 r_m),B(y,r_m)\big)}\int_{B(y,r_m)}|\nabla v(x)|^p\,dx\\&\le C(n,p,C_0)2^{-mp}r^p|B_i|^{-1}\int_{B_i}|\nabla v(x)|^p\,dx\\&\le C(n,p,C_0)2^{-mp}r^p\fint_{B_i}|\nabla v(x)|^p\,dx.\end{aligned}$$

The desired estimate (7.18) follows by substituting (7.20) and the estimate above to (7.19) and multiplying both sides by $|B_i|$. □

We turn to the proof of the local Hardy inequality. As in Section 6.5, we first prove a weighted version of the inequality. This is then bootstrapped in Theorem 7.31 to the unweighted inequality by choosing a suitable test function.

LEMMA 7.30. *Let $-\frac12<\beta<0$ and assume that $v\in\operatorname{Lip}_0(F^c)$. Then*

$$\int_{B(z,2r)\setminus F}|v(x)|^p d(x,F)^{\beta-p}\,dx\le C\int_{B(z,8r)\setminus F}|\nabla v(x)|^p d(x,F)^\beta\,dx,\tag{7.21}$$

with $C=\frac{1}{|\beta|}C(n,p,C_0)$.

PROOF. Since $v\in\operatorname{Lip}_0(F^c)$, we have $\nabla v(x)=0$ for almost every $x\in F$. By (7.18), (7.16) and (7.17), we obtain

$$\begin{aligned}\int_{A_m}|v(x)|^p\,dx&\le\sum_{i=1}^{N_m}\int_{B(x_i,2^{-m+2}r)}|v(x)|^p\,dx\\&\le C(n,p,C_0)2^{-mp}r^p\sum_{i=1}^{N_m}\int_{B(x_i,2^{-m+2}r)\setminus F}|\nabla v(x)|^p\,dx\\&\le C(n,p,C_0)2^{-mp}r^p\int_{\Omega_{m-2}}|\nabla v(x)|^p\,dx\\&\le C(n,p,C_0)2^{-mp}r^p\sum_{k=m-2}^{\infty}\int_{A_k}|\nabla v(x)|^p\,dx\end{aligned}\tag{7.22}$$

for every $m\in\mathbb{N}_0$.

Since

$$B(z,2r)\setminus F\subset\Omega_0=\big\{x\in B(z,8r):0<d(x,F)<2r\big\}$$

and $d(x,F)^{\beta-p} \le 2^{mp-m\beta} r^{\beta-p}$ for every $x \in A_m$, applying (7.22) we obtain

$$\begin{aligned}\int_{B(z,2r)\setminus F} |v(x)|^p d(x,F)^{\beta-p}\,dx &\le \int_{\Omega_0} |v(x)|^p d(x,F)^{\beta-p}\,dx \\ &\le \sum_{m=0}^{\infty} 2^{mp-m\beta} r^{\beta-p} \int_{A_m} |v(x)|^p\,dx \\ &= C(n,p,C_0) \sum_{m=0}^{\infty} 2^{-m\beta} r^{\beta} \sum_{k=m-2}^{\infty} \int_{A_k} |\nabla v(x)|^p\,dx \\ &= C(n,p,C_0) r^{\beta} \sum_{k=-2}^{\infty} \sum_{m=0}^{k+2} 2^{-m\beta} \int_{A_k} |\nabla v(x)|^p\,dx \\ &\le \frac{C(n,p,C_0)}{|\beta|} \sum_{k=-2}^{\infty} 2^{-k\beta} r^{\beta} \int_{A_k} |\nabla v(x)|^p\,dx \\ &\le \frac{C(n,p,C_0)}{|\beta|} \int_{B(z,8r)\setminus F} |\nabla v(x)|^p d(x,F)^{\beta}\,dx.\end{aligned}$$

In the penultimate step above, we used the estimate

$$\sum_{m=0}^{k+2} 2^{-m\beta} \le \frac{2^{-(k+2)\beta}}{1-2^{\beta}} \le C2^{-k\beta}|\beta|^{-1},$$

which holds for every $-\frac{1}{2} < \beta < 0$ with a constant C independent of k and β; compare to the end of the proof of Lemma 6.24. Inequality (7.21) follows. □

THEOREM 7.31. *Let $1 < p < \infty$ and assume that a closed set $E \subset \mathbb{R}^n$ satisfies the p-capacity density condition with a constant C_0. Let $z \in E$ and $r > 0$, and let F be as in Lemma 7.25 for $B(z,r)$. Assume that $u \in \operatorname{Lip}_c(F^c)$. Then*

$$\int_{B(z,8r)\setminus F} \frac{|u(x)|^p}{d(x,F)^p}\,dx \le C \int_{B(z,8r)} |\nabla u(x)|^p\,dx, \tag{7.23}$$

where $C = C(n,p,C_0)$.

PROOF. Let $-\frac{1}{2} < \beta < 0$ and define

$$v(x) = u(x) d(x,F)^{-\frac{\beta}{p}},$$

for every $x \in \mathbb{R}^n$. Then $v \in \operatorname{Lip}_c(F^c)$. By the Leibniz rule (2.4) and the chain rule (2.5), we have

$$|\nabla v(x)| \le |\nabla u(x)| d(x,F)^{-\frac{\beta}{p}} + \frac{|\beta|}{p} |u(x)| d(x,F)^{-\frac{\beta}{p}-1},$$

for almost every $x \in B(z,8r)\setminus F$. By (7.21) for v, we obtain

$$\begin{aligned}\int_{B(z,2r)\setminus F} \frac{|u(x)|^p}{d(x,F)^p}\,dx &= \int_{B(z,2r)\setminus F} |v(x)|^p d(x,F)^{\beta-p}\,dx \\ &\le \frac{C(n,p,C_0)}{|\beta|} \int_{B(z,8r)\setminus F} |\nabla v(x)|^p d(x,F)^{\beta}\,dx \\ &\le \frac{C(n,p,C_0)}{|\beta|} \int_{B(z,8r)} |\nabla u(x)|^p\,dx + C(n,p,C_0)|\beta|^{p-1} \int_{B(z,8r)\setminus F} \frac{|u(x)|^p}{d(x,F)^p}\,dx.\end{aligned}$$

We apply Lemma 7.27 with $\varsigma = 2$, $\sigma = 4$, $q = p$, $C_1 = C(n,p,C_0)|\beta|^{p-1}$ and $C_2 = \frac{1}{|\beta|}C(n,p,C_0)$. We choose $-\frac{1}{2} < \beta < 0$, depending on n, p and C_0, to be such that

$$C_3 = 1 - C_1(1 + C(n,p,\sigma,\varsigma)) \geq \frac{1}{2}.$$

Lemma 7.27 gives

$$\int_{B(z,8r)\setminus F} \frac{|u(x)|^p}{d(x,F)^p}\,dx \leq C(n,p,C_0)\int_{B(z,8r)} |\nabla u(x)|^p\,dx,$$

and this concludes the proof. □

7.8. Concluding argument

In the final step of the proof of Theorem 7.24 we improve the local Hardy inequality in Theorem 7.31. This is done by adapting the proof of Lemma 6.15 to the present setting and applying Lemma 7.27.

THEOREM 7.32. *Let $1 < p < \infty$ and assume that a closed set $E \subset \mathbb{R}^n$ satisfies the p-capacity density condition with a constant C_0. Let $z \in E$ and $r > 0$, let F be as in the statement of Lemma 7.25 for $B(z,r)$, and let $C_2 = C_2(n,p,C_0)$ be the constant in* (7.23)*, see Theorem* 7.31*. Then there exist $0 < \varepsilon = \varepsilon(n,p,C_2) < p-1$ and $C = C(n,p,C_2)$ such that the inequality*

$$\int_{B(z,12r)\setminus F} \frac{|u(x)|^{p-\varepsilon}}{d(x,F)^{p-\varepsilon}}\,dx \leq C\int_{B(z,12r)} |\nabla u(x)|^{p-\varepsilon}\,dx \tag{7.24}$$

holds for every $u \in \operatorname{Lip}_0(E^c)$.

PROOF. Without loss of generality, we may assume that $C_2 \geq 1$ in (7.23). We shall first prove (7.24) under the additional assumption that $u \in \operatorname{Lip}_c(E^c)$. Let $U = \mathbb{R}^n \setminus \operatorname{supp} u$. Then $u = 0$ in the open set U and $d(F,U^c) \geq d(E,U^c) > 0$.

Let $t > 0$ and define $F_t = G_t \cap H_t$, where

$$G_t = \left\{x \in B(z,8r) : M_{\frac{1}{2}d(x,F)}|\nabla u|(x) \leq t\right\}$$

and

$$H_t = \{x \in B(z,8r) : |u(x)| \leq t\,d(x,F)\}.$$

We show that the restriction of u to F_t is $C(n)t$-Lipschitz. To this end, fix $x, y \in F_t$, $x \neq y$, such that $d(y,F) \leq d(x,F)$. If $d(x,F) \geq 5|x-y|$, then

$$d(y,F) \geq d(x,F) - |x-y| \geq 4|x-y|.$$

Hence, by the pointwise estimate in Theorem 4.1, we obtain

$$\begin{aligned}|u(x) - u(y)| &\leq C(n)\,|x-y|\big(M_{2|x-y|}|\nabla u|(x) + M_{2|x-y|}|\nabla u|(y)\big)\\ &\leq C(n)\,|x-y|\big(M_{\frac{1}{2}d(x,F)}|\nabla u|(x) + M_{\frac{1}{2}d(y,F)}|\nabla u|(y)\big) \leq C(n)t|x-y|.\end{aligned}$$

On the other hand, if $d(y,F) \leq d(x,F) \leq 5|x-y|$, then

$$|u(x) - u(y)| \leq |u(x)| + |u(y)| \leq t\big(d(x,F) + d(y,F)\big) \leq 10t|x-y|.$$

Thus u is $C(n)t$-Lipschitz in F_t.

We apply Theorem 2.7 and extend the restriction $u|_{F_t}$ to a $C(n)t$-Lipschitz function $\widetilde{u}$ in $\mathbb{R}^n$. Then $\widetilde{u} = 0$ in an open set $\widetilde{U} \subset U$ with $d(F, \mathbb{R}^n \setminus \widetilde{U}) > 0$. Indeed, if

$$x \in \widetilde{U} = \{x \in B(z,8r) : d(x,F) < \tfrac{1}{2}d(F,U^c)\},$$

then $u(x)=0$. Since $|\nabla u|=0$ in U, we have $x\in F_t$, and so $\widetilde{u}(x)=0$. Define

$$v(x)=\widetilde{u}(x)\max\{0,1-d(x,B(z,8r))\},$$

for $x\in\mathbb{R}^n$. Then $v\in \operatorname{Lip}_c(F^c)$ and

$$|\nabla v(x)|=|\nabla\widetilde{u}(x)|\le \chi_{F_t}(x)|\nabla u(x)|+C(n)t\chi_{F_t^c}(x)$$

for almost every $x\in B(z,8r)$.

By applying Theorem 7.31 to the function $v\in \operatorname{Lip}_c(F^c)$, we obtain

$$\begin{aligned}\int_{(B(z,8r)\setminus F)\cap F_t}\frac{|u(x)|^p}{d(x,F)^p}\,dx &\le \int_{B(z,8r)\setminus F}\frac{|v(x)|^p}{d(x,F)^p}\,dx\\ &\le C_2\int_{F_t}|\nabla u(x)|^p\,dx+C_2C(n)^pt^p|B(z,8r)\setminus F_t|.\end{aligned}$$

Since $H_t=F_t\cup(H_t\setminus G_t)$ and $C_2\ge 1$, it follows that

$$\begin{aligned}&\int_{(B(z,8r)\setminus F)\cap H_t}\frac{|u(x)|^p}{d(x,F)^p}\,dx\\ &\quad\le C_2\int_{F_t}|\nabla u(x)|^p\,dx+C_2C(n)^pt^p|B(z,8r)\setminus F_t|\\ &\qquad+\int_{(H_t\setminus F)\setminus G_t}\frac{|u(x)|^p}{d(x,F)^p}\,dx\\ &\quad\le C_2\int_{G_t}|\nabla u(x)|^p\,dx+C_2C(n,p)t^p\big(|B(z,8r)\setminus F_t|+|H_t\setminus G_t|\big)\\ &\quad\le C_2\int_{G_t}|\nabla u(x)|^p\,dx+C_2C(n,p)t^p\big(|B(z,8r)\setminus H_t|+|B(z,8r)\setminus G_t|\big).\end{aligned}\tag{7.25}$$

Here $t>0$ was arbitrary, and thus we conclude that (7.25) holds for every $t>0$.

Next we multiply (7.25) by $t^{-1-\varepsilon}$, where $0<\varepsilon<p-1$, and integrate with respect to t over the set $(0,\infty)$. With a change of the order of integration on the left-hand side, this gives

$$\begin{aligned}\frac{1}{\varepsilon}\int_{B(z,8r)\setminus F}\left(\frac{|u(x)|}{d(x,F)}\right)^{p-\varepsilon}dx &\le C_2\int_0^\infty t^{-1-\varepsilon}\int_{G_t}|\nabla u(x)|^p\,dx\,dt\\ &\quad+C_2C(n,p)\int_0^\infty t^{p-1-\varepsilon}\big(|B(z,8r)\setminus H_t|+|B(z,8r)\setminus G_t|\big)\,dt.\end{aligned}$$

Since $|\nabla u(x)|\le M_{\frac12 d(x,F)}|\nabla u|(x)$ for almost every $x\in B(z,8r)$, see Corollary 1.23, we find by the definition of G_t that the first term on the right-hand side is dominated by

$$\frac{C_2}{\varepsilon}\int_{B(z,8r)}|\nabla u(x)|^{p-\varepsilon}\,dx.$$

Using the definitions of H_t and G_t, the second term on the right-hand side can be estimated from above by

$$\frac{C_2C(n,p)}{p-\varepsilon}\left(\int_{B(z,8r)\setminus F}\left(\frac{|u(x)|}{d(x,F)}\right)^{p-\varepsilon}dx+\int_{B(z,8r)}\big(M_{\frac12 d(x,F)}|\nabla u|(x)\big)^{p-\varepsilon}\,dx\right).$$

Fix $x\in B(z,8r)$ for a moment. Since $\frac12 d(x,F)\le 4r$, we have $B(x,\rho)\subset B(z,12r)$ for every $0<\rho\le\frac12 d(x,F)$. This implies

$$M_{\frac12 d(x,F)}|\nabla u|(x)=M_{\frac12 d(x,F)}\big(\chi_{B(z,12r)}|\nabla u|\big)(x)\le M\big(\chi_{B(z,12r)}|\nabla u|\big)(x)$$

for almost every $x \in B(z, 8r)$. By Theorem 1.15 (c), we have

$$\int_{B(z,8r)} \big(M_{\frac{1}{2}d(x,F)}|\nabla u|(x)\big)^{p-\varepsilon}\,dx \le \int_{\mathbb{R}^n} \big(M(\chi_{B(z,12r)}|\nabla u|)(x)\big)^{p-\varepsilon}\,dx$$
$$\le C(n,p,\varepsilon)\int_{B(z,12r)} |\nabla u(x)|^{p-\varepsilon}\,dx.$$

By combining the estimates above, we obtain

$$\int_{B(z,8r)\setminus F} \left(\frac{|u(x)|}{d(x,F)}\right)^{p-\varepsilon} dx$$
$$\le C_3 \int_{B(z,12r)\setminus F} \left(\frac{|u(x)|}{d(x,F)}\right)^{p-\varepsilon} dx + C_4 \int_{B(z,12r)} |\nabla u(x)|^{p-\varepsilon}\,dx,$$

where $C_3 = C_2 C(n,p)\frac{\varepsilon}{p-\varepsilon}$ and $C_4 = C_2\big(1 + C(n,p)\frac{\varepsilon}{p-\varepsilon}C(n,p,\varepsilon)\big)$. In order to apply Lemma 7.27, let $\varsigma = 8$, $\sigma = \frac{3}{2}$ and $q = p - \varepsilon$. We choose $0 < \varepsilon < p - 1$ depending on n, p and C_2 to be so small that $C_5 = 1 - C_3(1 + C(n,p,\sigma,\varsigma)) \ge \frac{1}{2}$. Lemma 7.27 implies

$$\int_{B(z,12r)\setminus F} \frac{|u(x)|^{p-\varepsilon}}{d(x,F)^{p-\varepsilon}}\,dx \le C(n,p,C_2)\int_{B(z,12r)} |\nabla u(x)|^{p-\varepsilon}\,dx,$$

and this proves inequality (7.24) when $u \in \operatorname{Lip}_c(E^c)$.

To prove the general case, let $u \in \operatorname{Lip}_0(E^c)$. Then u has a bounded support. Let

$$u_j(x) = \max\{|u(x)| - \tfrac{1}{j}, 0\},$$

for every $x \in \mathbb{R}^n$ and every $j \in \mathbb{N}$. Then $|\nabla u_j| \le |\nabla u|$ almost everywhere and u_j vanishes in the open neighborhood $U_j = \{|u| < \frac{1}{j}\}$ of E. In particular, $u_j \in \operatorname{Lip}_c(E^c)$ for every $j \in \mathbb{N}$. By Fatou's lemma and the special case of inequality (7.24) that was established above, we conclude that

$$\int_{B(z,12r)\setminus F} \frac{|u(x)|^{p-\varepsilon}}{d(x,F)^{p-\varepsilon}}\,dx \le \liminf_{j\to\infty} \int_{B(z,12r)\setminus F} \frac{|u_j(x)|^{p-\varepsilon}}{d(x,F)^{p-\varepsilon}}\,dx$$
$$\le C(n,p,C_2)\liminf_{j\to\infty}\int_{B(z,12r)} |\nabla u_j(x)|^{p-\varepsilon}\,dx$$
$$\le C(n,p,C_2)\int_{B(z,12r)} |\nabla u(x)|^{p-\varepsilon}\,dx.$$

This proofs the claim in the general case $u \in \operatorname{Lip}_0(E^c)$. □

We close this chapter by showing that the nonquantitative capacity condition

$$\operatorname{cap}_p\big(E \cap \overline{B}(x,r), B(x,2r)\big) > 0, \tag{7.26}$$

for every $x \in E$ and $r > 0$, is not self-improving.

EXAMPLE 7.33. Let $1 < p < 2$. We construct a set $E \subset \mathbb{R}^2$ which satisfies (7.26) with p, but for every $1 < q < p$ the corresponding condition does not hold since there exist a point $x \in E$ and a radius $r > 0$, depending on q, such that $\operatorname{cap}_q\big(E \cap \overline{B}(x,r), B(x,2r)\big) = 0$. We consider here only the planar case $n = 2$, but it is easy to generalize this example also to the higher dimensions.

For $i \in \mathbb{N}$, let $\lambda_i = 2 - p + \frac{1}{i}$ and let $E_i \subset [2^{-i}, 2^{-i+1}] \times [0, 2^{-i}]$ be a λ_i-regular compact set. For instance, E_i can be taken to be a Cantor-type fractal set. Let

$$E = \bigcup_{i=1}^{\infty} E_i \cup \{0\} \cup B(0,2)^c$$

where $0 = (0,0) \in \mathbb{R}^2$.

Let $x \in E$ and $r > 0$. If $x \in B(0,2)^c$ or $r > 2$, then $|E \cap \overline{B}(x,r)| > 0$, and Lemma 5.35 implies that $\operatorname{cap}_p\big(E \cap \overline{B}(x,r), B(x,2r)\big) > 0$. Thus we may assume that $x = 0$ or $x \in E_i$ for some $i \in \mathbb{N}$, and that $0 < r \le 2$. If $x \in E_i$, then by Lemma 7.8

$$\mathcal{H}^{\lambda_i}_\infty\big(E \cap \overline{B}(x,r)\big) \ge \mathcal{H}^{\lambda_i}_\infty\big(E_i \cap \overline{B}(x,r)\big) \ge C(\lambda_i, E_i) \min\{r, \operatorname{diam}(E_i)\}^{\lambda_i} > 0.$$

Theorem 5.41 implies

$$\operatorname{cap}_p\big(E \cap \overline{B}(x,r), B(x,2r)\big) \ge C(p, \lambda_i, E_i)\mathcal{H}^{\lambda_i}_\infty\big(E \cap \overline{B}(x,r)\big) r^{2-p-\lambda_i} > 0.$$

On the other hand, if $x = 0$, then we choose $i \in \mathbb{N}$ such that $E_i \subset B(0,r)$, and so

$$\operatorname{cap}_p\big(E \cap \overline{B}(0,r), B(0,2r)\big) \ge C(p, \lambda_i, E_i)\mathcal{H}^{\lambda_i}_\infty\big(E_i \cap \overline{B}(0,r)\big) r^{2-p-\lambda_i} > 0.$$

We conclude that (7.26), with p, holds for the set E.

This p is also the smallest exponent for which (7.26) is valid. Indeed, if $1 < q < p$, we choose $k \in \mathbb{N}$ to be so large that $q < p - \frac{1}{k}$, and consider $x = 0$ and $r < 2^{-k}$. Then, using Theorem 5.39 and countable subadditivity, we obtain

$$\begin{aligned}\operatorname{cap}_q\big(E \cap \overline{B}(0,r), B(0,2r)\big) &\le C\mathcal{H}^{2-q}_\infty\big(E \cap \overline{B}(0,r)\big) \\ &\le C\Big(\mathcal{H}^{2-q}_\infty(\{0\}) + \sum_{i=k}^{\infty} \mathcal{H}^{2-q}_\infty(E_i)\Big).\end{aligned}$$

Observe that $2 - q > 2 - p + \frac{1}{k} \ge 2 - p + \frac{1}{i} = \lambda_i$ for every $i \ge k$, and thus $\mathcal{H}^{2-q}_\infty(E_i) = 0$ for every $i \ge k$. Since $\mathcal{H}^{2-q}_\infty(\{0\}) = 0$, we conclude that

$$\operatorname{cap}_q\big(E \cap \overline{B}(0,r), B(0,2r)\big) = 0.$$

This shows that (7.26) does not hold for any $1 < q < p$.

7.9. Notes

References for Ahlfors–David regular sets include Brudnyi and Brudnyi [**59**,**60**], Ihnatsyeva and Vähäkangas [**202**], Jonsson and Wallin [**215**] and Triebel [**385**]. Lemma 7.8 has been considered in Egert, Haller-Dintelmann and Rehberg [**118**, Lemma 4.5].

For more information concerning the lower dimension we refer to Fraser [**136**]. This concept has also been called the lower Assouad dimension and denoted by $\underline{\dim}_{\mathrm{A}}(E)$, see for instance [**217**]. Essentially the same dimension was studied already by Larman [**255**]. The proof of Lemma 7.13 is inspired by Järvi and Vuorinen [**211**, Theorem 4.1], see also Lehrbäck [**259**, Lemma 4.1]. Corollary 7.15 was proved by Larman [**255**]. Lemma 7.19 has been considered by Käenmäki, Lehrbäck and Vuorinen [**217**], and the necessity part is essentially by Rajala and Vilppolainen [**351**, Remark 3.6(iii)].

Many of the implications in Theorem 7.20 can be found in Järvi and Vuorinen [**211**, Theorem 4.1] and Korte and Shanmugalingam [**242**, Theorem 1.1]. Ancona [**14**] proved that the capacity density condition is also necessary for the validity of Hardy's inequality when $p = n = 2$ and Lewis [**271**] generalized this for $p = n \geq 2$. See also Buckley and Koskela [**64**] for corresponding results in Orlicz–Sobolev spaces. Sugawa [**377**] proved that Hardy's inequality is equivalent to uniform perfectness of the complement for $p = n = 2$. Sugawa's paper also contains a survey on properties of uniformly perfect sets.

Theorem 7.21 has been proved by Lewis [**271**] for the Riesz capacities. The corresponding result for the variational capacity in a Euclidean case with weights is by Mikkonen [**327**]. This approach has been generalized to metric measure spaces by Björn, MacManus and Shanmugalingam [**36**]. These arguments are based on nonlinear potential theory. The direct proof of Theorem 7.21 in Section 7.6, Section 7.7 and Section 7.8 is an adaptation of the metric space proof in Lehrbäck, Tuominen and Vähäkangas [**264**].

It is also possible to consider restricted versions of the p-capacity density condition and the λ-Hausdorff content density condition. Then, instead of every $r > 0$, the conditions are required to hold for every $0 < r < r_0$, with some fixed $r_0 > 0$. It is straightforward to modify our results for these restricted conditions.

CHAPTER 8

Muckenhoupt Weights

This chapter discusses Muckenhoupt weights and related weighted norm inequalities for maximal functions. Self-improving phenomena are studied for weighted norm inequalities in Section 8.3, for Muckenhoupt weights in Section 8.4 and Section 8.5, and for reverse Hölder inequalities in Section 8.7 and Section 8.8. Instead of aiming at the most direct approach, we give several independent points of view to the self-improving phenomena. This causes some overlap, but these methods and results are applied in Chapter 9, Chapter 10 and Chapter 12.

8.1. Doubling weights

It is often more convenient to use cubes instead of balls. Unless specified otherwise, we apply half-open cubes whose sides are parallel to the coordinate axis.

DEFINITION 8.1. A cube in $\mathbb{R}^n$ is a set of the form

$$Q = [a_1, b_1) \times \ldots \times [a_n, b_n),$$

with $b_1 - a_1 = \cdots = b_n - a_n$. The side length of Q is $l(Q) = b_1 - a_1$. For $x \in \mathbb{R}^n$ and $r > 0$,

$$Q(x,r) = \{y \in \mathbb{R}^n : -r \le y_j - x_j < r \text{ for all } j = 1, \ldots, n\}$$

is the cube with center x and side length $2r$. Clearly,

$$|Q(x,r)| = (2r)^n \quad \text{and} \quad \operatorname{diam}(Q(x,r)) = (2\sqrt{n})r.$$

DEFINITION 8.2. A function $w \in L^1_{\mathrm{loc}}(\mathbb{R}^n)$, with $w(x) > 0$ for almost every $x \in \mathbb{R}^n$, is called a weight in $\mathbb{R}^n$. For a weight w and a measurable set $E \subset \mathbb{R}^n$, we write

$$w(E) = \int_E w(x)\,dx.$$

Note that the measure $w\,dx$ induced by a weight and the Lebesgue measure are mutually absolutely continuous. For $1 \le p < \infty$, the space $L^p(E; w\,dx)$ consists of measurable functions f on E satisfying

$$\|f\|_{L^p(E;w\,dx)} = \left(\int_E |f(x)|^p w(x)\,dx\right)^{\frac{1}{p}} < \infty.$$

In addition, $f \in L^p_{\mathrm{loc}}(E; w\,dx)$ if $f \in L^p(K; w\,dx)$ for every compact set $K \subset E$. We also write $L^p(w\,dx) = L^p(\mathbb{R}^n; w\,dx)$.

DEFINITION 8.3. A weight w in $\mathbb{R}^n$ is doubling if there exists a constant C such that $w(B(x,2r)) \le Cw(B(x,r))$ for every $x \in \mathbb{R}^n$ and $r > 0$.

It is immediate that the doubling condition in Definition 8.3 can be equivalently stated in the form that there exists a constant C such that $w(Q(x,2r)) \leq Cw(Q(x,r))$ for every $x \in \mathbb{R}^n$ and $r > 0$.

We collect definitions and properties of weighted maximal functions in the remark below.

REMARK 8.4. Let w be a doubling weight in $\mathbb{R}^n$, with a constant C_1.

(a) For a measurable function $f \in L^1_{\text{loc}}(w\,dx)$ the weighted maximal function $M^w f$ at $x \in \mathbb{R}^n$ is

$$M^w f(x) = \sup_{r>0} \frac{1}{w(B(x,r))} \int_{B(x,r)} |f(y)| w(y)\,dy.$$

Let $1 < p \leq \infty$. There exists a constant $C = C(n,p,C_1)$ such that

$$\|M^w f\|_{L^p(w\,dx)} \leq C \|f\|_{L^p(w\,dx)}$$

for every $f \in L^p(w\,dx)$. More precisely, we have $C = 1$ for $p = \infty$ and

$$C = 2 \cdot \left(\frac{C_1^3 p}{p-1} \right)^{\frac{1}{p}}$$

for $1 < p < \infty$. The proof of this statement is a straightforward adaptation of the proof of Theorem 1.15 (c).

(b) For a measurable function $f \in L^1_{\text{loc}}(w\,dx)$, we define

$$M^{c,w} f(x) = \sup_{Q \ni x} \frac{1}{w(Q)} \int_Q |f(y)| w(y)\,dy,$$

where the supremum is taken over all cubes $Q \subset \mathbb{R}^n$ with $x \in Q$. Let $1 < p \leq \infty$. There exists a constant $C = C(n,p,C_1)$ such that

$$\|M^{c,w} f\|_{L^p(w\,dx)} \leq C \|f\|_{L^p(w\,dx)}$$

for every $f \in L^p(w\,dx)$. This follows from (a) since $M^{c,w}$ and M^w are pointwise comparable by the doubling property of w, compare to Remark 1.2. For $w = 1$, we write $M^c f = M^{c,w} f$.

(c) Let $f \in L^1_{\text{loc}}(w\,dx)$. There exists a measurable set E with $w(E) = |E| = 0$ such that

$$\lim_{r \to 0} \frac{1}{w(B(x,r))} \int_{B(x,r)} |f(y) - f(x)| w(y)\,dy = 0$$

for every $x \in \mathbb{R}^n \setminus E$; compare to Theorem 1.19. As a consequence, if $x \in \mathbb{R}^n \setminus E$ and $(Q_j)_{j \in \mathbb{N}}$ is a sequence of cubes such that $x \in Q_j$ for every $j \in \mathbb{N}$ and $l(Q_j) \to 0$ as $j \to \infty$, then

$$f(x) = \lim_{j \to \infty} \frac{1}{w(Q_j)} \int_{Q_j} f(y) w(y)\,dy.$$

8.2. Dyadic cubes and the Calderón–Zygmund lemma

Let $Q_0 \subset \mathbb{R}^n$ be a cube. The dyadic decomposition $\mathcal{D}(Q_0)$ of Q_0 is defined as

$$\mathcal{D}(Q_0) = \bigcup_{j=0}^{\infty} \mathcal{D}_j(Q_0),$$

where each $\mathcal{D}_j(Q_0)$ consists of 2^{jn} pairwise disjoint half-open cubes Q, with side length $l(Q) = 2^{-j}l(Q_0)$, such that

$$Q_0 = \bigcup_{Q\in\mathcal{D}_j(Q_0)} Q$$

for every $j \in \mathbb{N}_0$. If $j \geq 1$ and $Q \in \mathcal{D}_j(Q_0)$, there exists a unique cube $\pi Q \in \mathcal{D}_{j-1}(Q_0)$, satisfying $Q \subset \pi Q$. The cube πQ is called the dyadic parent of Q, and Q is a dyadic child of πQ. The dyadic maximal function will be a useful tool for us.

DEFINITION 8.5. Let $Q_0 \subset \mathbb{R}^n$ be a cube and let $f \in L^1(Q_0)$. The dyadic (noncentered) maximal function $M^d_{Q_0}f$ is defined as

$$M^d_{Q_0}f(x) = \sup_{Q\ni x} \fint_Q |f(y)|\,dy, \tag{8.1}$$

where the supremum is taken over all dyadic cubes $Q \in \mathcal{D}(Q_0)$ with $x \in Q$.

Let $f, g \in L^1(Q_0)$ and $x \in Q_0$. Using the definition it is easy to show that $M^d_{Q_0}f(x) \geq 0$,

$$M^d_{Q_0}(f+g)(x) \leq M^d_{Q_0}f(x) + M^d_{Q_0}g(x),$$

and

$$M^d_{Q_0}(af)(x) = |a|M^d_{Q_0}f(x)$$

for every $a \in \mathbb{R}$, compare to Lemma 1.5.

The Calderón–Zygmund lemma and its variants will be used several times in the sequel.

LEMMA 8.6. *Let $Q_0 \subset \mathbb{R}^n$ be a cube. Assume that $f \in L^1(Q_0)$ is a nonnegative function and let*

$$E_t = \{x \in Q_0 : M^d_{Q_0}f(x) > t\} \tag{8.2}$$

for $t > 0$. For every $t \geq f_{Q_0}$, there exists a collection $\mathcal{D}_t$ of pairwise disjoint dyadic cubes $Q \in \mathcal{D}(Q_0)$ such that

$$E_t = \bigcup_{Q\in\mathcal{D}_t} Q \tag{8.3}$$

and

$$t < \fint_Q f(x)\,dx \leq 2^n t \tag{8.4}$$

for every $Q \in \mathcal{D}_t$. Moreover, $f(x) \leq t$ for almost every $x \in Q_0 \setminus E_t$.

PROOF. Let $t > 0$. If $E_t = \emptyset$, then the claim holds with $\mathcal{D}_t = \emptyset$. In this case the union in (8.3) is the empty set. Hence we may assume $E_t \neq \emptyset$.

For every $x \in E_t$ there exists a dyadic cube $Q \in \mathcal{D}(Q_0)$ with $x \in Q$ and $f_Q > t$. It follows that there exists a unique maximal dyadic cube $Q_x \in \mathcal{D}(Q_0)$ with $x \in Q_x$ and $f_{Q_x} > t$. Maximality means that if $Q_x \subsetneq Q \in \mathcal{D}(Q_0)$, then $f_Q \leq t$. The collection of all such maximal cubes is

$$\mathcal{D}_t = \{Q_x \in \mathcal{D}(Q_0) : x \in E_t\}.$$

If $Q_x = Q_0$ for some $x \in Q_0$, then $\mathcal{D}_t = \{Q_0\}$ and there are no cubes $Q \in \mathcal{D}(Q_0)$ with $Q_x \subsetneq Q$. This happens if and only if $t < f_{Q_0}$, which contradicts the assumption $t \geq f_{Q_0}$. Hence $\mathcal{D}_t \neq \{Q_0\}$.

Since either $Q \subset R$ or $R \subset Q$ for $Q, R \in \mathcal{D}(Q_0)$ with $Q \cap R \neq \emptyset$, by maximality the cubes in $\mathcal{D}_t$ are pairwise disjoint. Moreover, if $x \in E_t$, then $x \in Q_x \subset E_t$, and thus (8.3) holds.

Assume then that $Q \in \mathcal{D}_t \cap \mathcal{D}_j(Q_0)$, for some $j \in \mathbb{N}$, and let $\pi Q \in \mathcal{D}_{j-1}(Q_0)$ be the dyadic parent of Q. Then $Q \subsetneq \pi Q \subset Q_0$ and $|\pi Q| = 2^n|Q|$. By maximality of Q, we have $f_{\pi Q} \leq t$. Thus

$$t < f_Q = \frac{1}{|Q|}\int_Q f(x)\,dx \leq \frac{2^n}{|\pi Q|}\int_{\pi Q} f(x)\,dx \leq 2^n t.$$

Finally, by the Lebesgue differentiation theorem $f(x) \leq M^d_{Q_0} f(x) \leq t$ for almost every $x \in Q_0 \setminus E_t$. □

The Calderón–Zygmund lemma gives simple proofs for norm estimates for the dyadic maximal function. The next result is a weak type estimate, compare to Theorem 1.15 (b).

LEMMA 8.7. *Let $Q_0 \subset \mathbb{R}^n$ be a cube. Assume that $f \in L^1(Q_0)$ is a nonnegative function and let $E_t = \{x \in Q_0 : M^d_{Q_0} f(x) > t\}$. Then*

$$|E_t| \leq \frac{1}{t}\int_{E_t} f(x)\,dx$$

for every $t > 0$.

PROOF. Let $t > 0$. If $t < f_{Q_0}$, then $E_t = Q_0$ and thus

$$|E_t| = |Q_0| \leq \frac{1}{t}\int_{Q_0} f(x)\,dx = \frac{1}{t}\int_{E_t} f(x)\,dx.$$

Then assume that $t \geq f_{Q_0}$. Let $\mathcal{D}_t$ be the collection of dyadic subcubes of Q_0 given by Lemma 8.6. By using (8.3) and the facts that the cubes $Q \in \mathcal{D}_t$ are pairwise disjoint and $t < f_Q$ for every $Q \in \mathcal{D}_t$, we obtain

$$|E_t| = \Big|\bigcup_{Q\in\mathcal{D}_t} Q\Big| = \sum_{Q\in\mathcal{D}_t}|Q| \leq \frac{1}{t}\sum_{Q\in\mathcal{D}_t}\int_Q f(x)\,dx = \frac{1}{t}\int_{E_t} f(x)\,dx. \qquad \square$$

There exists a reverse weak type estimate for the dyadic maximal function, compare to Lemma 1.25.

LEMMA 8.8. *Let $Q_0 \subset \mathbb{R}^n$ be a cube. Assume that $f \in L^1(Q_0)$ is a nonnegative function and let $E_t = \{x \in Q_0 : M^d_{Q_0} f(x) > t\}$. Then*

$$\int_{E_t} f(x)\,dx \leq 2^n t|E_t|$$

for every $t \geq f_{Q_0}$.

PROOF. Without loss of generality, we may assume that $t > 0$. Let $\mathcal{D}_t$ be the collection of dyadic subcubes of Q_0 given by Lemma 8.6. Using (8.3) and the facts that the cubes $Q \in \mathcal{D}_t$ are pairwise disjoint and $f_Q \leq 2^n t$ for every $Q \in \mathcal{D}_t$, we obtain

$$\int_{E_t} f(x)\,dx = \sum_{Q\in\mathcal{D}_t}\int_Q f(x)\,dx \leq 2^n t\sum_{Q\in\mathcal{D}_t}|Q| = 2^n t|E_t|. \qquad \square$$

8.3. Self-improvement of weighted norm inequalities

We consider maximal function inequalities not only for the Lebesgue measure but also for weighted measures. The main result of this section, Theorem 8.11, gives a self-improvement property for weighted norm inequalities related to the Hardy–Littlewood–Wiener maximal function theorem.

We begin with a Whitney covering theorem. For the proof we refer to Stein [**369**, Chapter 6], see also Bennett and Sharpley [**28**, Section V.5] and Grafakos [**156**, Appendix J].

LEMMA 8.9. *Assume that $\Omega \subsetneq \mathbb{R}^n$ is a nonempty open set. There exist pairwise disjoint half-open dyadic cubes $Q_i \subset \Omega$, $i \in \mathbb{N}$, such that $\Omega = \bigcup_{i=1}^{\infty} Q_i$ and*

$$\operatorname{diam}(Q_i) \le d(Q_i, \partial\Omega) \le 4 \operatorname{diam}(Q_i) \tag{8.5}$$

for every $i \in \mathbb{N}$. Moreover, let $Q_i = Q(x_i, r_i)$ and $Q_i^* = Q(x_i, \frac{9}{8} r_i)$, for every $i \in \mathbb{N}$. There exists a constant $C = C(n)$ such that

$$\sum_{i=1}^{\infty} \chi_{Q_i^*}(x) \le C$$

for every $x \in \Omega$, and

$$\frac{3}{4} \operatorname{diam}(Q_i) \le d(x, \partial\Omega) \le 6 \operatorname{diam}(Q_i) \tag{8.6}$$

for every $x \in Q_i^*$.

A collection $\{Q_i : i \in \mathbb{N}\}$ of cubes $Q_i \subset \Omega$ satisfying the properties in Lemma 8.9 is called a Whitney decomposition of Ω.

Let $f \in L^\infty(\mathbb{R}^n)$ be a nonnegative function with a compact support. Let $t > 0$ and $E_t = \{x \in \mathbb{R}^n : M^* f(x) > t\}$, where $M^* f$ is the noncentered Hardy–Littlewood maximal function of f as in Remark 1.2. Then $E_t \subsetneq \mathbb{R}^n$ is an open set. Assume that $E_t \neq \emptyset$ and let $\{Q_i : i \in \mathbb{N}\}$ be a Whitney decomposition of E_t, see Lemma 8.9. Define

$$f_t(x) = \begin{cases} f(x), & x \in \mathbb{R}^n \setminus E_t, \\ \fint_{Q_i} f(y)\,dy, & x \in Q_i \text{ for some } i \in \mathbb{N}. \end{cases} \tag{8.7}$$

If $E_t = \emptyset$, then $f_t(x) = f(x)$ for every $x \in \mathbb{R}^n$. Since $f \in L^\infty(\mathbb{R}^n)$ is a nonnegative function and the support of f is compact, also f_t is nonnegative and has a compact support.

LEMMA 8.10. *Assume that $f \in L^\infty(\mathbb{R}^n)$ is a nonnegative function with a compact support. Let $t > 0$ and let E_t and f_t be as in (8.7). Then the following assertions hold.*

(a) *$f_t(x) \le C(n)t$ for almost every $x \in \mathbb{R}^n$.*
(b) *$M^* f(x) \le C(n) M^* f_t(x)$ for every $x \in \mathbb{R}^n \setminus E_t$.*

PROOF. By Corollary 1.23 we have

$$f_t(x) = f(x) \le M^* f(x) \le t$$

for almost every $x \in \mathbb{R}^n \setminus E_t$. On the other hand, if $x \in E_t$, then $x \in Q_i$ for some $i \in \mathbb{N}$. By (8.5), there exists a ball $B \subset \mathbb{R}^n$ that intersects $\mathbb{R}^n \setminus E_t$ and satisfies

$Q_i \subset B$ and $|B| \le C(n)|Q_i|$. For $z \in B \setminus E_t$ we have

$$f_t(x) = \fint_{Q_i} f(y)\,dy \le C(n) \fint_B f(y)\,dy \le C(n)M^*f(z) \le C(n)t.$$

Thus (a) holds for almost every $x \in \mathbb{R}^n$.

To prove part (b), let $x \in \mathbb{R}^n \setminus E_t$. Let $z \in \mathbb{R}^n$ and $r > 0$ be such that $x \in B(z,r)$. Then

$$\begin{aligned}\int_{B(z,r)} f(y)\,dy &= \int_{B(z,r)\setminus E_t} f(y)\,dy + \int_{B(z,r)\cap E_t} f(y)\,dy\\ &= \int_{B(z,r)\setminus E_t} f(y)\,dy + \sum_{i=1}^{\infty} \int_{B(z,r)\cap Q_i} f(y)\,dy.\end{aligned}$$

Consider a Whitney cube Q_i of E_t satisfying $B(z,r)\cap Q_i \ne \emptyset$. Since $x \in B(z,r)\setminus E_t$, by the lower bound in (8.5) we have $Q_i \subset B(z,4r)$. Thus (8.7) implies

$$\begin{aligned}\int_{B(z,r)\cap Q_i} f(y)\,dy &\le \int_{B(z,4r)\cap Q_i} f(y)\,dy = \int_{Q_i} f(y)\,dy\\ &= \int_{Q_i} f_t(y)\,dy = \int_{B(z,4r)\cap Q_i} f_t(y)\,dy.\end{aligned}$$

By collecting the estimates above, we obtain

$$\int_{B(z,r)} f(y)\,dy \le \int_{B(z,4r)\setminus E_t} f_t(y)\,dy + \sum_{i=1}^{\infty} \int_{B(z,4r)\cap Q_i} f_t(y)\,dy = \int_{B(z,4r)} f_t(y)\,dy,$$

and consequently

$$\fint_{B(z,r)} f(y)\,dy \le C(n) \fint_{B(z,4r)} f_t(y)\,dy \le C(n)M^*f_t(x).$$

By taking supremum over all balls $B(z,r)$ as above, we conclude that (b) holds. □

THEOREM 8.11. *Let $1 < p < \infty$ and let w be a weight in $\mathbb{R}^n$. Assume that there exists a constant C_1 such that*

$$\int_{\mathbb{R}^n} \big(M^*f(x)\big)^p w(x)\,dx \le C_1 \int_{\mathbb{R}^n} f(x)^p w(x)\,dx \tag{8.8}$$

for all nonnegative functions $f \in L^\infty(\mathbb{R}^n)$ with compact support. Then there exist $\varepsilon = \varepsilon(n,p,C_1)$, with $0 < \varepsilon < p-1$, and $C = C(n,p,C_1)$ such that

$$\int_{\mathbb{R}^n} \big(M^*f(x)\big)^{p-\varepsilon} w(x)\,dx \le C \int_{\mathbb{R}^n} f(x)^{p-\varepsilon} w(x)\,dx$$

for all nonnegative $f \in L^\infty(\mathbb{R}^n)$ with compact support.

PROOF. Fix a nonnegative function $f \in L^\infty(\mathbb{R}^n)$ with a compact support. Let $t > 0$, and let E_t and f_t be as in (8.7) for f. By the assumption

$$\int_{\mathbb{R}^n} \big(M^*f_t(x)\big)^p w(x)\,dx \le C_1 \int_{\mathbb{R}^n} f_t(x)^p w(x)\,dx.$$

By Lemma 8.10, we obtain

$$
\begin{aligned}
\int_{\mathbb{R}^n\setminus E_t} (M^* f(x))^p w(x)\,dx &\le C(n,p)\int_{\mathbb{R}^n\setminus E_t} (M^* f_t(x))^p w(x)\,dx \\
&\le C(n,p)\int_{\mathbb{R}^n} (M^* f_t(x))^p w(x)\,dx \\
&\le C(n,p,C_1)\Big(\int_{\mathbb{R}^n\setminus E_t} f(x)^p w(x)\,dx + t^p w(E_t)\Big).
\end{aligned}
\tag{8.9}
$$

This implies, for a fixed $t_0 > 0$,

$$
\begin{aligned}
&\int_{t_0}^{\infty} t^{-1-\varepsilon}\int_{\mathbb{R}^n\setminus E_t} (M^* f(x))^p w(x)\,dx\,dt \\
&\qquad \le C(n,p,C_1)\int_{t_0}^{\infty} t^{-1-\varepsilon}\int_{\mathbb{R}^n\setminus E_t} f(x)^p w(x)\,dx\,dt \\
&\qquad\quad + C(n,p,C_1)\int_{t_0}^{\infty} t^{p-1-\varepsilon} w(E_t)\,dt.
\end{aligned}
\tag{8.10}
$$

By Fubini's theorem, we have

$$
\begin{aligned}
&\int_{t_0}^{\infty} t^{-1-\varepsilon}\int_{\mathbb{R}^n\setminus E_t} (M^* f(x))^p w(x)\,dx\,dt \\
&\quad = \int_{\mathbb{R}^n} (M^* f(x))^p \Big(\int_{t_0}^{\infty} \chi_{\mathbb{R}^n\setminus E_t}(x) t^{-1-\varepsilon}\,dt\Big) w(x)\,dx \\
&\quad = \int_{\mathbb{R}^n} (M^* f(x))^p \Big(\int_{\max\{t_0, M^* f(x)\}}^{\infty} t^{-1-\varepsilon}\,dt\Big) w(x)\,dx \\
&\quad = \frac{1}{\varepsilon}\int_{\mathbb{R}^n} (M^* f(x))^p \big(\max\{t_0, M^* f(x)\}\big)^{-\varepsilon} w(x)\,dx.
\end{aligned}
$$

In a similar way, we obtain

$$
\begin{aligned}
\int_{t_0}^{\infty} t^{-1-\varepsilon}\int_{\mathbb{R}^n\setminus E_t} f(x)^p w(x)\,dx\,dt &= \frac{1}{\varepsilon}\int_{\mathbb{R}^n} f(x)^p \big(\max\{t_0, M^* f(x)\}\big)^{-\varepsilon} w(x)\,dx \\
&\le \frac{1}{\varepsilon}\int_{\mathbb{R}^n} f(x)^{p-\varepsilon} w(x)\,dx,
\end{aligned}
$$

where the inequality holds since $f(x) \le \max\{t_0, M^* f(x)\}$ for almost every $x \in \mathbb{R}^n$.

The last term in (8.10) can be estimated as

$$
\begin{aligned}
\int_{t_0}^{\infty} t^{p-1-\varepsilon} w(E_t)\,dt &= \int_{\mathbb{R}^n}\Big(\int_{t_0}^{\infty} \chi_{E_t}(x) t^{p-1-\varepsilon}\,dt\Big) w(x)\,dx \\
&= \int_{E_{t_0}}\Big(\int_{t_0}^{M^* f(x)} t^{p-1-\varepsilon}\,dt\Big) w(x)\,dx \\
&\le \frac{1}{p-\varepsilon}\int_{E_{t_0}} (M^* f(x))^{p-\varepsilon} w(x)\,dx \\
&\le \frac{1}{p-\varepsilon}\int_{\mathbb{R}^n} (M^* f(x))^p \big(\max\{t_0, M^* f(x)\}\big)^{-\varepsilon} w(x)\,dx.
\end{aligned}
$$

Multiplication of the obtained inequalities by $\varepsilon > 0$, together with (8.10), gives

$$\int_{\mathbb{R}^n} (M^*f(x))^p \big(\max\{t_0, M^*f(x)\}\big)^{-\varepsilon} w(x)\,dx \le C(n,p,C_1)\int_{\mathbb{R}^n} f(x)^{p-\varepsilon} w(x)\,dx + C(n,p,C_1)\frac{\varepsilon}{p-\varepsilon}\int_{\mathbb{R}^n} (M^*f(x))^p \big(\max\{t_0, M^*f(x)\}\big)^{-\varepsilon} w(x)\,dx.$$

Here the last term is finite by (8.8) and the fact that $t_0 > 0$. We choose $0 < \varepsilon < p-1$ to be so small that

$$C(n,p,C_1)\frac{\varepsilon}{p-\varepsilon} < \frac{1}{2}.$$

This allows us to absorb the last term to the left-hand side. Finally, by Fatou's lemma,

$$\begin{aligned}\int_{\mathbb{R}^n} (M^*f(x))^{p-\varepsilon} w(x)\,dx &\le \liminf_{t_0\to 0+}\int_{\mathbb{R}^n} (M^*f(x))^p \big(\max\{t_0, M^*f(x)\}\big)^{-\varepsilon} w(x)\,dx\\ &\le C(n,p,C_1)\int_{\mathbb{R}^n} f(x)^{p-\varepsilon} w(x)\,dx,\end{aligned}$$

and this concludes the proof. □

8.4. Muckenhoupt A_p weights for $1 < p < \infty$

Muckenhoupt weights are related to weighted norm inequalities for maximal functions and they are applied in harmonic analysis and partial differential equations.

DEFINITION 8.12. A weight w in $\mathbb{R}^n$ belongs to the Muckenhoupt class A_p, for $1 < p < \infty$, if there exists a constant C such that

$$\fint_Q w(x)\,dx \left(\fint_Q w(x)^{\frac{1}{1-p}}\,dx\right)^{p-1} \le C \tag{8.11}$$

for every cube $Q \subset \mathbb{R}^n$. The smallest possible constant C in (8.11) is called the A_p constant of w, and it is denoted by $[w]_{A_p}$.

The following basic properties of Muckenhoupt weights follow rather directly from the definitions and Hölder's inequality. The first assertion shows that the Muckenhoupt classes are nested and the second assertion is a duality statement.

LEMMA 8.13. *The following assertions hold.*

(a) *For $1 < p < q < \infty$, we have $A_p \subset A_q$.*
(b) *For $1 < p < \infty$, we have $w \in A_p$ if and only if $w^{1-p'} \in A_{p'}$, where $p' = \frac{p}{p-1}$.*

PROOF. To prove assertion (a), let $1 < p < q < \infty$, let $Q \subset \mathbb{R}^n$ be a cube, and assume that $w \in A_p$. By Hölder's inequality

$$\left(\fint_Q w(x)^{\frac{1}{1-q}}\,dx\right)^{q-1} \le \left(\fint_Q \left(w(x)^{\frac{1}{1-q}}\right)^{\frac{q-1}{p-1}}\,dx\right)^{(q-1)\frac{p-1}{q-1}} = \left(\fint_Q w(x)^{\frac{1}{1-p}}\,dx\right)^{p-1}.$$

This implies

$$\fint_Q w(x)\,dx\left(\fint_Q w(x)^{\frac{1}{1-q}}\,dx\right)^{q-1} \le \fint_Q w(x)\,dx\left(\fint_Q w(x)^{\frac{1}{1-p}}\,dx\right)^{p-1} \le [w]_{A_p},$$

and assertion (a) follows.

To prove assertion (b), let $1 < p < \infty$ and assume that $w \in A_p$. Then

$$\begin{aligned}
&\fint_Q w(x)^{1-p'}\,dx\left(\fint_Q \left(w(x)^{1-p'}\right)^{\frac{1}{1-p'}}\,dx\right)^{p'-1} \\
&\qquad = \fint_Q w(x)^{1-p'}\,dx\left(\fint_Q w(x)\,dx\right)^{\frac{1}{p-1}} \le [w]_{A_p}^{\frac{1}{p-1}}
\end{aligned}$$

for every cube $Q \subset \mathbb{R}^n$. This shows that $w^{1-p'} \in A_{p'}$. Conversely, if $w^{1-p'} \in A_{p'}$ then $w = (w^{1-p'})^{1-p} \in A_p$ by the argument above. □

Next, we consider a weighted Hölder inequality for integral averages.

LEMMA 8.14. *Let $1 < p < \infty$ and assume that $w \in A_p$. Let $Q \subset \mathbb{R}^n$ be a cube and $f \in L^1(Q)$ be a nonnegative function. Then*

$$\fint_Q f(x)\,dx \le [w]_{A_p}^{\frac{1}{p}}\left(\frac{1}{w(Q)}\int_Q f(x)^p w(x)\,dx\right)^{\frac{1}{p}}.$$

PROOF. Hölder's inequality and (8.11) imply

$$\begin{aligned}
\fint_Q f(x)\,dx &\le \frac{1}{|Q|}\left(\int_Q w(x)^{\frac{1}{1-p}}\,dx\right)^{\frac{p-1}{p}}\left(\int_Q f(x)^p w(x)\,dx\right)^{\frac{1}{p}} \\
&= \left(\frac{w(Q)}{|Q|}\right)^{\frac{1}{p}}\left(\fint_Q w(x)^{\frac{1}{1-p}}\,dx\right)^{\frac{p-1}{p}}\left(\frac{1}{w(Q)}\int_Q f(x)^p w(x)\,dx\right)^{\frac{1}{p}} \\
&\le [w]_{A_p}^{\frac{1}{p}}\left(\frac{1}{w(Q)}\int_Q f(x)^p w(x)\,dx\right)^{\frac{1}{p}}.
\end{aligned}$$

□

As an application of Lemma 8.14 we obtain that a Muckenhoupt weight is doubling, see Definition 8.3.

THEOREM 8.15. *Let $1 < p < \infty$ and assume that $w \in A_p$. Then w is doubling, that is, there exists a constant $C = C(n, p, w)$ such that $w(B(x, 2r)) \le Cw(B(x, r))$ for every $x \in \mathbb{R}^n$ and $r > 0$.*

PROOF. Let $x \in \mathbb{R}^n$ and $r > 0$. Since $B(x, 2r) \subset Q(x, 2r)$, Lemma 8.14 with $f = \chi_{B(x,r)}$ implies

$$\begin{aligned}
\frac{1}{2^n} = \frac{|B(x,r)|}{|B(x,2r)|} &\le C(n)\fint_{Q(x,2r)} f(y)\,dy \\
&\le C(n)[w]_{A_p}^{\frac{1}{p}}\left(\frac{1}{w(Q(x,2r))}\int_{Q(x,2r)} f(y)^p w(y)\,dy\right)^{\frac{1}{p}} \\
&\le C(n)[w]_{A_p}^{\frac{1}{p}}\left(\frac{w(B(x,r))}{w(B(x,2r))}\right)^{\frac{1}{p}}.
\end{aligned}$$

The claim follows by raising both sides to power p and reorganizing the terms. □

The class of Muckenhoupt weights is closed under maximum and minimum.

THEOREM 8.16. *Let $1 < p < \infty$ and assume that $v, w \in A_p$. Then $\max\{v, w\} \in A_p$ and $\min\{v, w\} \in A_p$.*

PROOF. The function $\max\{v,w\}$ is a weight since $\max\{v(x),w(x)\}>0$ for almost every $x\in\mathbb{R}^n$ and $\max\{v,w\}\in L^1_{\mathrm{loc}}(\mathbb{R}^n)$. By Definition 8.12 we have

$$\begin{aligned}\fint_Q \max\{v(x),w(x)\}\,dx &\le \fint_Q v(x)\,dx+\fint_Q w(x)\,dx\\ &\le [v]_{A_p}\Big(\fint_Q v(x)^{\frac{1}{1-p}}\,dx\Big)^{1-p}+[w]_{A_p}\Big(\fint_Q w(x)^{\frac{1}{1-p}}\,dx\Big)^{1-p}\\ &\le ([v]_{A_p}+[w]_{A_p})\Big(\fint_Q \max\{v(x),w(x)\}^{\frac{1}{1-p}}\,dx\Big)^{1-p}\end{aligned}$$

for every cube $Q\subset\mathbb{R}^n$. Thus $\max\{v,w\}\in A_p$.

Consider $\sigma(x)=\min\{v(x),w(x)\}$, for $x\in\mathbb{R}^n$. Clearly $\sigma(x)>0$ for almost every $x\in\mathbb{R}^n$ and $\sigma\in L^1_{\mathrm{loc}}(\mathbb{R}^n)$, and so σ is a weight. Lemma 8.13 (b) implies $v^{\frac{1}{1-p}},w^{\frac{1}{1-p}}\in A_{\frac{p}{p-1}}$. By the first part of the proof we have

$$\sigma^{\frac{1}{1-p}}=(\min\{v,w\})^{\frac{1}{1-p}}=\max\big\{v^{\frac{1}{1-p}},w^{\frac{1}{1-p}}\big\}\in A_{\frac{p}{p-1}},$$

and $\min\{v,w\}=\sigma\in A_p$ follows from Lemma 8.13 (b). □

A weighted norm inequality for the dyadic maximal function is related to Muckenhoupt weights.

LEMMA 8.17. *Let $1<p<\infty$ and assume that w is a weight such that*

$$\int_Q (M^d_Q f(x))^p w(x)\,dx\le C_1\int_{\mathbb{R}^n} f(x)^p w(x)\,dx$$

for every nonnegative $f\in L^\infty(\mathbb{R}^n)$ with compact support and for every cube $Q\subset\mathbb{R}^n$. Then $w\in A_p$ and $[w]_{A_p}\le C_1$.

PROOF. Fix a cube $Q\subset\mathbb{R}^n$. Let $\varepsilon>0$ and $f=(\varepsilon+w)^{\frac{1}{1-p}}\chi_Q$. Clearly

$$\fint_Q(\varepsilon+w(y))^{\frac{1}{1-p}}\,dy\le M^d_Q f(x)$$

for every $x\in Q$, and thus

$$\fint_Q w(x)\,dx\Big(\fint_Q(\varepsilon+w(y))^{\frac{1}{1-p}}\,dy\Big)^p\le\fint_Q(M^d_Q f(x))^p w(x)\,dx.$$

By the assumption, we have

$$\begin{aligned}\fint_Q(M^d_Q f(x))^p w(x)\,dx &\le C_1\frac{1}{|Q|}\int_{\mathbb{R}^n} f(x)^p w(x)\,dx\\ &\le C_1\fint_Q(\varepsilon+w(x))^{\frac{p}{1-p}}(\varepsilon+w(x))\,dx\\ &= C_1\fint_Q(\varepsilon+w(x))^{\frac{1}{1-p}}\,dx.\end{aligned}$$

This gives

$$\fint_Q w(x)\,dx\Big(\fint_Q(\varepsilon+w(y))^{\frac{1}{1-p}}\,dy\Big)^{p-1}\le C_1$$

for every $\varepsilon > 0$. Fatou's lemma implies

$$\begin{aligned}
&\fint_Q w(x)\,dx\left(\fint_Q w(x)^{\frac{1}{1-p}}\,dx\right)^{p-1}\\
&\qquad\le \liminf_{\varepsilon\to 0+}\fint_Q w(x)\,dx\left(\fint_Q (\varepsilon+w(x))^{\frac{1}{1-p}}\,dx\right)^{p-1}\le C_1,
\end{aligned}$$

and the claim follows. □

Theorem 8.19 gives a converse of Theorem 8.17, see also Theorem 9.9 and Theorem 8.28. First, we discuss a partial converse that provides a weighted weak type estimate for the dyadic maximal operator. The proof is based on the Calderón–Zygmund lemma and the argument is similar to the proof of Lemma 8.7.

THEOREM 8.18. *Let $Q_0 \subset \mathbb{R}^n$ be a cube, $f \in L^\infty(Q_0)$ be a nonnegative function and $E_t = \{x \in Q_0 : M^d_{Q_0} f(x) > t\}$ for $t > 0$. Let $1 < p < \infty$ and assume that $w \in A_p$. Then*

$$w(E_t) \le \frac{[w]_{A_p}}{t^p}\int_{Q_0} f(x)^p w(x)\,dx$$

for every $t > 0$.

PROOF. Let $t > 0$ and let $\mathcal{D}_t$ be the associated collection of maximal dyadic subcubes of Q_0 as in the proof of Lemma 8.6. Recall that in the case $0 < t < f_{Q_0}$ we have $\mathcal{D}_t = \{Q_0\}$. Since $t < f_Q$ for every $Q \in \mathcal{D}_t$, by (8.3) we have

$$\begin{aligned}
w(E_t) &= w\Bigg(\bigcup_{Q\in\mathcal{D}_t} Q\Bigg) \le \sum_{Q\in\mathcal{D}_t} w(Q)\\
&= \frac{1}{t^p}\sum_{Q\in\mathcal{D}_t} w(Q)t^p \le \frac{1}{t^p}\sum_{Q\in\mathcal{D}_t} w(Q)\left(\fint_Q f(x)\,dx\right)^p.
\end{aligned}$$

The integrals on the right-hand side can be estimated with Lemma 8.14. This, together with the pairwise disjointness of the cubes $Q \in \mathcal{D}_t$, implies

$$\begin{aligned}
w(E_t) &\le \frac{[w]_{A_p}}{t^p}\sum_{Q\in\mathcal{D}_t}\int_Q f(x)^p w(x)\,dx = \frac{[w]_{A_p}}{t^p}\int_{E_t} f(x)^p w(x)\,dx\\
&\le \frac{[w]_{A_p}}{t^p}\int_{Q_0} f(x)^p w(x)\,dx.
\end{aligned}$$

□

The following result shows that the Hardy–Littlewood maximal operator is bounded on $L^p(w\,dx)$ for $1 < p < \infty$ and $w \in A_p$. In particular, we obtain a converse of Lemma 8.17.

THEOREM 8.19. *Let $1 < p < \infty$ and assume that $w \in A_p$. There exists a constant $C = C(n, p, [w]_{A_p})$ such that*

$$\int_{\mathbb{R}^n} (M^* f(x))^p w(x)\,dx \le C\int_{\mathbb{R}^n} |f(x)|^p w(x)\,dx$$

for every $f \in L^p(w\,dx)$.

PROOF. Let $\sigma = w^{\frac{1}{1-p}}$. Lemma 8.13 (b) implies that $\sigma \in A_{p'}$, where $p' = \frac{p}{p-1}$. Let $f \in L^p(w\,dx)$, and fix $x \in \mathbb{R}^n$ and a ball $B = B(x_0, r)$ with $x \in B$. Then

$$\fint_B |f(y)|\,dy = A\left(\frac{|B|}{w(B(x_0,3r))}\left(\frac{1}{\sigma(B(x_0,3r))}\int_B |f(y)|\,dy\right)^{p-1}\right)^{\frac{1}{p-1}},$$

where

$$\begin{aligned} A &= \left(\frac{w(B(x_0,3r))\sigma(B(x_0,3r))^{p-1}}{|B|^p}\right)^{\frac{1}{p-1}} \\ &\le C(n,p)\left(\frac{w(Q(x_0,3r))\sigma(Q(x_0,3r))^{p-1}}{|Q(x_0,3r)|^p}\right)^{\frac{1}{p-1}} \\ &\le C(n,p)[w]_{A_p}^{\frac{1}{p-1}}. \end{aligned}$$

For $z \in B$, we have

$$\begin{aligned} \frac{1}{\sigma(B(x_0,3r))}\int_B |f(y)|\,dy &\le \frac{1}{\sigma(B(z,2r))}\int_{B(z,2r)} |f(y)\sigma(y)^{-1}|\sigma(y)\,dy \\ &\le M^\sigma(f\sigma^{-1})(z). \end{aligned}$$

Hence,

$$\begin{aligned} &\fint_B |f(y)|\,dy \\ &\quad\le C(n,p)[w]_{A_p}^{\frac{1}{p-1}}\left(\frac{1}{w(B(x_0,3r))}\int_B \left(M^\sigma(f\sigma^{-1})(z)\right)^{p-1} w(z)^{-1}w(z)\,dz\right)^{\frac{1}{p-1}} \\ &\quad\le C(n,p,[w]_{A_p})\left(\frac{1}{w(B(x,2r))}\int_{B(x,2r)} \left(M^\sigma(f\sigma^{-1})(z)\right)^{p-1} w(z)^{-1}w(z)\,dz\right)^{\frac{1}{p-1}} \\ &\quad\le C(n,p,[w]_{A_p})\big(M^w\big(M^\sigma(f\sigma^{-1})^{p-1}w^{-1}\big)(x)\big)^{\frac{1}{p-1}}. \end{aligned}$$

By taking supremum over all balls B as above, it follows that

$$M^*f(x) \le C(n,p,[w]_{A_p})\big(M^w\big(M^\sigma(f\sigma^{-1})^{p-1}w^{-1}\big)(x)\big)^{\frac{1}{p-1}}$$

for all $x \in \mathbb{R}^n$. Observe that $w \in A_p$ and $\sigma \in A_{p'}$ are doubling weights by Theorem 8.15, with constants depending on n, p and $[w]_{A_p}$. Remark 8.4 (a) implies

$$\begin{aligned} &\int_{\mathbb{R}^n} \left(M^*f(x)\right)^p w(x)\,dx \\ &\qquad\le C(n,p,[w]_{A_p})\int_{\mathbb{R}^n} \big(M^w\big(M^\sigma(f\sigma^{-1})^{p-1}w^{-1}\big)(x)\big)^{\frac{p}{p-1}} w(x)\,dx \\ &\qquad\le C(n,p,[w]_{A_p})\int_{\mathbb{R}^n} \left(M^\sigma(f\sigma^{-1})(x)\right)^p \sigma(x)\,dx \\ &\qquad\le C(n,p,[w]_{A_p})\int_{\mathbb{R}^n} |f(x)|^p w(x)\,dx, \end{aligned}$$

where the facts that $\sigma = w^{\frac{1}{1-p}}$ and $w = \sigma^{1-p}$ are used. □

Theorem 8.11 and Theorem 8.19 together imply the following Muckenhoupt A_p weighted norm inequality on $L^{p-\varepsilon}(w\,dx)$.

COROLLARY 8.20. *Let $1 < p < \infty$ and assume that $w \in A_p$. There exist $\varepsilon = \varepsilon(n, p, [w]_{A_p})$, with $0 < \varepsilon < p-1$, and $C = C(n, p, [w]_{A_p})$ such that*

$$\int_{\mathbb{R}^n} \big(M^* f(x)\big)^{p-\varepsilon} w(x)\,dx \le C \int_{\mathbb{R}^n} f(x)^{p-\varepsilon} w(x)\,dx$$

for all nonnegative $f \in L^\infty(\mathbb{R}^n)$ with compact support.

Next we obtain a self-improvement result for Muckenhoupt A_p weights for $1 < p < \infty$.

COROLLARY 8.21. *Let $1 < p < \infty$ and assume that $w \in A_p$. There exist $\varepsilon = \varepsilon(n, p, [w]_{A_p}) > 0$ and $C = C(n, p, [w]_{A_p})$ such that $w \in A_{p-\varepsilon}$ and $[w]_{A_{p-\varepsilon}} \le C[w]_{A_p}$.*

PROOF. Let $Q \subset \mathbb{R}^n$ be a cube. By Corollary 8.20 there exists $\varepsilon = \varepsilon(n, p, [w]_{A_p})$, with $0 < \varepsilon < p-1$, such that

$$\begin{aligned} \int_Q \big(M_Q^d f(x)\big)^{p-\varepsilon} w(x)\,dx &\le C(n,p) \int_{\mathbb{R}^n} \big(M^* f(x)\big)^{p-\varepsilon} w(x)\,dx \\ &\le C(n, p, [w]_{A_p}) \int_{\mathbb{R}^n} f(x)^{p-\varepsilon} w(x)\,dx \end{aligned}$$

for all nonnegative $f \in L^\infty(\mathbb{R}^n)$ with compact support. Lemma 8.17 implies that $w \in A_{p-\varepsilon}$ and $[w]_{A_{p-\varepsilon}} \le C(n, p, [w]_{A_p})$. □

8.5. Reverse Hölder inequalities for Muckenhoupt weights

We prove a reverse Hölder inequality for Muckenhoupt A_p weights, with $1 < p \le \infty$. For this purpose, we first discuss the limiting Muckenhoupt condition for $p = \infty$, which is the weakest A_p condition.

DEFINITION 8.22. A weight w in $\mathbb{R}^n$ belongs to the Muckenhoupt class A_∞ if there exist constants $C, \delta > 0$ such that

$$\frac{w(E)}{w(Q)} \le C \left(\frac{|E|}{|Q|} \right)^\delta \tag{8.12}$$

whenever $Q \subset \mathbb{R}^n$ is a cube and $E \subset Q$ is a measurable set.

An A_∞ weight is doubling, see Definition 8.3.

THEOREM 8.23. *Assume that $w \in A_\infty$. Then w is doubling, that is, there exists a constant $C = C(n, w)$ such that $w(B(x, 2r)) \le C w(B(x, r))$ for every $x \in \mathbb{R}^n$ and $r > 0$.*

PROOF. Let C_1 and δ be the constants in the A_∞ condition (8.12) for w. Choose $0 < \beta = \beta(w) < 1$ satisfying $0 < C_1\beta^\delta < 1$ and define $\alpha = 1 - C_1\beta^\delta > 0$. First assume that $Q \subset \mathbb{R}^n$ is a cube and $E \subset Q$ is a measurable set with $|E| \ge (1-\beta)|Q|$. Then $|Q \setminus E| \le \beta|Q|$ and (8.12) implies

$$w(Q \setminus E) \le C_1 \left(\frac{|Q \setminus E|}{|Q|} \right)^\delta w(Q) \le C_1\beta^\delta w(Q).$$

Consequently,

$$w(E) = w(Q) - w(Q \setminus E) \ge (1 - C_1\beta^\delta) w(Q) = \alpha w(Q). \tag{8.13}$$

Fix $x \in \mathbb{R}^n$ and $r > 0$. Define $r_1 = 2r$ and $r_{j+1} = (1-\beta)^{\frac{1}{n}} r_j$ for every $j = 1, \ldots, k-1$, where $k = k(n, \beta)$ is the smallest integer such that $Q(x, r_k) \subset B(x, r)$. Then $Q(x, r_{j+1}) \subset Q(x, r_j)$ and

$$|Q(x, r_{j+1})| \geq (1-\beta)|Q(x, r_j)|,$$

for every $j = 1, \ldots, k-1$. Hence we may apply (8.13) iteratively for these cubes, and obtain

$$\begin{aligned} w(B(x, 2r)) &\leq w(Q(x, 2r)) = w(Q(x, r_1)) \leq \alpha^{-1} w(Q(x, r_2)) \\ &\leq \cdots \leq \alpha^{1-k} w(Q(x, r_k)) \leq \alpha^{1-k} w(B(x, r)). \end{aligned}$$

This proves the claim with the constant $C(n, w) = \alpha^{1-k}$. $\square$

Let $Q \subset \mathbb{R}^n$ be a cube, σ be a weight and $f \in L^1(Q; \sigma\, dx)$ be a nonnegative function. The σ-weighted average of f on Q is defined as

$$f_{Q;\sigma} = \frac{1}{\sigma(Q)} \int_Q f(x)\sigma(x)\, dx. \tag{8.14}$$

In the next lemma a distributional inequality implies a reverse Hölder inequality.

LEMMA 8.24. *Let $\sigma \in A_\infty$ and let $w \geq 0$ be a measurable function such that $w\sigma \in L^1_{\mathrm{loc}}(\mathbb{R}^n)$. Assume that there exist $\alpha > 0$ and $0 < \beta < 1$ such that*

$$\sigma(\{x \in Q : w(x) > \beta w_{Q;\sigma}\}) \geq \alpha\sigma(Q)$$

for every cube $Q \subset \mathbb{R}^n$. Then there exist constants $q = q(n, \sigma, \alpha, \beta) > 1$ and $C = C(n, \sigma, \alpha, \beta)$ such that

$$\left(\frac{1}{\sigma(Q_0)} \int_{Q_0} w(x)^q \sigma(x)\, dx \right)^{\frac{1}{q}} \leq \frac{C}{\sigma(Q_0)} \int_{Q_0} w(x)\sigma(x)\, dx$$

for every cube $Q_0 \subset \mathbb{R}^n$.

PROOF. Let $Q_0 \subset \mathbb{R}^n$ be a cube. For $t > w_{Q_0;\sigma}$, let $\mathcal{D}_t \subset \mathcal{D}(Q_0)$ be the collection of maximal dyadic subcubes $Q \subset Q_0$ such that $w_{Q;\sigma} > t$, compare the proof of Lemma 8.6. Observe that $Q \subsetneq Q_0$ for every $Q \in \mathcal{D}_t$ since $t > w_{Q_0;\sigma}$.

Let $Q \in \mathcal{D}_t$ and let $\pi Q \in \mathcal{D}(Q_0)$ be the dyadic parent cube of Q, which satisfies the conditions $Q \subset \pi Q$ and $2^n|Q| = |\pi Q|$. Theorem 8.23 implies

$$\frac{\sigma(\pi Q)}{\sigma(Q)} \leq C(n, \sigma).$$

By the maximality of $Q \in \mathcal{D}_t$, we have $w_{\pi Q;\sigma} \leq t$ and

$$t < w_{Q;\sigma} = \frac{1}{\sigma(Q)} \int_Q w(x)\sigma(x)\, dx \leq \frac{\sigma(\pi Q)}{\sigma(Q)} \cdot \frac{1}{\sigma(\pi Q)} \int_{\pi Q} w(x)\sigma(x)\, dx \leq C(n, \sigma)t.$$

On the other hand, if $x \in Q_0 \setminus \bigcup_{Q \in \mathcal{D}_t} Q$, then $w_{Q';\sigma} \leq t$ whenever $Q' \in \mathcal{D}(Q_0)$ with $x \in Q'$; otherwise $Q' \subset Q$ for some $Q \in \mathcal{D}_t$, and this would contradict the choice of x. Since there are arbitrarily small such cubes Q' containing x, the Lebesgue differentiation theorem for the doubling measure $\sigma\, dx$, see Remark 8.4 (c), shows that $w(x) \leq t$ for almost every $x \in Q_0 \setminus \bigcup_{Q \in \mathcal{D}_t} Q$ with respect to the measure $\sigma\, dx$.

Let $d\mu = (\sigma w)\,dx$. Using the absolute continuity of μ with respect to $\sigma\,dx$ and the assumption, we obtain

$$
\begin{aligned}
\mu(\{x \in Q_0 : w(x) > t\}) &\le \sum_{Q\in\mathcal{D}_t} \mu(Q) \le C(n,\sigma)t \sum_{Q\in\mathcal{D}_t} \sigma(Q) \\
&\le \frac{C(n,\sigma)t}{\alpha} \sum_{Q\in\mathcal{D}_t} \sigma(\{x \in Q : w(x) > \beta w_{Q;\sigma}\}) \\
&\le \frac{C(n,\sigma)t}{\alpha} \sum_{Q\in\mathcal{D}_t} \sigma(\{x \in Q : w(x) > \beta t\}) \\
&\le \frac{C(n,\sigma)t}{\alpha} \sigma(\{x \in Q_0 : w(x) > \beta t\}),
\end{aligned}
\tag{8.15}
$$

where the final step holds since the cubes in $\mathcal{D}_t \subset \mathcal{D}(Q_0)$ are pairwise disjoint.

Next we multiply (8.15) by t^{q-2}, where $1 < q < 2$ is to be specified later, and then integrate the resulting estimate from $w_{Q_0;\sigma}$ to a fixed $t_0 > w_{Q_0;\sigma}$. This gives

$$
\begin{aligned}
&\int_{w_{Q_0;\sigma}}^{t_0} t^{q-2}\mu(\{x \in Q_0 : w(x) > t\})\,dt \\
&\quad\le \frac{C(n,\sigma)}{\alpha\beta^q} \int_0^{t_0} (\beta t)^q \sigma(\{x \in Q_0 : w(x) > \beta t\})\,\frac{dt}{t} \\
&\quad= \frac{C(n,\sigma)}{\alpha\beta^q} \int_0^{t_0} (\beta t)^q \sigma(\{x \in Q_0 : \min\{t_0, w(x)\} > \beta t\})\,\frac{dt}{t} \\
&\quad\le \frac{C(n,\sigma)}{q\alpha\beta^q} \int_{Q_0} w(x) \min\{t_0, w(x)\}^{q-1}\sigma(x)\,dx < \infty.
\end{aligned}
$$

By Fubini's theorem,

$$
\begin{aligned}
&\int_{w_{Q_0;\sigma}}^{t_0} t^{q-2}\mu(\{x \in Q_0 : w(x) > t\})\,dt \\
&\quad= \int_{\{x\in Q_0 : w(x) > w_{Q_0;\sigma}\}} w(x)\Big(\int_{w_{Q_0;\sigma}}^{\min\{t_0,w(x)\}} t^{q-2}\,dt\Big)\sigma(x)\,dx \\
&\quad= \int_{\{x\in Q_0 : w(x) > w_{Q_0;\sigma}\}} w(x)\Big(\frac{\min\{t_0, w(x)\}^{q-1}}{q-1} - \frac{w_{Q_0;\sigma}^{q-1}}{q-1}\Big)\sigma(x)\,dx \\
&\quad\ge \frac{1}{q-1}\int_{Q_0} w(x)\min\{t_0, w(x)\}^{q-1}\sigma(x)\,dx - \frac{2}{q-1} w_{Q_0;\sigma}^q \sigma(Q_0).
\end{aligned}
$$

Combining the estimates above, we arrive at

$$
\begin{aligned}
&\Big(\frac{1}{q-1} - \frac{C(n,\sigma)}{q\alpha\beta^q}\Big)^{\frac{1}{q}} \Big(\frac{1}{\sigma(Q_0)}\int_{Q_0} w(x)\min\{t_0, w(x)\}^{q-1}\sigma(x)\,dx\Big)^{\frac{1}{q}} \\
&\quad\le \Big(\frac{2}{q-1}\Big)^{\frac{1}{q}} w_{Q_0;\sigma}
\end{aligned}
$$

for every $t_0 > w_{Q_0;\sigma}$. The claim follows by first choosing $1 < q < 2$ to be so small that

$$
\frac{1}{q-1} - \frac{C(n,\sigma)}{q\alpha\beta^q} \ge 1,
$$

and then letting $t_0 \to \infty$ inside the final integral and applying Fatou's lemma. $\square$

The next result shows that $A_\infty = \bigcup_{1<p<\infty} A_p$.

THEOREM 8.25. *Assume that w is a weight in $\mathbb{R}^n$. The following conditions are equivalent:*

(a) $w \in A_\infty$,

(b) $w \in A_p$ *for some* $1 < p < \infty$,

(c) *there are constants C and $q > 1$ such that*

$$\left(\fint_Q w(x)^q \, dx \right)^{\frac{1}{q}} \le C \fint_Q w(x) \, dx$$

for every cube $Q \subset \mathbb{R}^n$.

Moreover, the constants and exponents in each of the conditions only depend on n, w and each other.

PROOF. First we show that (a) implies (b) by assuming that $\sigma \in A_\infty$ and showing that $\sigma \in A_p$ for some $1 < p < \infty$. Let $C(\sigma) > 0$ and $\delta = \delta(\sigma) > 0$ be the constants in the A_∞ condition (8.12) for σ. Choose $0 < \beta = \beta(\sigma) < 1$ satisfying $C(\sigma)\beta^\delta < 1$ and define $\alpha = 1 - C(\sigma)\beta^\delta > 0$.

Let $w = \sigma^{-1}$, let $Q \subset \mathbb{R}^n$ be a cube and define $E = \{x \in Q : w(x) > \beta w_{Q;\sigma}\}$. Since $w(x) \le \beta w_{Q;\sigma}$ for every $x \in Q \setminus E$, we obtain

$$\frac{|Q \setminus E|}{|Q|} = \frac{1}{|Q|} \int_{Q \setminus E} w(x)\sigma(x) \, dx \le \frac{\beta}{|Q|} \frac{|Q|}{\sigma(Q)} \sigma(Q \setminus E) \le \beta.$$

By (8.12), we have

$$1 = \frac{\sigma(E) + \sigma(Q \setminus E)}{\sigma(Q)} \le \frac{\sigma(E)}{\sigma(Q)} + C(\sigma)\left(\frac{|Q \setminus E|}{|Q|} \right)^\delta \le \frac{\sigma(E)}{\sigma(Q)} + C(\sigma)\beta^\delta, \tag{8.16}$$

and thus

$$\alpha\sigma(Q) = \big(1 - C(\sigma)\beta^\delta\big)\sigma(Q) \le \sigma(E).$$

From Lemma 8.24 we obtain constants $q = q(n, \sigma, \alpha, \beta) = q(n, \sigma) > 1$ and $C = C(n, \sigma, \alpha, \beta) = C(n, \sigma)$ such that

$$\begin{aligned} \frac{1}{\sigma(Q)^{\frac{1}{q}}} \left(\int_Q \sigma(x)^{1-q} \, dx \right)^{\frac{1}{q}} &= \left(\frac{1}{\sigma(Q)} \int_Q w(x)^q \sigma(x) \, dx \right)^{\frac{1}{q}} \\ &\le \frac{C}{\sigma(Q)} \int_Q w(x)\sigma(x) \, dx \\ &= C \frac{|Q|}{\sigma(Q)} = C \frac{|Q|^{\frac{1}{q}} |Q|^{\frac{q-1}{q}}}{\sigma(Q)^{\frac{1}{q}} \sigma(Q)^{\frac{q-1}{q}}}, \end{aligned}$$

where we also used the fact that $w(x)\sigma(x) = 1$ for almost every $x \in Q$. Reorganization of the terms gives

$$|Q|^{-\frac{q-1}{q}} \sigma(Q)^{\frac{q-1}{q}} |Q|^{-\frac{1}{q}} \left(\int_Q \sigma(x)^{1-q} \, dx \right)^{\frac{1}{q}} \le C(n, \sigma).$$

For $p = \frac{q}{q-1}$, the estimate above implies

$$\fint_Q \sigma(x) \, dx \left(\fint_Q \sigma(x)^{\frac{1}{1-p}} \, dx \right)^{p-1} \le C.$$

Hence $\sigma \in A_p$, with $p = p(n, \sigma)$ and the A_p constant $C = C(n, \sigma)$.

Next we show that (b) implies (c). Let $1 < p < \infty$ and assume that $w \in A_p$. Choose $0 < \beta = \beta(p, w) < 1$ satisfying $(\beta[w]_{A_p})^{\frac{1}{p-1}} < 1$ and define $\alpha = 1 - (\beta[w]_{A_p})^{\frac{1}{p-1}} > 0$. Let $Q \subset \mathbb{R}^n$ be a cube and $E = \{x \in Q : w(x) > \beta w_Q\}$. By (8.11), we have

$$\begin{aligned}\frac{1}{\beta}\left(\frac{|Q \setminus E|}{|Q|}\right)^{p-1} &= w_Q\left(\frac{1}{|Q|}\int_{Q\setminus E} (\beta w_Q)^{\frac{1}{1-p}}\,dx\right)^{p-1}\\ &\le w_Q\left(\fint_Q w(x)^{\frac{1}{1-p}}\,dx\right)^{p-1} \le [w]_{A_p}.\end{aligned}$$

As in (8.16), we obtain

$$|E| \ge (1 - (\beta[w]_{A_p})^{\frac{1}{p-1}})|Q| = \alpha|Q|.$$

Lemma 8.24, with $\sigma(x) = 1$ for every $x \in \mathbb{R}^n$, gives constants $q = q(n, \alpha, \beta) = q(n, p, w) > 1$ and $C = C(n, \alpha, \beta) = C(n, p, w) > 0$ such that

$$\left(\fint_Q w(x)^q\,dx\right)^{\frac{1}{q}} \le C \fint_Q w(x)\,dx$$

for every cube $Q \subset \mathbb{R}^n$, and thus assertion (c) holds.

Finally we show that (c) implies (a). Let w be a weight and assume that there are constants $C > 0$ and $q > 1$ such that the reverse Hölder condition (c) holds. Let $Q \subset \mathbb{R}^n$ be a cube and $E \subset Q$ be a measurable set. Using first Hölder's inequality and then the reverse Hölder inequality, we obtain

$$w(E) \le |E|^{\frac{q-1}{q}}\left(\int_Q w(x)^q\,dx\right)^{\frac{1}{q}} \le C|E|^{\frac{q-1}{q}}|Q|^{\frac{1}{q}}\fint_Q w(x)\,dx = C\left(\frac{|E|}{|Q|}\right)^{\frac{q-1}{q}} w(Q).$$

Thus w satisfies the A_∞ condition (8.12) with $\delta = \frac{q-1}{q} > 0$, and (a) holds. □

REMARK 8.26. If $1 < p < \infty$ and $w \in A_p$, then C and $q > 1$ in Theorem 8.25 (c) only depend on w through $[w]_{A_p}$. This follows from the proof of (b) implies (c) since α and β in the proof only depend on $[w]_{A_p}$ and p.

From Theorem 8.25 we obtain a second proof for the self-improvement property of Muckenhoupt weights, compare to Corollary 8.21. Yet another proof will be based on Theorem 8.39.

THEOREM 8.27. *Let $1 < p < \infty$ and assume that $w \in A_p$. There exists a constant $\varepsilon = \varepsilon(n, p, [w]_{A_p}) > 0$ such that $w \in A_{p-\varepsilon}$, with $[w]_{A_{p-\varepsilon}}$ only depending on n, p and $[w]_{A_p}$.*

PROOF. By Lemma 8.13 (b), we have $w^{\frac{1}{1-p}} \in A_{p'}$ with $p' = \frac{p}{p-1}$, and $[w]_{A_{p'}}$ only depends on p and $[w]_{A_p}$. Theorem 8.25 and Remark 8.26 imply that there exist $C = C(n, p, [w]_{A_p})$ and $q = q(n, p, [w]_{A_p}) > 1$ such that

$$\left(\fint_Q w(x)^{\frac{q}{1-p}}\,dx\right)^{\frac{1}{q}} \le C\fint_Q w(x)^{\frac{1}{1-p}}\,dx \tag{8.17}$$

for every cube $Q \subset \mathbb{R}^n$. Let $\varepsilon = \varepsilon(p,q) = \varepsilon(n,p,[w]_{A_p})$, with $0 < \varepsilon < p-1$, be determined by $\frac{q}{p-1} = \frac{1}{p-\varepsilon-1}$. By (8.17) and (8.11), we have

$$\begin{aligned}\left(\fint_Q w(x)^{\frac{1}{1-(p-\varepsilon)}}\,dx\right)^{p-\varepsilon-1} &= \left(\fint_Q w(x)^{\frac{q}{1-p}}\,dx\right)^{\frac{p-1}{q}} \\ &\le C(n,p,[w]_{A_p})\left(\fint_Q w(x)^{\frac{1}{1-p}}\,dx\right)^{p-1} \\ &\le C(n,p,[w]_{A_p})\left(\fint_Q w(x)\,dx\right)^{-1}\end{aligned}$$

for every cube $Q \subset \mathbb{R}^n$. This shows that $w \in A_{p-\varepsilon}$, with a constant $C(n,p,[w]_{A_p})$. □

By using the self-improvement of Muckenhoupt weights, we obtain a second proof for the fact that the Hardy–Littlewood maximal operator is bounded on $L^p(w\,dx)$ for $1 < p < \infty$ and $w \in A_p$, compare to Theorem 8.19.

THEOREM 8.28. *Let $1 < p < \infty$ and assume that $w \in A_p$. There exists a constant $C = C(n,p,[w]_{A_p})$ such that*

$$\int_{\mathbb{R}^n} Mf(x)^p w(x)\,dx \le C\int_{\mathbb{R}^n} |f(x)|^p w(x)\,dx$$

for every $f \in L^1_{\mathrm{loc}}(\mathbb{R}^n)$.

PROOF. Without loss of generality, we may assume that f is a nonnegative function for which the right-hand side is finite. By Theorem 8.27, there exists $1 < q = q(n,p,[w]_{A_p}) < p$ such that $w \in A_q$. Let $x \in \mathbb{R}^n$, $r > 0$ and $Q = Q(x,r)$. Lemma 8.14 implies

$$\begin{aligned}\fint_{B(x,r)} f(y)\,dy \le C(n)\fint_Q f(y)\,dy &\le C(n,q,w)\left(\frac{1}{w(Q)}\int_Q f(y)^q w(y)\,dy\right)^{\frac{1}{q}} \\ &\le C(n,q,w)\big(M^{c,w}(f^q)(x)\big)^{\frac{1}{q}}.\end{aligned}$$

By taking supremum over all $r > 0$, we obtain

$$Mf(x) \le C(n,q,w)\big(M^{c,w}(f^q)(x)\big)^{\frac{1}{q}}.$$

Using Remark 8.4 (b) we arrive at

$$\begin{aligned}\int_{\mathbb{R}^n} Mf(x)^p w(x)\,dx &\le C(n,p,q,w)\int_{\mathbb{R}^n}\big(M^{c,w}(f^q)(x)\big)^{\frac{p}{q}} w(x)\,dx \\ &\le C(n,p,w)\int_{\mathbb{R}^n} |f(x)|^p w(x)\,dx.\end{aligned}$$

Finally, the A_q constant of w depends on w only through $[w]_{A_p}$, see Theorem 8.27. The constant in Lemma 8.14 only depends on $[w]_{A_q}$ and q. The constant in Remark 8.4 (b) depends on n, $\frac{p}{q}$ and the doubling constant of w, which in turn only depends on $[w]_{A_q}$, n and q, see the proof of Theorem 8.15. Hence we conclude that the claim holds with $C = C(n,p,w) = C(n,p,[w]_{A_p})$. □

8.6. A_1 weights and Coifman–Rochberg lemma

This section discusses the Muckenhoupt condition with $p = 1$, which is the strongest A_p condition.

DEFINITION 8.29. A weight w in $\mathbb{R}^n$ belongs to the Muckenhoupt class A_1 if there exists a constant C such that

$$\fint_Q w(x)\,dx \operatorname*{ess\,sup}_{x\in Q} \frac{1}{w(x)} \le C, \tag{8.18}$$

for every cube $Q \subset \mathbb{R}^n$. The smallest possible constant C in (8.18) is called the A_1 constant of u, and it is denoted by $[w]_{A_1}$.

Assume that $w \in A_1$ and let $1 < p < \infty$. Then

$$\left(\fint_Q w(x)^{-\frac{1}{p-1}}\,dx\right)^{p-1} \le \operatorname*{ess\,sup}_{x\in Q} \frac{1}{w(x)} \le [w]_{A_1} \cdot \frac{|Q|}{w(Q)}$$

for every cube $Q \subset \mathbb{R}^n$, and thus $w \in A_p$ with $[w]_{A_p} \le [w]_{A_1}$. Together with Theorem 8.25 this implies

$$A_1 \subset A_p \subset A_\infty, \tag{8.19}$$

for $1 < p < \infty$.

REMARK 8.30. If w and σ are A_1 weights, then it is straightforward to show that $w\sigma^{1-p} \in A_p$ for every $1 < p < \infty$. Jones [**213**] proved that the converse claim holds as well, see also Coifman, Jones and Rubio de Francia [**88**] and Cruz-Uribe, Martell and Pérez [**92**]. In particular, this gives a characterization of A_p weights in terms of A_1 weights.

The Muckenhoupt A_1 condition in (8.18) can also be written as

$$\fint_Q w(x)\,dx \le C \operatorname*{ess\,inf}_{x\in Q} w(x), \tag{8.20}$$

for every cube $Q \subset \mathbb{R}^n$. This is useful in the proof of the following characterization.

LEMMA 8.31. *A weight w in $\mathbb{R}^n$ belongs to A_1 if and only if there exists a constant C_1 such that*

$$M^c w(x) \le C_1 w(x) \quad \textit{for almost every } x \in \mathbb{R}^n. \tag{8.21}$$

Furthermore, if $w \in A_1$, then (8.21) *holds with $C_1 = [w]_{A_1}$.*

PROOF. It is clear that (8.21) implies (8.20), with $C = C_1$, since

$$\fint_Q w(y)\,dy \le M^c w(x) \le C_1 w(x)$$

for almost every $x \in Q$ whenever $Q \subset \mathbb{R}^n$ is a cube.

Assume then that $w \in A_1$. Define $E = \{x \in \mathbb{R}^n : M^c w(x) > [w]_{A_1} w(x)\}$ and let $x \in E$. There exists a cube $Q \subset \mathbb{R}^n$ containing x such that

$$\fint_Q w(y)\,dy > [w]_{A_1} w(x). \tag{8.22}$$

For every $\varepsilon > 0$ there is a cube $\widetilde{Q}$, whose corners have rational coordinates, with $Q \subset \widetilde{Q}$ and $|\widetilde{Q} \setminus Q| < \varepsilon$. Then $|\widetilde{Q}| = |Q| + |\widetilde{Q} \setminus Q| < |Q| + \varepsilon$, and by choosing $\varepsilon > 0$ small enough, we have

$$\fint_{\widetilde{Q}} w(y)\, dy = \frac{1}{|\widetilde{Q}|} \int_{\widetilde{Q}} w(y)\, dy \geq \frac{1}{|Q| + \varepsilon} \int_Q w(y)\, dy > [w]_{A_1} w(x).$$

Hence we may assume that the corner points of the cube Q satisfying (8.22) are rational. The conditions in (8.20) and (8.22) imply

$$[w]_{A_1} w(x) < \fint_Q w(y)\, dy \leq [w]_{A_1} \operatorname*{ess\,inf}_{y \in Q} w(y).$$

Since $[w]_{A_1} > 0$, we conclude that

$$w(x) < \operatorname*{ess\,inf}_{y \in Q} w(y).$$

Let $(Q_i)_{i \in \mathbb{N}}$ be an enumeration of cubes in $\mathbb{R}^n$ with rational corners and define

$$E_i = \{x \in Q_i : w(x) < \operatorname*{ess\,inf}_{y \in Q_i} w(y)\}$$

for $i \in \mathbb{N}$. Then $|E_i| = 0$ for every $i \in \mathbb{N}$, and the argument above shows that $E \subset \bigcup_{i=1}^{\infty} E_i$. Thus $|E| = 0$, and the claim in (8.21) follows, with $C_1 = [w]_{A_1}$. □

The following lemma provides a weighted weak type estimate for the dyadic maximal function, compare to Theorem 8.18 and Lemma 8.7.

LEMMA 8.32. *Let $Q_0 \subset \mathbb{R}^n$ be a cube, $f \in L^\infty(Q_0)$ be a nonnegative function and $E_t = \{x \in Q_0 : M^d_{Q_0} f(x) > t\}$ for $t > 0$. Assume that $w \in A_1$. Then*

$$w(E_t) \leq \frac{[w]_{A_1}}{t} \int_{Q_0} f(x) w(x)\, dx$$

for every $t > 0$.

PROOF. Fix $t > 0$ and let $\mathcal{D}_t$ be the associated collection of maximal dyadic subcubes of Q_0 as in the proof of Lemma 8.6. Recall that in the case $0 < t < f_{Q_0}$ we have $\mathcal{D}_t = \{Q_0\}$. Since $t < f_Q$ for every $Q \in \mathcal{D}_t$, by (8.3) we have

$$\begin{aligned} w(E_t) &= \sum_{Q \in \mathcal{D}_t} w(Q) \leq \frac{1}{t} \sum_{Q \in \mathcal{D}_t} \frac{w(Q)}{|Q|} \int_Q f(x)\, dx \\ &= \frac{1}{t} \sum_{Q \in \mathcal{D}_t} \int_Q \left(\frac{w(Q)}{|Q|} \cdot \frac{1}{w(x)} \right) f(x) w(x)\, dx \leq \frac{[w]_{A_1}}{t} \int_{Q_0} f(x) w(x)\, dx. \end{aligned}$$ □

A lemma of Coifman and Rochberg gives a maximal function characterization for A_1 weights. Moreover, this result gives a method for constructing A_1 weights. Recall that $M^c f$ is the noncentered Hardy–Littlewood maximal function of $f \in L^1_{\mathrm{loc}}(\mathbb{R}^n)$ with respect to cubes, see Remark 8.4 (b).

THEOREM 8.33. *The following assertions hold.*

(a) *Let $0 < \delta < 1$ and assume that $f \in L^1_{\mathrm{loc}}(\mathbb{R}^n)$ satisfies $0 < M^c f(x) < \infty$ for almost every $x \in \mathbb{R}^n$. Then $w = (M^c f)^\delta$ is an A_1 weight and $[w]_{A_1} \leq \frac{C(n)}{1-\delta}$.*

(b) *Conversely, assume that $w \in A_1$. There exist $f \in L^1_{\mathrm{loc}}(\mathbb{R}^n)$, $0 < \delta < 1$, and a function g, with $g, \frac{1}{g} \in L^\infty(\mathbb{R}^n)$, such that $w(x) = g(x)(M^c f(x))^\delta$ for almost every $x \in \mathbb{R}^n$.*

PROOF. To prove part (a), it suffices to show that

$$\fint_Q (M^c f(y))^\delta\, dy \le \frac{C(n)}{1-\delta}(M^c f(x))^\delta$$

for every cube $Q \subset \mathbb{R}^n$ and for every $x \in Q$. Let $Q = Q(x_0, r)$ and write $f = f_1 + f_2$, where $f_1 = f\chi_{Q(x_0,2r)}$ and $f_2 = f\chi_{\mathbb{R}^n \setminus Q(x_0,2r)}$. We may consider $(M^c f_1)^\delta$ and $(M^c f_2)^\delta$ separately, since $M^c f \le M^c f_1 + M^c f_2$ and thus

$$(M^c f)^\delta \le (M^c f_1)^\delta + (M^c f_2)^\delta$$

for $0 < \delta < 1$. Hence it is enough to show that

$$\fint_Q (M^c f_i(y))^\delta\, dy \le \frac{C(n)}{1-\delta}(M^c f(x))^\delta, \tag{8.23}$$

for every $x \in Q$ with $i = 1, 2$. Moreover, we may assume that $\|f_i\|_{L^1(\mathbb{R}^n)} > 0$ for $i = 1, 2$, since otherwise $M^c f_i$ is identically zero and (8.23) follows immediately.

Let $t_0 > 0$ to be chosen later. By Theorem 1.15 (b), we have

$$\begin{aligned}
\fint_Q (M^c f_1(y))^\delta\, dy &= \frac{\delta}{|Q|}\int_0^\infty t^{\delta-1}\big|\{y \in Q : M^c f_1(y) > t\}\big|\, dt \\
&\le \frac{\delta}{|Q|}\int_0^\infty t^{\delta-1}\min\Big\{|Q|, \frac{C(n)}{t}\|f_1\|_{L^1(\mathbb{R}^n)}\Big\}\, dt \\
&= \frac{\delta}{|Q|}\bigg(\int_0^{t_0} + \int_{t_0}^\infty\bigg)t^{\delta-1}\min\Big\{|Q|, \frac{C(n)}{t}\|f_1\|_{L^1(\mathbb{R}^n)}\Big\}\, dt \\
&\le \frac{\delta}{|Q|}\bigg(|Q|\int_0^{t_0} t^{\delta-1}\, dt + C(n)\|f_1\|_{L^1(\mathbb{R}^n)}\int_{t_0}^\infty t^{\delta-2}\, dt\bigg) \\
&= t_0^\delta\bigg(1 + C(n)\frac{\|f_1\|_{L^1(\mathbb{R}^n)}}{|Q|}\frac{\delta}{1-\delta}t_0^{-1}\bigg).
\end{aligned}$$

By choosing $t_0 = \frac{\|f_1\|_{L^1(\mathbb{R}^n)}}{|Q|} > 0$, we obtain

$$\begin{aligned}
\fint_Q (M^c f_1(y))^\delta\, dy &\le \bigg(\frac{\|f_1\|_{L^1(\mathbb{R}^n)}}{|Q|}\bigg)^\delta\bigg(1 + C(n)\frac{\|f_1\|_{L^1(\mathbb{R}^n)}}{|Q|}\frac{\delta}{1-\delta}\frac{|Q|}{\|f_1\|_{L^1(\mathbb{R}^n)}}\bigg) \\
&= \bigg(1 + \frac{C(n)\delta}{1-\delta}\bigg)\bigg(\frac{1}{|Q|}\int_{Q(x_0,2r)}|f(y)|\, dy\bigg)^\delta \\
&= \bigg(2^{n\delta} + \frac{C(n)\delta 2^{n\delta}}{1-\delta}\bigg)\bigg(\frac{1}{|Q(x_0,2r)|}\int_{Q(x_0,2r)}|f(y)|\, dy\bigg)^\delta \\
&\le \frac{C(n)}{1-\delta}(M^c f(x))^\delta,
\end{aligned}$$

for every $x \in Q$. This proves (8.23) for $(M^c f_1)^\delta$.

To obtain an estimate for $(M^c f_2)^\delta$, we derive a pointwise estimate in Q. Let $y \in Q$ and let $Q' = Q(y', r')$ be any cube containing y. If $Q' \subset Q(x_0, 2r)$, then

$$\frac{1}{|Q'|}\int_{Q'}|f_2(z)|\, dz = 0.$$

Next we assume that Q' intersects the complement of $Q(x_0, 2r)$. Then $2r' \geq r$. Thus $Q \subset Q(y', 5r')$, and

$$\begin{aligned}\frac{1}{|Q'|}\int_{Q'}|f_2(z)|\,dz &\leq \frac{5^n}{|Q(y',5r')|}\int_{Q(y',5r')}|f_2(z)|\,dz \\ &\leq \frac{5^n}{|Q(y',5r')|}\int_{Q(y',5r')}|f(z)|\,dz \leq 5^n M^c f(x)\end{aligned}$$

for every $x \in Q$. The estimates above show that $M^c f_2(y) \leq 5^n M^c f(x)$ for every $x, y \in Q$, and this implies (8.23) for $(M^c f_2)^\delta$. The proof of (a) is complete.

We turn to part (b). Since $w \in A_1 \subset A_2$, by Theorem 8.25 there exist C and $q > 1$ such that

$$\left(\fint_Q w(y)^q\,dy\right)^{\frac{1}{q}} \leq C \fint_Q w(y)\,dy$$

for every cube $Q \subset \mathbb{R}^n$. Together with Lemma 8.31, this implies

$$\begin{aligned}(M^c(w^q)(x))^{\frac{1}{q}} &= \left(\sup_{Q\ni x}\fint_Q w(y)^q\,dy\right)^{\frac{1}{q}} \leq C\sup_{Q\ni x}\fint_Q w(y)\,dy \\ &= CM^c w(x) \leq C[w]_{A_1} w(x)\end{aligned}$$

for almost every $x \in \mathbb{R}^n$. On the other hand, by Hölder's inequality

$$w(x) \leq M^c w(x) = \sup_{Q\ni x}\fint_Q w(y)\,dy \leq \left(\sup_{Q\ni x}\fint_Q w(y)^q\,dy\right)^{\frac{1}{q}} = (M^c(w^q)(x))^{\frac{1}{q}}$$

for almost every $x \in \mathbb{R}^n$. Thus we have shown that

$$w(x) \leq (M^c f(x))^\delta \leq Cw(x)$$

for almost every $x \in \mathbb{R}^n$, where $f = w^q \in L^1_{\mathrm{loc}}(\mathbb{R}^n)$ and $\delta = \frac{1}{q}$. The claim in (b) follows by choosing

$$g(x) = \frac{w(x)}{(M^c f(x))^\delta},$$

and the proof is complete. □

REMARK 8.34. The claim in Theorem 8.33 (a) does not hold for $\delta = 1$. In fact, if f is a compactly supported measurable function with $0 < \|f\|_{L^\infty(\mathbb{R}^n)} < \infty$, then the function $M^c f$ cannot be an A_1 weight, or even an A_2 weight. To see this, assume that $M^c f \in A_2 \supset A_1$. Then also $w = (M^c f)^{-1}$ is an A_2 weight by Lemma 8.13 (b) for $p = 2$. By the boundedness of the maximal function on $L^2(w\,dx)$, see Theorem 8.28, and the fact that $w(x) \leq |f(x)|^{-1}$ almost everywhere, we obtain

$$\begin{aligned}\int_{\mathbb{R}^n} M^c f(x)\,dx &= \int_{\mathbb{R}^n}\left(M^c f(x)\right)^2 w(x)\,dx \leq C\int_{\mathbb{R}^n}|f(x)|^2 w(x)\,dx \\ &\leq C\int_{\mathbb{R}^n}|f(x)|\,dx < \infty.\end{aligned}$$

However, it can be shown that the left-hand side is finite only if $\|f\|_{L^\infty(\mathbb{R}^n)} = 0$, compare to Remark 1.12.

REMARK 8.35. The original result of Coifman and Rochberg in [**89**] is more general. Assume that μ is a positive Borel measure on $\mathbb{R}^n$ and define the maximal function

$$M\mu(x) = \sup_{Q \ni x} \frac{1}{|Q|} \int_Q 1 \, d\mu = \sup_{Q \ni x} \frac{\mu(Q)}{|Q|}$$

for every $x \in \mathbb{R}^n$. If $0 < M\mu(x) < \infty$ for almost every $x \in \mathbb{R}^n$ and $0 < \delta < 1$, then $w = (M\mu)^\delta$ is an A_1 weight and the A_1 constant only depends on n and δ. The claim (a) in Theorem 8.33 follows from this result by considering the measure $d\mu = |f(x)| \, dx$.

In particular, for $\mu = \delta$, where δ is the Dirac measure at the origin, we have

$$M\delta(x) = C(n)|x|^{-n}.$$

Thus $|x|^{-\alpha} \in A_1$ for $0 \le \alpha < n$. By Remark 8.30, we conclude that $|x|^{-\alpha} \in A_p$ whenever $1 < p < \infty$ and $n(1-p) < \alpha < n$.

In Chapter 10, we examine more carefully the A_p properties of distance weights $w(x) = d(x, E)^{-\alpha}$, for nonempty sets $E \subset \mathbb{R}^n$. In particular, we recover the above facts concerning the special case $E = \{0\} \subset \mathbb{R}^n$ from our more general results; see Example 10.27.

8.7. Self-improvement of reverse Hölder inequalities

A Muckenhoupt weight satisfies a reverse Hölder inequality by Theorem 8.25. The main goal in this section is to show that reverse Hölder inequalities are self-improving, compare to Theorem 8.27 for the corresponding result for Muckenhoupt weights. For a nonnegative function f on a set $E \subset \mathbb{R}^n$ and $t \ge 0$, we write

$$\{f > t\} = \{x \in E : f(x) > t\}. \tag{8.24}$$

LEMMA 8.36. *Let μ be a measure in $\mathbb{R}^n$. Let $E \subset \mathbb{R}^n$ be a μ-measurable set with $\mu(E) < \infty$ and let f be a nonnegative μ-measurable function on E. For $0 < q < \infty$ and $0 \le t_0 \le t_1 < \infty$, we have*

$$\begin{aligned} &\int_{\{t_0 < f \le t_1\}} f(x)^q \, d\mu(x) \\ &\quad = q \int_{t_0}^{t_1} t^{q-1} \mu(\{f > t\}) \, dt + t_0^q \mu(\{f > t_0\}) - t_1^q \mu(\{f > t_1\}). \end{aligned} \tag{8.25}$$

PROOF. Lemma 1.14 for the measure μ implies

$$\int_{\{t_0 < f \le t_1\}} f(x)^q \, d\mu(x) = q \int_0^\infty t^{q-1} \mu(\{t_0 < f \le t_1\} \cap \{f > t\}) \, dt,$$

where

$$\begin{aligned} &\int_0^\infty t^{q-1} \mu(\{t_0 < f \le t_1\} \cap \{f > t\}) \, dt \\ &\quad = \int_0^{t_0} t^{q-1} \mu(\{t_0 < f \le t_1\}) \, dt + \int_{t_0}^{t_1} t^{q-1} \mu(\{t < f \le t_1\}) \, dt \\ &\quad = \frac{t_0^q}{q} \mu(\{t_0 < f \le t_1\}) + \int_{t_0}^{t_1} t^{q-1} \mu(\{t < f \le t_1\}) \, dt. \end{aligned}$$

Since $\{t < f \le t_1\} = \{f > t\} \setminus \{f > t_1\}$ and the measures of the sets are finite by the assumption $\mu(E) < \infty$, we obtain

$$\mu(\{t < f \le t_1\}) = \mu(\{f > t\}) - \mu(\{f > t_1\})$$

for every $t_0 \le t \le t_1$. Consequently

$$\begin{aligned}\int_{t_0}^{t_1} t^{q-1}\mu(\{t < f \le t_1\})\,dt &= \int_{t_0}^{t_1} t^{q-1}\mu(\{f > t\})\,dt - \mu(\{f > t_1\})\int_{t_0}^{t_1} t^{q-1}\,dt \\ &= \int_{t_0}^{t_1} t^{q-1}\mu(\{f > t\})\,dt - \frac{t_1^q - t_0^q}{q}\mu(\{f > t_1\}).\end{aligned}$$

The claim follows by combining the equations above. □

The next lemma is a core of the self-improvement result for reverse Hölder inequalities.

LEMMA 8.37. *Let $1 < p < \infty$ and let $Q_0 \subset \mathbb{R}^n$ be a cube. Assume that $f, g \in L^p(Q_0)$ are nonnegative functions and that there exist $t_0 \ge 0$ and $C_1 > 1$ such that*

$$\int_{\{f>t\}} f(x)^p\,dx \le C_1\Big(t^{p-1}\int_{\{f>t\}} f(x)\,dx + \int_{\{g>t\}} g(x)^p\,dx\Big) \tag{8.26}$$

for every $t_0 \le t < \infty$. Let $q > p$ with $C_1\frac{q-p}{q-1} < 1$. Then there exists a constant $C = C(p, q, C_1)$ such that

$$\int_{Q_0} f(x)^q\,dx \le C\Big(t_0^{q-p}\int_{Q_0} f(x)^p\,dx + \int_{Q_0} g(x)^q\,dx\Big).$$

PROOF. Clearly

$$\begin{aligned}\int_{Q_0} f(x)^q\,dx &= \int_{\{f\le t_0\}} f(x)^q\,dx + \int_{\{f>t_0\}} f(x)^q\,dx \\ &\le t_0^{q-p}\int_{\{f\le t_0\}} f(x)^p\,dx + \int_{\{f>t_0\}} f(x)^q\,dx.\end{aligned} \tag{8.27}$$

It suffices to estimate the second integral on the right-hand side. Let $t_1 > t_0$. Using equation (8.25) with the exponent $q-p > 0$ and the measure $d\mu = f^p\,dx$, we obtain

$$\begin{aligned}\int_{\{t_0<f\le t_1\}} f(x)^q\,dx &= \int_{\{t_0<f\le t_1\}} f(x)^{q-p}\,d\mu(x) \\ &= (q-p)\int_{t_0}^{t_1} t^{q-p-1}\int_{\{f>t\}} f(x)^p\,dx\,dt \\ &\quad + t_0^{q-p}\int_{\{f>t_0\}} f(x)^p\,dx - t_1^{q-p}\int_{\{f>t_1\}} f(x)^p\,dx.\end{aligned}$$

Assumption (8.26) implies

$$\begin{aligned}&\int_{t_0}^{t_1} t^{q-p-1}\int_{\{f>t\}} f(x)^p\,dx\,dt \\ &\quad\le C_1\int_{t_0}^{t_1} t^{q-2}\int_{\{f>t\}} f(x)\,dx\,dt + C_1\int_{t_0}^{t_1} t^{q-p-1}\int_{\{g>t\}} g(x)^p\,dx\,dt.\end{aligned}$$

By (8.25), with the exponent $q-1>0$ and the measure $d\mu = f\,dx$, we obtain

$$\int_{t_0}^{t_1} t^{q-2}\int_{\{f>t\}} f(x)\,dx\,dt \le \frac{1}{q-1}\Big(\int_{\{t_0<f\le t_1\}} f(x)^q\,dx + t_1^{q-1}\int_{\{f>t_1\}} f(x)\,dx\Big).$$

On the other hand, with $d\mu = g^p\,dx$, we have

$$\begin{aligned}\int_{t_0}^{t_1} t^{q-p-1}\int_{\{g>t\}} g(x)^p\,dx\,dt &\le \int_0^\infty t^{q-p-1}\mu(\{g>t\})\,dt\\ &= \frac{1}{q-p}\int_{Q_0} g(x)^{q-p}\,d\mu(x) = \frac{1}{q-p}\int_{Q_0} g(x)^q\,dx.\end{aligned}$$

Consequently

$$\begin{aligned}\int_{\{t_0<f\le t_1\}} f(x)^q\,dx &\le C_1\frac{q-p}{q-1}\int_{\{t_0<f\le t_1\}} f(x)^q\,dx + t_0^{q-p}\int_{\{f>t_0\}} f(x)^p\,dx\\ &\quad + \Big(C_1\frac{q-p}{q-1}-1\Big)t_1^{q-p}\int_{\{f>t_1\}} f(x)^p\,dx + C_1\int_{Q_0} g(x)^q\,dx,\end{aligned} \tag{8.28}$$

where we also applied the estimate

$$\int_{\{f>t_1\}} f(x)\,dx \le \int_{\{f>t_1\}} f(x)\Big(\frac{f(x)}{t_1}\Big)^{p-1}dx = t_1^{1-p}\int_{\{f>t_1\}} f(x)^p\,dx.$$

Since

$$\int_{\{t_0<f\le t_1\}} f(x)^q\,dx \le t_1^q\big|\{t_0<f\le t_1\}\big| \le t_1^q|Q_0| < \infty,$$

we obtain from (8.28) that

$$\begin{aligned}\Big(1-C_1\frac{q-p}{q-1}\Big)\int_{\{t_0<f\le t_1\}} f(x)^q\,dx &\le t_0^{q-p}\int_{\{f>t_0\}} f(x)^p\,dx\\ &\quad + \Big(C_1\frac{q-p}{q-1}-1\Big)t_1^{q-p}\int_{\{f>t_1\}} f(x)^p\,dx + C_1\int_{Q_0} g(x)^q\,dx.\end{aligned}$$

Here $0<1-C_1\frac{q-p}{q-1}<1$, and thus

$$\begin{aligned}&\int_{\{t_0<f\le t_1\}} f(x)^q\,dx\\ &\quad\le Ct_0^{q-p}\int_{\{f>t_0\}} f(x)^p\,dx - t_1^{q-p}\int_{\{f>t_1\}} f(x)^p\,dx + C\int_{Q_0} g(x)^q\,dx\\ &\quad\le Ct_0^{q-p}\int_{\{f>t_0\}} f(x)^p\,dx + C\int_{Q_0} g(x)^q\,dx,\end{aligned}$$

with $C=C(p,q,C_1)\ge 1$. This upper bound does not depend on t_1, and by letting $t_1\to\infty$ and using Fatou's lemma, we obtain

$$\int_{\{f>t_0\}} f(x)^q\,dx \le Ct_0^{q-p}\int_{\{f>t_0\}} f(x)^p\,dx + C\int_{Q_0} g(x)^q\,dx.$$

Finally, by (8.27), we arrive at

$$\int_{Q_0} f(x)^q\,dx \le C(p,q,C_1)\Big(t_0^{q-p}\int_{Q_0} f(x)^p\,dx + \int_{Q_0} g(x)^q\,dx\Big),$$

which is the required estimate. □

We are ready for the main result of this section.

THEOREM 8.38. *Let $1 < p < \infty$ and let $Q_0 \subset \mathbb{R}^n$ be a cube. Assume that $f \in L^p(Q_0)$ is a nonnegative function and that there exists a constant C_1 such that*

$$\left(\fint_Q f(x)^p \, dx \right)^{\frac{1}{p}} \leq C_1 \fint_Q f(x) \, dx \tag{8.29}$$

for every cube $Q \subset Q_0$. Then there exist an exponent $q = q(n, p, C_1) > p$ and a constant $C = C(n, p, C_1)$ such that

$$\left(\fint_Q f(x)^q \, dx \right)^{\frac{1}{q}} \leq C \fint_Q f(x) \, dx \tag{8.30}$$

for every cube $Q \subset Q_0$.

PROOF. It suffices to prove that (8.30) holds for $Q = Q_0$. We may clearly assume that $f_{Q_0} > 0$. Let $M^d_{Q_0} f$ be the dyadic maximal function as in (8.1) and define

$$t_0 = \left(\fint_{Q_0} f(x)^p \, dx \right)^{\frac{1}{p}}. \tag{8.31}$$

Let $t \geq t_0 \geq f_{Q_0}$ and $E_t = \{x \in Q_0 : M^d_{Q_0} f(x) > t\}$. By Lemma 8.6, there exists a collection $\mathcal{D}_t$ of pairwise disjoint dyadic cubes $Q \in \mathcal{D}(Q_0)$ such that $E_t = \bigcup_{Q \in \mathcal{D}_t} Q$ and $t < f_Q \leq 2^n t$ for every $Q \in \mathcal{D}_t$. Let $F_t = \{x \in Q_0 : f(x) > t\}$. Since $f(x) \leq M^d_{Q_0} f(x)$ for almost every $x \in Q_0$ and f satisfies (8.29), we obtain

$$\begin{aligned}
\int_{F_t} f(x)^p \, dx \leq \int_{E_t} f(x)^p \, dx &= \sum_{Q \in \mathcal{D}_t} \int_Q f(x)^p \, dx \leq C_1^p \sum_{Q \in \mathcal{D}_t} |Q| \left(\fint_Q f(x) \, dx \right)^p \\
&\leq C_1^p (2^n t)^p \sum_{Q \in \mathcal{D}_t} |Q| = C(n, p, C_1) t^p |E_t|.
\end{aligned}$$

Let

$$\widetilde{f}(x) = \begin{cases} f(x), & \text{if } f(x) > \frac{t}{2}, \\ 0, & \text{otherwise}. \end{cases}$$

Then $f(x) \leq \widetilde{f}(x) + \frac{t}{2}$ for every $x \in Q_0$ and, as in the proof of Theorem 1.15, we have

$$M^d_{Q_0} f(x) \leq M^d_{Q_0} \widetilde{f}(x) + \frac{t}{2}$$

for every $x \in Q_0$. Lemma 8.32 implies

$$\begin{aligned}
|E_t| = \left| \{x \in Q_0 : M^d_{Q_0} f(x) > t\} \right| &\leq \left| \{x \in Q_0 : M^d_{Q_0} \widetilde{f}(x) > \tfrac{t}{2}\} \right| \\
&\leq \frac{2}{t} \int_{Q_0} \widetilde{f}(x) \, dx = \frac{2}{t} \int_{F_{\frac{t}{2}}} f(x) \, dx,
\end{aligned}$$

and thus

$$\int_{F_t} f(x)^p \, dx \leq C(n, p, C_1) t^p |E_t| \leq C(n, p, C_1) t^{p-1} \int_{F_{\frac{t}{2}}} f(x) \, dx.$$

On the other hand, we have

$$\int_{F_{\frac{t}{2}} \setminus F_t} f(x)^p \, dx = \int_{F_{\frac{t}{2}} \setminus F_t} f(x)^{p-1} f(x) \, dx \leq t^{p-1} \int_{F_{\frac{t}{2}}} f(x) \, dx,$$

and combination of the estimates above shows that

$$\int_{F_{\frac{t}{2}}} f(x)^p\,dx \le C(n,p,C_1)\Big(\frac{t}{2}\Big)^{p-1} \int_{F_{\frac{t}{2}}} f(x)\,dx$$

for $t \ge t_0$. We apply Lemma 8.37, with $g = 0$, and obtain $q = q(n,p,C_1) > p$ and a constant $C = C(n,p,C_1)$ such that

$$\int_{Q_0} f(x)^q\,dx \le C t_0^{q-p} \int_{Q_0} f(x)^p\,dx.$$

Finally (8.31) and (8.29) give

$$\begin{aligned} \int_{Q_0} f(x)^q\,dx &\le C\Big(\fint_{Q_0} f(x)^p\,dx\Big)^{\frac{q}{p}-1} \int_{Q_0} f(x)^p\,dx \\ &\le C\Big(\fint_{Q_0} f(x)^p\,dx\Big)^{\frac{q}{p}} |Q_0|\Big(\int_{Q_0} f(x)^p\,dx\Big)^{-1} \int_{Q_0} f(x)^p\,dx \qquad (8.32)\\ &= C\Big(\fint_{Q_0} f(x)^p\,dx\Big)^{\frac{q}{p}} |Q_0| \le C|Q_0|\Big(\fint_{Q_0} f(x)\,dx\Big)^q, \end{aligned}$$

proving (8.30) for Q_0. □

The next theorem gives an alternative way to show that every A_p weight satisfies a reverse Hölder inequality, compare to Theorem 8.25. Following the proof of Theorem 8.27, we obtain a third proof for the self-improvement of Muckenhoupt A_p weights.

THEOREM 8.39. *Let $1 \le p < \infty$ and assume that $w \in A_p$. There exist $\delta = \delta(n,p,[w]_{A_p}) > 0$ and $C = C(n,p,[w]_{A_p})$ such that*

$$\Big(\fint_Q w(x)^{1+\delta}\,dx\Big)^{\frac{1}{1+\delta}} \le C \fint_Q w(x)\,dx$$

for every cube $Q \subset \mathbb{R}^n$.

PROOF. By Theorem 8.13 (a) and (8.19) we have $A_p \subset A_q$ if $1 \le p < q < \infty$. Thus we may assume that $p > 2$. Fix a cube $Q \subset \mathbb{R}^n$. By Hölder's inequality, for every measurable function $f > 0$, we have

$$1 \le \fint_Q \frac{1}{f(x)}\,dx \fint_Q f(x)\,dx.$$

Inserting $f = w^{\frac{1}{p-1}}$ gives

$$1 \le \fint_Q w(x)^{\frac{1}{1-p}}\,dx \fint_Q w(x)^{\frac{1}{p-1}}\,dx.$$

Since $w \in A_p$, we obtain

$$\begin{aligned} \fint_Q w(x)\,dx \Big(\fint_Q w(x)^{\frac{1}{1-p}}\,dx\Big)^{p-1} &\le [w]_{A^p} = [w]_{A^p}\cdot 1^{p-1} \\ &\le [w]_{A^p}\Big(\fint_Q w(x)^{\frac{1}{1-p}}\,dx\Big)^{p-1}\Big(\fint_Q w(x)^{\frac{1}{p-1}}\,dx\Big)^{p-1}. \end{aligned}$$

By dividing both sides by

$$0 < \left(\fint_Q w(x)^{\frac{1}{1-p}}\, dx \right)^{p-1} < \infty$$

we obtain

$$\fint_Q w(x)\, dx \le [w]_{A^p} \left(\fint_Q w(x)^{\frac{1}{p-1}}\, dx \right)^{p-1}$$

and thus

$$\left(\fint_Q \left(w(x)^{\frac{1}{p-1}} \right)^{p-1} dx \right)^{\frac{1}{p-1}} \le [w]_{A_p}^{\frac{1}{p-1}} \fint_Q w(x)^{\frac{1}{p-1}}\, dx$$

for every cube $Q \subset \mathbb{R}^n$. Since $p - 1 > 1$, this shows that $w^{\frac{1}{p-1}}$ satisfies a uniform reverse Hölder inequality with respect to cubes in $\mathbb{R}^n$. By Theorem 8.38, there exist $q = q(n, p, [w]_{A_p}) > p - 1$ and $C = C(n, p, [w]_{A_p})$ such that

$$\left(\fint_Q w(x)^{\frac{q}{p-1}}\, dx \right)^{\frac{1}{q}} \le C \fint_Q w(x)^{\frac{1}{p-1}}\, dx \le C \left(\fint_Q w(x)\, dx \right)^{\frac{1}{p-1}}$$

for every cube $Q \subset \mathbb{R}^n$. The last inequality follows from Hölder's inequality. Thus we arrive at

$$\left(\fint_Q w(x)^{\frac{q}{p-1}}\, dx \right)^{\frac{p-1}{q}} \le C \fint_Q w(x)\, dx$$

for every cube $Q \subset \mathbb{R}^n$. The claim follows by choosing $\delta > 0$ such that $1 + \delta = \frac{q}{p-1}$. □

8.8. General self-improvement result for reverse Hölder inequalities

There exists a more general version of Theorem 8.38, which is relevant in the regularity theory for nonlinear partial differential equations in Chapter 12. First we introduce an appropriate maximal function on cubes.

DEFINITION 8.40. Let $1 \le p < \infty$, let $Q_0 \subset \mathbb{R}^n$ be a cube, and assume that $f \in L^p(Q_0)$ is a nonnegative function. The noncentered maximal function $M_{Q_0}^p f$ on Q_0 is defined as

$$M_{Q_0}^p f(x) = \sup_{Q \ni x} \left(\fint_Q f(y)^p\, dy \right)^{\frac{1}{p}},$$

where the supremum is taken over all cubes $Q \subset Q_0$ with $x \in Q$.

We begin with a reverse weak type inequality for this maximal operator.

LEMMA 8.41. *Let $1 \le p < \infty$ and let $Q_0 \subset \mathbb{R}^n$ be a cube. Assume that $f \in L^p(Q_0)$ is a nonnegative function. There exists a constant $C = C(n)$ such that*

$$\int_{\{x \in Q_0 : f(x) > t\}} f(x)^p\, dx \le C t^p \left| \{ x \in Q_0 : M_{Q_0}^p f(x) > t \} \right| \tag{8.33}$$

whenever $t^p \ge (f^p)_{Q_0}$.

PROOF. We apply Lemma 8.6 for f^p at the level t^p and obtain a collection $\mathcal{D}_{t^p}$ of pairwise disjoint dyadic cubes $Q \in \mathcal{D}(Q_0)$ such that $t^p < (f^p)_Q \le 2^n t^p$ for every $Q \in \mathcal{D}_{t^p}$ and

$$\{ x \in Q_0 : M_{Q_0}^d f^p(x) > t^p \} = \bigcup_{Q \in \mathcal{D}_{t^p}} Q.$$

Since $f(x)^p \le M^d_{Q_0} f^p(x)$ for almost every $x \in Q_0$, we have

$$\begin{aligned}\int_{\{x\in Q_0 : f(x)>t\}} f(x)^p\,dx &\le \int_{\{x\in Q_0 : M^d_{Q_0} f^p(x)>t^p\}} f(x)^p\,dx = \sum_{Q\in\mathcal{D}_{t^p}} \int_Q f(x)^p\,dx \\ &= \sum_{Q\in\mathcal{D}_{t^p}} |Q| \fint_Q f(x)^p\,dx \le 2^n t^p \sum_{Q\in\mathcal{D}_{t^p}} |Q| \\ &\le 2^n t^p \big|\{x\in Q_0 : M^d_Q f^p(x) > t^p\}\big|.\end{aligned}$$

Since $\left(M^d_{Q_0} f^p(x)\right)^{\frac{1}{p}} \le M^p_{Q_0} f(x)$ for every $x \in Q_0$, inequality (8.33) follows. □

The next result is a weak type estimate for the maximal operator $M^p_{Q_0}$. Recall that Mf is the usual centered Hardy–Littlewood maximal function of $f \in L^1_{\text{loc}}(\mathbb{R}^n)$.

LEMMA 8.42. *Let $1 \le p < \infty$ and let $Q_0 \subset \mathbb{R}^n$ be a cube. Assume that $f \in L^p(Q_0)$ is a nonnegative function. There exist constants $C_1 = C_1(n)$ and $C_2 = C_2(n)$ such that*

$$\big|\{x\in Q_0 : M^p_{Q_0} f(x) > t\}\big| \le \frac{C_1}{t^p} \int_{\{x\in Q_0 : f(x)>\frac{t}{C_2}\}} f(x)^p\,dx \tag{8.34}$$

for every $t > 0$.

PROOF. Let $g = f^p \chi_{Q_0} \in L^1(\mathbb{R}^n)$. We have

$$M^p_Q f(x) = M^p_Q(f\chi_{Q_0})(x) \le \left(C_2(n) Mg(x)\right)^{\frac{1}{p}}$$

for every $x \in Q_0$, where $C_2(n) \ge 1$. The proof of Theorem 1.15 shows that

$$\begin{aligned}\big|\{x\in Q_0 : M^p_{Q_0} f(x) > t\}\big| &\le \Big|\Big\{x\in\mathbb{R}^n : Mg(x) > \tfrac{t^p}{C_2(n)}\Big\}\Big| \\ &\le \frac{C_1(n)}{t^p} \int_{\{x\in\mathbb{R}^n : g(x)\ge \frac{t^p}{2C_2(n)}\}} g(x)\,dx \\ &\le \frac{C_1(n)}{t^p} \int_{\{x\in Q_0 : f(x)\ge \frac{t}{2C_2(n)}\}} f(x)^p\,dx.\end{aligned}$$

In the final step we used the inequality $(2C_2(n))^{\frac{1}{p}} \le 2C_2(n)$. □

THEOREM 8.43. *Let $1 < p < \infty$ and $C_1 > 0$, and let $\Omega \subset \mathbb{R}^n$ be an open set. There exist $\theta = \theta(n,p) > 0$, $q = q(n,p,C_1) > p$ and $C = C(n,p,C_1) \ge 1$ such that if $f, g \in L^p_{\text{loc}}(\Omega)$ are nonnegative functions satisfying*

$$\begin{aligned}\left(\fint_{Q(z,r)} f(x)^p\,dx\right)^{\frac{1}{p}} &\le C_1\left[\fint_{Q(z,2r)} f(x)\,dx + \left(\fint_{Q(z,2r)} g(x)^p\,dx\right)^{\frac{1}{p}}\right] \\ &\quad + \theta\left(\fint_{Q(z,2r)} f(x)^p\,dx\right)^{\frac{1}{p}},\end{aligned}\tag{8.35}$$

for every cube $Q(z,r)$ with $Q(z,2r) \subset \Omega$, then

$$\left(\fint_{Q(z,r)} f(x)^q\,dx\right)^{\frac{1}{q}} \le C\left(\fint_{Q(z,2r)} f(x)^p\,dx\right)^{\frac{1}{p}} + C\left(\fint_{Q(z,2r)} g(x)^q\,dx\right)^{\frac{1}{q}}$$

for every cube $Q(z,r)$ with $\overline{Q(z,2r)} \subset \Omega$.

PROOF. Let $Q_0 = Q(x_0, r_0)$ be a cube with $\overline{Q}_0 \subset \Omega$. We begin by constructing a specific Whitney type decomposition $\mathcal{W}$ of Q_0. For $i \in \mathbb{N}$ let

$$Q_i = Q\big(x_0, (1 - 2^{-i})r_0\big).$$

We divide each Q_i into $(2^{i+1} - 2)^n$ pairwise disjoint Q_0-dyadic half-open cubes, with common side length $2^{-i}r_0$, which cover Q_i. Denote this collection by $\mathcal{L}_i$. We define recursively a collection $\mathcal{W}_i$ of pairwise disjoint cubes by setting $\mathcal{W}_1 = \mathcal{L}_1$ and

$$\mathcal{W}_{i+1} = \big\{Q \in \mathcal{L}_{i+1} : Q \cap \widetilde{Q} = \emptyset \text{ for every } \widetilde{Q} \in \mathcal{W}_i\big\}$$

for every $i \in \mathbb{N}$. Let $\mathcal{W} = \bigcup_{i=1}^{\infty} \mathcal{W}_i$. The cubes in $\mathcal{W}$ are pairwise disjoint and they cover the interior of Q_0, and hence they cover Q_0 up to measure zero. Moreover, if $Q = Q(z, r) \in \mathcal{W}$, then the doubled cube $Q(z, 2r)$ is a subset of Q_0.

Let $f, g \in L^p_{\mathrm{loc}}(\Omega)$ be nonnegative functions such that (8.35) holds for some $0 < \theta < 1$, to be specified later, and define

$$t_0 = \left(\fint_{Q_0} f(x)^p \, dx\right)^{\frac{1}{p}} < \infty. \tag{8.36}$$

Without loss of generality, we may assume that $t_0 > 0$.

Let $t \geq t_0$. For $Q \in \mathcal{W}$, we have

$$\fint_Q f(x)^p \, dx \leq \frac{1}{|Q|} \int_{Q_0} f(x)^p \, dx = \frac{|Q_0|}{|Q|} \fint_{Q_0} f(x)^p \, dx \leq a_Q t^p, \tag{8.37}$$

where $a_Q = \frac{|Q_0|}{|Q|} > 1$. Define functions $\widetilde{f}$ and $\widetilde{g}$ in the interior of Q_0 by setting

$$\widetilde{f}(x) = a_Q^{-\frac{1}{p}} f(x) \quad \text{and} \quad \widetilde{g}(x) = a_Q^{-\frac{1}{p}} g(x),$$

for every $x \in Q \in \mathcal{W}$. Clearly $0 \leq \widetilde{f} \leq f$ and $0 \leq \widetilde{g} \leq g$ almost everywhere in Q_0, and thus $\widetilde{f}, \widetilde{g} \in L^p(Q_0)$.

Let $Q \in \mathcal{W}$. By (8.37), we have

$$\fint_Q \widetilde{f}(x)^p \, dx \leq t^p,$$

and Lemma 8.41 gives

$$\int_{\{x \in Q : \widetilde{f}(x) > t\}} \widetilde{f}(x)^p \, dx \leq C(n) t^p \big|\{x \in Q : M_Q^p \widetilde{f}(x) > t\}\big|. \tag{8.38}$$

To estimate the right-hand side of (8.38), let $x \in Q$ and let $Q_x = Q(z_x, r_x) \subset Q$ be a subcube of Q containing x. Then the construction above guarantees that the doubled cube $Q(z_x, 2r_x)$ is contained in $Q_0 \subset \Omega$, and (8.35) implies

$$\begin{aligned}
\left(\fint_{Q_x} \widetilde{f}(y)^p \, dy\right)^{\frac{1}{p}} &= a_Q^{-\frac{1}{p}} \left(\fint_{Q_x} f(y)^p \, dy\right)^{\frac{1}{p}} \\
&\leq C_1 a_Q^{-\frac{1}{p}} \left[\fint_{Q(z_x, 2r_x)} f(y) \, dy + \left(\fint_{Q(z_x, 2r_x)} g(y)^p \, dy\right)^{\frac{1}{p}}\right] \\
&\quad + \theta a_Q^{-\frac{1}{p}} \left(\fint_{Q(z_x, 2r_x)} f(y)^p \, dy\right)^{\frac{1}{p}}.
\end{aligned} \tag{8.39}$$

It is easy to see that the cube $Q(z_x, 2r_x)$ intersects at most those cubes in $\mathcal{W}$ which touch Q, that is, those Whitney type cubes whose closures intersect the closure of Q. In particular, there exists a cube $Q' = Q(x', r') \in \mathcal{W}$ which touches Q and satisfies

$$\widetilde{f}(y) \geq (a_{Q'})^{-\frac{1}{p}} f(y) \quad \text{and} \quad \widetilde{g}(y) \geq (a_{Q'})^{-\frac{1}{p}} g(y)$$

for almost every $y \in Q(z_x, 2r_x)$. Moreover, by the construction of the cubes it holds that $Q \subset Q(x', 5r')$. This implies $|Q| \leq |Q(x', 5r')| = 5^n|Q'|$ and $a_{Q'} \leq 5^n a_Q$. Concluding from above, we obtain

$$\fint_{Q(z_x,2r_x)} f(y)\,dy \leq (5^n a_Q)^{\frac{1}{p}} \fint_{Q(z_x,2r_x)} \widetilde{f}(y)\,dy.$$

A similar reasoning shows that

$$\biggl(\fint_{Q(z_x,2r_x)} g(y)^p\,dy\biggr)^{\frac{1}{p}} \leq (5^n a_Q)^{\frac{1}{p}} \biggl(\fint_{Q(z_x,2r_x)} \widetilde{g}(y)^p\,dy\biggr)^{\frac{1}{p}}$$

and

$$\biggl(\fint_{Q(z_x,2r_x)} f(y)^p\,dy\biggr)^{\frac{1}{p}} \leq (5^n a_Q)^{\frac{1}{p}} \biggl(\fint_{Q(z_x,2r_x)} \widetilde{f}(y)^p\,dy\biggr)^{\frac{1}{p}}.$$

By substituting the estimates above to (8.39) and taking supremum over all cubes Q_x as above, we have

$$M^p_Q \widetilde{f}(x) \leq C(n, C_1)\bigl(M_{Q_0}\widetilde{f}(x) + M^p_{Q_0}\widetilde{g}(x)\bigr) + C(n)\theta M^p_{Q_0}\widetilde{f}(x)$$

for every $x \in Q$. This implies

$$\{x \in Q : M^p_Q \widetilde{f}(x) > t\} \subset Q \cap (E \cup F \cup G),$$

where

$$E = \Bigl\{x \in Q_0 : M_{Q_0}\widetilde{f}(x) > \frac{t}{3C(n, C_1)}\Bigr\},$$

$$F = \Bigl\{x \in Q_0 : M^p_{Q_0}\widetilde{g}(x) > \frac{t}{3C(n, C_1)}\Bigr\},$$

$$G = \Bigl\{x \in Q_0 : M^p_{Q_0}\widetilde{f}(x) > \frac{t}{3C(n)\theta}\Bigr\}.$$

By (8.38) we have

$$\int_{\{x\in Q:\widetilde{f}(x)>t\}} \widetilde{f}(x)^p\,dx \leq C(n)t^p|Q \cap (E \cup F \cup G)|. \tag{8.40}$$

By summing (8.40) over all Whitney cubes $Q \in \mathcal{W}$ and applying Lemma 8.42, we obtain

$$\begin{aligned}
\int_{\{\widetilde{f}>t\}} \widetilde{f}(x)^p\,dx &\leq C(n)t^p|E \cup F \cup G| \leq C(n)t^p(|E| + |F| + |G|)\\
&\leq C(n, C_1)t^{p-1} \int_{\{\widetilde{f}>\tau t\}} \widetilde{f}(x)\,dx + C(n, p, C_1) \int_{\{\widetilde{g}>\tau t\}} \widetilde{g}(x)^p\,dx \qquad (8.41)\\
&\quad + C(n, p)\theta^p \int_{\{\widetilde{f}>\tau t\}} \widetilde{f}(x)^p\,dx,
\end{aligned}$$

where $0 < \tau = \tau(n, C_1) < 1$, and notation as in (8.24) is used for the set Q_0. On the other hand, we have

$$\int_{\{\tau t<\widetilde{f}\leq t\}} \widetilde{f}(x)^p\,dx \leq t^{p-1} \int_{\{\widetilde{f}>\tau t\}} \widetilde{f}(x)\,dx. \tag{8.42}$$

By adding (8.41) and (8.42) and reorganizing terms, we arrive at

$$\begin{aligned}(1 - C(n,p)\theta^p) &\int_{\{\widetilde{f}>\tau t\}} \widetilde{f}(x)^p\,dx \\ &\le C(n,p,C_1)\Big((\tau t)^{p-1} \int_{\{\widetilde{f}>\tau t\}} \widetilde{f}(x)\,dx + \int_{\{\widetilde{g}>\tau t\}} \widetilde{g}(x)^p\,dx \Big).\end{aligned}$$

Recall that here $t \ge t_0$ was arbitrary. Also note that the term that is absorbed into the left-hand side is finite since $\widetilde{f} \in L^p(Q_0)$.

Let $0 < \theta = \theta(n,p) < 1$ be so small that $1 - C(n,p)\theta^p \ge \frac{1}{2}$. Lemma 8.37, applied for the functions $\widetilde{f}, \widetilde{g} \in L^p(Q_0)$, and the trivial estimates $\widetilde{f} \le f$, $\widetilde{g} \le g$ imply the existence of $q = q(n,p,C_1) > p$ such that

$$\int_{Q_0} \widetilde{f}(x)^q\,dx \le C(n,p,C_1)\Big((\tau t_0)^{q-p} \int_{Q_0} f(x)^p\,dx + \int_{Q_0} g(x)^q\,dx \Big).$$

By applying (8.36) and estimating as in (8.32), we obtain

$$\Big(\fint_{Q_0} \widetilde{f}(x)^q\,dx \Big)^{\frac{1}{q}} \le C\Big(\fint_{Q_0} f(x)^p\,dx \Big)^{\frac{1}{p}} + C\Big(\fint_{Q_0} g(x)^q\,dx \Big)^{\frac{1}{q}}$$

with $C = C(n,p,C_1)$.

Finally, it follows from the construction of $\mathcal{W}$ that $Q_1 = Q(x_0, \frac{r_0}{2})$ is divided to 2^n cubes $Q \in \mathcal{W}$, each having side length $\frac{r_0}{2}$. Since $a_Q = \frac{|Q_0|}{|Q|} = 4^n$ for these cubes, it holds by the definition of $\widetilde{f}$ that $\widetilde{f}(x) = 4^{-\frac{n}{p}} f(x)$ for every $x \in Q_1$. Hence we conclude that

$$\Big(\fint_{Q_1} f(x)^q\,dx \Big)^{\frac{1}{q}} \le C\Big(\fint_{Q_0} f(x)^p\,dx \Big)^{\frac{1}{p}} + C\Big(\fint_{Q_0} g(x)^q\,dx \Big)^{\frac{1}{q}}$$

with $C = C(n,p,C_1)$. This is the desired inequality for the cube $Q(z,r) = Q_1$, and the proof is complete. □

REMARK 8.44. By covering cubes in (8.35) with balls by Lemma 1.13, and covering balls in the final inequality by dyadic cubes, we obtain the following variant of Theorem 8.43. Let $1 < p < \infty$, $C_1 > 0$, $1 < \tau_1, \tau_2 < \infty$, and let $\Omega \subset \mathbb{R}^n$ be an open set. There exist $\theta = \theta(n,p,\tau_1) > 0$, $q = q(n,p,C_1,\tau_1) > p$ and $C = C(n,p,C_1,\tau_1,\tau_2) \ge 1$ such that if $f, g \in L^p_{\mathrm{loc}}(\Omega)$ are nonnegative functions satisfying

$$\begin{aligned}\Big(\fint_{B(z,r)} f(x)^p\,dx \Big)^{\frac{1}{p}} &\le C_1\Big[\fint_{B(z,\tau_1 r)} f(x)\,dx + \Big(\fint_{B(z,\tau_1 r)} g(x)^p\,dx \Big)^{\frac{1}{p}} \Big] \\ &\quad + \theta\Big(\fint_{B(z,\tau_1 r)} f(x)^p\,dx \Big)^{\frac{1}{p}},\end{aligned}$$

for every ball $B(z,r)$, with $B(z,\tau_1 r) \subset \Omega$, then

$$\Big(\fint_{B(z,r)} f(x)^q\,dx \Big)^{\frac{1}{q}} \le C\Big(\fint_{B(z,\tau_2 r)} f(x)^p\,dx \Big)^{\frac{1}{p}} + C\Big(\fint_{B(z,\tau_2 r)} g(x)^q\,dx \Big)^{\frac{1}{q}}$$

for every ball $B(z,r)$ with $B(z,\tau_2 r) \subset \Omega$.

REMARK 8.45. By tracking the constants in the proof of Theorem 8.43 and using covering arguments, we also obtain the following variant of Theorem 8.43 for balls. Let $1 < p < \infty$, $C_1 > 0$, $1 < \tau_1, \tau_2 < \infty$, and let $\Omega \subset \mathbb{R}^n$ be an open set. There exist $\theta = \theta(n, p, \tau_1) > 0$, $\varepsilon_0 = \varepsilon_0(n, p, C_1, \tau_1) > 0$ and $C = C(n, p, C_1, \tau_1, \tau_2) \geq 1$ such that the following claim holds for every ε with $0 < \varepsilon < \varepsilon_0$: For nonnegative functions $f, g \in L^{p-\varepsilon}_{\mathrm{loc}}(\Omega)$ satisfying

$$\left(\fint_{B(z,r)} f(x)^{p-\varepsilon}\,dx\right)^{\frac{1}{p-\varepsilon}} \leq C_1\left[\fint_{B(z,\tau_1 r)} f(x)\,dx + \left(\fint_{B(z,\tau_1 r)} g(x)^{p-\varepsilon}\,dx\right)^{\frac{1}{p-\varepsilon}}\right] + \theta\left(\fint_{B(z,\tau_1 r)} f(x)^{p-\varepsilon}\,dx\right)^{\frac{1}{p-\varepsilon}}$$

for every ball $B(z,r)$ with $B(z,\tau_1 r) \subset \Omega$, we have

$$\left(\fint_{B(z,r)} f(x)^p\,dx\right)^{\frac{1}{p}} \leq C\left(\fint_{B(z,\tau_2 r)} f(x)^{p-\varepsilon}\,dx\right)^{\frac{1}{p-\varepsilon}} + C\left(\fint_{B(z,\tau_2 r)} g(x)^p\,dx\right)^{\frac{1}{p}}$$

for every ball $B(z,r)$ with $B(z,\tau_2 r) \subset \Omega$.

8.9. Notes

For the dyadic maximal function in Section 8.2 we refer to Journé [**216**, Chapter 1]. The proof of Theorem 8.11 is based on an unpublished note on Muckenhoupt weights by Stephen Keith and Xiao Zhong.

Weighted norm inequalities have been studied by Muckenhoupt [**333, 334**] and by Coifman and Fefferman [**87**]. Other references for Muckenhoupt weights include Cruz-Uribe, Martell and Pérez [**92**], Duoandikoetxea [**107**, Chapter 7], [**108**], García-Cuerva and Rubio de Francia [**139**, Chapter 4], Garnett [**142**, Chapter 6], Genebashvili, Gogatishvili, Kokilashvili and Krbec [**145**], Grafakos [**157**, Chapter 9], Heinonen, Kilpeläinen and Martio [**187**, Chapter 15], Journé [**216**, Chapter 2], Lu, Ding and Yan [**286**, Section 1.4], Stein [**371**, Chapter 5], Strömberg and Torchinsky [**376**, Chapter 1], Torchinsky [**383**, Chapter 9] and Turesson [**388**, Section 1.2]. Theorem 8.16 can be found in Kilpeläinen, Koskela and Masaoka [**221**, Proposition 4.3]. The elementary proof of Theorem 8.19 is from Lerner [**267**]. Theorem 8.25 is in Coifman and Fefferman [**87**], see also Turesson [**388**, Section 1.2.4]. Theorem 8.33 is from Coifman and Rochberg [**89**].

Theorem 8.38 has been studied by Gehring [**144**] and Theorem 8.43 by Meyers and Elcrat [**325**]. The proof of Theorem 8.38 is based on Bojarski and Iwaniec [**52**], Garnett [**142**, Chapter 6] and Kinnunen [**223**]. See also Duoandikoetxea [**108**, Section 4.2]. For the proof of Theorem 8.43 we refer to Björn and Björn [**31**, Section 3.5], Giaquinta [**147**, Chapter 5], [**148**, Section 6.2], Giaquinta and Martinazzi [**149**], Giaquinta and Modica [**150**], Giusti [**153**, Section 6.4], Iwaniec and Martin [**207**, Chapter 14], Martín and Milman [**309**], Stredulinsky [**373, 374**], Strömberg and Torchinsky [**376**, Chapter 1] and Zatorska-Goldstein [**402**].

CHAPTER 9

Weighted Maximal and Poincaré Inequalities

This chapter begins with Poincaré inequalities on cubes and, more generally, on open sets satisfying a chaining condition. In Section 9.2, we study weighted norm inequalities for dyadic maximal operators, and as a consequence obtain a weighted Poincaré inequality on cubes. Two weight Poincaré inequalities are discussed in Section 9.3, Section 9.4 and Section 9.5. Weighted Poincaré inequalities are proved using the sparse domination method in Lemma 9.13 and Lemma 9.19. In Section 9.6 we investigate two weight Poincaré inequalities on open sets satisfying a chaining condition. The sparse domination method is applied in Section 9.7 to prove the John–Nirenberg inequality for functions of bounded mean oscillation (BMO), and at the end of the section we obtain weighted Sobolev inequalities for $p = n$ as a consequence of the John–Nirenberg inequality. Finally, in Section 9.8 we examine connections between Muckenhoupt weights and BMO functions and prove a boundedness result for a maximal operator on BMO.

9.1. Poincaré inequalities on cubes

We consider Poincaré inequalities for more general open sets than balls, compare to Section 3.3. The proof is based on a chaining argument and a local-to-global inequality, which allows us to reduce the problem to Poincaré inequalities on balls. As in Lemma 8.9, we consider a Whitney decomposition $\mathcal{W}(\Omega)$ of an open set Ω and write $Q^* = Q(x, \frac{9}{8}r)$ for $Q = Q(x,r) \in \mathcal{W}(\Omega)$.

DEFINITION 9.1. Let $\Omega \subset \mathbb{R}^n$ be a connected open set and let $\mathcal{W}(\Omega)$ be a Whitney decomposition of Ω. We say that

$$\mathcal{C}(Q) = (Q_0, \dots, Q_k) \subset \mathcal{W}(\Omega)$$

is chain in Ω joining Q_0 to $Q = Q_k$, if $Q_i \neq Q_j$ whenever $i \neq j$, and for each $j \in \{1, \dots, k\}$ there exists a cube $Q(z,r) \subset Q_j^* \cap Q_{j-1}^*$ satisfying

$$r \geq C(n) \max\{l(Q_j^*), l(Q_{j-1}^*)\}. \tag{9.1}$$

For a fixed cube $Q_0 \in \mathcal{W}(\Omega)$, the collection $\{\mathcal{C}(Q) : Q \in \mathcal{W}(\Omega)\}$ is called a chain decomposition of Ω. The shadow of a Whitney cube $Q \in \mathcal{W}(\Omega)$ is the set $\mathcal{S}(Q) = \{Q' \in \mathcal{W}(\Omega) : Q \in \mathcal{C}(Q')\}$.

Notice the duality of the concepts of chain and shadow: $Q \in \mathcal{C}(Q')$ if and only if $Q' \in \mathcal{S}(Q)$.

DEFINITION 9.2. An open set $\Omega \subset \mathbb{R}^n$ satisfies the Boman chain condition if Ω is connected and there exist $\tau > 1$ and a chain decomposition $\{\mathcal{C}(Q) : Q \in \mathcal{W}(\Omega)\}$ of Ω, with a fixed cube $Q_0 \in \mathcal{W}(\Omega)$, such that if $Q = Q(x,r) \in \mathcal{W}(\Omega)$ and $Q' \in \mathcal{S}(Q)$, then

$$Q' \subset Q(x, \tau r). \tag{9.2}$$

Note that if an open set $\Omega \subset \mathbb{R}^n$ satisfies the Boman chain condition, then Ω is bounded and connected.

REMARK 9.3. All open cubes and balls satisfy the Boman chain condition, with $\tau = \tau(n)$ independent of the cube. More generally, if $\Omega \subset \mathbb{R}^n$ is a so-called John domain, then Ω satisfies the Boman chain condition, see Buckley, Koskela and Lu [**65**]. The class of John domains includes bounded and connected open sets with a smooth or a Lipschitz boundary.

The following lemma provides a special case of a local-to-global inequality for open sets satisfying the Boman chain condition, including all cubes. See also Theorem 9.28 for a more general local-to-global inequality.

LEMMA 9.4. *Assume that $\Omega \subset \mathbb{R}^n$ is an open set satisfying the Boman chain condition, with $\tau > 1$. There exists a constant $C = C(n, \tau)$ such that*

$$\int_\Omega |f(x) - f_\Omega|\,dx \le C \sum_{Q \in \mathcal{W}(\Omega)} \int_{Q^*} |f(x) - f_{Q^*}|\,dx$$

for every $f \in L^1(\Omega)$.

PROOF. Observe that

$$\begin{aligned}
\int_\Omega |f(x) - f_\Omega|\,dx &\le \int_\Omega |f(x) - f_{Q_0^*}|\,dx + \int_\Omega |f_\Omega - f_{Q_0^*}|\,dx \\
&\le 2\int_\Omega |f(x) - f_{Q_0^*}|\,dx \qquad (9.3)\\
&\le 2\sum_{Q \in \mathcal{W}(\Omega)} \int_{Q^*} |f(x) - f_{Q^*}|\,dx + 2\sum_{Q \in \mathcal{W}(\Omega)} |Q||f_{Q^*} - f_{Q_0^*}|.
\end{aligned}$$

The first term on the right-hand side of (9.3) is already of the desired form. To estimate the second term, we use the chain $\mathcal{C}(Q) = (Q_0, \dots, Q_k)$ joining the cube Q_0 to $Q_k = Q \in \mathcal{W}(\Omega)$, and obtain

$$\sum_{Q \in \mathcal{W}(\Omega)} |Q||f_{Q^*} - f_{Q_0^*}| \le \sum_{Q \in \mathcal{W}(\Omega)} |Q| \sum_{i=1}^{k} |f_{Q_i^*} - f_{Q_{i-1}^*}|.$$

By (9.1), we have

$$\begin{aligned}
|f_{Q_i^*} - f_{Q_{i-1}^*}| &\le |f_{Q_i^*} - f_{Q_i^* \cap Q_{i-1}^*} + f_{Q_i^* \cap Q_{i-1}^*} - f_{Q_{i-1}^*}| \\
&\le \fint_{Q_i^* \cap Q_{i-1}^*} |f(x) - f_{Q_i^*}|\,dx + \fint_{Q_i^* \cap Q_{i-1}^*} |f(x) - f_{Q_{i-1}^*}|\,dx \\
&\le C(n) \sum_{j=i-1}^{i} \fint_{Q_j^*} |f(x) - f_{Q_j^*}|\,dx
\end{aligned}$$

for every $i \in \{1, \dots, k\}$. Using the estimates above, the assumption $Q_{i_1} \neq Q_{i_2}$ for $i_1 \neq i_2$, and the duality between chains and shadows, we arrive at

$$\begin{aligned}
\sum_{Q\in\mathcal{W}(\Omega)} |Q||f_{Q^*} - f_{Q_0^*}| &\le C(n) \sum_{Q\in\mathcal{W}(\Omega)} |Q| \sum_{i=1}^{k} \sum_{j=i-1}^{i} \fint_{Q_j^*} |f(x) - f_{Q_j^*}|\,dx \\
&\le C(n) \sum_{Q\in\mathcal{W}(\Omega)} |Q| \sum_{R\in\mathcal{C}(Q)} \fint_{R^*} |f(x) - f_{R^*}|\,dx \\
&\le C(n) \sum_{R\in\mathcal{W}(\Omega)} \sum_{Q\in\mathcal{S}(R)} |Q| \fint_{R^*} |f(x) - f_{R^*}|\,dx \\
&\le C(n,\tau) \sum_{R\in\mathcal{W}(\Omega)} \int_{R^*} |f(x) - f_{R^*}|\,dx,
\end{aligned}$$

where the final inequality follows from (9.2). □

THEOREM 9.5. *Assume that $\Omega \subset \mathbb{R}^n$ is an open set satisfying the Boman chain condition, with $\tau > 1$. There exists a constant $C = C(n, \tau)$ such that*

$$\int_\Omega |u(x) - u_\Omega|\,dx \le C \operatorname{diam}(\Omega) \int_\Omega |\nabla u(x)|\,dx$$

for every $u \in W^{1,1}(\Omega)$.

PROOF. Let $\mathcal{W}(\Omega) = \{Q_i : i \in \mathbb{N}\}$ be a Whitney decomposition of Ω, where $Q_i = Q(x_i, r_i)$. Let $B_i = B\big(x_i, \frac{9\sqrt{n}}{8} r_i\big)$ for every $i \in \mathbb{N}$. Then $B_i \Subset \Omega$, the interior of Q_i^* is contained in B_i, and

$$\sum_{i=1}^{\infty} \chi_{B_i} \le C(n)\chi_\Omega. \tag{9.4}$$

Lemma 9.4 and the 1-Poincaré inequality for balls, see Theorem 3.14, imply

$$\begin{aligned}
\int_\Omega |u(x) - u_\Omega|\,dx &\le C(n,\tau) \sum_{i=1}^{\infty} \int_{Q_i^*} |u(x) - u_{Q_i^*}|\,dx \\
&\le C(n,\tau) \sum_{i=1}^{\infty} \int_{Q_i^*} |u(x) - u_{B_i}|\,dx \\
&\le C(n,\tau) \sum_{i=1}^{\infty} \int_{B_i} |u(x) - u_{B_i}|\,dx \\
&\le C(n,\tau) \sum_{i=1}^{\infty} r_i \int_{B_i} |\nabla u(x)|\,dx \\
&\le C(n,\tau) \operatorname{diam}(\Omega) \int_\Omega |\nabla u(x)|\,dx,
\end{aligned}$$

where the final inequality follows from (9.4). □

COROLLARY 9.6. *Let $Q \subset \mathbb{R}^n$ be a cube. There exists a constant $C = C(n)$ such that*

$$\int_Q |u(x) - u_Q|\,dx \le C l(Q) \int_Q |\nabla u(x)|\,dx$$

for every $u \in W^{1,1}(Q)$.

PROOF. The interior of the cube Q satisfies the Boman chain condition with $\tau = \tau(n) > 1$, and hence the claimed inequality, with a constant $C = C(n)$, follows from Theorem 9.5. $\square$

9.2. Single weight maximal and Poincaré inequalities

We prove weighted Poincaré inequalities by applying weighted norm inequalities for dyadic maximal functions. We begin by defining a weighted dyadic maximal function. The corresponding unweighted dyadic maximal function $M^d_{Q_0} f$ was defined in (8.1). For simplicity, we only consider functions $f \in L^\infty(Q_0)$.

DEFINITION 9.7. Let w be a weight in $\mathbb{R}^n$ and let $Q_0 \subset \mathbb{R}^n$ be a cube. For $f \in L^\infty(Q_0)$, the weighted dyadic maximal function $M^{d,w}_{Q_0} f$ is defined at $x \in Q_0$ by

$$M^{d,w}_{Q_0} f(x) = \sup_{Q \ni x} \frac{1}{w(Q)} \int_Q |f(y)| w(y)\, dy,$$

where the supremum is taken over all dyadic cubes $Q \in \mathcal{D}(Q_0)$ with $x \in Q$.

The proof of the following weighted norm inequality is a variant of the proof of Theorem 1.15; compare also to the proof of Lemma 8.32. Observe that the result holds for an arbitrary weight. In particular, the weight is not assumed to be doubling nor to belong to a Muckenhoupt class.

THEOREM 9.8. *Let $1 < p < \infty$ and let $Q_0 \subset \mathbb{R}^n$ be a cube. Assume that w is a weight in $\mathbb{R}^n$. Then*

$$\int_{Q_0} \left(M^{d,w}_{Q_0} f(x)\right)^p w(x)\, dx \le \frac{p2^p}{p-1} \int_{Q_0} |f(x)|^p w(x)\, dx$$

for every $f \in L^\infty(Q_0)$.

PROOF. Without loss of generality, we may assume that $f \in L^\infty(Q_0)$ is nonnegative. For $t > 0$, let $E_t = \{x \in Q_0 : M^{d,w}_{Q_0} f(x) > t\}$ and let $\mathcal{D}_t$ be the collection of maximal dyadic cubes $Q \in \mathcal{D}(Q_0)$ satisfying

$$\frac{1}{w(Q)} \int_Q f(x) w(x)\, dx > t,$$

as in the proof of Lemma 8.6. Then $E_t = \bigcup_{Q \in \mathcal{D}_t} Q$ and the cubes $Q \in \mathcal{D}_t$ are pairwise disjoint. Thus we obtain the weak type inequality

$$w(E_t) = \sum_{Q \in \mathcal{D}_t} w(Q) \le \frac{1}{t} \sum_{Q \in \mathcal{D}_t} \int_Q f(x) w(x)\, dx \le \frac{1}{t} \int_{Q_0} f(x) w(x)\, dx, \tag{9.5}$$

for every $t > 0$.

Let

$$\widetilde{f}(x) = \begin{cases} f(x), & \text{if } f(x) > \frac{t}{2}, \\ 0, & \text{otherwise,} \end{cases}$$

for every $x \in Q_0$. Then $f(x) \le \widetilde{f}(x) + \frac{t}{2}$ for every $x \in Q_0$ and we obtain

$$M^{d,w}_{Q_0} f(x) \le M^{d,w}_{Q_0} \widetilde{f}(x) + \frac{t}{2}$$

for every $x \in Q_0$. The weak type estimate in (9.5) implies

$$\begin{aligned} w(\{x \in Q_0 : M^{d,w}_{Q_0} f(x) > t\}) &\le w(\{x \in Q_0 : M^{d,w}_{Q_0} \widetilde{f}(x) > \tfrac{t}{2}\}) \\ &\le \frac{2}{t} \int_{Q_0} \widetilde{f}(x) w(x)\,dx \\ &= \frac{2}{t} \int_{\{x\in Q_0 : f(x) > \frac{t}{2}\}} f(x) w(x)\,dx. \end{aligned}$$

As in the proof of Theorem 1.15, we obtain

$$\begin{aligned} \int_{Q_0} \big(M^{d,w}_{Q_0} f(x)\big)^p w(x)\,dx &= p \int_0^\infty t^{p-1} w(\{x \in Q_0 : M^{d,w}_{Q_0} f(x) > t\})\,dt \\ &\le 2p \int_0^\infty t^{p-2} \bigg(\int_{\{x\in Q_0 : f(x) > \frac{t}{2}\}} f(x) w(x)\,dx \bigg)\,dt \\ &= 2p \int_{Q_0} f(x) \bigg(\int_0^\infty t^{p-2} \chi_{\{x\in Q_0 : f(x) > \frac{t}{2}\}}(x)\,dt \bigg) w(x)\,dx \\ &= 2p \int_{Q_0} f(x) \bigg(\int_0^{2f(x)} t^{p-2}\,dt \bigg) w(x)\,dx \\ &= 2^p \frac{p}{p-1} \int_{Q_0} f(x)^p w(x)\,dx. \end{aligned}$$

□

The following result is a strong type version of Theorem 8.18 for the dyadic maximal function. The proof is based on the self-improvement of Muckenhoupt weights.

THEOREM 9.9. *Let $1 < p < \infty$, let $Q_0 \subset \mathbb{R}^n$ be a cube, and assume that $w \in A_p$. There exists a constant $C = C(n,p,w)$ such that*

$$\int_{Q_0} \big(M^d_{Q_0} f(x)\big)^p w(x)\,dx \le C \int_{Q_0} |f(x)|^p w(x)\,dx$$

for every $f \in L^\infty(Q_0)$.

PROOF. Without loss of generality, we may assume that $f \in L^\infty(Q_0)$ is non-negative. By Theorem 8.27, there exists $0 < \varepsilon = \varepsilon(n,p,w) < p-1$ such that $w \in A_{p-\varepsilon}$. Let $x \in Q_0$ and let $Q \in \mathcal{D}(Q_0)$ with $x \in Q$. Theorem 8.14 implies

$$\fint_Q f(y)\,dy \le C(n,p,w,\varepsilon) \bigg(\frac{1}{w(Q)} \int_Q f(y)^{p-\varepsilon} w(y)\,dy \bigg)^{\frac{1}{p-\varepsilon}}.$$

By taking supremums, we obtain

$$M^d_{Q_0} f(x) \le C(n,p,w,\varepsilon) \big(M^{d,w}_{Q_0} f^{p-\varepsilon}(x) \big)^{\frac{1}{p-\varepsilon}}.$$

Theorem 9.8 with the exponent $\frac{p}{p-\varepsilon} > 1$ implies

$$\begin{aligned} \int_{Q_0} \big(M^d_{Q_0} f(x)\big)^p\, w(x)\,dx &\le C(n,p,w,\varepsilon) \int_{Q_0} \big(M^{d,w}_{Q_0} f^{p-\varepsilon}(x)\big)^{\frac{p}{p-\varepsilon}}\, w(x)\,dx \\ &\le C(n,p,w,\varepsilon) \int_{Q_0} \big(f(x)^{p-\varepsilon}\big)^{\frac{p}{p-\varepsilon}} w(x)\,dx \\ &= C(n,p,w,\varepsilon) \int_{Q_0} f(x)^p w(x)\,dx. \end{aligned}$$

□

We continue with a pointwise estimate, which is a counterpart of Theorem 3.4 for cubes.

LEMMA 9.10. *Let $Q \subset \mathbb{R}^n$ be a cube and assume that $u \in \operatorname{Lip}(Q)$. There exists a constant $C = C(n)$ such that*

$$|u(x) - u_Q| \le C\, l(Q) M_Q^d |\nabla u|(x)$$

for every $x \in Q$.

PROOF. Let $Q_0 = Q$ and $x \in Q_0$. For each $j \in \mathbb{N}$, we choose recursively the dyadic child $Q_j \subset Q_{j-1}$ with $x \in Q_j$ and $2^n |Q_j| = |Q_{j-1}|$. Since u is continuous at x, we have

$$\begin{aligned}|u(x) - u_Q| &= \left| \sum_{j=1}^{\infty} u_{Q_j} - u_{Q_{j-1}} \right| \le \sum_{j=1}^{\infty} |u_{Q_j} - u_{Q_{j-1}}| \\ &\le \sum_{j=1}^{\infty} \fint_{Q_j} |u(y) - u_{Q_{j-1}}|\, dy \le \sum_{j=1}^{\infty} 2^n \fint_{Q_{j-1}} |u(y) - u_{Q_{j-1}}|\, dy.\end{aligned}$$

By Corollary 9.6, we obtain

$$\begin{aligned}|u(x) - u_Q| &\le \sum_{j=1}^{\infty} C(n) \big(2^{-j+1} l(Q)\big) \fint_{Q_{j-1}} |\nabla u(y)|\, dy \\ &\le C(n) l(Q) M_{Q_0}^d |\nabla u|(x) \sum_{j=1}^{\infty} 2^{-j+1} = C(n) l(Q) M_{Q_0}^d |\nabla u|(x). \qquad \square\end{aligned}$$

Weighted Poincaré inequalities for cubes follow as in Lemma 3.9.

THEOREM 9.11. *Let $1 < p < \infty$ and assume that $w \in A_p$. Let $Q \subset \mathbb{R}^n$ be a cube and let $u \in \operatorname{Lip}(Q)$. There exists a constant $C = C(n, p, w)$ such that*

$$\int_Q |u(x) - u_Q|^p w(x)\, dx \le C l(Q)^p \int_Q |\nabla u(x)|^p w(x)\, dx. \tag{9.6}$$

PROOF. By Lemma 9.10 and Theorem 9.9, we have

$$\begin{aligned}\int_Q |u(x) - u_Q|^p w(x)\, dx &\le C(n,p) l(Q)^p \int_Q \big(M_Q^d |\nabla u|(x)\big)^p w(x)\, dx \\ &\le C(n,p,w) l(Q)^p \int_Q |\nabla u(x)|^p w(x)\, dx. \qquad \square\end{aligned}$$

REMARK 9.12. For $n = 1$, the validity of the weighted Poincaré inequality (9.6) for all cubes in $\mathbb{R}$ implies that $w \in A_p(\mathbb{R})$, see Björn, Buckley and Keith [**35**]. For $n \ge 2$, this implication is not true. For instance, let $\alpha > -n$ and let $w(x) = |x|^\alpha$, for every $x \in \mathbb{R}^n \setminus \{0\}$. Then the Poincaré inequality in (9.6) with respect to w holds for any $1 < p < \infty$, see Heinonen, Kilpeläinen and Martio [**187**, p. 10]. However, w is an A_p weight if and only if $p > 1 + \frac{\alpha}{n}$ or $\alpha = 0$, see Remark 8.35 and Theorem 10.26. See also Gong [**155**] for absolute continuity of doubling measures supporting a Poincaré inequality.

9.3. Weighted local Fefferman–Stein inequalities

The following sparse domination lemma is a fundamental tool here and also in Section 9.7. It can be regarded as a refined version of Lemma 9.10 and Lemma 4.13. This technique allows to change simple nested sequences of cubes, as used for instance in the proof of Lemma 9.10, to an appropriate sparse collection of dyadic subcubes, with a controlled overlap property. Notice that the sparse collection of cubes depends on the function in question.

LEMMA 9.13. *Assume that $w \in A_\infty$. Let $Q_0 \subset \mathbb{R}^n$ be a cube and let $f \in L^1(Q_0)$. There exists a collection $\mathcal{S} \subset \mathcal{D}(Q_0)$ of dyadic subcubes of Q_0 satisfying the following conditions.*

(a) *There exists a constant $C = C(n, w)$ such that*

$$|f(x) - f_{Q_0}| \leq C \sum_{Q \in \mathcal{S}} \chi_Q(x) \fint_Q |f(y) - f_Q| \, dy,$$

for every Lebesgue point $x \in Q_0$ of f.

(b) *There exist a constant $0 < \kappa = \kappa(w) < 1$ and a collection $\{E_Q : Q \in \mathcal{S}\}$ of pairwise disjoint measurable sets $E_Q \subset Q$ with $w(E_Q) \geq \kappa w(Q)$ for every $Q \in \mathcal{S}$. Moreover, if $R, S \in \mathcal{S}$ and $R \subsetneq S$, then $R \subset S \setminus E_S$.*

PROOF. We may assume that

$$\fint_{Q_0} |f(y) - f_{Q_0}| \, dy > 0,$$

since otherwise there is nothing to prove. Let $C(w) > 0$ and $\delta(w) > 0$ be the constants in the A_∞ condition (8.12) for w, and choose $\rho > 1$ such that $C(w)\rho^{-\delta(w)} < 1$.

The collection $\mathcal{S}$ is constructed recursively. Let $Q_0 \in \mathcal{S}$. For $S \in \mathcal{S}$, we add to $\mathcal{S}$ the maximal dyadic cubes $S' \subsetneq S$ that satisfy the stopping condition

$$\fint_{S'} |f(y) - f_S| \, dy > \rho \fint_S |f(y) - f_S| \, dy.$$

By iterating this process, we obtain a collection $\mathcal{S} \subset \mathcal{D}(Q_0)$ satisfying the following properties.

(S1) For $Q \in \mathcal{D}(Q_0)$, let $\pi_{\mathcal{S}} Q$ be the minimal cube in $\mathcal{S}$ with $Q \subset \pi_{\mathcal{S}} Q$. If $\pi_{\mathcal{S}}(Q) = S \in \mathcal{S}$, then

$$\fint_Q |f(y) - f_S| \, dy \leq \rho \fint_S |f(y) - f_S| \, dy.$$

(S2) For $S \in \mathcal{S}$, let $\operatorname{ch}_{\mathcal{S}}(S)$ denote the collection of maximal cubes in $\mathcal{S}$ that are strictly contained in S; these are the children of S with respect to $\mathcal{S}$. For every $S' \in \operatorname{ch}_{\mathcal{S}}(S)$ we have

$$\frac{1}{|S'|} \int_{S'} |f(y) - f_S| \, dy > \frac{\rho}{|S|} \int_S |f(y) - f_S| \, dy.$$

Moreover, the sets $S' \in \operatorname{ch}_{\mathcal{S}}(S)$ are pairwise disjoint, and thus

$$\sum_{S' \in \operatorname{ch}_{\mathcal{S}}(S)} |S'| \leq \frac{|S|}{\rho} \frac{\sum_{S' \in \operatorname{ch}_{\mathcal{S}}(S)} \int_{S'} |f(y) - f_S| \, dy}{\int_S |f(y) - f_S| \, dy} \leq \frac{|S|}{\rho}.$$

(S3) Let $S \in \mathcal{S}$. From the pairwise disjointness of the sets $S' \in \operatorname{ch}_{\mathcal{S}}(S)$, condition (S2), the A_∞ property of w, and the choice of ρ, we obtain

$$\frac{1}{w(S)} \sum_{S' \in \operatorname{ch}_{\mathcal{S}}(S)} w(S') \leq C(w) \Big(\frac{1}{|S|} \sum_{S' \in \operatorname{ch}_{\mathcal{S}}(S)} |S'| \Big)^{\delta(w)} \leq C(w) \rho^{-\delta(w)} < 1.$$

For every $S \in \mathcal{S}$, let

$$E_S = S \setminus \bigcup_{S' \in \operatorname{ch}_{\mathcal{S}}(S)} S'.$$

If $S, R \in \mathcal{S}$ and $S \neq R$, then $E_S \cap E_R = \emptyset$. This is clear if $S \cap R = \emptyset$. On the other hand, if $R \subsetneq S$, there exists $S' \in \operatorname{ch}_{\mathcal{S}}(S)$ with $R \subset S'$, and thus

$$E_R \cap E_S \subset R \cap E_S \subset S' \cap E_S = \emptyset.$$

If $S \subsetneq R$, then $E_R \cap E_S = \emptyset$, and we conclude that the sets E_S, for $S \in \mathcal{S}$, are pairwise disjoint. Moreover, we see that $R \subset S \setminus E_S$ if $R, S \in \mathcal{S}$ and $R \subsetneq S$.

Let $S \in \mathcal{S}$. To complete the proof of assertion (b), it suffices to show that $w(E_S) \geq \kappa w(S)$. To this end, condition (S3) implies

$$\sum_{S' \in \operatorname{ch}_{\mathcal{S}}(S)} w(S') \leq C(w) \rho^{-\delta(w)} w(S),$$

where $C(w)\rho^{-\delta(w)} < 1$. Hence

$$w(E_S) = w(S) - \sum_{S' \in \operatorname{ch}_{\mathcal{S}}(S)} w(S') \geq \big(1 - C(w)\rho^{-\delta(w)}\big) w(S),$$

and we conclude that (b) holds with $0 < \kappa = 1 - C(w)\rho^{-\delta(w)} < 1$.

In the proof of assertion (a) we apply the operators

$$\Delta_Q f(x) = \sum_{Q' \in \operatorname{ch}(Q)} \chi_{Q'}(x)(f_{Q'} - f_Q),$$

where $x \in Q_0$ and $Q \in \mathcal{D}(Q_0)$ is a dyadic cube and $\operatorname{ch}(Q)$ is the collection of its 2^n dyadic children. For $S \in \mathcal{S}$ and $Q \in \mathcal{D}(Q_0)$, we write $Q \in \mathcal{C}(S)$ if $\pi_{\mathcal{S}} Q = S$. Let $x \in Q_0$ be a Lebesgue point of f. Then

$$\begin{aligned} |f(x) - f_{Q_0}| &= \Big| \sum_{Q \in \mathcal{D}(Q_0)} \Delta_Q f(x) \Big| = \Big| \sum_{S \in \mathcal{S}} \sum_{Q \in \mathcal{C}(S)} \Delta_Q f(x) \Big| \\ &\leq \sum_{S \in \mathcal{S}} \Big| \sum_{Q \in \mathcal{C}(S)} \Delta_Q f(x) \Big|. \end{aligned} \tag{9.7}$$

Let $S \in \mathcal{S}$ and assume that $x \in S$. Then either $x \in S \setminus E_S$ or $x \in E_S$. First we consider the case $x \in S \setminus E_S$. By the definition of E_S there exists $S' \in \operatorname{ch}_{\mathcal{S}}(S)$ with $x \in S'$, and by telescoping we obtain

$$\sum_{Q \in \mathcal{C}(S)} \Delta_Q f(x) = f_{S'} - f_S = \fint_{S'} (f(y) - f_S)\, dy.$$

Let $\pi S' \in \mathcal{D}(Q_0)$ be the dyadic parent of S'. Then $\pi_{\mathcal{S}}(\pi S') = S \in \mathcal{S}$, and condition (S1) implies

$$\begin{aligned}\Big|\sum_{Q\in\mathcal{C}(S)} \Delta_Q f(x)\Big| &= \Big|\fint_{S'} (f(y) - f_S)\,dy\Big| \le \fint_{S'} |f(y) - f_S|\,dy \\ &\le 2^n \fint_{\pi S'} |f(y) - f_S|\,dy \le \rho 2^n \fint_S |f(y) - f_S|\,dy.\end{aligned}$$

Assume then that $x \in E_S$. For $k \in \mathbb{N}_0$, let $Q_k \subset S$ be the unique dyadic cube $Q_k \in \mathcal{D}(Q_0)$ with $x \in Q_k$ and $l(S) = 2^k l(Q_k)$. Then $Q_k \in \mathcal{C}(S)$ for every $k \in \mathbb{N}_0$ since otherwise $x \in S'$ for some $S' \in \operatorname{ch}_{\mathcal{S}}(S)$, but this is not possible since $x \in E_S$. By telescoping, we obtain

$$\sum_{Q\in\mathcal{C}(S)} \Delta_Q f(x) = \lim_{k\to\infty} (f_{Q_k} - f_S).$$

We remark that the limit exists, and is equal to $f(x) - f_S$, since x is assumed to be a Lebesgue point of f. Since $x \in E_S$, we have $\pi_{\mathcal{S}}(Q_k) = S$ for every $k \in \mathbb{N}_0$. By property (S1),

$$|f_{Q_k} - f_S| = \Big|\fint_{Q_k} (f(y) - f_S)\,dy\Big| \le \fint_{Q_k} |f(y) - f_S|\,dy \le \rho \fint_S |f(y) - f_S|\,dy,$$

and thus

$$\Big|\sum_{Q\in\mathcal{C}(S)} \Delta_Q f(x)\Big| \le \rho \fint_S |f(y) - f_S|\,dy \le \rho 2^n \fint_S |f(y) - f_S|\,dy$$

for every $x \in E_S$.

By combining the estimates above, and noticing that $\sum_{Q\in\mathcal{C}(S)} \Delta_Q f(x) = 0$ if $x \notin S$, we obtain

$$\Big|\sum_{Q\in\mathcal{C}(S)} \Delta_Q f(x)\Big| \le \rho 2^n \chi_S(x) \fint_S |f(y) - f_S|\,dy$$

for every Lebesgue point $x \in Q_0$ of f. This, together with (9.7), concludes the proof of assertion (a). □

The next result can be viewed as a weighted local Fefferman–Stein inequality. It is not clear how to modify the proofs of Lemma 9.10 or Lemma 4.13 to obtain a more direct proof of this theorem. Instead, we make use of the sparse domination lemma, L^p-duality, and boundedness results for weighted maximal operators. We need the following dyadic sharp maximal function, compare to Definition 4.12.

DEFINITION 9.14. Let $Q_0 \subset \mathbb{R}^n$ be a cube. For $f \in L^1(Q_0)$, the dyadic sharp maximal function $M^{d,\sharp}_{Q_0} f$ is defined at $x \in Q_0$ by

$$M^{d,\sharp}_{Q_0} f(x) = \sup_{Q\ni x} \fint_Q |f(y) - f_Q|\,dy,$$

where the supremum is taken over all dyadic cubes $Q \in \mathcal{D}(Q_0)$ with $x \in Q$.

THEOREM 9.15. *Let $1 < p < \infty$ and assume that $w \in A_\infty$. Let $Q_0 \subset \mathbb{R}^n$ be a cube and let $f \in L^1(Q_0)$. There exists a constant $C = C(n,p,w)$ such that*

$$\int_{Q_0} |f(x) - f_{Q_0}|^p w(x)\,dx \le C \int_{Q_0} \big(M^{d,\sharp}_{Q_0} f(x)\big)^p w(x)\,dx. \tag{9.8}$$

PROOF. Without loss of generality, we may assume that the right-hand side of (9.8) is finite. Let $\mathcal{S} \subset \mathcal{D}(Q_0)$ be the sparse collection of cubes associated to function f given by Lemma 9.13, and define

$$F(x) = \sum_{Q\in\mathcal{S}} \chi_Q(x) \fint_Q |f(y) - f_Q|\,dy,$$

for every $x \in Q_0$. Lemma 9.13 implies

$$\int_{Q_0} |f(x) - f_{Q_0}|^p w(x)\,dx \le C(n,p,w) \int_{Q_0} F(x)^p w(x)\,dx.$$

We estimate the integral on the right-hand side by the L^p-duality

$$\Big(\int_{Q_0} F(x)^p w(x)\,dx\Big)^{\frac{1}{p}} = \sup_g \int_{Q_0} F(x) g(x) w(x)\,dx,$$

where the supremum is taken over all nonnegative $g \in L^\infty(Q_0) \cap L^q(Q_0; w\,dx)$ with $q = \frac{p}{p-1}$ and $\|g\|_{L^q(Q_0;w\,dx)} \le 1$; this fact follows from density of simple functions in $L^q(Q_0; w\,dx)$. Let g be a function as above. By the definition of F, we have

$$\int_{Q_0} F(x) g(x) w(x)\,dx = \sum_{Q\in\mathcal{S}} \fint_Q |f(y) - f_Q|\,dy \int_Q g(x) w(x)\,dx. \tag{9.9}$$

From Lemma 9.13 we obtain for every $Q \in \mathcal{S}$ a subset $E_Q \subset Q$ such that

$$\begin{aligned}\int_Q g(x)w(x)\,dx &\le \kappa^{-1} \int_{E_Q} \Big(\frac{1}{w(Q)} \int_Q g(x) w(x)\,dx\Big) w(z)\,dz \\ &\le \kappa^{-1} \int_{E_Q} M^{d,w}_{Q_0} g(z) w(z)\,dz,\end{aligned}$$

where $M^{d,w}_{Q_0} g$ is as in Definition 9.7 and $\kappa = \kappa(w)$.

From (9.9) we obtain, by pairwise disjointness of the sets E_Q and Hölder's inequality, that

$$\begin{aligned}\int_{Q_0} F(x) g(x) w(x)\,dx &\le \kappa^{-1} \sum_{Q\in\mathcal{S}} \int_{E_Q} \Big(\fint_Q |f(y) - f_Q|\,dy\Big) M^{d,w}_{Q_0} g(z) w(z)\,dz \\ &\le \kappa^{-1} \sum_{Q\in\mathcal{S}} \int_{E_Q} M^{d,\sharp}_{Q_0} f(z) M^{d,w}_{Q_0} g(z) w(z)\,dz \\ &\le \kappa^{-1} \int_{Q_0} M^{d,\sharp}_{Q_0} f(z) M^{d,w}_{Q_0} g(z) w(z)\,dz \\ &\le \kappa^{-1} \|M^{d,\sharp}_{Q_0} f\|_{L^p(Q_0;w\,dx)} \|M^{d,w}_{Q_0} g\|_{L^q(Q_0;w\,dx)}.\end{aligned}$$

Finally, Theorem 9.8 implies

$$\|M^{d,w}_{Q_0} g\|_{L^q(Q_0;w\,dx)} \le C(p) \|g\|_{L^q(Q_0;w\,dx)} \le C(p),$$

and the claim follows by taking supremum over all functions g as above. $\square$

9.4. Two weight maximal inequalities

In this section we consider two weight inequalities for the following dyadic fractional maximal function, compare to Definition 1.38.

DEFINITION 9.16. Let $Q_0 \subset \mathbb{R}^n$ be a cube and let $0 \le \alpha < n$. For $f \in L^1(Q_0)$, the dyadic fractional maximal function $M^d_{\alpha,Q_0} f$ at $x \in Q_0$ is

$$M^d_{\alpha,Q_0} f(x) = \sup_{Q \ni x} \frac{1}{|Q|^{1-\frac{\alpha}{n}}} \int_Q |f(y)|\, dy,$$

where the supremum is taken over all dyadic cubes $Q \in \mathcal{D}(Q_0)$ with $x \in Q$.

From the Poincaré inequality on cubes we obtain pointwise estimates in terms of the dyadic fractional maximal function, compare to Section 4.3.

LEMMA 9.17. *Let $Q_0 \subset \mathbb{R}^n$ be a cube and let $u \in \operatorname{Lip}(Q_0)$. Then*

$$M^{d,\sharp}_{Q_0} u(x) \le C(n) M^d_{1,Q_0} |\nabla u|(x)$$

for every $x \in Q_0$.

PROOF. Let $x \in Q_0$ and let $Q \in \mathcal{D}(Q_0)$ with $x \in Q$. By Corollary 9.6, we have

$$\begin{aligned} \fint_Q |u(y) - u_Q|\, dy &\le C(n) l(Q) \fint_Q |\nabla u(y)|\, dy \\ &= \frac{C(n)}{|Q|^{1-\frac{1}{n}}} \int_Q |\nabla u(y)|\, dy \le C(n) M^d_{1,Q_0} |\nabla u|(x). \end{aligned}$$

The claim follows by taking supremum over all cubes Q as above. □

The next lemma gives a necessary condition for a weighted Hardy–Littlewood–Sobolev theorem for the dyadic fractional maximal function, compare to Theorem 1.42. This result is also a generalization of Lemma 8.17. The corresponding sufficient condition will be discussed in Theorem 9.20.

LEMMA 9.18. *Let $1 < p \le q < \infty$ and $0 \le \alpha < n$, and let $Q_0 \subset \mathbb{R}^n$ be a cube. Assume that (w, v) is a pair of weights such that*

$$\Big(\int_{Q_0} (M^d_{\alpha,Q_0} f(x))^q w(x)\, dx \Big)^{\frac{1}{q}} \le C_1 \Big(\int_{Q_0} |f(x)|^p v(x)\, dx \Big)^{\frac{1}{p}}$$

for every $f \in L^\infty(Q_0)$, and write $\sigma = v^{\frac{1}{1-p}}$. Then

$$\Big(\frac{1}{|Q|^{1-\frac{\alpha}{n}}} \Big)^p w(Q)^{\frac{p}{q}} \sigma(Q)^{p-1} \le C_1^p \tag{9.10}$$

for every dyadic cube $Q \in \mathcal{D}(Q_0)$.

PROOF. Let $\varepsilon > 0$ and $Q \in \mathcal{D}(Q_0)$, and define

$$f(x) = (\varepsilon + v(x))^{\frac{1}{1-p}} \chi_Q(x)$$

for every $x \in Q_0$. Then

$$\frac{1}{|Q|^{1-\frac{\alpha}{n}}} \int_Q (\varepsilon + v(y))^{\frac{1}{1-p}}\, dy \le M^d_{\alpha,Q_0} f(x)$$

for every $x \in Q$, and thus

$$\int_Q w(x)\,dx \left(\frac{1}{|Q|^{1-\frac{\alpha}{n}}} \int_Q (\varepsilon + v(y))^{\frac{1}{1-p}}\,dy \right)^q \leq \int_Q \left(M^d_{\alpha,Q_0} f(x) \right)^q w(x)\,dx.$$

Since $Q \subset Q_0$ and $f \in L^\infty(Q_0)$, we have

$$\begin{aligned} \int_Q \left(M^d_{\alpha,Q_0} f(x) \right)^q w(x)\,dx &\leq C_1^q \left(\int_{Q_0} f(x)^p v(x)\,dx \right)^{\frac{q}{p}} \\ &\leq C_1^q \left(\int_Q (\varepsilon + v(x))^{\frac{p}{1-p}} (\varepsilon + v(x))\,dx \right)^{\frac{q}{p}} \\ &= C_1^q \left(\int_Q (\varepsilon + v(x))^{\frac{1}{1-p}}\,dx \right)^{\frac{q}{p}}, \end{aligned}$$

and thus

$$\left(\frac{1}{|Q|^{1-\frac{\alpha}{n}}} \right)^q w(Q) \left(\int_Q (\varepsilon + v(y))^{\frac{1}{1-p}}\,dy \right)^{q-\frac{q}{p}} \leq C_1^q$$

for every $\varepsilon > 0$. By Fatou's lemma, we conclude that

$$\begin{aligned} &\left(\frac{1}{|Q|^{1-\frac{\alpha}{n}}} \right)^p w(Q)^{\frac{p}{q}} \sigma(Q)^{p-1} \\ &\qquad \leq \liminf_{\varepsilon \to 0+} \left(\frac{1}{|Q|^{1-\frac{\alpha}{n}}} \right)^p w(Q)^{\frac{p}{q}} \left(\int_Q (\varepsilon + v(y))^{\frac{1}{1-p}}\,dy \right)^{p-1} \leq C_1^p, \end{aligned}$$

and the claim follows. □

For the converse of Lemma 9.18 we need another variant of the sparse domination lemma. Here we write

$$f_{\alpha,Q} = \frac{1}{|Q|^{1-\frac{\alpha}{n}}} \int_Q |f(y)|\,dy$$

for $Q \in \mathcal{D}(Q_0)$ and $f \in L^1(Q_0)$.

LEMMA 9.19. *Let $0 \leq \alpha < n$ and assume that $\sigma \in A_\infty$. Let $Q_0 \subset \mathbb{R}^n$ be a cube and let $f \in L^\infty(Q_0)$. There exists a collection $\mathcal{S} \subset \mathcal{D}(Q_0)$ of dyadic subcubes of Q_0 satisfying the following conditions.*

(a) *For $1 \leq p < \infty$, there exists a constant $C = C(n, p, \sigma)$ such that*

$$\left(M^d_{\alpha,Q_0} f(x) \right)^p \leq C \sum_{Q \in \mathcal{S}} (f_{\alpha,Q})^p \chi_Q(x)$$

for every $x \in Q_0$.

(b) *There exist a constant $\kappa = \kappa(n, \sigma) > 0$ and a collection $\{E_Q : Q \in \mathcal{S}\}$ of pairwise disjoint measurable sets $E_Q \subset Q$ with $\sigma(E_Q) \geq \kappa \sigma(Q)$ for every $Q \in \mathcal{S}$.*

PROOF. Let $C(\sigma) > 0$ and $\delta(\sigma) > 0$ be the constants in the A_∞ condition (8.12) for w. Choose $a > 2^n$ such that

$$\rho = C(\sigma) \left(\frac{2^n}{a} \right)^{\delta(\sigma)} < 1.$$

Without loss of generality, we may assume that $f_{\alpha,Q_0} > 0$. Let k_0 be the smallest integer satisfying $a^{k_0} \geq f_{\alpha,Q_0}$, and let

$$D_k = \{x \in Q_0 : a^k < M^d_{\alpha,Q_0} f(x) \leq a^{k+1}\}$$

and

$$S_k = \{x \in Q_0 : a^k < M^d_{\alpha,Q_0} f(x)\}$$

for every integer $k > k_0$.

Let $\mathcal{S}_{k_0} = \{Q_0\}$. For $k > k_0$, let $\mathcal{S}_k$ be the collection of pairwise disjoint maximal dyadic cubes $Q \in \mathcal{D}(Q_0)$ satisfying

$$a^k < \frac{1}{|Q|^{1-\frac{\alpha}{n}}} \int_Q |f(y)|\, dy. \tag{9.11}$$

Since $k > k_0$, each of these maximal cubes is strictly contained in Q_0. Observe that $S_k = \bigcup_{Q \in \mathcal{S}_k} Q$ for every integer $k > k_0$.

For every $k \geq k_0$ and every $R \in \mathcal{S}_{k+1}$, there exists a unique $Q \in \mathcal{S}_k$ with $R \subset Q$. In particular, the sets

$$E_{k,Q} = Q \setminus \bigcup_{R \in \mathcal{S}_{k+1}} R = Q \setminus \bigcup_{\substack{R \in \mathcal{S}_{k+1} \\ R \subset Q}} R$$

in the collection $\{E_{k,Q} : k \geq k_0 \text{ and } Q \in \mathcal{S}_k\}$ are pairwise disjoint.

We claim that

$$\sigma(E_{k,Q}) \geq (1-\rho)\sigma(Q) \tag{9.12}$$

for every $k \geq k_0$ and $Q \in \mathcal{S}_k$. In order to prove this inequality, we first apply (9.11) and obtain

$$\sum_{\substack{R \in \mathcal{S}_{k+1} \\ R \subset Q}} |R| \leq \sum_{\substack{R \in \mathcal{S}_{k+1} \\ R \subset Q}} \frac{|R|^{\frac{\alpha}{n}}}{a^{k+1}} \int_R |f(y)|\, dy \leq \frac{|Q|^{\frac{\alpha}{n}}}{a^{k+1}} \int_Q |f(y)|\, dy \leq \frac{2^n}{a}|Q|,$$

whenever $Q \in \mathcal{S}_k$; recall here that $\alpha \geq 0$. The final step above follows from the maximality of the collection $\mathcal{S}_k$ if $k > k_0$ and from inequality $a^{k_0} \geq f_{\alpha,Q_0}$ if $k = k_0$. Since $\sigma \in A_\infty$, we have

$$\frac{\sigma\Big(\bigcup_{R \in \mathcal{S}_{k+1} : R \subset Q} R\Big)}{\sigma(Q)} \leq C(\sigma)\bigg(\frac{1}{|Q|}\Big|\bigcup_{R \in \mathcal{S}_{k+1} : R \subset Q} R\Big|\bigg)^{\delta(\sigma)} \leq C(\sigma)\Big(\frac{2^n}{a}\Big)^{\delta(\sigma)} = \rho.$$

From this we deduce that

$$\sigma(E_{k,Q}) \geq \sigma(Q) - \rho\sigma(Q) = (1-\rho)\sigma(Q),$$

and (9.12) follows. Moreover, since $E_{k,Q} \neq \emptyset$ for every $k \geq k_0$ and $Q \in \mathcal{S}_k$, we have $\mathcal{S}_k \cap \mathcal{S}_j = \emptyset$ for $k \neq j$. We conclude from above that condition (b) holds for the collection $\mathcal{S} = \bigcup_{k \geq k_0} \mathcal{S}_k$, with $\kappa = 1 - \rho > 0$.

Observe that

$$M^d_{\alpha,Q_0} f(x) \leq |Q_0|^{\frac{\alpha}{n}} M^d_{Q_0} f(x) \leq |Q_0|^{\frac{\alpha}{n}} \|f\|_{L^\infty(Q_0)} < \infty$$

for every $x \in Q_0$, and thus

$$Q_0 = \{x \in Q_0 : M^d_{Q_0} f(x) \leq a^{k_0+1}\} \cup \bigcup_{k > k_0} D_k.$$

Since $a^{k_0-1} < f_{\alpha,Q_0}$, we obtain

$$\begin{aligned}
\left(M^d_{\alpha,Q_0}f(x)\right)^p &\le a^{2p}(f_{\alpha,Q_0})^p\chi_{Q_0}(x) + \sum_{k>k_0}\left(M^d_{\alpha,Q_0}f(x)\right)^p\chi_{D_k}(x)\\
&\le a^{2p}(f_{\alpha,Q_0})^p\chi_{Q_0}(x) + a^p\sum_{k>k_0}a^{kp}\chi_{S_k}(x)\\
&\le a^{2p}(f_{\alpha,Q_0})^p\chi_{Q_0}(x) + a^p\sum_{k>k_0}\sum_{Q\in\mathcal{S}_k}(f_{\alpha,Q})^p\chi_Q(x)\\
&\le a^{2p}\sum_{k\ge k_0}\sum_{Q\in\mathcal{S}_k}(f_{\alpha,Q})^p\chi_Q(x)\\
&= a^{2p}\sum_{Q\in\mathcal{S}}(f_{\alpha,Q})^p\chi_Q(x)
\end{aligned}$$

for every $x \in Q_0$, and this proves condition (a). □

The following result is an extension of Theorem 9.9 for two weights and two exponents. Indeed, if $\alpha = 0$, $p = q$ and $w = v$, then (9.13) reduces to a localized A_p condition for w.

THEOREM 9.20. *Let $1 < p \le q < \infty$ and $0 \le \alpha < n$, and assume that (w, v) is a pair of weights such that $\sigma = v^{\frac{1}{1-p}} \in A_\infty$. Let $Q_0 \subset \mathbb{R}^n$ be a cube and assume that there exists a constant $C_1 > 0$ such that*

$$\left(\frac{1}{|Q|^{1-\frac{\alpha}{n}}}\right)^p w(Q)^{\frac{p}{q}}\sigma(Q)^{p-1} \le C_1 \tag{9.13}$$

for all dyadic cubes $Q \in \mathcal{D}(Q_0)$. Then there exists a constant $C = C(n, p, q, \sigma, C_1)$ such that

$$\left(\int_{Q_0}\left(M^d_{\alpha,Q_0}f(x)\right)^q w(x)\,dx\right)^{\frac{1}{q}} \le C\left(\int_{Q_0}|f(x)|^p v(x)\,dx\right)^{\frac{1}{p}}$$

for every $u \in L^\infty(Q_0)$.

PROOF. Let $f \in L^\infty(Q_0)$ and let $\mathcal{S}$ be the associated sparse collection of cubes, given by Lemma 9.19 with $\sigma \in A_\infty$ and exponent p. Then

$$\begin{aligned}
\left(\int_{Q_0}\left(M^d_{\alpha,Q_0}f(x)\right)^q w(x)\,dx\right)^{\frac{p}{q}} &= \left\|\left(M^d_{\alpha,Q_0}f\right)^p\right\|_{L^{\frac{q}{p}}(Q_0;w\,dx)}\\
&\le C\left\|\sum_{Q\in\mathcal{S}}(f_{\alpha,Q})^p\chi_Q\right\|_{L^{\frac{q}{p}}(Q_0;w\,dx)}\\
&\le C\sum_{Q\in\mathcal{S}}\left\|(f_{\alpha,Q})^p\chi_Q\right\|_{L^{\frac{q}{p}}(Q_0;w\,dx)}\\
&\le C\sum_{Q\in\mathcal{S}}\left(\frac{1}{|Q|^{1-\frac{\alpha}{n}}}\int_Q|f(y)|\,dy\right)^p w(Q)^{\frac{p}{q}},
\end{aligned}$$

where $C = C(n,p,q,\sigma)$. On the other hand, by (9.13),

$$\begin{aligned}
&\sum_{Q\in\mathcal{S}}\left(\frac{1}{|Q|^{1-\frac{\alpha}{n}}}\int_Q |f(y)|\,dy\right)^p w(Q)^{\frac{p}{q}} \\
&\quad = \sum_{Q\in\mathcal{S}}\left(\frac{1}{|Q|^{1-\frac{\alpha}{n}}}\right)^p w(Q)^{\frac{p}{q}}\sigma(Q)^{p-1}\left(\frac{1}{\sigma(Q)}\int_Q |f(y)|\sigma^{-1}(y)\sigma(y)\,dy\right)^p \sigma(Q) \\
&\quad \le C_1 \sum_{Q\in\mathcal{S}}\left(\frac{1}{\sigma(Q)}\int_Q |f(y)|\sigma^{-1}(y)\sigma(y)\,dy\right)^p \sigma(Q).
\end{aligned}$$

We use the properties of the collection $\mathcal{S}$ and apply Lemma 9.8 to bound the final term by

$$\begin{aligned}
&\frac{C_1}{\kappa}\sum_{Q\in\mathcal{S}}\left(\frac{1}{\sigma(Q)}\int_Q |f(y)|\,\sigma^{-1}(y)\sigma(y)\,dy\right)^p \sigma(E_Q) \\
&\quad = \frac{C_1}{\kappa}\sum_{Q\in\mathcal{S}}\int_{E_Q}\left(\frac{1}{\sigma(Q)}\int_Q |f(y)|\,\sigma^{-1}(y)\sigma(y)\,dy\right)^p \sigma(x)\,dx \\
&\quad \le \frac{C_1}{\kappa}\int_{Q_0}(M^{d,\sigma}_{Q_0}(f\sigma^{-1})(x))^p\sigma(x)\,dx \\
&\quad \le \frac{C(p)C_1}{\kappa}\int_{Q_0}|f(y)\sigma^{-1}(y)|^p\sigma(y)\,dy.
\end{aligned}$$

This concludes the proof since $\kappa = \kappa(n,\sigma)$ and $v = \sigma^{1-p}$. □

9.5. Two weight Poincaré inequalities

The following result is a generalization of Theorem 9.11 and it gives a sufficient condition for the validity of a two weight Poincaré inequality.

THEOREM 9.21. *Let $1 < p \le q < \infty$ and let (w,v) be a pair of weights with $w \in A_\infty$ and $\sigma = v^{\frac{1}{1-p}} \in A_\infty$. Let $Q_0 \subset \mathbb{R}^n$ be a cube and assume that there exists a constant C_1 such that*

$$\left(\frac{1}{|Q|^{1-\frac{1}{n}}}\right)^p w(Q)^{\frac{p}{q}}\sigma(Q)^{p-1} \le C_1 \tag{9.14}$$

for all dyadic cubes $Q \in \mathcal{D}(Q_0)$. Then there exists a constant $C = C(n,p,q,w,\sigma,C_1)$ such that

$$\left(\int_{Q_0}|u(x)-u_{Q_0}|^q w(x)\,dx\right)^{\frac{1}{q}} \le C\left(\int_{Q_0}|\nabla u(x)|^p v(x)\,dx\right)^{\frac{1}{p}}$$

for every $u \in \operatorname{Lip}(Q_0)$.

PROOF. Let $u \in \operatorname{Lip}(Q_0)$. Theorem 9.15 and Lemma 9.17 imply

$$\begin{aligned}
\int_{Q_0}|u(x)-u_{Q_0}|^q w(x)\,dx &\le C(n,q,w)\int_{Q_0}(M^{d,\sharp}_{Q_0}u(x))^q w(x)\,dx \\
&\le C(n,q,w)\int_{Q_0}(M^d_{1,Q_0}|\nabla u|(x))^q w(x)\,dx.
\end{aligned}$$

By Theorem 9.20, we obtain

$$\int_{Q_0} (M^d_{1,Q_0}|\nabla u|(x))^q w(x)\,dx \le C(n,p,q,\sigma,C_1)\Big(\int_{Q_0} |\nabla u(x)|^p v(x)\,dx\Big)^{\frac{q}{p}},$$

and the claim follows. □

REMARK 9.22. We show that Theorem 9.21 is a generalization of Theorem 9.11. Let $1 < p < \infty$ and let $Q_0 \subset \mathbb{R}^n$ be a cube. Assume that $w \in A_p \subset A_\infty$ and let $v = l(Q_0)^p w$ and $\sigma = v^{\frac{1}{1-p}}$. Lemma 8.13 (b) implies $\sigma \in A_{p'} \subset A_\infty$. For $q = p$ and every $Q \in \mathcal{D}(Q_0)$, we have

$$\begin{aligned}\Big(\frac{1}{|Q|^{1-\frac{1}{n}}}\Big)^p w(Q)^{\frac{p}{q}}\sigma(Q)^{p-1} &= l(Q)^p|Q|^{-p}w(Q)\Big(l(Q_0)^{-\frac{p}{p-1}}\int_Q w(x)^{\frac{1}{1-p}}\,dx\Big)^{p-1}\\ &= \Big(\frac{l(Q)}{l(Q_0)}\Big)^p \fint_Q w(x)\,dx\Big(\fint_Q w(x)^{\frac{1}{1-p}}\,dx\Big)^{p-1}\\ &\le [w]_{A_p},\end{aligned}$$

where the final step holds since $l(Q) \le l(Q_0)$. This shows that the assumption of Theorem 9.21 is satisfied, and the claim of Theorem 9.11 follows from Theorem 9.21.

REMARK 9.23. Let $1 < p < n$ and $q = p^* = \frac{np}{n-p}$, and let $w(x) = v(x) = 1$ for every $x \in \mathbb{R}^n$. Then $\sigma(x) = 1$ for every $x \in \mathbb{R}^n$ and

$$\Big(\frac{1}{|Q|^{1-\frac{1}{n}}}\Big)^p w(Q)^{\frac{p}{q}}\sigma(Q)^{p-1} = |Q|^{\frac{p}{n}-p}|Q|^{1-\frac{p}{n}}|Q|^{p-1} = 1$$

for every cube $Q \subset \mathbb{R}^n$. By Theorem 9.21, we find that the Sobolev–Poincaré inequality

$$\Big(\int_Q |u(x)-u_Q|^{p^*}\,dx\Big)^{\frac{1}{p^*}} \le C(n,p)\Big(\int_Q |\nabla u(x)|^p\,dx\Big)^{\frac{1}{p}} \tag{9.15}$$

holds for all cubes $Q \subset \mathbb{R}^n$ and every $u \in \operatorname{Lip}(Q)$. Note that the 1-Poincaré inequality for cubes is needed in the proof of Theorem 9.21 via the proof of Lemma 9.17.

The following Poincaré inequality for cubes follows from Remark 9.23 by adapting the proof of Theorem 3.14. We remark that inequalities (9.15) and (9.16) are obtained as special cases of Theorem 9.21, but there are also direct proofs, for instance by following the lines of the proofs of the corresponding inequalities for balls in Chapter 3.

COROLLARY 9.24. *Let $1 < p < \infty$ and assume that $u \in W^{1,p}_{\mathrm{loc}}(\mathbb{R}^n)$. Let $1 \le q \le \frac{np}{n-p}$ for $1 < p < n$, and $1 \le q < \infty$ for $n \le p < \infty$. There exists a constant $C = C(n,p,q)$ such that*

$$\Big(\fint_Q |u(x)-u_Q|^q\,dx\Big)^{\frac{1}{q}} \le Cl(Q)\Big(\fint_Q |\nabla u(x)|^p\,dx\Big)^{\frac{1}{p}} \tag{9.16}$$

for every cube $Q \subset \mathbb{R}^n$.

For compactly supported Lipschitz functions we obtain the following two weight Sobolev–Gagliardo–Nirenberg inequality; compare to Theorem 3.6 and Remark 3.16.

COROLLARY 9.25. *Let $1 < p \le q < \infty$ and let (w, v) be a pair of weights with $w \in A_\infty$ and $\sigma = v^{\frac{1}{1-p}} \in A_\infty$. Assume that there exists a constant C_1 such that*

$$\left(\frac{1}{|Q|^{1-\frac{1}{n}}}\right)^p w(Q)^{\frac{p}{q}} \sigma(Q)^{p-1} \le C_1$$

for all cubes $Q \subset \mathbb{R}^n$. Then there exists a constant $C = C(n, p, q, w, \sigma, C_1)$ such that

$$\left(\int_{\mathbb{R}^n} |u(x)|^q w(x)\,dx\right)^{\frac{1}{q}} \le C\left(\int_{\mathbb{R}^n} |\nabla u(x)|^p v(x)\,dx\right)^{\frac{1}{p}}$$

for every $u \in \operatorname{Lip}_c(\mathbb{R}^n)$.

PROOF. Let $u \in \operatorname{Lip}_c(\mathbb{R}^n)$ and let $Q_k = Q(0, k)$ for every $k \in \mathbb{N}$. Since $u \in L^\infty(\mathbb{R}^n)$, we have

$$\lim_{k\to\infty} |u_{Q_k}| \le \lim_{k\to\infty} \frac{\|u\|_{L^\infty(\mathbb{R}^n)}}{|Q_k|} = 0.$$

Fatou's lemma and Theorem 9.21 imply

$$\begin{aligned}\left(\int_{\mathbb{R}^n} |u(x)|^q w(x)\,dx\right)^{\frac{1}{q}} &\le \liminf_{k\to\infty}\left(\int_{Q_k} |u(x) - u_{Q_k}|^q w(x)\,dx\right)^{\frac{1}{q}} \\ &\le C \liminf_{k\to\infty}\left(\int_{Q_k} |\nabla u(x)|^p\, v(x)\,dx\right)^{\frac{1}{p}} \\ &= C\left(\int_{\mathbb{R}^n} |\nabla u(x)|^p\, v(x)\,dx\right)^{\frac{1}{p}},\end{aligned}$$

where $C = C(n, p, q, w, \sigma, C_1)$. □

9.6. Local-to-global inequalities on open sets

The covering lemma below is proved by maximal function arguments.

LEMMA 9.26. *Let $1 \le p < \infty$ and $1 \le \tau < \infty$ and assume that w is a doubling weight in $\mathbb{R}^n$. Let $\Omega \subset \mathbb{R}^n$ be an open set with a Whitney decomposition $\mathcal{W}(\Omega) = \{Q_i : i \in \mathbb{N}\}$, where $Q_i = Q(x_i, r_i)$. There exists a constant $C = C(n, p, \tau, w)$ such that*

$$\left\|\sum_{i=1}^\infty a_i \chi_{Q(x_i,\tau r_i)}\right\|_{L^p(w\,dx)} \le C\left\|\sum_{i=1}^\infty a_i \chi_{Q_i}\right\|_{L^p(w\,dx)}$$

whenever $a_i \ge 0$ for every $i \in \mathbb{N}$.

PROOF. The case $p = 1$ follows from the fact that w is a doubling weight. Hence, we may assume that $1 < p < \infty$. By duality and density of compactly supported continuous functions in $L^{p'}(w\,dx)$, with $p' = \frac{p}{p-1}$, it suffices to verify that there exists a constant $C = C(n, p, \tau, w)$ such that

$$\left|\int_{\mathbb{R}^n} \sum_{i=1}^\infty a_i \chi_{Q(x_i,\tau r_i)}(y) g(y) w(y)\,dy\right| \le C\left\|\sum_{i=1}^\infty a_i \chi_{Q_i}\right\|_{L^p(w\,dx)} \tag{9.17}$$

for every $g \in C_0(\mathbb{R}^n)$ with $\|g\|_{L^{p'}(w\,dx)} \le 1$. Let $i \in \mathbb{N}$ and assume that g is as above. Then

$$\int_{Q(x_i,\tau r_i)} |g(y)| w(y)\,dy \le w(Q(x_i, \tau r_i)) M^{c,w} g(x) \tag{9.18}$$

for every $x \in Q(x_i, \tau r_i)$. Here $M^{c,w}$ is the weighted maximal operator as in Remark 8.4 (b). By integrating (9.18) over the cube $Q_i \subset Q(x_i, \tau r_i)$ and using the fact that w is doubling, we obtain

$$\begin{aligned}\int_{Q(x_i,\tau r_i)} |g(y)|w(y)\,dy &\le \frac{w(Q(x_i,\tau r_i)}{w(Q_i)} \int_{Q_i} M^{c,w}g(x)w(x)\,dx \\ &\le C(w,\tau)\int_{Q_i} M^{c,w}g(x)w(x)\,dx.\end{aligned}$$

By the triangle inequality we conclude

$$\begin{aligned}\left|\int_{\mathbb{R}^n} \sum_{i=1}^{\infty} a_i \chi_{Q(x_i,\tau r_i)}(y)g(y)w(y)\,dy\right| &\le \sum_{i=1}^{\infty} a_i \int_{Q(x_i,\tau r_i)} |g(y)|w(y)\,dy \\ &\le C(w,\tau)\sum_{i=1}^{\infty} a_i \int_{Q_i} M^{c,w}g(x)w(x)\,dx \\ &= C(w,\tau)\int_{\mathbb{R}^n} \sum_{i=1}^{\infty} a_i\chi_{Q_i}(x)M^{c,w}g(x)w(x)\,dx.\end{aligned}$$

The required estimate (9.17) follows by applying Hölder's inequality to the right hand side and using the boundedness of $M^{c,w}$ on $L^{p'}(w\,dx)$, see Remark 8.4 (b). $\square$

For a weight w in $\mathbb{R}^n$ and a measurable set $E \subset \mathbb{R}^n$ with $0 < w(E) < \infty$, we write

$$f_{E;w} = \frac{1}{w(E)}\int_E f(x)w(x)\,dx$$

for every $f \in L^1(E; w\,dx)$; compare to (8.14).

LEMMA 9.27. *Assume that w is a doubling weight in $\mathbb{R}^n$ and that $\Omega \subset \mathbb{R}^n$ is an open set with a Whitney decomposition $\mathcal{W}(\Omega)$. Let $Q_0, Q \in \mathcal{W}(\Omega)$ and let $\mathcal{C}(Q)$ be a chain in Ω joining Q_0 to Q. Then*

$$|f_{Q^*;w} - f_{Q_0^*;w}| \le C(n,w) \sum_{R\in\mathcal{C}(Q)} \frac{1}{w(R^*)}\int_{R^*} |f(x) - f_{R^*;w}|w(x)\,dx$$

for every $f \in L^1_{\mathrm{loc}}(\Omega; w\,dx)$.

PROOF. Let $\mathcal{C}(Q) = (Q_0, \dots, Q_k)$, with $Q = Q_k$, and let $f \in L^1_{\mathrm{loc}}(\Omega; w\,dx)$. Then

$$\left|f_{Q^*;w} - f_{Q_0^*;w}\right| \le \sum_{i=1}^{k} \left|f_{Q_i^*;w} - f_{Q_{i-1}^*;w}\right|.$$

Let $i \in \{1, \dots, k\}$. By (9.1) and the assumption that w is doubling, we obtain

$$\max\{w(Q_i^*), w(Q_{i-1}^*)\} \le C(n,w)w(Q_i^* \cap Q_{i-1}^*).$$

Thus

$$\begin{aligned}|f_{Q_i^*;w} - f_{Q_{i-1}^*;w}| &= |f_{Q_i^*;w} - f_{Q_i^*\cap Q_{i-1}^*;w} + f_{Q_i^*\cap Q_{i-1}^*;w} - f_{Q_{i-1}^*;w}| \\ &\le \frac{1}{w(Q_i^*\cap Q_{i-1}^*)}\int_{Q_i^*\cap Q_{i-1}^*} |f(x) - f_{Q_i^*;w}|w(x)\,dx \\ &\quad + \frac{1}{w(Q_i^*\cap Q_{i-1}^*)}\int_{Q_i^*\cap Q_{i-1}^*} |f(x) - f_{Q_{i-1}^*;w}|w(x)\,dx \\ &\le C(n,w)\sum_{j=i-1}^{i}\frac{1}{w(Q_j^*)}\int_{Q_j^*}|f(x) - f_{Q_j^*;w}|w(x)\,dx.\end{aligned}$$

Since $Q_{i_1} \neq Q_{i_2}$ for $i_1 \neq i_2$, we conclude

$$|f_{Q^*;w} - f_{Q_0^*;w}| \le 2C(n,w)\sum_{R\in\mathcal{C}(Q)}\frac{1}{w(R^*)}\int_{R^*}|f(x) - f_{R^*;w}|w(x)\,dx,$$

and the proof is complete. □

We are ready for a weighted local-to-global inequality.

THEOREM 9.28. *Let $1 \le p < \infty$ and let w be a doubling weight in $\mathbb{R}^n$. Assume that $\Omega \subset \mathbb{R}^n$ is an open set satisfying the Boman chain condition, with $\tau > 1$. There exists a constant $C = C(n,p,w,\tau)$ such that*

$$\inf_{c\in\mathbb{R}}\int_\Omega |f(x) - c|^p w(x)\,dx \le C\sum_{Q\in\mathcal{W}(\Omega)}\int_{Q^*}|f(x) - f_{Q^*;w}|^p w(x)\,dx \tag{9.19}$$

for every $f \in L^1_{\mathrm{loc}}(\Omega; w\,dx)$.

PROOF. Throughout this proof C is a constant that depends at most on the parameters p, n, w and τ. Let Q_0 be the fixed Whitney cube in the chain decomposition of Ω, with $\tau > 1$, see Definition 9.2. By the triangle inequality,

$$\begin{aligned}|f(x) - f_{Q_0^*;w}| &\le \Big|f(x) - \sum_{Q\in\mathcal{W}(\Omega)} f_{Q^*;w}\chi_Q(x)\Big| + \Big|\sum_{Q\in\mathcal{W}(\Omega)} f_{Q^*;w}\chi_Q(x) - f_{Q_0^*;w}\Big| \\ &= g_1(x) + g_2(x),\end{aligned}$$

for every $x \in \Omega$. This implies

$$\int_\Omega |f(x) - f_{Q_0^*;w}|^p w(x)\,dx \le 2^p\int_\Omega g_1(x)^p w(x)\,dx + 2^p\int_\Omega g_2(x)^p w(x)\,dx.$$

We estimate the integrals on the right-hand side separately. By the properties of the Whitney cubes we have

$$\begin{aligned}\int_\Omega g_1(x)^p w(x)\,dx &= \sum_{Q\in\mathcal{W}(\Omega)}\int_Q g_1(x)^p w(x)\,dx = \sum_{Q\in\mathcal{W}(\Omega)}\int_Q |f(x) - f_{Q^*;w}|^p w(x)\,dx \\ &\le \sum_{Q\in\mathcal{W}(\Omega)}\int_{Q^*}|f(x) - f_{Q^*;w}|^p w(x)\,dx.\end{aligned}$$

The second integral is more difficult to estimate. We begin with

$$\begin{aligned}\int_\Omega g_2(x)^p w(x)\,dx &= \int_\Omega \left| \sum_{Q\in\mathcal{W}(\Omega)} (f_{Q^*;w} - f_{Q_0^*;w})\chi_Q(x)\right|^p w(x)\,dx \\ &\le \int_{\mathbb{R}^n} \left(\sum_{Q\in\mathcal{W}(\Omega)} |f_{Q^*;w} - f_{Q_0^*;w}|\chi_Q(x)\right)^p w(x)\,dx.\end{aligned}$$

Define

$$a_R = \left(\frac{1}{w(R^*)} \int_{R^*} |f(x) - f_{R^*;w}|^p w(x)\,dx \right)^{\frac{1}{p}} \ge 0$$

for cubes $R = Q(x_R, r_R) \in \mathcal{W}(Q)$. By Lemma 9.27 and Hölder's inequality, we have

$$|f_{Q^*;w} - f_{Q_0^*;w}|\chi_Q \le C \sum_{R\in\mathcal{C}(Q)} a_R \chi_Q$$

for every $Q \in \mathcal{W}(\Omega)$. By summing these estimates over all Whitney cubes and applying (9.2), we obtain

$$\begin{aligned}\sum_{Q\in\mathcal{W}(\Omega)} |f_{Q^*;w} - f_{Q_0^*;w}|\chi_Q &\le C \sum_{Q\in\mathcal{W}(\Omega)} \sum_{R\in\mathcal{C}(Q)} a_R\chi_Q = C \sum_{R\in\mathcal{W}(\Omega)} a_R \sum_{Q\in\mathcal{S}(R)} \chi_Q \\ &\le C \sum_{R\in\mathcal{W}(\Omega)} a_R \chi_{Q(x_R,\tau r_R)}.\end{aligned}$$

An application of Lemma 9.26 shows that

$$\begin{aligned}\int_\Omega g_2(x)^p w(x)\,dx &\le C \int_{\mathbb{R}^n} \left(\sum_{R\in\mathcal{W}(\Omega)} a_R \chi_{Q(x_R,\tau r_R)}(x) \right)^p w(x)\,dx \\ &\le C \int_{\mathbb{R}^n} \left(\sum_{R\in\mathcal{W}(\Omega)} a_R \chi_R(x) \right)^p w(x)\,dx \\ &\le C \sum_{R\in\mathcal{W}(\Omega)} a_R^p \int_{\mathbb{R}^n} \chi_R(x) w(x)\,dx \\ &\le C \sum_{R\in\mathcal{W}(\Omega)} \frac{w(R)}{w(R^*)} \int_{R^*} |f(x) - f_{R^*;w}|^p w(x)\,dx \\ &\le C \sum_{R\in\mathcal{W}(\Omega)} \int_{R^*} |f(x) - f_{R^*;w}|^p w(x)\,dx,\end{aligned}$$

and the proof is complete. □

REMARK 9.29. Theorem 9.28 applies for $f \in L^1_{\text{loc}}(\Omega; w\,dx)$ even if $f_{\Omega;w}$ is not defined. However, if $f \in L^1(\Omega; w\,dx)$ and $1 \le p < \infty$, then

$$\int_\Omega |f(x) - f_{\Omega;w}|^p w(x)\,dx \le 2^p \inf_{c\in\mathbb{R}} \int_\Omega |f(x) - c|^p w(x)\,dx. \tag{9.20}$$

This follows, since

$$\begin{aligned}\Big(\int_\Omega |f(x)-f_{\Omega;w}|^p w(x)\,dx\Big)^{\frac1p} &\le \Big(\int_\Omega |f(x)-c|^p w(x)\,dx\Big)^{\frac1p} + w(\Omega)^{\frac1p}|f_{\Omega;w}-c|\\ &= \Big(\int_\Omega |f(x)-c|^p w(x)\,dx\Big)^{\frac1p} + w(\Omega)^{\frac1p}\Big|\frac{1}{w(\Omega)}\int_\Omega (f(x)-c)w(x)\,dx\Big|\\ &\le \Big(\int_\Omega |f(x)-c|^p w(x)\,dx\Big)^{\frac1p} + w(\Omega)^{\frac1p}\Big(\frac{1}{w(\Omega)}\int_\Omega |f(x)-c|^p w(x)\,dx\Big)^{\frac1p}\end{aligned}$$

for every $c \in \mathbb{R}$.

Recall that the dilated cubes Q^* appearing in Theorem 9.30 have a bounded overlap property.

THEOREM 9.30. *Let $1 \le p < \infty$ and $w \in A_\infty$. Assume that $\Omega \subset \mathbb{R}^n$ is an open set satisfying the Boman chain condition, with $\tau > 1$. There exists a constant $C = C(n,p,w,\tau)$ such that*

$$\inf_{c\in\mathbb{R}} \int_\Omega |f(x)-c|^p w(x)\,dx \le C \sum_{Q\in\mathcal{W}(\Omega)} \int_{Q^*} (M^{d,\sharp}_{Q^*} f(x))^p w(x)\,dx$$

for every $f \in \operatorname{Lip}_{\mathrm{loc}}(\Omega)$.

PROOF. Theorem 8.23 implies that w is a doubling weight. By Theorem 9.28, inequality (9.20), and Theorem 9.15, we obtain

$$\begin{aligned}\inf_{c\in\mathbb{R}} \int_\Omega |f(x)-c|^p w(x)\,dx &\le C \sum_{Q\in\mathcal{W}(\Omega)} \int_{Q^*} |f(x)-f_{Q^*;w}|^p w(x)\,dx\\ &\le 2^p C \sum_{Q\in\mathcal{W}(\Omega)} \int_{Q^*} |f(x)-f_{Q^*}|^p w(x)\,dx \qquad (9.21)\\ &\le 2^p C \sum_{Q\in\mathcal{W}(\Omega)} \int_{Q^*} (M^{d,\sharp}_{Q^*} f(x))^p w(x)\,dx,\end{aligned}$$

where $C = C(n,p,w,\tau)$. □

The following two-weight Poincaré inequality holds for open sets satisfying the Boman chain condition.

THEOREM 9.31. *Let $1 < p \le q < \infty$ and let (w,v) be a pair of weights with $w \in A_\infty$ and $\sigma = v^{\frac{1}{1-p}} \in A_\infty$. Assume that $\Omega \subset \mathbb{R}^n$ is an open set satisfying the Boman chain condition, with $\tau > 1$, and that there exists a constant C_1 such that*

$$\Big(\frac{1}{|Q'|^{1-\frac1n}}\Big)^p w(Q')^{\frac pq}\sigma(Q')^{p-1} \le C_1$$

for all cubes Q' satisfying $Q' \subset Q^*$ for some Whitney cube $Q \in \mathcal{W}(\Omega)$. Then there exists a constant $C = C(n,p,q,w,\sigma,\tau,C_1)$ such that

$$\inf_{c\in\mathbb{R}}\Big(\int_\Omega |u(x)-c|^q w(x)\,dx\Big)^{\frac1q} \le C\Big(\int_\Omega |\nabla u(x)|^p v(x)\,dx\Big)^{\frac1p}$$

for every $u \in \operatorname{Lip}_{\mathrm{loc}}(\Omega)$.

PROOF. Theorem 8.23 implies that $w \in A_\infty$ is a doubling weight. As in (9.21), we have

$$\inf_{c\in\mathbb{R}} \int_\Omega |u(x)-c|^q w(x)\,dx \le 2^p C(n,p,w,\tau) \sum_{Q\in\mathcal{W}(\Omega)} \int_{Q*} |u(x)-u_{Q^*}|^q w(x)\,dx.$$

Theorem 9.21 and the assumption $q \ge p$ imply

$$\begin{aligned}\inf_{c\in\mathbb{R}} \int_\Omega |u(x)-c|^q w(x)\,dx &\le C \sum_{Q\in\mathcal{W}(\Omega)} \left(\int_{Q^*} |\nabla u(x)|^p v(x)\,dx\right)^{\frac{q}{p}} \\ &\le C\left(\sum_{Q\in\mathcal{W}(\Omega)} \int_{Q^*} |\nabla u(x)|^p v(x)\,dx\right)^{\frac{q}{p}} \\ &\le C\left(\int_\Omega |\nabla u(x)|^p v(x)\,dx\right)^{\frac{q}{p}},\end{aligned}$$

where $C = C(n,p,q,w,\sigma,\tau,C_1)$. □

As a consequence of Theorem 9.31, we obtain a Sobolev–Poincaré inequality on open sets satisfying the Boman chain condition.

THEOREM 9.32. *Let $1 < p < n$ and $p^* = \frac{np}{n-p}$, and assume that $\Omega \subset \mathbb{R}^n$ is an open set satisfying the Boman chain condition, with $\tau > 1$. There exists a constant $C = C(n,p,\tau)$ such that*

$$\left(\int_\Omega |u(x)-u_\Omega|^{p^*}\,dx\right)^{\frac{1}{p^*}} \le C\left(\int_\Omega |\nabla u(x)|^p\,dx\right)^{\frac{1}{p}},$$

for every $u \in W^{1,p}(\Omega)$.

PROOF. Assume first that $u \in W^{1,p}(\Omega) \cap C^\infty(\Omega) \subset \mathrm{Lip}_{\mathrm{loc}}(\Omega)$. Then

$$\inf_{c\in\mathbb{R}} \left(\int_\Omega |u(x)-c|^{p^*}\,dx\right)^{\frac{1}{p^*}} \le C\left(\int_\Omega |\nabla u(x)|^p\,dx\right)^{\frac{1}{p}}$$

by Remark 9.23 and Theorem 9.31. Since $W^{1,p}(\Omega) \subset L^1(\Omega)$, in this case we obtain the claimed inequality from inequality (9.20). In the general case $u \in W^{1,p}(\Omega)$ the claim follows by approximation in $W^{1,p}(\Omega)$ and Fatou's lemma, compare to the proof of Theorem 3.14. □

9.7. BMO and John–Nirenberg inequality

As another application of the sparse domination technique in Lemma 9.13, we consider functions of bounded mean oscillation. The central result is the John–Nirenberg inequality, Theorem 9.39, which implies a self-improvement property for BMO functions.

DEFINITION 9.33. A function $f \in L^1_{\mathrm{loc}}(\mathbb{R}^n)$ belongs to the space of bounded mean oscillation $\mathrm{BMO}(\mathbb{R}^n)$ if

$$\|f\|_{\mathrm{BMO}(\mathbb{R}^n)} = \sup_Q \fint_Q |f(x)-f_Q|\,dx < \infty,$$

where the supremum is taken over all cubes $Q \subset \mathbb{R}^n$.

REMARK 9.34. If $f \in \mathrm{BMO}(\mathbb{R}^n)$, then $|f| \in \mathrm{BMO}(\mathbb{R}^n)$ and

$$\||f|\|_{\mathrm{BMO}(\mathbb{R}^n)} \le 2\|f\|_{\mathrm{BMO}(\mathbb{R}^n)}. \tag{9.22}$$

Indeed, if $Q \subset \mathbb{R}^n$ is a cube, then by (9.20) and the triangle inequality

$$\begin{aligned}\fint_Q \big||f(x)| - |f|_Q\big|\,dx &\le 2 \fint_Q \big||f(x)| - |f_Q|\big|\,dx \\ &\le 2 \fint_Q |f(x) - f_Q|\,dx \le 2\|f\|_{\mathrm{BMO}(\mathbb{R}^n)}.\end{aligned}$$

Inequality (9.22) follows by taking supremum over all cubes $Q \subset \mathbb{R}^n$.

REMARK 9.35. We note that $L^\infty(\mathbb{R}^n) \subset \mathrm{BMO}(\mathbb{R}^n)$ and

$$\|f\|_{\mathrm{BMO}(\mathbb{R}^n)} \le 2\|f\|_{L^\infty(\mathbb{R}^n)}.$$

This follows, since

$$\begin{aligned}\fint_Q |f(x) - f_Q|\,dx &\le \fint_Q (|f(x)| + |f_Q|)\,dx = \fint_Q |f(x)|\,dx + |f_Q| \\ &\le \fint_Q |f(x)|\,dx + |f|_Q \le 2 \fint_Q |f(x)|\,dx \le 2\|f\|_{L^\infty(\mathbb{R}^n)}\end{aligned}$$

for every cube $Q \subset \mathbb{R}^n$.

The following example gives an unbounded BMO function. Thus $L^\infty(\mathbb{R}^n)$ is a proper subset of $\mathrm{BMO}(\mathbb{R}^n)$.

EXAMPLE 9.36. We show that $\log|x| \in \mathrm{BMO}(\mathbb{R}^n)$. By Remark 9.29 it is enough to show that for every cube $Q(x,r)$, with $x \in \mathbb{R}^n$ and $r > 0$, there exists a constant $c_{Q(x,r)} \in \mathbb{R}$ such that

$$\sup_{Q(x,r)} \fint_{Q(x,r)} \big|\log|y| - c_{Q(x,r)}\big|\,dy < \infty.$$

We consider two cases.

First assume that $|x| < 2r\sqrt{n}$. In this case, we choose $c_{Q(x,r)} = \log(2r)$. By a change of variables $y = 2rz$, $dy = (2r)^n\,dz$, we obtain

$$\begin{aligned}\frac{1}{(2r)^n} \int_{Q(x,r)} \big|\log|y| - c_{Q(x,r)}\big|\,dy &= \int_{Q(\frac{x}{2r},\frac{1}{2})} \big|\log(2r|z|) - \log(2r)\big|\,dz \\ &= \int_{Q(\frac{x}{2r},\frac{1}{2})} \big|\log|z|\big|\,dz \le \int_{B(0,2\sqrt{n})} \big|\log|z|\big|\,dz = C(n) < \infty.\end{aligned}$$

The final estimate holds since if $z \in Q(\frac{x}{2r}, \frac{1}{2})$, then

$$|z| \le \frac{|x|}{2r} + \frac{\sqrt{n}}{2} < \sqrt{n} + \frac{\sqrt{n}}{2} \le 2\sqrt{n},$$

and thus $Q(\frac{x}{2r}, \frac{1}{2}) \subset B(0, 2\sqrt{n})$.

Assume then that $|x| \ge 2r\sqrt{n}$. In this case, we choose $c_{Q(x,r)} = \log|x|$. A change of variables, as above, gives

$$\begin{aligned}\frac{1}{(2r)^n} \int_{Q(x,r)} \big|\log|y| - c_{Q(x,r)}\big|\,dy &= \int_{Q(\frac{x}{2r},\frac{1}{2})} \big|\log|2rz| - \log|x|\big|\,dz \\ &= \int_{Q(\frac{x}{2r},\frac{1}{2})} \left|\log \frac{|2rz|}{|x|}\right|\,dz.\end{aligned} \tag{9.23}$$

Let $z \in Q(\frac{x}{2r}, \frac{1}{2})$. Then

$$\frac{|x|}{2r} - \frac{\sqrt{n}}{2} \le |z| \le \frac{|x|}{2r} + \frac{\sqrt{n}}{2}.$$

Thus

$$\frac{|2rz|}{|x|} \ge 1 - \frac{2r}{|x|}\frac{\sqrt{n}}{2} \ge 1 - \frac{1}{2} = \frac{1}{2}$$

and

$$\frac{|2rz|}{|x|} \le 1 + \frac{2r}{|x|}\frac{\sqrt{n}}{2} \le 1 + \frac{1}{2} < 2.$$

These estimates show that

$$-\log 2 = \log\frac{1}{2} \le \log\frac{|2rz|}{|x|} \le \log 2$$

and consequently

$$\left|\log\frac{|2rz|}{|x|}\right| \le \log 2$$

for every $z \in Q(\frac{x}{2r}, \frac{1}{2})$. This, together with (9.23), gives

$$\frac{1}{(2r)^n}\int_{Q(x,r)} \left|\log|y| - c_{Q(x,r)}\right| dy \le \int_{Q(\frac{x}{2r},\frac{1}{2})} \left|\log\frac{|2rz|}{|x|}\right| dz \le \log 2 < \infty.$$

We conclude that the mean oscillation of $\log|x|$ is uniformly bounded and thus $\log|x| \in \mathrm{BMO}(\mathbb{R}^n)$. Observe however that $\log|x| \notin L^\infty(\mathbb{R}^n)$.

We also need a local dyadic variant of BMO.

DEFINITION 9.37. Let $Q_0 \subset \mathbb{R}^n$ be a cube. A function $f \in L^1(Q_0)$ belongs to the space $\mathrm{BMO}^d(Q_0)$ of dyadic bounded mean oscillation if

$$\|f\|_{\mathrm{BMO}^d(Q_0)} = \sup_{Q\in\mathcal{D}(Q_0)} \fint_Q |f(y) - f_Q|\,dy < \infty,$$

where the supremum is taken over all dyadic subcubes of Q_0.

Let $M^{d,\sharp}_{Q_0} f$ be the dyadic sharp maximal function of $f \in L^1(Q_0)$, see Definition 9.14. Then

$$\|f\|_{\mathrm{BMO}^d(Q_0)} = \sup_{Q\in\mathcal{D}(Q_0)} \fint_Q |f(y) - f_Q|\,dy = \left\|M^{d,\sharp}_{Q_0} f\right\|_{L^\infty(Q_0)},$$

and thus $f \in \mathrm{BMO}^d(Q_0)$ if and only if $M^{d,\sharp}_{Q_0} f \in L^\infty(Q_0)$.

Functions in the Sobolev space $W^{1,n}$ are examples of BMO functions.

EXAMPLE 9.38. Let $Q_0 \subset \mathbb{R}^n$ be a cube and let $u \in W^{1,n}(Q_0)$. Then by Hölder's inequality $u \in W^{1,1}(Q_0)$. The 1-Poincaré inequality on cubes, see Corollary 9.6, and Hölder's inequality give

$$\begin{aligned}\fint_Q |u(x) - u_Q|\,dx &\le C(n)l(Q)\left(\fint_Q |\nabla u(x)|^n\,dx\right)^{\frac{1}{n}}\\ &= C(n)l(Q)|Q|^{-\frac{1}{n}}\left(\int_Q |\nabla u(x)|^n\,dx\right)^{\frac{1}{n}}\end{aligned}$$

for every $Q \in \mathcal{D}(Q_0)$. Since $l(Q)|Q|^{-\frac{1}{n}} = 1$ and $Q \subset Q_0$ if $Q \in \mathcal{D}(Q_0)$, we have

$$\begin{aligned}\|u\|_{\mathrm{BMO}^d(Q_0)} &= \sup_{Q\in\mathcal{D}(Q_0)} \fint_Q |u(x) - u_Q|\,dx \\ &\le C(n)\left(\int_{Q_0} |\nabla u(x)|^n\,dx\right)^{\frac{1}{n}} \le C(n)\|u\|_{W^{1,n}(Q_0)}.\end{aligned}$$

Thus $W^{1,n}(Q_0) \subset \mathrm{BMO}^d(Q_0)$, and the inclusion is a bounded linear operator. In a similar fashion, it can be shown that $W^{1,n}(\mathbb{R}^n) \subset \mathrm{BMO}(\mathbb{R}^n)$.

We prove a weighted John–Nirenberg inequality for the dyadic BMO.

THEOREM 9.39. *Assume that $w \in A_\infty$. Let $Q_0 \subset \mathbb{R}^n$ be a cube and assume that $f \in \mathrm{BMO}^d(Q_0)$. There exists a constant $C = C(n, w)$ such that*

$$\frac{w(\{x \in Q_0 : |f(x) - f_{Q_0}| > t\})}{w(Q_0)} \le C \exp\left(-\frac{t}{C\|f\|_{\mathrm{BMO}^d(Q_0)}}\right)$$

for every $t > 0$.

PROOF. We may assume that $\|f\|_{\mathrm{BMO}^d(Q_0)} > 0$. Let $\mathcal{S} \subset \mathcal{D}(Q_0)$ be the sparse collection of cubes given by Lemma 9.13, associated to the function $f \in \mathrm{BMO}^d(Q_0)$. For $j \in \mathbb{N}_0$, let $\mathcal{S}_j$ denote the collection of those pairwise disjoint cubes $S \in \mathcal{S}$ for which there is a sequence $S = S_j \subsetneq \cdots \subsetneq S_0 = Q_0$ such that $S_i \in \mathrm{ch}_{\mathcal{S}}(S_{i-1})$ for every $i = 1, \ldots, j$. Observe that

$$\mathcal{S}_0 = \{Q_0\}, \qquad \mathcal{S}_1 = \mathrm{ch}_{\mathcal{S}}(Q_0), \qquad \text{and} \qquad \mathcal{S}_j = \bigcup_{Q\in\mathcal{S}_{j-1}} \mathrm{ch}_{\mathcal{S}}(Q)$$

for $j \in \mathbb{N}$. Let $x \in Q_0$ be a Lebesgue point of f. If $|f(x) - f_{Q_0}| > t$, then by Lemma 9.13 (a),

$$t < |f(x) - f_{Q_0}| \le C_1 \sum_{Q\in\mathcal{S}} \chi_Q(x) \fint_Q |f(y) - f_Q|\,dy \le C_1\|f\|_{\mathrm{BMO}^d(Q_0)} \sum_{Q\in\mathcal{S}} \chi_Q(x),$$

where $C_1 = C_1(n, w) > 0$. Let $j \in \mathbb{N}_0$ be such that

$$j < \frac{t}{C_1\|f\|_{\mathrm{BMO}^d(Q_0)}} \le j + 1.$$

Then $j < \sum_{Q\in\mathcal{S}} \chi_Q(x)$, and thus x belongs to at least $j + 1$ cubes in $\mathcal{S}$. It follows that $x \in S$ for some $S \in \mathcal{S}_j$, and therefore

$$w(\{x \in Q_0 : |f(x) - f_{Q_0}| > t\}) \le w\left(\bigcup_{S\in\mathcal{S}_j} S\right) = \sum_{S\in\mathcal{S}_j} w(S).$$

Here we also used the fact that $w\,dx$-almost every point in Q_0 is a Lebesgue point of f. If $j > 0$ then, by Lemma 9.13 (b),

$$\sum_{S\in\mathcal{S}_j} w(S) = \sum_{Q\in\mathcal{S}_{j-1}} \sum_{S\in\mathrm{ch}_{\mathcal{S}}(Q)} w(S) \le \sum_{Q\in\mathcal{S}_{j-1}} w(Q \setminus E_Q) \le (1-\kappa) \sum_{Q\in\mathcal{S}_{j-1}} w(Q).$$

By iterating this estimate, and using the fact that $\mathcal{S}_0 = \{Q_0\}$, we obtain

$$\sum_{S\in\mathcal{S}_j} w(S) \le (1-\kappa)^j w(Q_0) = \frac{\exp\left(-(j+1)\log\frac{1}{1-\kappa}\right)}{1-\kappa} w(Q_0).$$

The desired conclusion, with $C = \max\left\{\frac{1}{1-\kappa}, \frac{C_1}{\log\frac{1}{1-\kappa}}\right\}$, follows by combining the estimates above. □

The John–Nirenberg inequality implies a self-improvement property for the dyadic BMO. This result holds for all A_∞-weights. We remark that inequality (9.24) also follows from Theorem 9.15.

THEOREM 9.40. *Let $1 < p < \infty$ and assume that $w \in A_\infty$. Let $Q_0 \subset \mathbb{R}^n$ be a cube and let $f \in \mathrm{BMO}^d(Q_0)$. There exists a constant $C = C(n, p, w)$ such that*

$$\frac{1}{w(Q_0)} \int_{Q_0} |f(x) - f_{Q_0}|^p \, w(x) dx \le C \|f\|^p_{\mathrm{BMO}^d(Q_0)}. \tag{9.24}$$

PROOF. Let $C_1 = C_1(n, w) > 0$ be the constant as in Theorem 9.39. By Theorem 9.39 and Fubini's theorem,

$$\begin{aligned}
&\frac{1}{w(Q_0)} \int_{Q_0} |f(x) - f_{Q_0}|^p w(x)\, dx \\
&\quad \le \frac{p}{w(Q_0)} \int_0^\infty t^{p-1} w(\{x \in Q_0 : |f(x) - f_{Q_0}| > t\})\, dt \\
&\quad \le pC_1 \int_0^\infty t^{p-1} \exp\left(- \frac{t}{C_1 \|f\|_{\mathrm{BMO}^d(Q_0)}} \right) dt \\
&\quad \le pC_1^{1+p} \|f\|^p_{\mathrm{BMO}^d(Q_0)} \int_0^\infty t^{p-1} e^{-t}\, dt,
\end{aligned}$$

and the claim follows. □

We also obtain a weighted exponential integrability result for dyadic BMO functions.

THEOREM 9.41. *Assume that $w \in A_\infty$. Let Q_0 be a cube and let $f \in \mathrm{BMO}^d(Q_0)$. There exists a constant $C = C(n, w)$ such that*

$$\frac{1}{w(Q_0)} \int_{Q_0} \exp\left(\frac{|f(x) - f_{Q_0}|}{C \|f\|_{\mathrm{BMO}^d(Q_0)}} \right) w(x)\, dx \le C.$$

PROOF. Let $C_1 = C_1(n, w) > 0$ be the constant in Theorem 9.39 and define

$$g(x) = \frac{|f(x) - f_{Q_0}|}{2C_1 \|f\|_{\mathrm{BMO}^d(Q_0)}}$$

for every $x \in Q_0$. By Fubini's theorem and Theorem 9.39,

$$\begin{aligned}
\frac{1}{w(Q_0)} \int_{Q_0} e^{g(x)} w(x)\, dx &= 1 + \frac{1}{w(Q_0)} \int_{Q_0} (e^{g(x)} - 1) w(x)\, dx \\
&= 1 + \frac{1}{w(Q_0)} \int_0^\infty e^t w(\{x \in Q_0 : g(x) > t\})\, dt \\
&\le 1 + C_1 \int_0^\infty e^{-t}\, dt = C_2 < \infty.
\end{aligned}$$

The proof is complete, with $C = \max\{2C_1, C_2\}$. □

By Example 9.38 we immediately obtain the following corollary of Theorem 9.40 and Theorem 9.41. Part (b) of Corollary 9.42 is a version of Trudinger's inequality.

COROLLARY 9.42. *Assume that $w \in A_\infty$. Let Q_0 be a cube and let $u \in W^{1,n}(Q_0)$.*

(a) *Let $1 < p < \infty$. There exists a constant $C = C(n,p,w)$ such that*

$$\left(\frac{1}{w(Q_0)} \int_{Q_0} |u(x) - u_{Q_0}|^p \, w(x) dx\right)^{\frac{1}{p}} \le C\left(\int_{Q_0} |\nabla u(x)|^n \, dx\right)^{\frac{1}{n}}.$$

(b) *There exists a constant $C = C(n,w)$ such that*

$$\frac{1}{w(Q_0)} \int_{Q_0} \exp\left(\frac{|u(x) - u_{Q_0}|}{C\|\nabla u\|_{L^n(Q_0)}}\right) w(x) \, dx \le C.$$

9.8. Maximal functions and BMO

The main result of this section is a boundedness result for a maximal operator on $\mathrm{BMO}(\mathbb{R}^n)$, see Theorem 9.46. For this purpose, we first study the connection between Muckenhoupt weights and BMO functions.

LEMMA 9.43. *Assume that $w \in A_\infty$. Then $\log w \in \mathrm{BMO}(\mathbb{R}^n)$. Moreover, if $w \in A_2$, then*

$$\|\log w\|_{\mathrm{BMO}(\mathbb{R}^n)} \le \log(2[w]_{A_2}).$$

PROOF. By Theorem 8.25, we have $w \in A_p$ for some $1 < p < \infty$. By using Lemma 8.13 and replacing w with $w^{\frac{1}{1-p}} \in A_{p'}$ if $p > 2$, we may assume that $w \in A_2$. Let $Q \subset \mathbb{R}^n$ be a cube and let $f = \log w$. Then $\exp f = w \in A_2$, and thus

$$\fint_Q \exp(f(x)) \, dx \fint_Q \exp(-f(x)) \, dx \le [w]_{A_2}.$$

Multiplying this inequality by $\exp(-f_Q)\exp(f_Q) = 1$ gives

$$\fint_Q \exp(f(x) - f_Q) \, dx \fint_Q \exp(f_Q - f(x)) \, dx \le [w]_{A_2}.$$

By Jensen's inequality

$$1 = \exp(0) = \exp\left(\fint_Q (f_Q - f(x)) \, dx\right) \le \fint_Q \exp(f_Q - f(x)) \, dx,$$

and hence

$$\fint_Q \exp(f(x) - f_Q) \, dx \le [w]_{A_2}\left(\fint_Q \exp(f_Q - f(x)) \, dx\right)^{-1} \le [w]_{A_2}.$$

In a similar way we obtain

$$\fint_Q \exp(f_Q - f(x)) \, dx \le [w]_{A_2}.$$

Thus,

$$\begin{aligned}\fint_Q \exp|f(x) - f_Q| \, dx &= \fint_Q \max\{\exp(f(x) - f_Q), \exp(f_Q - f(x))\} \, dx \\ &\le \fint_Q \exp(f(x) - f_Q) \, dx + \fint_Q \exp(f_Q - f(x)) \, dx \le 2[w]_{A_2},\end{aligned}$$

and Jensen's inequality implies

$$\exp\left(\fint_Q |f(x) - f_Q| \, dx\right) \le \fint_Q \exp|f(x) - f_Q| \, dx \le 2[w]_{A_2}.$$

After taking logarithms and then supremum over all cubes $Q \subset \mathbb{R}^n$, we conclude that $f = \log w \in \mathrm{BMO}(\mathbb{R}^n)$ with $\|f\|_{\mathrm{BMO}(\mathbb{R}^n)} \le \log(2[w]_{A_2})$. □

The following result is a converse of Lemma 9.43. The proof is based on the John–Nirenberg inequality.

THEOREM 9.44. *Assume that* $f \in \mathrm{BMO}(\mathbb{R}^n)$. *There exists* $\delta = \delta(n, f) > 0$ *such that* $\exp(\delta f) \in A_2$.

PROOF. We may assume that $\|f\|_{\mathrm{BMO}(\mathbb{R}^n)} > 0$. Let $C = C(n)$ be the constant in Theorem 9.41 with $w = 1$ in $\mathbb{R}^n$. Let $Q \subset \mathbb{R}^n$ be a cube and define $g = \delta f$ with

$$\delta = \frac{1}{C\|f\|_{\mathrm{BMO}(\mathbb{R}^n)}}.$$

Since $\|f\|_{\mathrm{BMO}^d(Q)} \le \|f\|_{\mathrm{BMO}(\mathbb{R}^n)}$, Theorem 9.41 implies

$$\fint_Q \exp(|g(x) - g_Q|)\,dx = \fint_Q \exp(\delta|f(x) - f_Q|)\,dx \le C.$$

Thus

$$\fint_Q \exp(g(x) - g_Q)\,dx \fint_Q \exp(g_Q - g(x))\,dx \le C^2,$$

and multiplying this inequality by $\exp(g_Q)\exp(-g_Q) = 1$ gives

$$\fint_Q \exp(g(x))\,dx \fint_Q \exp(-g(x))\,dx \le C^2.$$

This is the A_2 condition for $\exp g$ on the cube Q. The claim follows by taking supremum over all cubes $Q \subset \mathbb{R}^n$. □

REMARK 9.45. The parameter $\delta > 0$ is necessary in Theorem 9.44. Indeed, we have $-n\log|x| \in \mathrm{BMO}(\mathbb{R}^n)$ but $|x|^{-n} = \exp(-n\log|x|) \notin A_2$.

Using Lemma 9.43, Theorem 9.41, which was based on the John–Nirenberg inequality, and the Coifman–Rochberg lemma, see Theorem 8.33, we obtain the following boundedness result for the maximal operator on $\mathrm{BMO}(\mathbb{R}^n)$.

THEOREM 9.46. *Let* $f \in \mathrm{BMO}(\mathbb{R}^n)$ *and assume that* $M^c f$ *is finite almost everywhere in* $\mathbb{R}^n$. *Then* $M^c f \in \mathrm{BMO}(\mathbb{R}^n)$ *and there exists a constant* $C = C(n)$ *such that*

$$\|M^c f\|_{\mathrm{BMO}(\mathbb{R}^n)} \le C\|f\|_{\mathrm{BMO}(\mathbb{R}^n)}.$$

PROOF. We may assume that $\|f\|_{\mathrm{BMO}(\mathbb{R}^n)} > 0$, and by Remark 9.34, we may also assume that $f \ge 0$. Let $C_1 = C_1(n)$ be the constant in Theorem 9.41 with $w = 1$ in $\mathbb{R}^n$. Let $Q \subset \mathbb{R}^n$ be a cube and define $g = \delta f$ with

$$\delta = \frac{1}{C_1\|f\|_{\mathrm{BMO}(\mathbb{R}^n)}}.$$

Theorem 9.41 implies

$$\fint_Q \exp(g(x) - g_Q)\,dx \le \fint_Q \exp(|g(x) - g_Q|)\,dx = \fint_Q \exp(\delta|f(x) - f_Q|)\,dx \le C_1.$$

By Jensen's inequality,

$$\exp(g_Q) \le \fint_Q \exp(g(x))\,dx \le C_1 \exp(g_Q).$$

This holds for all cubes $Q \subset \mathbb{R}^n$, and thus

$$\exp(M^c g(x)) \le M^c(\exp(g))(x) \le C_1 \exp(M^c g(x)) \tag{9.25}$$

for every $x \in \mathbb{R}^n$. Since $\exp(g) \in L^1_{\mathrm{loc}}(\mathbb{R}^n)$ and $M^c g(x) = \delta M^c f(x) < \infty$ for almost every $x \in \mathbb{R}^n$, Theorem 8.33 (a) shows that

$$w = \big(M^c(\exp(g))\big)^{\frac{1}{2}}$$

is an A_1 weight with $[w]_{A_1} \le C_2 = C_2(n)$. Let

$$v = (\exp(M^c g))^{\frac{1}{2}} = \exp\big(\tfrac{1}{2} M^c g\big).$$

The inequalities in (9.25) imply

$$\begin{aligned}\left(\fint_Q v(x)\,dx\right) \operatorname*{ess\,sup}_{x\in Q} \frac{1}{v(x)} &\le C_1^{\frac{1}{2}} \left(\fint_Q w(x)\,dx\right) \operatorname*{ess\,sup}_{x\in Q} \frac{1}{w(x)} \\ &\le C_1^{\frac{1}{2}} [w]_{A_1} \le C(C_1, C_2),\end{aligned}$$

for every cube $Q \subset \mathbb{R}^n$. Thus v is an A_1 weight and $[v]_{A_1} \le C(C_1, C_2)$. By Lemma 9.43 and the fact that $v \in A_2$ with $[v]_{A_2} \le [v]_{A_1}$, we obtain

$$\frac{\|M^c g\|_{\mathrm{BMO}(\mathbb{R}^n)}}{2} = \|\log v\|_{\mathrm{BMO}(\mathbb{R}^n)} \le \log(2[v]_{A_2}) \le \log(2[v]_{A_1}) \le C(C_1, C_2).$$

Hence,

$$\|M^c f\|_{\mathrm{BMO}(\mathbb{R}^n)} = \frac{\|M^c g\|_{\mathrm{BMO}(\mathbb{R}^n)}}{\delta} \le \frac{2C(C_1, C_2)}{\delta} = C(n)\|f\|_{\mathrm{BMO}(\mathbb{R}^n)}. \qquad \square$$

9.9. Notes

Sobolev–Poincaré type inequalities on open subsets of $\mathbb{R}^n$ have been studied by Bojarski [**49**], Gol′dshteĭn and Reshetnyak [**154**], Martio [**310**] and Reshetnyak [**353**]. See also Maz′ya [**318, 321**]. Chaining arguments can be found in Bojarski [**49**], Boman [**53**], Hajłasz [**167**], Hajłasz and Koskela [**171**], Hurri-Syrjänen [**194, 195**], Iwaniec and Nolder [**208**], Jones [**212, 214**], Ohtsuka [**338**], Smith and Stegenga [**364**] and Staples [**366, 367**]. Section 9.1 is based on Chua [**85**] and Iwaniec and Nolder [**208**]. Theorem 9.11 has been studied by Fabes, Kenig and Serapioni [**124**]. The argument in Section 9.2 is based on Chiarenza and Frasca [**83**]. See also Heinonen, Kilpeläinen and Martio [**187**, 15.20]. Characterization of local-to-global inequality in Theorem 9.32 on domains satisfying a separation property has been given by Buckley and Koskela [**63**], see also Dyda, Ihnatsyeva and Vähäkangas [**112**].

A nonweighted version of Lemma 9.13 can be found in Lerner, Ombrosi and Rivera-Ríos [**269**]. Theorem 9.15 is a version of an inequality of Fefferman and Stein [**128**]. See also Duoandikoetxea [**107**, Lemma 6.9], García-Cuerva and Rubio de Francia [**139**, Theorem 3.6], Grafakos [**157**, Theorem 7.4.5], Journé [**216**, p. 41], Stein [**371**, Corollary 1, p. 154]. For the corresponding result with a good lambda inequality, we refer to Strömberg and Torchinsky [**376**, Chapter 3]. A local version of the Fefferman–Stein inequality has been studied in Iwaniec [**206**], Lerner [**268**] and Wik [**399**]. We follow Kurki and Vähäkangas [**252**].

The necessity in Lemma 9.18 is discussed in Cruz-Uribe [**91**] and Muckenhoupt and Wheeden [**335**]. The proof of Theorem 9.20 is adapted from Pérez [**344**]. The A_∞-condition in Theorem 9.20 can be relaxed, see Pérez [**345**] and Cruz-Uribe and Moen [**93**]. Section 9.5 and Section 9.6 are based on Kurki and Vähäkangas [**252**];

see also Chanillo, Strömberg and Wheeden [**80**], Chanillo and Wheeden [**81**], Chua [**85**] and Cruz-Uribe, Martell and Pérez [**92**]. For applications of weighted norm inequalities in the theory of Schrödinger equations, see Chang, Wilson and Wolff [**79**] and Fefferman [**127**]. For Lemma 9.26 we refer to Strömberg and Torchinsky [**375**], [**376**, Lemma 4, p. 115]. See also Bojarski [**49**, Lemma 4.2], Boman [**53**], Chua [**85**], Iwaniec and Nolder [**208**, Lemma 4] and Staples [**367**]. For related local-to-global results we refer to Hurri-Syrjänen [**196**], Hurri-Syrjänen, Marola and Vähäkangas [**197**], Maasalo [**294**], Marola and Saari [**308**], Reimann and Rychener [**352**], Smith and Stegenga [**364**] and Staples [**366, 367**].

The John–Nirenberg lemma can be found in Bennet and Sharpley [**28**], Duoandikoetxea [**107**], García-Cuerva and Rubio de Francia [**139**], Garnett [**142**], Grafakos [**157**], Journé [**216**], Lu, Ding and Yan [**286**], Stein [**371**] and Torchinsky [**383**]. The weighted John–Nirenberg theorem has been studied by Muckenhoupt and Wheeden [**336**]. See also Ombrosi, Pérez, Rivera-Ríos and Rela [**339**] and Canto and Pérez [**72**]. Corollary 9.42 (b) in the unweighted case can be found in Bojarski and Iwaniec [**52**]. See also Hajłasz and Koskela [**171**, Section 6]. John–Nirenberg lemma in more general contexts has been discussed in Buckley [**61**], Hytönen [**199**], Kronz [**247**] and Mateu, Mattila, Nicolau and Orobitg [**313**]. See also Björn and Björn [**31**, Section 3.3], Heinonen, Kilpeläinen and Martio [**187**] and Strömberg and Torchinsky [**376**, Chapter 3].

Boundedness of the Hardy–Littlewood maximal function in BMO has been studied by Bennett, DeVore and Sharpley [**27**]. See also Bennett and Sharpley [**28**, Theorem 7.18, p. 389]. The proof of Theorem 9.46 is by Chiarenza and Frasca [**84**]. See also Aalto and Kinnunen [**1**].

CHAPTER 10

Distance Weights and Hardy–Sobolev Inequalities

This chapter discusses topics related to the distance function. We begin with analytic and geometric conditions which guarantee that a power of the distance function is a Muckenhoupt weight. By an application of general weighted Poincaré inequalities from Chapter 9, we obtain Hardy–Sobolev and Hardy–Sobolev–Poincaré inequalities for distance weights. We give sufficient and necessary conditions for this kind of inequalities through dimensional conditions and we also characterize Hardy–Sobolev inequalities in terms of the properties of the variational capacity.

10.1. Aikawa condition

The following integrability condition for the distance function is closely related to Muckenhoupt properties.

DEFINITION 10.1. A nonempty set $E \subset \mathbb{R}^n$ satisfies the Aikawa condition for $\alpha \in \mathbb{R}$ if there exists a constant C such that

$$\fint_{B(x,r)} d(y,E)^{-\alpha}\,dy \le Cr^{-\alpha} \tag{10.1}$$

for every $x \in E$ and $r > 0$. We use here the conventions $0^0 = 1$ and $0^{-\alpha} = \infty$ for $\alpha > 0$. We denote by $\mathcal{A}(E)$ the set of $\alpha \in \mathbb{R}$ for which E satisfies the Aikawa condition.

Since $d(y,E) = d(y,\overline{E})$ for every $y \in \mathbb{R}^n$, it is immediate that $\alpha \in \mathcal{A}(E)$ if and only if $\alpha \in \mathcal{A}(\overline{E})$. Note that if $0 < \alpha \in \mathcal{A}(E)$, then $|\overline{E}| = 0$.

EXAMPLE 10.2. Let $E = \{0\} = \{(0,0,\dots,0)\} \subset \mathbb{R}^n$, $\alpha \in \mathbb{R}$ and $r > 0$. By integrating over spheres, see (3.1), we obtain

$$\begin{aligned}\int_{B(0,r)} d(y,E)^{-\alpha}\,dy &= \int_{B(0,r)} |y|^{-\alpha}\,dy = C(n)\int_0^r \rho^{-\alpha}\rho^{n-1}\,d\rho\\ &= \begin{cases} C(n,\alpha)r^{n-\alpha}, & \text{if } \alpha < n,\\ \infty, & \text{if } \alpha \ge n.\end{cases}\end{aligned}$$

This shows that $\alpha \in \mathcal{A}(E)$ if and only if $\alpha < n$.

REMARK 10.3. Let $E \subset \mathbb{R}^n$ be a nonempty set. It is easy to see that E satisfies the Aikawa condition for every $\alpha \le 0$. On the other hand, for $\alpha \ge n$, we have

$$\fint_{B(x,r)} d(y,E)^{-\alpha}\,dy \ge \fint_{B(x,r)} |y-x|^{-\alpha}\,dy = \infty$$

for every $x \in E$ and $r > 0$, as in Example 10.2. This shows that E does not satisfy the Aikawa condition for any $\alpha \ge n$. Thus we may restrict our attention to the range $0 < \alpha < n$ in the Aikawa condition.

For a bounded set it is sufficient to require the Aikawa condition for small radii only.

LEMMA 10.4. *Let $0 < \alpha < n$ and let $E \subset \mathbb{R}^n$ be a nonempty bounded set. Assume that* (10.1) *holds with a constant C_1 whenever $x \in E$ and $0 < r < \frac{1}{4}\operatorname{diam}(E)$. Then* (10.1) *holds for every $x \in E$ and $r > 0$ with a constant $C = C(n, \alpha, C_1)$.*

PROOF. By Example 10.2, we may assume that $\operatorname{diam}(E) > 0$. Let $x \in E$ and write $R = \frac{1}{4}\operatorname{diam}(E)$. We show that (10.1) holds for every $r \geq R$ with a constant $C = C(n, \alpha, C_1)$. First assume that $R \leq r \leq 8R = 2\operatorname{diam}(E)$ and let $A = \{y \in \mathbb{R}^n : d(y, E) < \frac{R}{10}\}$. By the boundedness of E and Lemma 1.13, there exist $N = N(n) \in \mathbb{N}$ and points $x_1, \ldots, x_N \in E$ such that $A \subset \bigcup_{j=1}^{N} B(x_j, \frac{R}{2})$. Since (10.1) holds for all radii up to R, we obtain

$$\int_{B(x,r)\cap A} d(y,E)^{-\alpha}\,dy \leq \sum_{j=1}^{N} \int_{B(x_j,\frac{R}{2})} d(y,E)^{-\alpha}\,dy$$
$$\leq C(n) N C_1 2^{\alpha-n} R^{n-\alpha} \leq C(n,\alpha,C_1) r^{n-\alpha}.$$

On the other hand, for $y \in B(x,r) \setminus A$ we have $d(y,E) \geq \frac{R}{10}$, and thus

$$\int_{B(x,r)\setminus A} d(y,E)^{-\alpha}\,dy \leq C(n,\alpha) r^n R^{-\alpha} \leq C(n,\alpha) r^{n-\alpha}.$$

A combination of the two estimates above implies

$$\fint_{B(x,r)} d(y,E)^{-\alpha}\,dy \leq C(n,\alpha,C_1) r^{-\alpha},$$

for every $R \leq r \leq 8R$.

Then assume that $8R < r < \infty$. By the beginning of the proof we have

$$\int_{B(x,8R)} d(y,E)^{-\alpha}\,dy \leq C(n,\alpha,C_1)(8R)^{n-\alpha} \leq C(n,\alpha,C_1) r^{n-\alpha}.$$

If $y \in B(x,r) \setminus B(x,8R)$, then $d(y,E) \geq 4R = \operatorname{diam}(E)$, and thus

$$|x-y| \leq \operatorname{diam}(E) + d(y,E) \leq 2d(y,E).$$

As in Example 10.2, we obtain

$$\int_{B(x,r)\setminus B(x,8R)} d(y,E)^{-\alpha}\,dy \leq C(\alpha) \int_{B(x,r)} |x-y|^{-\alpha}\,dy \leq C(n,\alpha) r^{n-\alpha}.$$

We conclude that (10.1) holds for every $r \geq R$, with a constant $C = C(n,\alpha,C_1)$, and the proof is complete. □

In certain cases it is more convenient to use cubes instead of balls in the Aikawa condition.

LEMMA 10.5. *Let $\alpha \in \mathbb{R}$ and assume that $E \subset \mathbb{R}^n$ is a nonempty set. Then $\alpha \in \mathcal{A}(E)$ if and only if there exists a constant C such that*

$$\fint_{Q(x,r)} d(y,E)^{-\alpha}\,dy \leq C r^{-\alpha} \tag{10.2}$$

for all cubes $Q(x,r) \subset \mathbb{R}^n$ that intersect E. Moreover, the constants in (10.1) *and* (10.2) *only depend on each other, n and α.*

PROOF. Assume that (10.1) holds with a constant C_1. Let $Q(x,r)$ be a cube that intersects E and let $z \in Q(x,r) \cap E$. Then $Q(x,r) \subset B(z, 2\sqrt{n}r)$ and $|Q(x,r)| = C(n)|B(z, 2\sqrt{n}r)|$. By (10.1) we have

$$\begin{aligned}\fint_{Q(x,r)} d(y,E)^{-\alpha}\,dy &\le C(n) \fint_{B(z,2\sqrt{n}r)} d(y,E)^{-\alpha}\,dy \\ &\le C(n)C_1(2\sqrt{n}r)^{-\alpha} = C(n,\alpha,C_1)r^{-\alpha}.\end{aligned}$$

This shows that (10.2) holds.

Then assume that (10.2) holds with a constant C_1, and let $B(x,r)$ be a ball with $x \in E$ and $r > 0$. Since $B(x,r) \subset Q(x,r)$ and $|Q(x,r)| = C(n)|B(x,r)|$, we have

$$\fint_{B(x,r)} d(y,E)^{-\alpha}\,dy \le C(n) \fint_{Q(x,r)} d(y,E)^{-\alpha}\,dy \le C(n,C_1)r^{-\alpha},$$

and thus (10.1) holds. □

REMARK 10.6. If $\alpha \in \mathcal{A}(E)$, then $\alpha' \in \mathcal{A}(E)$ for every $\alpha' < \alpha$. To see this, assume that (10.2) holds with a constant C_1 and let $Q(x,r) \subset \mathbb{R}^n$ be a cube that intersects E. Then

$$\begin{aligned}\fint_{Q(x,r)} d(y,E)^{-\alpha'}\,dy &= \fint_{Q(x,r)} d(y,E)^{(\alpha-\alpha')-\alpha}\,dy \\ &\le C(n,\alpha,\alpha')r^{\alpha-\alpha'} \fint_{Q(x,r)} d(y,E)^{-\alpha}\,dy \le C(n,\alpha,\alpha',C_1)r^{-\alpha'}.\end{aligned}$$

It is significantly more difficult to show that if $\alpha \in \mathcal{A}(E)$ for some $0 < \alpha < n$, then there exists α', with $\alpha < \alpha' < n$, such that $\alpha' \in \mathcal{A}(E)$. The proof of this self-improvement result for the Aikawa condition is based on the self-improvement properties of reverse Hölder inequalities.

LEMMA 10.7. *Let $E \subset \mathbb{R}^n$ and assume that $\alpha \in \mathcal{A}(E)$ for some $0 < \alpha < n$ with a constant C_1. If $0 < \beta < \alpha$, then there exists a constant $C = C(n,\alpha,\beta,C_1)$ such that*

$$\left(\fint_{Q(x,r)} d(y,E)^{-\alpha}\,dy\right)^{\frac{\beta}{\alpha}} \le C \fint_{Q(x,r)} d(y,E)^{-\beta}\,dy \tag{10.3}$$

for every cube $Q(x,r) \subset \mathbb{R}^n$.

PROOF. Let $0 < \beta < \alpha$ and assume first that $Q(x,2r) \cap E \neq \emptyset$. Then $d(y,E)^{\beta} \le C(n,\beta)r^{\beta}$ for every $y \in Q(x,r)$. By (10.2), we obtain

$$\begin{aligned}\fint_{Q(x,r)} d(y,E)^{-\alpha}\,dy &\le C(n) \fint_{Q(x,2r)} d(y,E)^{-\alpha}\,dy \le C(n,\alpha,C_1)\left(r^{-\beta}\right)^{\frac{\alpha}{\beta}} \\ &\le C(n,\alpha,\beta,C_1)\left(\fint_{Q(x,r)} d(y,E)^{-\beta}\,dy\right)^{\frac{\alpha}{\beta}},\end{aligned}$$

and (10.3) follows. On the other hand, if $Q(x,2r) \cap E = \emptyset$ and $y \in Q(x,r)$, then

$$C(n)d(y,E) \le d(Q(x,r),E) \le d(y,E).$$

This implies (10.3) in the case $Q(x,2r) \cap E = \emptyset$. □

The following result shows that the Aikawa condition is self-improving.

THEOREM 10.8. *Let $E \subset \mathbb{R}^n$ and $0 < \alpha < n$. Assume that $\alpha \in \mathcal{A}(E)$ with a constant C_1. There exists $\alpha' = \alpha'(n, \alpha, C_1)$, with $\alpha < \alpha' < n$, such that $\alpha' \in \mathcal{A}(E)$ with a constant $C = C(n, \alpha, C_1)$.*

PROOF. Let $0 < \beta < \alpha$, $p = \frac{\alpha}{\beta} > 1$ and $w(y) = d(y, E)^{-\beta}$ for every $y \in \mathbb{R}^n$. By Lemma 10.7, we have

$$\left(\fint_{Q(x,r)} w(y)^p \, dy\right)^{\frac{1}{p}} \le C(n, \alpha, \beta, C_1) \fint_{Q(x,r)} w(y) \, dy$$

for every cube $Q(x, r) \subset \mathbb{R}^n$. Since reverse Hölder inequalities are self-improving, see Theorem 8.38, there exists $q = q(n, \alpha, \beta, C_1) > p$ such that

$$\left(\fint_{Q(x,r)} w(y)^q \, dy\right)^{\frac{1}{q}} \le C \fint_{Q(x,r)} w(y) \, dy \le C\left(\fint_{Q(x,r)} w(y)^p \, dy\right)^{\frac{1}{p}} \tag{10.4}$$

for every cube $Q(x, r) \subset \mathbb{R}^n$ with $C = C(n, \alpha, \beta, C_1)$. Let $\alpha' = q\beta > p\beta = \alpha$. Then $w(y)^q = d(y, E)^{-\alpha'}$ and $w(y)^p = d(y, E)^{-\alpha}$. Let $Q(x, r) \subset \mathbb{R}^n$ be a cube with $Q(x, r) \cap E \neq \emptyset$. By (10.4) and $\alpha \in \mathcal{A}(E)$, we obtain

$$\fint_{Q(x,r)} d(y, E)^{-\alpha'} \, dy \le C\left(\fint_{Q(x,r)} d(y, E)^{-\alpha} \, dy\right)^{\frac{q}{p}} \le C(r^{-\alpha})^{\frac{q}{p}} = Cr^{-\alpha'},$$

where $C = C(n, \alpha, \beta, C_1)$. Thus $\alpha' \in \mathcal{A}(E)$, and by Remark 10.3 we have $\alpha' < n$. Finally, by choosing $\beta = \frac{\alpha}{2}$, we conclude that the corresponding α' and the constant in the Aikawa condition only depend on n, α and C_1. □

Since always $0 \in \mathcal{A}(E)$ and $n \notin \mathcal{A}(E)$, the observations in Remark 10.6 and Theorem 10.8 show that the set $\mathcal{A}(E)$ is either an interval of the form $(-\infty, \alpha_0)$, with $0 < \alpha_0 \le n$, or $\mathcal{A}(E) = (-\infty, 0]$.

10.2. A_p properties of distance functions

The distance weight satisfies a reverse Hölder inequality under an appropriate Aikawa condition, see Definition 10.1.

THEOREM 10.9. *Let $E \subset \mathbb{R}^n$ and $0 \le \alpha < n$, and let $w(y) = d(y, E)^{-\alpha}$ for every $y \in \mathbb{R}^n$. If $\alpha \in \mathcal{A}(E)$, then $w \in A_\infty$.*

PROOF. If $\alpha = 0$, then $w(y) = 1$ for every $y \in \mathbb{R}^n$, and thus $w \in A_\infty$. We may hence assume that $0 < \alpha < n$ and that (10.1) holds with a constant C_1 for every $x \in E$ and $r > 0$. This implies that w is locally integrable. Since $\alpha \in \mathcal{A}(E)$ and $\alpha > 0$, we have $|\overline{E}| = 0$. Therefore $w(x) > 0$ for almost every $x \in \mathbb{R}^n$, and thus w is a weight, see Definition 8.2.

By Theorem 10.8 there exists $\alpha' = \alpha'(n, \alpha, C_1)$, with $\alpha < \alpha' < n$, such that $\alpha' \in \mathcal{A}(E)$ with a constant $C = C(n, \alpha, C_1)$. Let $p = \frac{\alpha'}{\alpha} > 1$. Lemma 10.7 implies

$$\begin{aligned}\left(\fint_{Q(x,r)} w(y)^p \, dy\right)^{\frac{1}{p}} &= \left(\fint_{Q(x,r)} d(y, E)^{-\alpha'} \, dy\right)^{\frac{\alpha}{\alpha'}} \\ &\le C \fint_{Q(x,r)} d(y, E)^{-\alpha} \, dy = C \fint_{Q(x,r)} w(y) \, dy\end{aligned}$$

for every cube $Q(x, r) \subset \mathbb{R}^n$, where $C = C(n, \alpha, C_1)$. The claim $w \in A_\infty$ follows from Theorem 8.25. □

Theorem 8.25 implies that, under the assumptions of Theorem 10.9, we have $w \in A_p$ for some $1 < p < \infty$. Next we strengthen this result by showing that $w \in A_1$. Recall that $A_1 \subset A_p \subset A_\infty$ whenever $1 < p < \infty$, see (8.19). We also extend the range of the Aikawa condition.

THEOREM 10.10. *Let $E \subset \mathbb{R}^n$ and $\alpha \in \mathbb{R}$, and let $w(y) = d(y,E)^{-\alpha}$ for every $y \in \mathbb{R}^n$. Then the following assertions hold.*

(a) *If $0 \le \alpha \in \mathcal{A}(E)$, then $w \in A_p$ for every $1 \le p \le \infty$.*
(b) *If $\alpha < 0$ and $1 < p < \infty$ are such that $\frac{-\alpha}{p-1} \in \mathcal{A}(E)$, then $w \in A_p$.*

Moreover, the A_p constant of w, in (a) *and* (b), *only depends on n, p, α and the constant in the Aikawa condition.*

PROOF. As in the proof of Theorem 10.9 we see that w is a weight in (a) and (b). Since $A_1 \subset A_p$ for every $1 \le p \le \infty$ by (8.19), in assertion (a) it suffices to show that $w \in A_1$. This is clear if $\alpha = 0$, and thus we may assume that $0 < \alpha \in \mathcal{A}(E)$. By Lemma 10.5 there is a constant C_1 such that (10.2) holds for all cubes $Q(x,r)$ intersecting E.

Let $Q(x,r) \subset \mathbb{R}^n$ be a cube and assume first that $Q(x,2r) \cap E \neq \emptyset$. Then there is $z \in E$ such that $Q(x,r) \subset Q(z,3r)$. By Lemma 10.5 we obtain

$$\fint_{Q(x,r)} w(y)\,dy \le C(n) \fint_{Q(z,3r)} d(y,E)^{-\alpha}\,dy \le C(n,\alpha,C_1) r^{-\alpha}.$$

On the other hand, since $\alpha > 0$, we have

$$\frac{1}{w(y)} = d(y,E)^{\alpha} \le |y-z|^{\alpha} \le C(n,\alpha) r^{\alpha}$$

for every $y \in Q(x,r) \setminus \overline{E}$. Since $|\overline{E}| = 0$, by combining the estimates above we obtain

$$\fint_{Q(x,r)} w(y)\,dy \operatorname*{ess\,sup}_{y \in Q(x,r)} \frac{1}{w(y)} \le C(n,\alpha,C_1).$$

This shows that the A_1 condition (8.18) holds with a constant $C(n,\alpha,C_1)$ for the cube $Q(x,r)$ when $Q(x,2r) \cap E \neq \emptyset$.

Assume then that $Q(x,2r) \cap E = \emptyset$. In this case

$$C(n) d(y,E) \le d(Q(x,r),E) \le d(y,E)$$

for every $y \in Q(x,r)$, and thus

$$\fint_{Q(x,r)} w(y)\,dy \operatorname*{ess\,sup}_{y \in Q(x,r)} \frac{1}{w(y)} \le C(n,\alpha) d\big(Q(x,r),E\big)^{-\alpha} d\big(Q(x,r),E\big)^{\alpha} \le C(n,\alpha).$$

Hence (8.18) holds and the proof of (a) is complete.

In part (b), let

$$\sigma(y) = w(y)^{1-p'} = w(y)^{-\frac{1}{p-1}} = d(y,E)^{\frac{\alpha}{p-1}}$$

for every $y \in \mathbb{R}^n$. By part (a) we have $\sigma \in A_1 \subset A_{\frac{p}{p-1}}$, and the claim $w \in A_p$ follows from the duality property of A_p weights, see Theorem 8.13 (b). $\square$

Our next goal is to obtain partial converses to assertions (a) and (b) in Theorem 10.10 for porous sets.

DEFINITION 10.11. A set $E \subset \mathbb{R}^n$ is porous if there exists a constant C such that for every $x \in \mathbb{R}^n$ and $r > 0$ there exists $y \in \mathbb{R}^n$ satisfying $B(y,Cr) \subset B(x,r) \setminus E$.

REMARK 10.12. Assume that $E \subset \mathbb{R}^n$ is porous. Then also $\overline{E}$ is porous. Let $x \in \overline{E}$ and $r > 0$, and let $y \in \mathbb{R}^n$ be such that $B(y, Cr) \subset B(x, r) \setminus \overline{E}$. Then

$$\frac{|\overline{E} \cap B(x,r)|}{|B(x,r)|} \leq \frac{|B(x,r) \setminus B(y,Cr)|}{|B(x,r)|} = \frac{r^n - (Cr)^n}{r^n} = 1 - C^n < 1.$$

The Lebesgue density theorem, see Theorem 1.24, implies that $|\overline{E}| = 0$, and hence also $|E| = 0$.

It is easy to characterize porosity in terms of cubes instead of balls.

LEMMA 10.13. *A set $E \subset \mathbb{R}^n$ is porous if and only if there exists a constant C such that for all cubes $Q_0 \subset \mathbb{R}^n$ there is a cube $Q \subset Q_0 \setminus E$ with $l(Q) = Cl(Q_0)$ and $d(Q, E) \geq \frac{l(Q)}{2}$. Moreover, constant C and the constant C_1 in the definition of porosity for E only depend on each other and n.*

PROOF. Assume first that $E \subset \mathbb{R}^n$ is porous with constant C_1 and let $Q_0 = Q(x, r) \subset \mathbb{R}^n$ be a cube. There exists a point $y \in \mathbb{R}^n$ such that $B(y, C_1 r) \subset B(x, r) \setminus E \subset Q_0 \setminus E$. Hence $Q = Q\big(y, \frac{C_1 r}{4\sqrt{n}}\big) \subset B\big(y, \frac{C_1 r}{2}\big) \subset Q_0 \setminus E$. It follows that $l(Q) = \frac{C_1}{4\sqrt{n}} l(Q_0)$ and

$$d(Q, E) \geq \frac{C_1 r}{2} \geq \frac{C_1 r}{4\sqrt{n}} = \frac{l(Q)}{2},$$

as desired.

Conversely, assume that the condition concerning cubes holds. Let $x \in \mathbb{R}^n$, $r > 0$ and $Q_0 = Q\big(x, \frac{r}{2\sqrt{n}}\big)$, and let $Q = Q(y, r') \subset Q_0 \setminus E$ be as in the assumption. Then $B(y, r') \subset Q \subset Q_0 \setminus E \subset B(x, r) \setminus E$. Since

$$r' = \frac{l(Q)}{2} = \frac{Cl(Q_0)}{2} = \frac{Cr}{2\sqrt{n}},$$

we conclude that E is porous. □

The following lemma gives one direction of the connection between porosity and the Aikawa condition; see also Theorem 10.25 below.

LEMMA 10.14. *Assume that $E \subset \mathbb{R}^n$ satisfies the Aikawa condition for some $\alpha > 0$ with a constant C_1. Then E is porous with a constant $C = C(n, \alpha, C_1)$.*

PROOF. Let $Q_0 \subset \mathbb{R}^n$ be a cube. Let j be the smallest integer satisfying

$$j > 1 + \frac{1}{\alpha} \log_2(n^{\frac{\alpha}{2}} C_2),$$

where $C_2 = C_2(n, \alpha, C_1) \geq 1$ is a constant for which (10.2) holds.

Let $\mathcal{D}_j(Q_0) = \{Q_1, \ldots, Q_{2^{jn}}\}$ be the collection of pairwise disjoint half-open dyadic subcubes of Q_0, with side-length $l(Q_i) = 2^{-j} l(Q_0)$ for each $i = 1, \ldots, 2^{jn}$. We claim that there is a cube $Q_i \in \mathcal{D}_j(Q_0)$ that does not intersect E. For a contradiction, assume that E intersects all cubes $Q_i \in \mathcal{D}_j(Q_0)$. For every $i = 1, \ldots, 2^{jn}$ and every $y \in Q_i$,

$$d(y, E) \leq \operatorname{diam}(Q_i) = \sqrt{n} l(Q_i). \tag{10.5}$$

By (10.5) and (10.2),

$$\begin{aligned}(\sqrt{n})^{-\alpha}2^{j\alpha}l(Q_0)^{n-\alpha} &= \sum_{i=1}^{2^{jn}}\int_{Q_i}\big(\sqrt{n}l(Q_i)\big)^{-\alpha}\,dy\\ &\le \int_{Q_0} d(y,E)^{-\alpha}\,dy \le C_2 2^{\alpha} l(Q_0)^{n-\alpha}.\end{aligned}$$

From this we obtain

$$j \le 1 + \frac{1}{\alpha}\log_2(n^{\frac{\alpha}{2}}C_2),$$

which contradicts the original choice of j. Hence, there is $Q_i = Q(x_i, 2^{-j-1}l(Q_0)) \in \mathcal{D}_j(Q_0)$ such that $Q_i \cap E = \emptyset$. Let $Q = Q(x_i, 2^{-j-2}l(Q_0))$. Then $Q \subset Q_i \subset Q_0 \setminus E$ and $d(Q,E) \ge \frac{l(Q)}{2}$. Thus Lemma 10.13 implies that E is porous, with a constant $C = C(n,\alpha,C_1)$. □

We have the following partial converse of Theorem 10.10 (a). This result also contains a strong self-improvement property for the distance weight under the assumption that the set is porous.

THEOREM 10.15. *Assume that $E \subset \mathbb{R}^n$ is a nonempty porous set. Let $\alpha > 0$ and let $w(y) = d(y,E)^{-\alpha}$ for every $y \in \mathbb{R}^n$. The following conditions are equivalent:*

(a) $\alpha \in \mathcal{A}(E)$,
(b) $w \in A_1$,
(c) $w \in A_\infty$.

Moreover, the constants in each of the conditions only depend on each other, the porosity constant, n, w and α.

PROOF. By Theorem 10.10, condition (a) implies both conditions (b) and (c), and the implication from (b) to (c) follows from the inclusion $A_1 \subset A_\infty$, see (8.19). Thus it suffices to show that (c) implies (a).

To this end, assume that $w \in A_\infty$ and let $Q_0 \subset \mathbb{R}^n$ be a cube. Theorem 8.25 implies that $w \in A_p$ for some $1 < p = p(n,w) < \infty$, and so

$$\fint_{Q_0} w(y)\,dy\Big(\fint_{Q_0} w(y)^{\frac{1}{1-p}}\,dy\Big)^{p-1} \le C(n,w).$$

Since E is porous, by Lemma 10.13 there is a cube $Q \subset Q_0 \setminus E$ satisfying $l(Q) = C_1 l(Q_0)$ and $d(Q,E) \ge \frac{l(Q)}{2}$. Thus

$$\begin{aligned}\Big(\fint_{Q_0} w(y)^{\frac{1}{1-p}}\,dy\Big)^{p-1} &\ge C(n,p,C_1)\Big(\fint_{Q} w(y)^{\frac{1}{1-p}}\,dy\Big)^{p-1}\\ &\ge C(n,p,\alpha,C_1)\Big(\fint_{Q} l(Q)^{\frac{\alpha}{p-1}}\,dy\Big)^{p-1}\\ &\ge C(n,p,\alpha,C_1)l(Q)^{\alpha} \ge C(n,\alpha,p,C_1)l(Q_0)^{\alpha}.\end{aligned}$$

By combining the two estimates above we obtain

$$\fint_{Q_0} d(y,E)^{-\alpha}\,dy = \fint_{Q_0} w(y)\,dy \le C\Big(\fint_{Q_0} w(y)^{\frac{1}{1-p}}\,dy\Big)^{1-p} \le Cl(Q_0)^{-\alpha},$$

where $C = C(n,p,\alpha,w,C_1)$, and the claim $\alpha \in \mathcal{A}(E)$ follows from Lemma 10.5. □

There is also a partial converse of assertion (b) in Theorem 10.10.

THEOREM 10.16. *Assume that $E \subset \mathbb{R}^n$ is a nonempty porous set. Let $\alpha < 0$ and let $w(y) = d(y,E)^{-\alpha}$ for every $y \in \mathbb{R}^n$. If $1 < p < \infty$ and $w \in A_p$, then $\frac{-\alpha}{p-1} \in \mathcal{A}(E)$.*

PROOF. Let $1 < p < \infty$ and $w \in A_p$. The duality property of A_p weights, see Theorem 8.13 (b), implies

$$d(\cdot,E)^{-(\frac{-\alpha}{p-1})} = d(\cdot,E)^{\frac{\alpha}{p-1}} = w^{-\frac{1}{p-1}} = w^{1-p'} \in A_{\frac{p}{p-1}} \subset A_\infty.$$

Hence the claim follows from the implication from (c) to (a) in Theorem 10.15. □

10.3. Assouad dimension

Next we connect the Aikawa condition with geometric properties of the sets. The Assouad dimension below is a natural dual of the lower dimension that was discussed in Section 7.3. Recall from Definition 7.11 that for a bounded set $A \subset \mathbb{R}^n$ and $r > 0$, we let $N(A,r)$ denote the minimal number of open balls of radius r that are needed to cover the set A.

DEFINITION 10.17. Let $E \subset \mathbb{R}^n$. The Assouad dimension $\dim_{\mathrm{A}}(E)$ is the infimum of the numbers $\lambda \geq 0$ for which there exists a constant C such that the

$$N\big(E \cap \overline{B}(x,R), r\big) \leq C\Big(\frac{R}{r}\Big)^\lambda \tag{10.6}$$

for every $x \in E$ and $0 < r < R$.

In particular, the estimate in (10.6) holds whenever $\lambda > \dim_{\mathrm{A}}(E)$, and possibly also when $\lambda = \dim_{\mathrm{A}}(E)$. It is clear that $\dim_{\mathrm{A}}(E) \leq \dim_{\mathrm{A}}(E')$ for $E \subset E' \subset \mathbb{R}^n$.

REMARK 10.18. Let $E \subset \mathbb{R}^n$, $x \in E$ and $0 < r < R$. The collection

$$\{B(z,\tfrac{r}{5}) : z \in E \cap \overline{B}(x,R)\}$$

is a cover of $E \cap \overline{B}(x,R)$. By Lemma 1.13, there exist points $x_i \in E \cap \overline{B}(x,R)$, $i = 1,\dots,N$, such that the balls $B(x_i,\frac{r}{5}) \subset B(x,2R)$ are pairwise disjoint and $E \cap \overline{B}(x,R) \subset \bigcup_{i=1}^N B(x_i,r)$. This implies

$$Nr^n = C(n)\sum_{i=1}^N \big|B\big(x_i,\tfrac{r}{5}\big)\big| = C(n)\Big|\bigcup_{i=1}^N B\big(x_i,\tfrac{r}{5}\big)\Big| \leq C(n)|B(x,2R)| = C(n)R^n,$$

and thus

$$N\big(E \cap \overline{B}(x,R), r\big) \leq N \leq C(n)\Big(\frac{R}{r}\Big)^n.$$

From this we conclude that $0 \leq \dim_{\mathrm{A}}(E) \leq n$ for every $E \subset \mathbb{R}^n$.

The Hausdorff dimension always gives a lower bound for the Assouad dimension. Recall from Corollary 7.15 that for the lower dimension we have $\dim_{\mathrm{L}}(E) \leq \dim_{\mathrm{H}}(E)$ for all closed sets $E \subset \mathbb{R}^n$.

LEMMA 10.19. *Let $E \subset \mathbb{R}^n$. Then* $\dim_{\mathrm{H}}(E) \leq \dim_{\mathrm{A}}(E)$.

PROOF. The Hausdorff dimension is countably stable, see Falconer [**125**, p. 29], and thus $\dim_{\mathrm{H}}(E) = \sup_{k\in\mathbb{N}} \dim_{\mathrm{H}}(E \cap \overline{B}(x,k))$ for every $x \in \mathbb{R}^n$. Hence it suffices to show that $\dim_{\mathrm{H}}(E \cap \overline{B}(x,R)) \leq \dim_{\mathrm{A}}(E)$ for every $x \in E$ and $R > 0$. Let

$s > \dim_{\mathrm{A}}(E)$ and let $\dim_{\mathrm{A}}(E) < \lambda < s$. Let C_1 be the constant in (10.6) for λ, and let $R > 0$ and $x \in E$. Then $E \cap \overline{B}(x, R)$ can be covered by

$$N \leq C_1 \left(\frac{R}{r} \right)^{\lambda}$$

balls of radius r, for every $0 < r < R$. By Definition 1.43, we have

$$\mathcal{H}^s_r\big(E \cap \overline{B}(x, R)\big) \leq N r^s \leq C_1 R^{\lambda} r^{s-\lambda}.$$

Letting $r \to 0$ shows that $\mathcal{H}^s\big(E \cap \overline{B}(x, R)\big) = 0$. Since this holds for every $s > \dim_{\mathrm{A}}(E)$, we conclude that $\dim_{\mathrm{H}}(E \cap \overline{B}(x, R)) \leq \dim_{\mathrm{A}}(E)$. □

In general, and even for closed sets, the inequality in Lemma 10.19 can be strict, as the following example illustrates.

EXAMPLE 10.20. Let $E = \{\frac{1}{k} : k \in \mathbb{N}\} \cup \{0\} \subset \mathbb{R}$. The set E is countable, and thus $\dim_{\mathrm{H}}(E) = 0$. We claim that $\dim_{\mathrm{A}}(E) = 1$. To see this, let $R = \frac{1}{K}$ for $K \in \mathbb{N} \setminus \{1\}$ and $0 < r = \frac{1}{K^2} < R$. The distance between two consecutive points $\frac{1}{k}$ and $\frac{1}{k+1}$ in $E \cap \overline{B}(0, R)$ is

$$\frac{1}{k} - \frac{1}{k+1} = \frac{1}{k(k+1)} < \frac{1}{K^2} = r.$$

If intervals $B_i = B(x_i, r) = (x_i - r, x_i + r)$, $i = 1, 2, \ldots, N$, cover the set $E \cap \overline{B}(0, R)$, then all subintervals of $[0, R]$ that are not covered by the intervals B_i have length less than r. On the other hand, we may clearly assume that $x_i \in [0, R]$ for every $i = 1, 2, \ldots, N$, and thus each of the intervals B_i covers a subinterval of $[0, R]$ of length at least r. On both sides of a subinterval of $[0, R]$ that is not covered by the intervals B_i, there are intervals B_{i_1} and B_{i_2} for some $i_1, i_2 \in \{1, 2, \ldots N\}$. It follows that the intervals B_i cover at least half of the interval $[0, R]$. Hence their total length is at least $\frac{R}{2}$, and thus $N 2r \geq \frac{R}{2}$. From this we conclude that there exists a constant $C > 0$, independent of r and R, such that

$$N\big(E \cap \overline{B}(0, R), r\big) \geq C \left(\frac{R}{r} \right).$$

Since $\frac{R}{r} = K$, letting $K \to \infty$ shows that that $\dim_{\mathrm{A}}(E) \geq 1$. On the other hand, $\dim_{\mathrm{A}}(E) \leq 1$ by Remark 10.18, and thus $\dim_{\mathrm{A}}(E) = 1$. By considering the natural embedding of $\mathbb{R}$ to $\mathbb{R}^n$, this example can be easily lifted to any $\mathbb{R}^n$.

If $E \subset \mathbb{R}^n$ is Ahlfors–David λ-regular, see Section 7.2, then

$$\dim_{\mathrm{H}}(E) = \dim_{\mathrm{L}}(E) = \lambda,$$

by Theorem 7.12. Next we show that for a λ-regular set $E \subset \mathbb{R}^n$ also the Assouad dimension is equal to λ, and thus

$$\dim_{\mathrm{L}}(E) = \dim_{\mathrm{A}}(E) = \dim_{\mathrm{H}}(E) = \lambda.$$

THEOREM 10.21. *Let $0 < \lambda \leq n$ and assume that a closed set $E \subset \mathbb{R}^n$ is Ahlfors–David λ-regular. Then* $\dim_{\mathrm{A}}(E) = \lambda$.

PROOF. By Lemma 10.19 we have $\lambda = \dim_{\mathrm{H}}(E) \leq \dim_{\mathrm{A}}(E)$. Hence it suffices to show that $\dim_{\mathrm{A}}(E) \leq \lambda$. This can be shown essentially in the same way as the bound $\dim_{\mathrm{L}}(E) \leq \lambda$ in Theorem 7.12.

Let $x \in E$ and $0 < r < R$. Assume first that $2R \le \operatorname{diam}(E)$. By Lemma 1.13 and boundedness, there exist $N \in \mathbb{N}$ and points $x_i \in E \cap \overline{B}(x,R)$, $i = 1,2,\dots,N$, such that $E \cap \overline{B}(x,R) \subset \bigcup_{i=1}^N B(x_i,r)$ and the balls $B(x_i,\frac{r}{5})$ are pairwise disjoint. Hence, by the Ahlfors–David regularity and the additivity and monotonicity of the Hausdorff measure,

$$N r^\lambda = \sum_{i=1}^N r^\lambda \le C \sum_{i=1}^N \mathcal{H}^\lambda\big(E \cap \overline{B}(x_i,\tfrac{r}{6})\big) \le C\mathcal{H}^\lambda\big(E \cap \overline{B}(x,2R)\big) \le CR^\lambda.$$

Thus

$$N\big(E \cap \overline{B}(x,R), r\big) \le N \le C\Big(\frac{R}{r}\Big)^\lambda. \tag{10.7}$$

On the other hand, if $2R > \operatorname{diam}(E) \ge r$, then a similar argument as above, for $E = E \cap \overline{B}(x,\operatorname{diam}(E))$, shows that

$$N\big(E \cap \overline{B}(x,R), r\big) \le N\big(E \cap \overline{B}(x,\operatorname{diam}(E)), r\big) \le C\Big(\frac{\operatorname{diam}(E)}{r}\Big)^\lambda \le C\Big(\frac{R}{r}\Big)^\lambda.$$

Finally, if $r > \operatorname{diam}(E)$, then $N\big(E \cap \overline{B}(x,R), r\big) = 1$. Hence (10.7) holds in all cases, and the claim $\dim_{\mathrm{A}}(E) \le \lambda$ follows. □

Our next results are concerned with the close connection between the Assouad dimension and the Aikawa condition, see Definition 10.1.

THEOREM 10.22. *Let $E \subset \mathbb{R}^n$ and assume that $\alpha \in \mathbb{R}$ is such that $\alpha \in \mathcal{A}(E)$. Then* $\dim_{\mathrm{A}}(E) \le n - \alpha$.

PROOF. For $\alpha \le 0$, the claim is clear by Remark 10.18. Hence we may assume that $\alpha > 0$, and by Remark 10.3 we have $\alpha < n$. Let $x \in E$ and $0 < r < R$, and write $F = E \cap \overline{B}(x,R)$. By Lemma 1.13 and the boundedness of F, there exist $N \in \mathbb{N}$ and a collection $\{B(x_i,\frac{r}{5}) : i = 1,\dots,N\}$ of pairwise disjoint open balls, with $x_i \in F$, such that $F \subset \bigcup_{i=1}^N B(x_i,r)$.

Let $F_r = \{y \in \mathbb{R}^n : d(y,F) < r\} \subset B(x,2R)$ and let C_1 be the constant in the Aikawa condition (10.1) for α. Using pairwise disjointness of the balls $B(x_i,\frac{r}{5}) \subset F_r$ and the fact that $d(y,E) \le d(y,F) < r$ for every $y \in F_r$, we obtain

$$\begin{aligned} NC(n)r^n &\le \sum_{i=1}^N \big|B(x_i,\tfrac{r}{5})\big| \le |F_r| \le r^\alpha \int_{F_r} d(y,E)^{-\alpha}\,dy \\ &\le r^\alpha \int_{B(x,2R)} d(y,E)^{-\alpha}\,dy \le r^\alpha |B(x,2R)| C_1 R^{-\alpha} \\ &= C(n,C_1) r^n \Big(\frac{R}{r}\Big)^{n-\alpha}. \end{aligned}$$

Thus

$$N\big(E \cap \overline{B}(x,R), r\big) = N(F,r) \le N \le C(n,C_1)\Big(\frac{R}{r}\Big)^{n-\alpha},$$

and the claim $\dim_{\mathrm{A}}(E) \le n - \alpha$ follows since $n - \alpha > 0$. □

Notice that if $\alpha > 0$, then by the self-improvement of the Aikawa condition, see Theorem 10.8, we obtain a strict inequality in Theorem 10.22. See also the proof of

Theorem 10.24 below. For the converse direction we assume a strict upper bound for the dimension.

THEOREM 10.23. *Let $E \subset \mathbb{R}^n$ be a nonempty set and assume that $\alpha \in \mathbb{R}$ with $\dim_{\mathrm{A}}(E) < n - \alpha$. Then $\alpha \in \mathcal{A}(E)$.*

PROOF. The claim is clear if $\alpha \leq 0$, and so we may assume that $\alpha > 0$. Let $\lambda > 0$ with $\dim_{\mathrm{A}}(E) < \lambda < n - \alpha$, and let $x \in E$ and $r > 0$. Define

$$F_j = \{y \in B(x,r) : d(y,E) < 2^{-j+1}r\} \quad \text{and} \quad A_j = F_j \setminus F_{j+1},$$

for $j \in \mathbb{N}$. Since $\lambda > \dim_{\mathrm{A}}(E)$, there is a constant C_1 such that the set $E \cap \overline{B}(x,2r)$ can be covered by $N_j \leq C_1 2^{j\lambda}$ balls of radius $2^{1-j}r$, for every $j \in \mathbb{N}$. It follows that each F_j can be covered by at most N_j balls of radius $2^{2-j}r$. If B_i^j, $i = 1, \ldots, N_j$, are such balls, then

$$|F_j| \leq \sum_{i=1}^{N_j} |B_i^j| \leq N_j C(n)(2^{2-j}r)^n \leq C(n, C_1)(2^{-j})^{n-\lambda} r^n. \tag{10.8}$$

Since $\overline{E} \cap B(x,r) \subset F_j$ for every $j \in \mathbb{N}$ and $\lambda < n - \alpha < n$, by letting $j \to \infty$ we see in particular that $|\overline{E} \cap B(x,r)| = 0$. Here $r > 0$ is arbitrary, and thus $|\overline{E}| = 0$.

If $y \in A_j$, then $2^{-j}r \leq d(y,E) < 2^{-j+1}r$. In addition, $A_j \subset F_j$ for every $j \in \mathbb{N}$ and the sets A_j cover $B(x,r)$ up to the set $\overline{E} \cap B(x,r)$, which has measure zero. By (10.8) we obtain

$$\begin{aligned}
\fint_{B(x,r)} d(y,E)^{-\alpha}\,dy &\leq C(n) r^{-n} \sum_{j=1}^{\infty} \int_{A_j} d(y,E)^{-\alpha}\,dy \\
&\leq C(n) r^{-n} \sum_{j=1}^{\infty} |F_j| (2^{-j}r)^{-\alpha} \\
&\leq C(n, C_1) r^{-\alpha} \sum_{j=1}^{\infty} (2^{-j})^{n-\lambda-\alpha} \\
&\leq C(n, \alpha, \lambda, C_1) r^{-\alpha},
\end{aligned}$$

where the geometric series converges since $\lambda < n - \alpha$. This shows that $\alpha \in \mathcal{A}(E)$. □

Combining the two results above and the self-improvement of the Aikawa condition, see Theorem 10.8, we obtain a characterization connecting the Assouad dimension and the Aikawa condition.

THEOREM 10.24. *Let $E \subset \mathbb{R}^n$ be a nonempty set and let $\alpha > 0$. Then $\alpha \in \mathcal{A}(E)$ if and only if $\dim_{\mathrm{A}}(E) < n - \alpha$.*

PROOF. If $\dim_{\mathrm{A}}(E) < n - \alpha$, then $\alpha \in \mathcal{A}(E)$ by Theorem 10.23. Conversely, assume that $0 < \alpha \in \mathcal{A}(E)$. By Remark 10.3 we have $\alpha < n$, and by Theorem 10.8 there is $\alpha' > \alpha$ with $\alpha' \in \mathcal{A}(E)$. Thus Theorem 10.22 implies $\dim_{\mathrm{A}}(E) \leq n - \alpha' < n - \alpha$, as desired. □

Notice that the assumption $\alpha > 0$ in Theorem 10.24 is essential: if $E \subset \mathbb{R}^n$ and $\dim_{\mathrm{A}}(E) = n$, then $0 \in \mathcal{A}(E)$, but $\dim_{\mathrm{A}}(E) \not< n - 0$. For instance, this holds if $E = B(0,1)$, see Lemma 10.19 and Remark 10.18.

Recall from Theorem 7.22 that if $E \subset \mathbb{R}^n$ is an unbounded closed set and $1 < p \leq n$, then E satisfies the p-capacity density condition in Definition 6.17 if and only if $\dim_{\mathrm{L}}(E) > n-p$. Similarly, for $\alpha > 0$, Theorem 10.24 shows that a nonempty set $E \subset \mathbb{R}^n$ satisfies the Aikawa condition for α if and only if $\dim_{\mathrm{A}}(E) < n - \alpha$. In the limiting case $p = n$, the capacity density condition and the positivity of the lower dimension have a simple geometric characterization through uniform perfectness, see Theorem 7.20. In a corresponding manner, porosity of $E \subset \mathbb{R}^n$ gives a geometric characterization for the properties that the Aikawa condition holds for some $\alpha > 0$ and that the Assouad dimension is strictly less than the dimension of the space.

THEOREM 10.25. *Let $E \subset \mathbb{R}^n$ be a nonempty set. The following conditions are equivalent:*

(a) $\dim_{\mathrm{A}}(E) < n$,
(b) *there is $\alpha > 0$ such that $\alpha \in \mathcal{A}(E)$,*
(c) *E is porous.*

Moreover, the constants and parameters in each of the conditions only depend on each other and n, and if any of these conditions holds, then $|\overline{E}| = 0$.

PROOF. The implication from (a) to (b) follows from Theorem 10.23 by choosing $0 < \alpha < n - \dim_{\mathrm{A}}(E)$, and the implication from (b) is (c) is established in Lemma 10.14. Hence it suffices to show that (c) implies (a).

To this end, let $x \in E$ and $R > 0$, and write $Q_0 = Q(x, R)$. As before, let $\mathcal{D}_j(Q_0)$ be the collection of pairwise disjoint half-open dyadic subcubes of Q_0, with side length $2^{-j}l(Q_0) = 2^{-j}2R$. Since E is porous, we obtain from Lemma 10.13 that there is $j_0 \in \mathbb{N}$, independent of Q_0, and a cube $Q \in \mathcal{D}_{j_0}(Q_0)$ such that $E \cap \overline{Q} = \emptyset$. Hence at most $2^{nj_0} - 1$ closed cubes of side length $2^{-j_0}2R$ are needed to cover $E \cap \overline{Q}_0$. By a simple iteration, we conclude that if $k \in \mathbb{N}$, then at most $(2^{nj_0} - 1)^k$ closed cubes of side length $2^{-kj_0}2R$ are needed to cover $E \cap \overline{Q}_0$.

Let $0 < r < R$ and choose $k \in \mathbb{N}$ satisfying

$$(\sqrt{n})^{-1}2^{(k-1)j_0} \leq \frac{R}{r} < (\sqrt{n})^{-1}2^{kj_0}.$$

Since $\overline{B}(x, R) \subset \overline{Q}_0$ and $r > \sqrt{n}2^{-kj_0}R$, the set $E \cap \overline{B}(x, R)$ can be covered by at most $(2^{nj_0} - 1)^k$ open balls of radius r; indeed, if $Q = Q(z, 2^{-kj_0}R) \in \mathcal{D}_{kj_0}(Q_0)$, then $\overline{Q} \subset B(z, r)$. Since

$$\log_2(2^{nj_0} - 1) < \log_2(2^{nj_0}) = nj_0,$$

we may choose $\lambda > 0$ satisfying

$$\frac{1}{j_0}\log_2(2^{nj_0} - 1) < \lambda < n.$$

Then $2^{nj_0} - 1 < 2^{\lambda j_0}$, and we obtain from the considerations above and the choice of k that

$$N\big(E \cap \overline{B}(x, R), r\big) \leq (2^{nj_0} - 1)^k < \big(2^{\lambda j_0}\big)^k \leq C(n, j_0, \lambda)\Big(\frac{R}{r}\Big)^{\lambda}.$$

Hence $\dim_{\mathrm{A}}(E) \leq \lambda < n$, as claimed.

Finally, all conditions in the theorem imply that $\overline{E}$ is porous, and hence $|\overline{E}| = 0$ by Remark 10.12. □

We restate results concerning the Muckenhoupt properties of the distance function in terms of the Assouad dimension. This leads to the following characterizations for a porous set.

THEOREM 10.26. *Let $1 < p < \infty$ and assume that $E \subset \mathbb{R}^n$ is a nonempty porous set. Let $\alpha \in \mathbb{R}$ and define $w(y) = d(y,E)^{-\alpha}$ for every $y \in \mathbb{R}^n$. Then the following assertions hold.*

(a) *$w \in A_1$ if and only if $0 \le \alpha < n - \dim_{\mathrm{A}}(E)$.*
(b) *$w \in A_p$ if and only if*

$$(1-p)\big(n - \dim_{\mathrm{A}}(E)\big) < \alpha < n - \dim_{\mathrm{A}}(E). \tag{10.9}$$

PROOF. Since E is porous, $\dim_{\mathrm{A}}(E) < n$ by Theorem 10.25.

We consider first part (b). If $0 \le \alpha < n - \dim_{\mathrm{A}}(E)$, we have $\alpha \in \mathcal{A}(E)$ by Theorem 10.23 and thus Theorem 10.10 (a) implies $w \in A_p$. On the other hand, if

$$(1-p)\big(n - \dim_{\mathrm{A}}(E)\big) < \alpha < 0,$$

then

$$0 < \frac{-\alpha}{p-1} < n - \dim_{\mathrm{A}}(E).$$

From Theorem 10.23 we obtain $\frac{-\alpha}{p-1} \in \mathcal{A}(E)$ and so $w \in A_p$ by Theorem 10.10 (b).

Conversely, assume that $w \in A_p \subset A_\infty$. If $\alpha > 0$, Theorem 10.15 implies $\alpha \in \mathcal{A}(E)$, and so $\alpha < n - \dim_{\mathrm{A}}(E)$ by Theorem 10.24. Note that porosity is needed in Theorem 10.15. If $\alpha = 0$, then (10.9) holds since $\dim_{\mathrm{A}}(E) < n$ by porosity. Finally, if $\alpha < 0$, then $\frac{-\alpha}{p-1} \in \mathcal{A}(E)$ by Theorem 10.16. Theorem 10.24 gives

$$0 < \frac{-\alpha}{p-1} < n - \dim_{\mathrm{A}}(E),$$

showing that (10.9) also holds in this case. The proof of (b) is complete.

Consider then part (a). If $0 \le \alpha < n - \dim_{\mathrm{A}}(E)$, the claim $w \in A_1$ follows from Theorem 10.23 and Theorem 10.10 (a) just as in case (b). Conversely, if $w \in A_1$ and $\alpha > 0$, then $\alpha < n - \dim_{\mathrm{A}}(E)$ by Theorem 10.15 and Theorem 10.24. If $\alpha = 0$ then $0 \le \alpha < n - \dim_{\mathrm{A}}(E)$ holds since $\dim_{\mathrm{A}}(E) < n$ by porosity. It remains to show that the condition $\alpha \ge 0$ is necessary in (a). To this end, assume that $w \in A_1$ and let $Q = Q(x,r)$ be a cube with $x \in E$. Since $w(y) > 0$ for almost every $y \in Q$, we have $⨍_Q w(y)\,dy > 0$. If $\alpha < 0$, then

$$\operatorname*{ess\,sup}_{y\in Q} \frac{1}{w(y)} = \operatorname*{ess\,sup}_{y\in Q} d(y,E)^{\alpha} = \infty,$$

and this clearly violates the A_1 condition (8.18). We conclude that $w \in A_1$ is possible only when $\alpha \ge 0$. □

EXAMPLE 10.27. Consider the porous set $E = \{0\} \subset \mathbb{R}^n$, for which $\dim_{\mathrm{A}}(E) = 0$, and let $w(x) = d(x,E)^{-\alpha} = |x|^{-\alpha}$ for every $x \in \mathbb{R}^n \setminus \{0\}$. Theorem 10.26 implies that $w \in A_1$ if and only if $0 \le \alpha < n$, and $w \in A_p$, for $1 < p < \infty$, if and only if $(1-p)n < \alpha < n$. Compare to Remark 8.35.

10.4. Distance weighted Poincaré inequalities

This section combines the abstract weighted Poincaré inequalities from Chapter 9 with the Muckenhoupt properties of the distance functions. The following result follows by specializing Theorem 9.11 to a power of the distance function.

THEOREM 10.28. *Let $E \subset \mathbb{R}^n$ be a nonempty set and assume that $1 < p < n - \dim_A(E)$. Let $Q \subset \mathbb{R}^n$ be a cube and assume that $u \in \mathrm{Lip}(Q)$. There exists a constant $C = C(n, p, E)$ such that*

$$\Big(\int_Q |u(x) - u_Q|^p d(x,E)^{-p}\,dx\Big)^{\frac{1}{p}} \le C l(Q)\Big(\int_Q |\nabla u(x)|^p d(x,E)^{-p}\,dx\Big)^{\frac{1}{p}}. \tag{10.10}$$

PROOF. Let $w(x) = d(x,E)^{-p}$ for every $x \in \mathbb{R}^n$. We have $p \in \mathcal{A}(E)$ by Theorem 10.23 and thus Theorem 10.10 (a) implies $w \in A_p$. The claim follows from Theorem 9.11. □

By applying Theorem 9.21 with suitable weights instead of Theorem 9.11, we obtain an improvement of Theorem 10.28. Note that for $q = p$ the left-hand side of (10.11) is the same as in (10.10), while the right-hand side of (10.11) is significantly smaller on cubes Q which intersect E or are close to E.

THEOREM 10.29. *Let $1 < p < n$ and $p \le q \le \frac{np}{n-p}$ and assume that $E \subset \mathbb{R}^n$ is a nonempty set with $\dim_A(E) < \frac{q}{p}(n-p)$. Let $Q \subset \mathbb{R}^n$ be a cube and let $u \in \mathrm{Lip}(Q)$. There exists a constant $C = C(n, p, q, E)$ such that*

$$\Big(\int_Q |u(x) - u_Q|^q d(x,E)^{\frac{q}{p}(n-p)-n}\,dx\Big)^{\frac{1}{q}} \le C\Big(\int_Q |\nabla u(x)|^p\,dx\Big)^{\frac{1}{p}}. \tag{10.11}$$

PROOF. Let

$$w(x) = d(x,E)^{\frac{q}{p}(n-p)-n} \quad \text{and} \quad v(x) = 1 = \sigma(x)$$

for every $x \in \mathbb{R}^n$. For $\alpha = n - \frac{q}{p}(n-p) \ge 0$ we have $\dim_A(E) < n - \alpha$. By Theorem 10.23, Theorem 10.10 (a) and (8.19), we conclude $w \in A_1 \subset A_\infty$.

Let $Q_0 \subset \mathbb{R}^n$ be a cube. It suffices to show that there is a constant $C_1 = C_1(n, p, q, E) > 0$ such that

$$\Big(\frac{1}{|Q|^{1-\frac{1}{n}}}\Big)^p w(Q)^{\frac{p}{q}} \sigma(Q)^{p-1} \le C_1 \tag{10.12}$$

holds for all dyadic cubes $Q \in \mathcal{D}(Q_0)$ since then inequality (10.11) follows from Theorem 9.21. To this end, let $Q = Q(x,r) \in \mathcal{D}(Q_0)$ be a dyadic cube.

Assume first that $Q(x,2r) \cap E \neq \emptyset$. Then there exists $z \in E$ such that $Q \subset Q(z,3r)$. Since $\dim_A(E) < \frac{q}{p}(n-p)$, Theorem 10.23 implies that E satisfies the Aikawa condition (10.2) for $\alpha = n - \frac{q}{p}(n-p)$, with a constant $C = C(n,p,q,E)$. Thus

$$w(Q)^{\frac{p}{q}} \le \Big(\int_{Q(z,3r)} d(y,E)^{\frac{q}{p}(n-p)-n}\,dy\Big)^{\frac{p}{q}} \le C\big(r^{\frac{q}{p}(n-p)}\big)^{\frac{p}{q}} = Cr^{n-p} = C|Q|^{1-\frac{p}{n}},$$

where $C = C(n,p,q,E)$. Since $\sigma(Q)^{p-1} = |Q|^{p-1}$, we obtain

$$\Big(\frac{1}{|Q|^{1-\frac{1}{n}}}\Big)^p w(Q)^{\frac{p}{q}} \sigma(Q)^{p-1} \le C|Q|^{-p+\frac{p}{n}+1-\frac{p}{n}+p-1} = C,$$

with $C = C(n,p,q,E)$. This proves (10.12) in the case $Q(x,2r) \cap E \neq \emptyset$.

Assume then that $Q(x,2r)\cap E=\emptyset$. In this case $C(n)d(y,E)\le d(Q,E)\le d(y,E)$ for every $y\in Q=Q(x,r)$. Hence

$$\begin{aligned} w(Q)^{\frac{p}{q}}\sigma(Q)^{p-1} &\le C(n,p,q)|Q|^{\frac{p}{q}}d(Q,E)^{n-p-\frac{np}{q}}|Q|^{p-1}\\ &\le C(n,p,q)|Q|^{\frac{p}{q}+p-1}d(Q,E)^{n-p-\frac{np}{q}}. \end{aligned}$$

By assumption $n-p-\frac{np}{q}\le 0$, and thus

$$\begin{aligned} \Big(\frac{1}{|Q|^{1-\frac{1}{n}}}\Big)^p w(Q)^{\frac{p}{q}}\sigma(Q)^{p-1} &\le C|Q|^{-p+\frac{p}{n}+\frac{p}{q}+p-1}d(Q,E)^{n-p-\frac{np}{q}}\\ &\le C|Q|^{-p+\frac{p}{n}+\frac{p}{q}+p-1+1-\frac{p}{n}-\frac{p}{q}}=C, \end{aligned}$$

where $C=C(n,p,q)$. This shows that (10.12) also holds when $Q(x,2r)\cap E=\emptyset$, and the proof is complete. $\square$

The inequality in (10.11) is a Hardy–Sobolev–Poincaré inequality, compare to Definition 10.32 below. By applying the weighted local-to-global inequality, see Theorem 9.28, we obtain a variant of the Hardy–Sobolev–Poincaré inequality on open sets satisfying the Boman chain condition.

THEOREM 10.30. *Let $1<p<n$ and $p\le q\le\frac{np}{n-p}$. Assume that $\Omega\subset\mathbb{R}^n$ is an open set satisfying the Boman chain condition, with $\tau>1$, and that $\dim_{\mathrm{A}}(\partial\Omega)<\frac{q}{p}(n-p)$. There exists a constant $C=C(n,p,q,\tau)$ such that*

$$\inf_{c\in\mathbb{R}}\Big(\int_\Omega|u(x)-c|^q d(x,\partial\Omega)^{\frac{q}{p}(n-p)-n}\,dx\Big)^{\frac{1}{q}}\le C\Big(\int_\Omega|\nabla u(x)|^p\,dx\Big)^{\frac{1}{p}}$$

for every $u\in\operatorname{Lip}_{\mathrm{loc}}(\Omega)$.

PROOF. Let $\alpha=n-\frac{q}{p}(n-p)\ge 0$ and let $w(x)=d(x,\partial\Omega)^{-\alpha}$ for every $x\in\mathbb{R}^n$. Since $\dim_{\mathrm{A}}(\partial\Omega)<n-\alpha$, we have $w\in A_1\subset A_p$ by Theorems 10.23 and 10.10 (a). Theorem 8.15 implies that w is a doubling weight. Using Theorem 9.28 and (9.20) we obtain

$$\begin{aligned} \inf_{c\in\mathbb{R}}\int_\Omega|u(x)-c|^q w(x)\,dx &\le C\sum_{Q\in\mathcal{W}(\Omega)}\int_{Q^*}|u(x)-u_{Q^*;w}|^q w(x)\,dx\\ &\le C\sum_{Q\in\mathcal{W}(\Omega)}\int_{Q^*}|u(x)-u_{Q^*}|^q w(x)\,dx \qquad (10.13)\\ &\le C\sum_{Q\in\mathcal{W}(\Omega)}l(Q)^{-\alpha}\int_{Q^*}|u(x)-u_{Q^*}|^q\,dx, \end{aligned}$$

where $C=C(n,p,q,\tau)$. By Corollary 9.24 we have

$$\int_{Q^*}|u(x)-u_{Q^*}|^q\,dx\le C(n,p,q)l(Q^*)^\alpha\Big(\int_{Q^*}|\nabla u(x)|^p\,dx\Big)^{\frac{q}{p}}$$

for every $Q \in \mathcal{W}(\Omega)$; observe here that $q + n - \frac{nq}{p} = \alpha$. Since $q \geq p$, substitution of these estimates for Whitney cubes to (10.13) gives

$$\begin{aligned}\inf_{c\in\mathbb{R}} \int_\Omega |u(x) - c|^q w(x)\,dx &\leq C \sum_{Q\in\mathcal{W}(\Omega)} \Big(\int_{Q^*} |\nabla u(x)|^p\,dx\Big)^{\frac{q}{p}}\\ &\leq C\Big(\sum_{Q\in\mathcal{W}(\Omega)} \int_{Q^*} |\nabla u(x)|^p\,dx\Big)^{\frac{q}{p}},\end{aligned}$$

with $C = C(n, p, q, \tau)$. The desired estimate follows since the dilated Whitney cubes $Q^* \subset \Omega$ have a bounded overlap. □

REMARK 10.31. Let $\Omega \subset \mathbb{R}^n$ be a nonempty open set satisfying the assumptions of Theorem 10.30. Then Ω is necessarily bounded, and thus

$$n - 1 \leq \dim_{\mathrm{H}}(\partial\Omega) \leq \dim_{\mathrm{A}}(\partial\Omega) < \frac{q}{p}(n-p)$$

by Lemma 10.19. This poses additional restrictions for the use of Theorem 10.30. However, for instance the domain inside the von Koch snowflake curve of dimension $n - 1 < \lambda < n$ satisfies the Boman chain condition, and hence Theorem 10.30 can be applied for such open sets.

10.5. Hardy–Sobolev inequalities

Hardy–Sobolev inequalities generalize both Hardy and Sobolev inequalities.

DEFINITION 10.32. Let $1 \leq p < \infty$ and $1 \leq q < \infty$, and let $\Omega \subsetneq \mathbb{R}^n$ be an open set. We say that the (q, p)-Hardy–Sobolev inequality holds in Ω, with a constant C, if

$$\Big(\int_\Omega |u(x)|^q d(x, \partial\Omega)^{\frac{q}{p}(n-p)-n}\,dx\Big)^{\frac{1}{q}} \leq C\Big(\int_\Omega |\nabla u(x)|^p\,dx\Big)^{\frac{1}{p}} \tag{10.14}$$

for every $u \in \operatorname{Lip}_0(\Omega)$.

REMARK 10.33. Let $1 \leq p < \infty$ and $1 \leq q < \infty$, and let $\Omega \subsetneq \mathbb{R}^n$ be an open set. Approximation and a straightforward application of Fatou's lemma, as in the proof of Theorem 3.6, show that if (10.14) holds for every $u \in \operatorname{Lip}_0(\Omega)$, then (10.14) holds for every $u \in W_0^{1,p}(\Omega)$.

For $q = p^* = \frac{np}{n-p}$ inequality (10.14) reduces to the Sobolev inequality (3.6), and for $q = p$ it reduces to the p-Hardy inequality (6.1). If $|\Omega^c| = 0$, then integrals in (10.14) can be taken over the whole $\mathbb{R}^n$. We first examine the validity of such inequalities for functions $u \in \operatorname{Lip}_c(\mathbb{R}^n)$, that is, we do not require u to be zero on Ω^c. In this case (10.14) is called a global Hardy–Sobolev inequality. We have the following characterization.

THEOREM 10.34. *Let $1 < p < n$ and $p \leq q < \frac{np}{n-p}$ and assume that $E \subset \mathbb{R}^n$ is a nonempty closed set. The following conditions are equivalent.*

(a) $\dim_{\mathrm{A}}(E) < \frac{q}{p}(n-p)$.

(b) *There exists a constant* $C = C(n,p,q,E)$ *such that*

$$\left(\int_{\mathbb{R}^n} |u(x)|^q d(x,E)^{\frac{q}{p}(n-p)-n}\,dx\right)^{\frac{1}{q}} \le C\left(\int_{\mathbb{R}^n} |\nabla u(x)|^p\,dx\right)^{\frac{1}{p}}, \tag{10.15}$$

for every $u \in \operatorname{Lip}_c(\mathbb{R}^n)$.

PROOF. Assume first that (a) holds. Let w, v and σ be as in the proof of Theorem 10.29. Then the same proof shows that the assumptions of Corollary 9.25 are valid, and it follows that (10.15) holds for every $u \in \operatorname{Lip}_c(\mathbb{R}^n)$.

Assume then that (10.15) holds for every $u \in \operatorname{Lip}_c(\mathbb{R}^n)$ with a constant C_1. Let $x \in E$, $r > 0$, and

$$u(y) = \max\{0, 1 - \tfrac{1}{r} d(y, B(x,r))\}$$

for every $y \in \mathbb{R}^n$. Then $u \in \operatorname{Lip}_c(\mathbb{R}^n)$, $u(y) = 1$ for every $y \in B(x,r)$, u is supported in $\overline{B}(x,2r)$ and $|\nabla u(y)| \le \frac{1}{r}$ for almost every $y \in \overline{B}(x,2r)$. Using (10.15) for u, we obtain

$$\begin{aligned}
\int_{B(x,r)} d(y,E)^{\frac{q}{p}(n-p)-n}\,dy &\le \int_{\mathbb{R}^n} |u(y)|^q d(y,E)^{\frac{q}{p}(n-p)-n}\,dy \\
&\le C_1^q\left(\int_{\mathbb{R}^n} |\nabla u(y)|^p\,dy\right)^{\frac{q}{p}} \\
&\le C r^{-q} \big|B(x,2r)\big|^{\frac{q}{p}} = C r^{\frac{q}{p}(n-p)},
\end{aligned}$$

where $C = C(n,p,q,C_1)$. This shows that the Aikawa condition (10.1) holds for $\alpha = n - \frac{q}{p}(n-p) > 0$. By Theorem 10.24 we have $\dim_{\mathrm{A}}(E) < n - \alpha = \frac{q}{p}(n-p)$. □

Notice that the strict inequality $q < \frac{np}{n-p}$ is essential in the necessity part of Theorem 10.34 since the case $q = \frac{np}{n-p}$ reduces to the Sobolev–Gagliardo–Nirenberg inequality, see Theorem 3.6, and the role of the set E becomes redundant. In particular, the dimensional restriction $\dim_{\mathrm{A}}(E) < n$ does not follow from the Sobolev inequality.

REMARK 10.35. Let $1 < p < n$ and $p \le q < \frac{np}{n-p}$ and assume that $\Omega \subsetneq \mathbb{R}^n$ is an open set with $\dim_{\mathrm{A}}(\Omega^c) < \frac{q}{p}(n-p)$. Then the (q,p)-Hardy–Sobolev inequality holds in Ω by Theorem 10.34.

The following special case $q = p$ can be regarded as a dual pair of Theorem 7.17.

COROLLARY 10.36. *Let* $1 < p < n$ *and assume that* $\Omega \subsetneq \mathbb{R}^n$ *is an open set with* $\dim_{\mathrm{A}}(\Omega^c) < n - p$. *Then the* p*-Hardy inequality* (6.1) *holds in* Ω. *Moreover, the* p*-Hardy inequality holds for every* $u \in \operatorname{Lip}_0(\mathbb{R}^n)$, *not only for* $u \in \operatorname{Lip}_0(\Omega)$.

PROOF. The claim follows from Theorem 10.34 for $q = p$ and $E = \Omega^c$. Theorem 10.25 implies that $|\Omega^c| = 0$ and thus the integrals in (10.15) can be taken over Ω for $u \in \operatorname{Lip}_0(\mathbb{R}^n)$. □

Hardy–Sobolev inequalities, for $p \le q \le p^* = \frac{np}{n-p}$, can be obtained by interpolating between the Hardy and Sobolev inequalities. See also Remark 10.53.

THEOREM 10.37. *Let* $1 \le p < n$ *and assume that the* p*-Hardy inequality holds in an open set* $\Omega \subset \mathbb{R}^n$. *Then the* (q,p)*-Hardy–Sobolev inequality holds in* Ω *for every* $p \le q \le \frac{np}{n-p}$.

PROOF. For $q = p$ we have the p-Hardy inequality and for $q = \frac{np}{n-p}$ we have the Sobolev inequality (3.6), and both of these hold in Ω. Hence we may assume that $p < q < \frac{np}{n-p} = p^*$. Let

$$\alpha = \frac{p^2}{np - nq + qp} > 1 \quad \text{and} \quad \alpha' = \frac{\alpha}{\alpha - 1},$$

and let $u \in \operatorname{Lip}_0(\Omega)$. By Hölder's inequality, we have

$$\begin{aligned}\Big(\int_\Omega |u(x)|^q d(x,\partial\Omega)^{\frac{q}{p}(n-p)-n}\,dx\Big)^{\frac{1}{q}} &= \Big(\int_\Omega |u(x)|^{\frac{p}{\alpha}+\frac{p^*}{\alpha'}} d(x,\partial\Omega)^{\frac{-p}{\alpha}}\,dx\Big)^{\frac{1}{q}} \\ &\le \Big(\int_\Omega |u(x)|^p d(x,\partial\Omega)^{-p}\,dx\Big)^{\frac{1}{q\alpha}}\Big(\int_\Omega |u(x)|^{p^*}\,dx\Big)^{\frac{1}{q\alpha'}}.\end{aligned}$$

The first integral on the right-hand side can be estimated with the p-Hardy inequality and the second with the Sobolev inequality. Since $\frac{1}{q\alpha} + \frac{n}{n-p}\frac{1}{q\alpha'} = \frac{1}{p}$, we conclude that

$$\begin{aligned}&\Big(\int_\Omega |u(x)|^q d(x,\partial\Omega)^{\frac{q}{p}(n-p)-n}\,dx\Big)^{\frac{1}{q}} \\ &\qquad\le C\Big(\int_\Omega |\nabla u(x)|^p\,dx\Big)^{\frac{1}{q\alpha}}\Big(\int_\Omega |\nabla u(x)|^p\,dx\Big)^{\frac{n}{n-p}\frac{1}{q\alpha'}} \\ &\qquad= C\Big(\int_\Omega |\nabla u(x)|^p\,dx\Big)^{\frac{1}{p}}.\end{aligned}$$

□

REMARK 10.38. Let $1 < p \le q < \frac{np}{n-p} < \infty$. If $\Omega \subsetneq \mathbb{R}^n$ is an open set and $\dim_{\mathrm{A}}(\Omega^c) < \frac{q}{p}(n-p)$, then the (q,p)-Hardy–Sobolev inequality holds in Ω by Remark 10.35. In this case Theorem 10.37 does not provide new information since for the p-Hardy inequality we need the assumption $\dim_{\mathrm{A}}(\Omega^c) < n - p$, which is stronger than the assumption for the (q,p)-Hardy–Sobolev inequality. On the other hand, we know from Theorem 7.17 that if $\dim_{\mathrm{L}}(\Omega^c) > n - p$ and Ω^c is unbounded, then the p-Hardy inequality holds in Ω. By Theorem 10.37 also the (q,p)-Hardy–Sobolev inequality holds in Ω.

The punctured ball $\Omega = B(0,1) \setminus \{0\}$, considered in Example 6.13, gives a simple model case where a suitable combination of thick and thin parts in the complement Ω^c implies that a p-Hardy inequality holds in Ω. Theorem 10.40 below is a generalization of this idea. In the proof we apply the following lemma concerning Hardy–Sobolev inequalities in intersections of open sets.

LEMMA 10.39. *Let $1 < p < n$ and $p \le q \le \frac{np}{n-p}$, and let $\Omega_1, \Omega_2 \subset \mathbb{R}^n$ be open sets with $\Omega = \Omega_1 \cap \Omega_2 \neq \emptyset$. Assume that the (q,p)-Hardy–Sobolev inequality holds in Ω_1 and Ω_2, with constants C_1 and C_2, respectively. Then the (q,p)-Hardy–Sobolev inequality holds in the open set Ω with a constant $C = C(C_1, C_2)$.*

PROOF. Let $u \in \operatorname{Lip}_0(\Omega)$. Since $\partial\Omega \subset \partial\Omega_1 \cup \partial\Omega_2$ and $\Omega \subset \Omega_1$ and $\Omega \subset \Omega_2$, we have $d(x,\partial\Omega) = \min\{d(x,\partial\Omega_1), d(x,\partial\Omega_2)\}$ for every $x \in \Omega$. This and the assumed

Hardy–Sobolev inequalities for Ω_1 and Ω_2 imply

$$\begin{aligned}
&\int_\Omega |u(x)|^q d(x,\partial\Omega)^{\frac{q}{p}(n-p)-n}\,dx \\
&\quad\le \int_\Omega |u(x)|^q d(x,\partial\Omega_1)^{\frac{q}{p}(n-p)-n}\,dx + \int_\Omega |u(x)|^q d(x,\partial\Omega_2)^{\frac{q}{p}(n-p)-n}\,dx \\
&\quad\le \int_{\Omega_1} |u(x)|^q d(x,\partial\Omega_1)^{\frac{q}{p}(n-p)-n}\,dx + \int_{\Omega_2} |u(x)|^q d(x,\partial\Omega_2)^{\frac{q}{p}(n-p)-n}\,dx \\
&\quad\le C_1^q\Big(\int_{\Omega_1} |\nabla u(x)|^p\,dx\Big)^{\frac{q}{p}} + C_2^q\Big(\int_{\Omega_2} |\nabla u(x)|^p\,dx\Big)^{\frac{q}{p}} \\
&\quad\le (C_1^q + C_2^q)\Big(\int_\Omega |\nabla u(x)|^p\,dx\Big)^{\frac{q}{p}},
\end{aligned}$$

where the final inequality holds since $|\nabla u| = 0$ almost everywhere in Ω^c. The claim follows, with $C = \big(C_1^q + C_2^q\big)^{\frac{1}{q}} \le C_1 + C_2$. □

THEOREM 10.40. *Let $1 < p < n$ and $p \le q \le \frac{np}{n-p}$. Assume that an open set $\Omega_1 \subset \mathbb{R}^n$ and a closed set $F \subset \mathbb{R}^n$ are such that Ω_1^c is unbounded,*

$$\dim_{\mathrm{L}}(\Omega_1^c) > n-p \quad \text{and} \quad \dim_{\mathrm{A}}(F) < \frac{q}{p}(n-p).$$

Then the (q,p)-Hardy–Sobolev inequality holds in the open set $\Omega = \Omega_1 \setminus F$.

PROOF. Notice that $\Omega \neq \emptyset$ since $\dim_{\mathrm{A}}(F) < n$ by the assumptions. Let $\Omega_2 = F^c$. Then $\dim_{\mathrm{A}}(\Omega_2^c) = \dim_{\mathrm{A}}(F) < \frac{q}{p}(n-p)$, and Theorem 10.34 implies that the (q,p)-Hardy–Sobolev inequality holds in the open set Ω_2. On the other hand, since $\dim_{\mathrm{L}}(\Omega_1^c) > n-p$ and Ω_1^c is unbounded, the p-Hardy inequality holds in Ω_1 by Theorem 7.17. Theorem 10.37 implies that the (q,p)-Hardy–Sobolev inequality holds in Ω_1, and the (q,p)-Hardy–Sobolev inequality for $\Omega = \Omega_1 \cap \Omega_2$ follows from Lemma 10.39. □

See Example 10.49 for an application of Theorem 10.40.

10.6. Necessary conditions for Hardy–Sobolev inequalities

Let $1 < p < n$ and $p \le q < \frac{np}{n-p}$. By Remark 10.38, the requirement

$$\dim_{\mathrm{L}}(\Omega^c) > n-p \quad \text{or} \quad \dim_{\mathrm{A}}(\Omega^c) < \frac{q}{p}(n-p) \tag{10.16}$$

is sufficient for the (q,p)-Hardy–Sobolev inequality in Ω, the former condition with the additional assumption that Ω^c is unbounded. Our main goal in this section is to show that conversely, if the (q,p)-Hardy–Sobolev inequality holds in Ω, then

$$\dim_{\mathrm{H}}(\Omega^c) \ge n-p \quad \text{or} \quad \dim_{\mathrm{A}}(\Omega^c) < \frac{q}{p}(n-p),$$

see Theorem 10.45. In addition, we discuss a localized version of this dimensional dichotomy and prove that in the case $q = p$ the inequality for the Hausdorff dimension can be made strict. However, it is not possible to replace $\dim_{\mathrm{H}}$ with the smaller quantity $\dim_{\mathrm{L}}$ in these necessary conditions, see Example 10.49. Hence the dimensional conditions for the complement do not quite give a geometric characterization for the validity of Hardy–Sobolev inequalities.

Before the proofs of the necessary conditions we need some preparations. The next lemma shows that if a part of the complement of Ω is small enough, then the functions for the Hardy–Sobolev inequality need not vanish in that part of Ω^c.

LEMMA 10.41. *Let $1 \le p < n$ and $1 \le q < \infty$ and assume that the (q,p)-Hardy–Sobolev inequality* (10.14) *holds in an open set $\Omega \subsetneq \mathbb{R}^n$ with a constant C_1. Assume that $U \subset \mathbb{R}^n$ is a bounded open set satisfying $\mathcal{H}^{n-p}_\infty(\Omega^c \cap \overline{U}) = 0$. Then there exists a constant $C = C(n,p,C_1)$ such that* (10.14) *holds for every $u \in \operatorname{Lip}_0(\Omega \cup U)$ with the constant C.*

PROOF. Let $u \in \operatorname{Lip}_0(\Omega \cup U)$ and $j \in \mathbb{N}$. Without loss of generality, we may assume that $\|u\|_\infty > 0$. By the definition of the Hausdorff content, see Definition 1.43, and the compactness of $\Omega^c \cap \overline{U}$, there are balls $B(z_i, r_i)$, $i = 1, \dots, N$, such that $\Omega^c \cap \overline{U} \subset \bigcup_{i=1}^N B(z_i, r_i)$ and

$$\sum_{i=1}^N r_i^{n-p} \le \|u\|_\infty^{-p} 2^{-j}.$$

Let

$$\psi_j(x) = \min_{1 \le i \le N} \min\{1, r_i^{-1} d(x, B(z_i, 2r_i))\},$$

for every $x \in \Omega$. Then

$$|\nabla \psi_j(x)|^p \le \sum_{i=1}^N r_i^{-p} \chi_{B(z_i,3r_i)}(x)$$

for almost every $x \in \Omega$. Let $u_j(x) = \psi_j(x) u(x)$ for $x \in \Omega$. Then $u_j \in \operatorname{Lip}_0(\Omega)$ and, by the Leibniz rule (2.4), we have

$$|\nabla u_j(x)| \le |\nabla \psi_j(x)||u(x)| + |\nabla u(x)|$$

for almost every $x \in \Omega$. Using the (q,p)-Hardy–Sobolev inequality for u_j, we obtain

$$\begin{aligned}
&\left(\int_\Omega |u_j(x)|^q d(x, \partial\Omega)^{\frac{q}{p}(n-p)-n} \, dx \right)^{\frac{p}{q}} \\
&\quad \le 2^p C_1^p \left(\|u\|_\infty^p \int_\Omega |\nabla \psi_j(x)|^p \, dx + \int_\Omega |\nabla u(x)|^p \, dx \right) \\
&\quad \le C \left(\|u\|_\infty^p \sum_{i=1}^N r_i^{n-p} + \int_\Omega |\nabla u(x)|^p \, dx \right) \le C 2^{-j} + C \int_\Omega |\nabla u(x)|^p \, dx,
\end{aligned}$$

where $C = C(n,p,C_1)$. The claim follows by Fatou's Lemma since $u_j(x) \to u(x)$ for almost every $x \in \Omega$ as $j \to \infty$. □

From Lemma 10.41 and Theorem 10.8, we obtain the following result.

LEMMA 10.42. *Let $1 < p < n$ and $p \le q < \frac{np}{n-p}$ and assume that the (q,p)-Hardy–Sobolev inequality holds in an open set $\Omega \subsetneq \mathbb{R}^n$ with a constant C_1. Assume that $B(x,R) \subset \mathbb{R}^n$ is an open ball such that $\mathcal{H}^{n-p}_\infty\big(\Omega^c \cap \overline{B}(x,2R)\big) = 0$. There exists $\delta = \delta(n,p,q,C_1) > 0$ such that*

$$\dim_{\mathrm{A}}\big(\Omega^c \cap \overline{B}(x,R)\big) \le \frac{q}{p}(n-p) - \delta.$$

PROOF. Let $z \in \Omega^c \cap B(x,R)$ and $0 < r < \frac{R}{2}$. Then $B(z,2r) \subset B(x,2R)$. Let

$$u(y) = \max\{0, 1 - \tfrac{1}{r} d(y, B(z,r))\}$$

for every $y \in \mathbb{R}^n$. Then $u \in \operatorname{Lip}_0(\Omega \cup B(x,2R))$, $u(y) = 1$ for every $y \in B(z,r)$, and $|\nabla u(y)| \le \frac{1}{r}\chi_{B(z,2r)}(y)$ for almost every $y \in \mathbb{R}^n$. By Lemma 10.41, the (q,p)-Hardy–Sobolev inequality holds in Ω for u with a constant $C = C(n,p,C_1)$. Since $d(y,\Omega^c) \le d(y, \Omega^c \cap \overline{B}(x,R))$ for every $y \in B(z,2r)$ and $|\Omega^c \cap \overline{B}(x,2R)| = 0$, we obtain

$$\begin{aligned}\int_{B(z,r)} d\big(y, \Omega^c \cap \overline{B}(x,R)\big)^{\frac{q}{p}(n-p)-n}\,dy &\le \int_{\Omega\cap B(z,2r)} |u(y)|^q d(y,\Omega^c)^{\frac{q}{p}(n-p)-n}\,dy \\ &\le C\Big(\int_\Omega |\nabla u(y)|^p\,dy\Big)^{\frac{q}{p}} \\ &\le Cr^{-q}\big|B(z,2r)\big|^{\frac{q}{p}} = Cr^{\frac{q}{p}(n-p)},\end{aligned}$$

where $C = C(n,p,q,C_1)$. This shows that the set $\Omega^c \cap \overline{B}(x,R)$ satisfies the Aikawa condition (10.1) for $\alpha = n - \frac{q}{p}(n-p) > 0$ whenever $0 < r < \frac{R}{2}$. Lemma 10.4 implies that $\alpha \in \mathcal{A}(\Omega^c \cap \overline{B}(x,R))$, with a constant $C = C(n,p,q,C_1)$. By Theorem 10.8, there is $\delta = \delta(n,p,q,C_1) > 0$ such that $\alpha + \delta \in \mathcal{A}(\Omega^c \cap \overline{B}(x,R))$. The claim $\dim_{\mathrm{A}}(\Omega^c \cap \overline{B}(x,R)) \le \frac{q}{p}(n-p) - \delta$ follows from Theorem 10.22. □

REMARK 10.43. In the case $q = p$ we have the following uniformity concerning $\delta > 0$ in Lemma 10.42. Let $1 < p_1 < p_2 < \infty$ and assume that the p-Hardy inequality holds in Ω for every $p_1 \le p \le p_2$ with the same constant C_1. If $p_1 \le p \le p_2$ and $\mathcal{H}^{n-p}_\infty(\Omega^c \cap \overline{B}(x,2R)) = 0$, then we can choose $\delta > 0$ in Lemma 10.42 to be independent of p, that is, $\delta = \delta(n,p_1,p_2,C_1) > 0$. This can be seen by tracing the dependence of δ on p explicitly via Theorem 10.8 and Theorem 8.38 to Lemma 8.37.

We have now established enough tools to prove the necessity of a local dimensional dichotomy.

THEOREM 10.44. *Let $1 < p < n$ and $p \le q < \frac{np}{n-p}$ and assume that the (q,p)-Hardy–Sobolev inequality holds in an open set $\Omega \subsetneq \mathbb{R}^n$. If $x \in \mathbb{R}^n$ and $R > 0$, then*

$$\dim_{\mathrm{H}}\big(\Omega^c \cap \overline{B}(x,2R)\big) \ge n-p \quad \text{or} \quad \dim_{\mathrm{A}}\big(\Omega^c \cap \overline{B}(x,R)\big) < \frac{q}{p}(n-p).$$

PROOF. Let $x \in \mathbb{R}^n$ and $R > 0$. The claim holds if $\dim_{\mathrm{H}}\big(\Omega^c \cap \overline{B}(x,2R)\big) \ge n-p$. Hence we may assume that $\dim_{\mathrm{H}}(\Omega^c \cap \overline{B}(x,2R)) < n-p$, and consequently $\mathcal{H}^{n-p}_\infty\big(\Omega^c \cap \overline{B}(x,2R)\big) = 0$. The assumptions of Lemma 10.42 are satisfied, and we conclude that

$$\dim_{\mathrm{A}}\big(\Omega^c \cap \overline{B}(x,R)\big) < \frac{q}{p}(n-p). \qquad \square$$

Along the same lines we obtain a global dimension dichotomy result.

THEOREM 10.45. *Let $1 < p < n$ and $p \le q < \frac{np}{n-p}$ and assume that the (q,p)-Hardy–Sobolev inequality holds in an open set $\Omega \subsetneq \mathbb{R}^n$. Then*

$$\dim_{\mathrm{H}}(\Omega^c) \ge n-p \quad \text{or} \quad \dim_{\mathrm{A}}(\Omega^c) < \frac{q}{p}(n-p).$$

PROOF. Let C_1 be the constant in the (q,p)-Hardy–Sobolev inequality for Ω. We may assume that $\dim_H(\Omega^c) < n-p$. Let $z \in \Omega^c$ and $r > 0$. Arguing as in the proof of Lemma 10.42, we obtain

$$\int_{B(z,r)} d(y,\Omega^c)^{\frac{q}{p}(n-p)-n}\,dy \le C r^{\frac{q}{p}(n-p)},$$

where $C = C(n,p,q,C_1)$. Thus $0 < n - \frac{q}{p}(n-p) \in \mathcal{A}(\Omega^c)$, and the claim follows from Theorem 10.24. □

In the case $q = p$ we obtain even better dichotomy results, due to the self-improvement property of Hardy inequalities, see Theorem 6.16.

THEOREM 10.46. *Let $1 < p < \infty$ and assume that the p-Hardy inequality holds in an open set $\Omega \subsetneq \mathbb{R}^n$ with a constant C_1. Let $x \in \mathbb{R}^n$ and $R > 0$. There exists $\varepsilon = \varepsilon(n,p,C_1) > 0$ such that either*

$$\dim_H\bigl(\Omega^c \cap \overline{B}(x,2R)\bigr) > n-p+\varepsilon \quad \text{or} \quad \dim_A\bigl(\Omega^c \cap \overline{B}(x,R)\bigr) < n-p-\varepsilon.$$

PROOF. By Theorem 6.16, there are $\varepsilon_1 > 0$ and $C_2 > 0$, both only depending on C_1, n and p, such that the q-Hardy inequality holds in Ω with the constant C_2 for every q with $p - \varepsilon_1 < q \le p$.

Let $x \in \mathbb{R}^n$ and $R > 0$, and let $0 < \varepsilon < \frac{1}{2}\varepsilon_1$ to be specified later. If

$$\dim_H\bigl(\Omega^c \cap \overline{B}(x,2R)\bigr) > n-p+\varepsilon,$$

the claim holds. Hence we may assume that $\dim_H\bigl(\Omega^c \cap \overline{B}(x,2R)\bigr) \le n-p+\varepsilon$, and consequently $\mathcal{H}^{n-q}_\infty\bigl(\Omega^c \cap \overline{B}(x,2R)\bigr) = 0$ for $q = p-2\varepsilon$. As the $(p-2\varepsilon)$-Hardy inequality holds in Ω and $p-2\varepsilon > p-\varepsilon_1$, we may use Lemma 10.42 and Remark 10.43 to conclude that there exists $\delta > 0$, independent of ε with $0 < \varepsilon < \frac{1}{2}\varepsilon_1$, such that

$$\dim_A\bigl(\Omega^c \cap \overline{B}(x,R)\bigr) \le n-p+2\varepsilon-\delta.$$

The claim follows by choosing $0 < \varepsilon < \min\bigl\{\frac{1}{2}\varepsilon_1, \frac{1}{3}\delta\bigr\}$. □

Along the same lines we obtain a global dimension dichotomy result for the p-Hardy inequality.

THEOREM 10.47. *Let $1 < p < \infty$ and assume that the p-Hardy inequality holds in an open set $\Omega \subsetneq \mathbb{R}^n$ with a constant C_1. There exists $\varepsilon = \varepsilon(n,p,C_1) > 0$ such that either*

$$\dim_H(\Omega^c) > n-p+\varepsilon \quad \text{or} \quad \dim_A(\Omega^c) < n-p-\varepsilon.$$

In particular, either $\dim_H(\Omega^c) > n-p$ or $\dim_A(\Omega^c) < n-p$.

PROOF. If $\dim_H(\Omega^c) \le n-p+\varepsilon$, then, as in the proof of Lemma 10.42, there are $\delta = \delta(n,p,C_1) > 0$ and $C = C(n,p,C_1)$ such that

$$\int_{B(z,r)} d(x,\Omega^c)^{-p+2\varepsilon-\delta}\,dx \le C r^{n-p+2\varepsilon-\delta},$$

for every $z \in \Omega^c$ and $r > 0$. Moreover, both C and δ are independent of ε. Choosing ε as in the proof of Theorem 10.46 proves the claim. □

EXAMPLE 10.48. Let $0 < \lambda < n-1$ and let $E \subset \mathbb{R}^n$ be an unbounded λ-regular set. By Theorem 7.12 and Theorem 10.21 we have $\dim_{\mathrm{L}}(E) = \dim_{\mathrm{H}}(E) = \dim_{\mathrm{A}}(E) = \lambda$. Let $\Omega = \mathbb{R}^n \setminus E$. The p-Hardy inequality holds in Ω for every $p > n - \lambda$, by Theorem 7.17, and for every $1 < p < n - \lambda$, by Corollary 10.36. On the other hand, it follows from Theorem 10.47 that for $p = n - \lambda$ the p-Hardy inequality does not hold in Ω. Notice that the (q,p)-Hardy–Sobolev inequality holds in Ω whenever $1 < p < q \leq \frac{np}{n-p}$, by Remark 10.38, since at least one of the conditions $\lambda > n - p$ or $\lambda < \frac{q}{p}(n-p)$ holds.

EXAMPLE 10.49. Consider the case $1 < p = q < \infty$ and let $0 < \lambda < n-1$. Let $\Omega_1 = B(0,1)$ and let $F \subset \Omega_1$ be a closed λ-regular set. Then $\dim_{\mathrm{L}}(\Omega_1^c) = n$ and $\dim_{\mathrm{L}}(F) = \dim_{\mathrm{H}}(F) = \dim_{\mathrm{A}}(F) = \lambda$, by Theorem 7.12 and Theorem 10.21. Define $\Omega = \Omega_1 \setminus F$. Then $\dim_{\mathrm{L}}(\Omega^c) = \lambda$, and hence the p-Hardy inequality holds in Ω for every $p > n - \lambda$, by Theorem 7.17. On the other hand, if $1 < p < n - \lambda$, then the p-Hardy inequality holds in Ω by Theorem 10.40. Finally, from Theorem 10.46 it follows that the $(n-\lambda)$-Hardy inequality does not hold in Ω. In conclusion, the p-Hardy inequality holds in Ω if and only if $1 < p < n - \lambda$ or $n - \lambda < p < \infty$. Notice that here $\dim_{\mathrm{A}}(\Omega^c) = n$, and hence the mixed case in Theorem 10.40 is really needed in order to obtain the p-Hardy inequality for $1 < p < n - \lambda$.

Observe that for $1 < p < n - \lambda$ the p-Hardy inequality holds in Ω but $\dim_{\mathrm{L}}(\Omega^c) = \lambda < n - p$ and $\dim_{\mathrm{A}}(\Omega^c) = n > n - p$. Hence neither of the conditions in (10.16), with $q = p$, is valid. This shows that (10.16) is not necessary for the p-Hardy inequality in Ω and consequently it is not possible to replace $\dim_{\mathrm{H}}(\Omega^c)$ by $\dim_{\mathrm{L}}(\Omega^c)$ in Theorem 10.47. On the other hand, for $p = n - \lambda$ we have $\dim_{\mathrm{H}}(\Omega^c) = n > n - p = \lambda$, but the p-Hardy inequality does not hold in Ω. Hence the condition $\dim_{\mathrm{H}}(\Omega^c) > n - p$ is not sufficient for the p-Hardy inequality in Ω, unlike the condition $\dim_{\mathrm{L}}(\Omega^c) > n - p$ when Ω^c is unbounded, see Theorem 7.17.

10.7. Testing conditions

In this section we discuss characterizations of Hardy–Sobolev inequalities in terms of capacities. These conditions do not have an immediate connection to the geometry of an open set or its complement, but they give convenient tests for the validity of Hardy–Sobolev inequalities.

THEOREM 10.50. *Let $1 < p \leq q < \infty$ and let $\Omega \subsetneq \mathbb{R}^n$ be an open set. The (q,p)-Hardy–Sobolev inequality holds in Ω if and only if there exists a constant C_1 such that*

$$\int_E d(x,\partial\Omega)^{\frac{q}{p}(n-p)-n}\,dx \leq C_1 \operatorname{cap}_p(E,\Omega)^{\frac{q}{p}} \tag{10.17}$$

for every compact set $E \subset \Omega$.

PROOF. Assume first that the (q,p)-Hardy–Sobolev inequality holds in Ω, with a constant C. Let $E \subset \Omega$ be a compact set and let $u \in \operatorname{Lip}_0(\Omega)$ be such that $u(x) = 1$ for every $x \in E$. Then the (q,p)-Hardy–Sobolev inequality implies

$$\begin{aligned}\int_E d(x,\partial\Omega)^{\frac{q}{p}(n-p)-n}\,dx &\leq \int_\Omega |u(x)|^q d(x,\partial\Omega)^{\frac{q}{p}(n-p)-n}\,dx \\ &\leq C^q\left(\int_\Omega |\nabla u(x)|^p\,dx\right)^{\frac{q}{p}}.\end{aligned}$$

Inequality (10.17) with the constant $C_1 = C^q$ follows by taking infimum over all such functions u, see Theorem 5.44.

Assume then that (10.17) holds with a constant C_1 for all compact sets $E \subset \Omega$, and let $u \in \operatorname{Lip}_0(\Omega)$. For $j \in \mathbb{Z}$, let $E_j = \{x \in \Omega : |u(x)| > 2^j\}$. By (10.17) and the fact that the set $\overline{E}_j \subset \Omega$ is compact, we have

$$\begin{aligned}
&\int_\Omega |u(x)|^q d(x,\partial\Omega)^{\frac{q}{p}(n-p)-n}\,dx \\
&\qquad\le \sum_{j=-\infty}^{\infty} 2^{(j+2)q} \int_{E_{j+1}\setminus E_{j+2}} d(x,\partial\Omega)^{\frac{q}{p}(n-p)-n}\,dx \\
&\qquad\le \sum_{j=-\infty}^{\infty} 2^{(j+2)q} \int_{\overline{E}_{j+1}} d(x,\partial\Omega)^{\frac{q}{p}(n-p)-n}\,dx \\
&\qquad\le 4^q C_1 \sum_{j=-\infty}^{\infty} 2^{jq} \operatorname{cap}_p(\overline{E}_{j+1},\Omega)^{\frac{q}{p}}
\end{aligned} \tag{10.18}$$

for every $j \in \mathbb{Z}$. Define $u_j \colon \mathbb{R}^n \to [0,1]$ by

$$u_j(x) = \begin{cases} 1, & \text{if } |u(x)| \ge 2^{j+1}, \\ 2^{-j}|u(x)| - 1, & \text{if } 2^j < |u(x)| < 2^{j+1}, \\ 0, & \text{if } |u(x)| \le 2^j. \end{cases}$$

Then $u_j \in \operatorname{Lip}_0(\Omega)$, $u_j = 1$ in $\overline{E}_{j+1}$ and $u_j = 0$ in $\mathbb{R}^n \setminus E_j$. By Theorem 5.44, we can test the capacity $\operatorname{cap}_p(\overline{E}_{j+1},\Omega)$ with u_j, and obtain

$$\begin{aligned}
\operatorname{cap}_p(\overline{E}_{j+1},\Omega) &\le \int_{E_j\setminus E_{j+1}} |\nabla u_j(x)|^p\,dx \\
&\le 2^{-jp} \int_{E_j\setminus E_{j+1}} |\nabla u(x)|^p\,dx.
\end{aligned}$$

Since $\frac{q}{p} \ge 1$, it follows that

$$\begin{aligned}
\sum_{j=-\infty}^{\infty} 2^{jq} \operatorname{cap}_p(\overline{E}_{j+1},\Omega)^{\frac{q}{p}} &\le \sum_{j=-\infty}^{\infty} \Big(\int_{E_j\setminus E_{j+1}} |\nabla u(x)|^p\,dx \Big)^{\frac{q}{p}} \\
&\le \Big(\sum_{j=-\infty}^{\infty} \int_{E_j\setminus E_{j+1}} |\nabla u(x)|^p\,dx \Big)^{\frac{q}{p}} \\
&= \Big(\int_\Omega |\nabla u(x)|^p\,dx \Big)^{\frac{q}{p}}.
\end{aligned} \tag{10.19}$$

From (10.18) and (10.19) we conclude that the (q,p)-Hardy–Sobolev inequality in (10.14) holds in Ω with the constant $4C_1^{\frac{1}{q}}$. □

The variational capacity is countably subadditive by Theorem 5.34 (e). However, the corresponding countable additivity on pairwise disjoint sets does not hold. Next we prove a weaker converse of the subadditivity, where we only consider intersections of a compact set with the Whitney cubes, and allow a constant in the inequality. This is called the quasiadditivity property of the variational capacity.

THEOREM 10.51. *Let $1 < p \le q < \infty$. Assume that the (q,p)-Hardy–Sobolev inequality (10.14) holds in an open set $\Omega \subset \mathbb{R}^n$ with a constant C_1, and let $\mathcal{W}(\Omega) = \{Q_i : i \in \mathbb{N}\}$ be a Whitney decomposition of Ω. There exists a constant $C = C(n,p,q,C_1)$ such that*

$$\sum_{i=1}^{\infty} \operatorname{cap}_p(E \cap \overline{Q}_i, \Omega)^{\frac{q}{p}} \le C \operatorname{cap}_p(E,\Omega)^{\frac{q}{p}} \tag{10.20}$$

for every compact set $E \subset \Omega$.

PROOF. As in Lemma 8.9 we write $Q_i = Q(x_i, r_i) \in \mathcal{W}(\Omega)$ and $Q_i^* = Q(x_i, \frac{9}{8}r_i)$ with $\sum_{i=1}^{\infty} \chi_{Q_i^*}(x) \le C(n)$ for every $x \in \Omega$.

Let $E \subset \Omega$ be a compact set and $u \in \operatorname{Lip}_0(\Omega)$ be such that $u(x) = 1$ for every $x \in E$. For every $i \in \mathbb{N}$ we choose a Lipschitz function φ_i satisfying $\varphi_i(x) = 1$ for every $x \in \overline{Q}_i$, and $|\nabla \varphi_i(x)| \le C(n) r_i^{-1}$ and $0 \le \varphi_i(x) \le \chi_{Q_i^*}(x)$ for every $x \in \mathbb{R}^n$. By Theorem 5.44, we can test the capacity $\operatorname{cap}_p(E \cap \overline{Q}_i, \Omega)$ with $u_i = u\varphi_i$ for every $i \in \mathbb{N}$. By the Leibniz rule (2.4), we have

$$|\nabla u_i(x)| \le C\big(|\nabla u(x)| + r_i^{-1}|u(x)|\big)\chi_{Q_i^*}(x)$$

for almost every $x \in \Omega$. Here $r_i^{-1} \le C(n) d(x,\partial\Omega)^{-1}$ for every $x \in Q_i^*$. Thus the left-hand side in (10.20) is bounded from above by

$$\begin{aligned}
&\sum_{i=1}^{\infty}\Big(\int_{\Omega} |\nabla u_i(x)|^p dx\Big)^{\frac{q}{p}} \\
&\le C \sum_{i=1}^{\infty}\Big(\int_{Q_i^*} |\nabla u(x)|^p \, dx + \int_{Q_i^*} |u(x)|^p d(x,\partial\Omega)^{-p} \, dx\Big)^{\frac{q}{p}} \qquad (10.21)\\
&\le C \sum_{i=1}^{\infty}\Big(\int_{Q_i^*} |\nabla u(x)|^p \, dx\Big)^{\frac{q}{p}} + C \sum_{i=1}^{\infty}\Big(\int_{Q_i^*} |u(x)|^p d(x,\partial\Omega)^{-p} \, dx\Big)^{\frac{q}{p}}.
\end{aligned}$$

The first term on the right-hand side of (10.21) can be estimated, by the bounded overlap of the cubes Q_i^*, as

$$\sum_{i=1}^{\infty}\Big(\int_{Q_i^*} |\nabla u(x)|^p \, dx\Big)^{\frac{q}{p}} \le \Big(\sum_{i=1}^{\infty}\int_{Q_i^*} |\nabla u(x)|^p \, dx\Big)^{\frac{q}{p}} \le C\Big(\int_{\Omega} |\nabla u(x)|^p \, dx\Big)^{\frac{q}{p}}.$$

For the second term on the right-hand side of (10.21), we apply Hölder's inequality, the estimates $|Q_i^*| \le C d(x,\partial\Omega)^n$ for every $x \in Q_i^*$ and $\sum_{i=1}^{\infty} \chi_{Q_i^*}(x) \le C$ for every $x \in \Omega$, and inequality (10.14). This gives

$$\begin{aligned}
\sum_{i=1}^{\infty}\Big(\int_{Q_i^*} |u(x)|^p d(x,\partial\Omega)^{-p} \, dx\Big)^{\frac{q}{p}} &\le \sum_{i=1}^{\infty} |Q_i^*|^{\frac{q}{p}\frac{q-p}{q}} \int_{Q_i^*} |u(x)|^q d(x,\partial\Omega)^{-q} \, dx \\
&\le C \sum_{i=1}^{\infty} \int_{Q_i^*} |u(x)|^q d(x,\partial\Omega)^{-q+n\frac{q}{p}-n} \, dx \\
&\le C \int_{\Omega} |u(x)|^q d(x,\partial\Omega)^{\frac{q}{p}(n-p)-n} \, dx \\
&\le C\Big(\int_{\Omega} |\nabla u(x)|^p \, dx\Big)^{\frac{q}{p}}.
\end{aligned}$$

After collecting the estimates above and taking infimum over all functions u as above, the claim follows from Theorem 5.44. $\square$

The quasiadditivity property in Theorem 10.51 gives the following characterization for Hardy–Sobolev inequalities. Notice that for $q = \frac{np}{n-p}$ the claim in part (a) of Theorem 10.52 holds for every open set $\Omega \subsetneq \mathbb{R}^n$, and thus also (b) and (c) hold for every open set $\Omega \subsetneq \mathbb{R}^n$. An important ingredient in the proof below is the use of the maximal function $M_\Omega u \in W_0^{1,p}(\Omega)$ as a capacity test function.

THEOREM 10.52. *Let $1 < p < n$ and $p \le q \le \frac{np}{n-p}$. Assume that $\Omega \subsetneq \mathbb{R}^n$ is an open set and let $\mathcal{W}(\Omega) = \{Q_i : i \in \mathbb{N}\}$ be a Whitney decomposition of Ω. The following conditions are equivalent.*

(a) *The (q,p)-Hardy–Sobolev inequality holds in Ω.*
(b) *There exists a constant C such that inequality*

$$\sum_{i=1}^{\infty} \operatorname{cap}_p(E \cap \overline{Q}_i, \Omega)^{\frac{q}{p}} \le C \operatorname{cap}_p(E, \Omega)^{\frac{q}{p}}$$

holds for every compact set $E \subset \Omega$.
(c) *There exists a constant C such that inequality*

$$\sum_{i \in I} \operatorname{cap}_p(\overline{Q}_i, \Omega)^{\frac{q}{p}} \le C \operatorname{cap}_p\Bigl(\bigcup_{i \in I} \overline{Q}_i, \Omega\Bigr)^{\frac{q}{p}}$$

holds whenever $I \subset \mathbb{N}$ is a finite set.

Moreover, the constants in each of the conditions only depend on each other, n, p and q.

PROOF. The implication from (a) to (b) is proved in Theorem 10.51. The implication from (b) to (c) follows by considering $E = \bigcup_{i \in I} \overline{Q}_i$.

Then assume that condition (c) holds with a constant C_1 and fix a compact set $E \subset \Omega$. To show that (a) holds, by Theorem 10.50 it suffices to prove that there exists a constant $C = C(n, p, C_1)$ such that

$$\int_E d(x, \partial\Omega)^{\frac{q}{p}(n-p)-n} \, dx \le C \operatorname{cap}_p(E, \Omega)^{\frac{q}{p}}. \tag{10.22}$$

Let $u \in \operatorname{Lip}_0(\Omega)$ with $u(x) = 1$ for every $x \in E$. We partition the Whitney decomposition $\mathcal{W}(\Omega) = \{Q_i : i \in \mathbb{N}\}$ as $\mathcal{W}_1 = \{Q_i \in \mathcal{W}(\Omega) : u_{Q_i} < \frac{1}{2}\}$ and $\mathcal{W}_2 = \mathcal{W}(\Omega) \setminus \mathcal{W}_1$. Then the left-hand side of (10.22) can be written as

$$\Bigl(\sum_{Q \in \mathcal{W}_1} + \sum_{Q \in \mathcal{W}_2}\Bigr) \int_{E \cap Q} d(x, \partial\Omega)^{\frac{q}{p}(n-p)-n} \, dx. \tag{10.23}$$

To estimate the first sum, we observe that, for every $Q \in \mathcal{W}_1$ and every $x \in E \cap Q$,

$$\tfrac{1}{2} = 1 - \tfrac{1}{2} < u(x) - u_Q = |u(x) - u_Q|.$$

By the (q,p)-Poincaré inequality on cubes, see Corollary 9.24, we obtain

$$
\begin{aligned}
\sum_{Q\in\mathcal{W}_1}\int_{E\cap Q} d(x,\partial\Omega)^{\frac{q}{p}(n-p)-n}\,dx &\le C\sum_{Q\in\mathcal{W}_1} l(Q)^{\frac{q}{p}(n-p)-n}\int_Q |u(x)-u_Q|^q\,dx\\
&\le C\sum_{Q\in\mathcal{W}_1} l(Q)^{\frac{q}{p}(n-p)}l(Q)^q\Big(\fint_Q |\nabla u(x)|^p\,dx\Big)^{\frac{q}{p}}\\
&\le C\sum_{Q\in\mathcal{W}_1}\Big(\int_Q |\nabla u(x)|^p\,dx\Big)^{\frac{q}{p}}\\
&\le C\Big(\int_\Omega |\nabla u(x)|^p\,dx\Big)^{\frac{q}{p}},
\end{aligned}
$$

where $C = C(n,p,q)$.

The second sum in (10.23) can be estimated by using the boundedness of the local maximal operator M_Ω. Let $Q\in\mathcal{W}_2$ and $x\in Q$. Then $Q\subset B(x,\operatorname{diam}(Q))\subset\Omega$. The inequalities in (8.5) and the definition of $\mathcal{W}_2$ imply

$$
M_\Omega u(x)\ge \fint_{B(x,\operatorname{diam}(Q))}|u(y)|\,dy \ge C(n)\fint_Q |u(y)|\,dy\ge C(n)u_Q\ge \frac{C(n)}{2}. \tag{10.24}
$$

Since $u\in \operatorname{Lip}_0(\Omega)\subset W_0^{1,p}(\Omega)$, we have by Corollary 4.28 that $M_\Omega u\in W_0^{1,p}(\Omega)$. Moreover, $M_\Omega u$ is continuous in Ω by Lemma 1.31, and by (10.24) there exists $C_2=C_2(n)>0$ such that $C_2M_\Omega u\ge 1$ in $\bigcup_{Q\in\mathcal{W}_2}\overline{Q}$. It follows from Theorem 5.46 that we may test the capacity $\operatorname{cap}_p\big(\bigcup_{Q\in\mathcal{W}_2}\overline{Q},\Omega\big)$ with the function $C_2M_\Omega u$. Note that the set $\{x\in\Omega : u(x)\ge\frac12\}$ is compact since $u\in\operatorname{Lip}_0(\Omega)$, and so $\mathcal{W}_2$ is finite.

Let $Q=Q(z,r)\in\mathcal{W}_2$. As $1<p<n$, we have, by the monotonicity properties of the capacity and Remark 5.37, that

$$
\operatorname{cap}_p(\overline{Q},\Omega)\ge\operatorname{cap}_p(\overline{Q},\mathbb{R}^n)\ge\operatorname{cap}_p(\overline{B}(z,r),\mathbb{R}^n)\ge C(n,p)r^{n-p}.
$$

Thus

$$
\int_{E\cap Q} d(x,\partial\Omega)^{\frac{q}{p}(n-p)-n}\,dx\le C(n,p,q)r^{\frac{q}{p}(n-p)}\le C(n,p,q)\operatorname{cap}_p(\overline{Q},\Omega)^{\frac{q}{p}}.
$$

From the assumed condition (c), Theorem 4.25, and the maximal function theorem, see Theorem 1.15, we hence obtain

$$
\begin{aligned}
\sum_{Q\in\mathcal{W}_2}\int_{E\cap Q} d(x,\partial\Omega)^{\frac{q}{p}(n-p)-n}\,dx &\le C\sum_{Q\in\mathcal{W}_2}\operatorname{cap}_p(\overline{Q},\Omega)^{\frac{q}{p}}\\
&\le C\operatorname{cap}_p\Big(\bigcup_{Q\in\mathcal{W}_2}\overline{Q},\Omega\Big)^{\frac{q}{p}}\\
&\le C\Big(\int_\Omega |\nabla M_\Omega u(x)|^p\,dx\Big)^{\frac{q}{p}}\\
&\le C\Big(\int_\Omega |\nabla u(x)|^p\,dx\Big)^{\frac{q}{p}},
\end{aligned}
$$

where $C = C(n, p, q, C_1)$. By combining the estimates for $\mathcal{W}_1$ and $\mathcal{W}_2$ we conclude that

$$\int_E d(x, \partial\Omega)^{\frac{q}{p}(n-p)-n}\,dx = \sum_{Q\in\mathcal{W}} \int_{E\cap Q} d(x, \partial\Omega)^{\frac{q}{p}(n-p)-n}\,dx$$
$$\leq C(n, p, q, C_1)\biggl(\int_\Omega |\nabla u(x)|^p\,dx\biggr)^{\frac{q}{p}},$$

and the desired estimate (10.22) follows from Theorem 5.44 by taking infimum over all Lipschitz functions u as above. □

REMARK 10.53. Theorem 10.52 can be used to give an alternative proof for Theorem 10.37, and in fact we obtain the following more general statement. Let $1 < p \leq q < p^* = \frac{np}{n-p} < \infty$ and assume that the (q, p)-Hardy–Sobolev inequality holds in an open set $\Omega \subset \mathbb{R}^n$. Then the (q', p)-Hardy–Sobolev inequality holds in Ω for every $q \leq q' \leq p^*$. This follows from Theorem 10.52 and the simple observation that if $q \leq q'$, then

$$\biggl(\sum_{i=1}^{\infty} \operatorname{cap}_p(E \cap \overline{Q}_i, \Omega)^{\frac{q'}{p}}\biggr)^{\frac{p}{q'}} \leq \biggl(\sum_{i=1}^{\infty} \operatorname{cap}_p(E \cap \overline{Q}_i, \Omega)^{\frac{q}{p}}\biggr)^{\frac{p}{q}}$$

whenever $\mathcal{W}(\Omega) = \{Q_i : i \in \mathbb{N}\}$ is a Whitney decomposition of Ω and $E \subset \Omega$ is a compact set.

10.8. Notes

The Aikawa condition was introduced and applied in connection with quasi-additivity properties of Riesz capacities by Aikawa [**7**, **8**]. Self-improvement of the Aikawa condition is implicit in the proof of Lemma 2.4 in Koskela and Zhong [**246**]. This condition was applied for weighted Hardy inequalities by Lehrbäck [**258**]. Section 10.2 follows the ideas in Dyda, Ihnatsyeva, Lehrbäck, Tuominen and Vähäkangas [**111**], but the connection between the Aikawa condition and the A_p properties of distance functions is mentioned already by Aikawa in [**8**, p. 151]. Horiuchi [**193**] has similar results in terms of the so-called $P(s)$-condition, which can also be characterized using the Assouad dimension, see Lehrbäck and Vähäkangas [**265**, Theorem 3.4].

The Assouad dimension was studied by Assouad [**19**, **20**], but essentially the same concept has appeared earlier for instance in Larman [**255**]. See Luukkainen [**293**] and Fraser [**136**] for more information concerning the history and theory of Assouad dimension. To emphasize the duality with the lower dimension, sometimes the Assouad dimension is called the upper Assouad dimension and denoted by $\overline{\dim}_{\mathrm{A}}(E)$.

Assouad dimension of porous sets is discussed in Luukkainen [**293**, Section 5]. Theorem 10.24 was considered in Lehrbäck and Tuominen [**263**]. Theorem 10.26 is from Dyda, Ihnatsyeva, Lehrbäck, Tuominen and Vähäkangas [**111**, Corollary 3.8]. See also Durán and López García [**109**, Lemma 3.3] and Aimar, Carena, Durán and Toschi [**9**] concerning A_p properties of distance functions.

Theorem 10.29 has been studied by Cruz-Uribe, Moen and Rodney [**94**] and Kurki and Vähäkangas [**252**]. For Theorem 10.30, see Kurki and Vähäkangas [**252**]. Conditions for Hardy inequalities in terms of the lower dimension and the Assouad dimension were obtained in Lehrbäck [**261**]. Dyda and Vähäkangas [**113**] have corresponding results for fractional Hardy inequalities. Similar ideas are also present in the unpublished manuscript by Wannebo [**394**], see for instance [**394**, Definition 6.6].

Hardy–Sobolev inequalities have been considered for instance in Dyda, Ihnatsyeva, Lehrbäck, Tuominen and Vähäkangas [**111**], Lehrbäck and Vähäkangas [**265**], Maz′ya [**318**, Section 2.1.6], [**321**, Section 2.1.7] and Opic and Kufner [**340**, Section 21]. Horiuchi [**192, 193**] has closely related results for embeddings between weighted Sobolev spaces. It is also possible to deduce a weighted version of the global Hardy–Sobolev inequality (10.15) using Theorem 9.21, see Dyda, Ihnatsyeva, Lehrbäck, Tuominen and Vähäkangas [**111**, Section 6]. The mixed case in Theorem 10.40, for $p = q$, has been considered in Lehrbäck [**258, 261**] for weighted inequalities. The necessary conditions in Section 10.6 are based on Koskela and Zhong [**246**], see also Lehrbäck [**258, 261**] and Lehrbäck and Tuominen [**263**].

The proof of Theorem 10.50 is based on a truncation argument that can be found in Maz′ya [**318**, p. 110]. For more information about this kind of characterizations, we refer to Maz′ya [**318–320**]. The truncation method has been applied, for example, in Adams and Hedberg [**4**, Theorem 7.2.1], Bakry, Coulhon, Ledoux and Saloff-Coste [**21**], Biroli and Mosco [**30**], Björn and Björn [**31**, Lemma 4.20], Capogna, Danielli and Garofalo [**73**], Coulhon [**90**], Franchi, Gutiérrez and Wheeden [**131**], Garofalo and Nhieu [**143**], Hajłasz and Koskela [**171**], Heinonen and Koskela [**188, 189**], Kinnunen and Korte [**225**], Long and Nie [**285**], Maheux and Saloff-Coste [**297**], Saloff-Coste [**357**, Section 3.2], Semmes [**358**], Tartar [**379**] and Turesson [**388**].

Quasiadditivity of capacity has been studied by Aikawa [**7, 8**], Dyda and Vähäkangas [**114**], Hurri-Syrjänen and Vähäkangas [**198**] and Lehrbäck and Shanmugalingam [**262**]. For $q > p$, the quasiadditivity results in Section 10.7 have not appeared previously, but their counterparts for fractional Hardy inequalities have been obtained by Hurri-Syrjänen and Vähäkangas [**198**].

CHAPTER 11

The p-Laplace Equation

Let $1 < p < \infty$ and let $\Omega \subset \mathbb{R}^n$ be an open set. Assume that $u \in C^2(\Omega)$ with $\nabla u(x) \neq 0$ for every $x \in \Omega$. We consider the p-Laplace equation

$$\operatorname{div}(|\nabla u(x)|^{p-2}\nabla u(x)) = 0$$

for every $x \in \Omega$. Here the divergence of a differentiable vector field $F\colon \Omega \to \mathbb{R}^n$, $F(x) = (F_1(x), \dots, F_n(x))$, at $x \in \Omega$ is

$$\operatorname{div} F(x) = \sum_{i=1}^{n} \frac{\partial F_i(x)}{\partial x_i}.$$

A direct computation shows that

$$\begin{aligned}&\operatorname{div}(|\nabla u(x)|^{p-2}\nabla u(x))\\ &\quad= |\nabla u(x)|^{p-4}\Big(|\nabla u(x)|^2 \Delta u(x) + (p-2)\sum_{i,j=1}^{n} \frac{\partial u}{\partial x_i}(x)\frac{\partial u}{\partial x_j}(x)\frac{\partial^2 u}{\partial x_i \partial x_j}(x)\Big)\end{aligned}$$

for every $x \in \Omega$. This gives a formula for a classical solution to the p-Laplace equation in Ω. We remark that the assumption $\nabla u(x) \neq 0$ is not needed when $p \geq 2$. For $p = 2$ we have the Laplace equation, which is linear. When $p \neq 2$, the p-Laplace equation is nonlinear and a sum of two solutions is not necessarily a solution. However, a constant can be added to a solution and a solution can be multiplied by a constant without destroying the property of being a solution. This chapter discusses existence and regularity of a weak solution to the p-Laplace equation.

11.1. Weak solutions

Assume that $u \in C^2(\Omega)$, with $\nabla u(x) \neq 0$ for every $x \in \Omega$, is a classical solution to the p-Laplace equation. An integration by parts gives

$$0 = -\int_\Omega \operatorname{div}(|\nabla u(x)|^{p-2}\nabla u(x))\varphi(x)\,dx = \int_\Omega |\nabla u(x)|^{p-2}\nabla u(x) \cdot \nabla\varphi(x)\,dx$$

for every $\varphi \in C_0^\infty(\Omega)$. Observe that there is no boundary term since φ is compactly supported in Ω. Conversely, if $u \in C^2(\Omega)$ is such that $\nabla u(x) \neq 0$ for every $x \in \Omega$ and

$$\int_\Omega |\nabla u(x)|^{p-2}\nabla u(x) \cdot \nabla\varphi(x)\,dx = 0$$

for every $\varphi \in C_0^\infty(\Omega)$, then by the computation above

$$\int_\Omega \operatorname{div}(|\nabla u(x)|^{p-2}\nabla u(x))\varphi(x)\,dx = 0$$

for every $\varphi \in C_0^\infty(\Omega)$. This implies

$$\operatorname{div}(|\nabla u(x)|^{p-2}\nabla u(x)) = 0$$

for every $x \in \Omega$. We will use from now on the convention $|0|^{p-2}0 = 0$, and with this we may omit the assumption that $\nabla u(x) \neq 0$ for every $x \in \Omega$.

The considerations above motivate the definition of a weak solution. Second order derivatives are needed in the definition of a classical solution to the p-Laplace equation, but in the following definition it is enough to only assume that first order weak derivatives exist.

DEFINITION 11.1. Let $1 < p < \infty$ and let $\Omega \subset \mathbb{R}^n$ be an open set. A function $u \in W^{1,p}_{\text{loc}}(\Omega)$ is a weak solution to the p-Laplace equation in Ω if

$$\int_\Omega |\nabla u(x)|^{p-2}\nabla u(x) \cdot \nabla\varphi(x)\, dx = 0 \tag{11.1}$$

for every $\varphi \in C_0^\infty(\Omega)$.

We show that being a weak solution is a local property.

LEMMA 11.2. *Let $1 < p < \infty$ and let $\Omega \subset \mathbb{R}^n$ be an open set. A function $u \in W^{1,p}_{\text{loc}}(\Omega)$ is a weak solution to the p-Laplace equation in Ω if and only if for every $x \in \Omega$ there exists $r_x > 0$ such that $B(x, r_x) \Subset \Omega$ and u is a weak solution to the p-Laplace equation in $B(x, r_x)$.*

PROOF. Assume first that $u \in W^{1,p}_{\text{loc}}(\Omega)$ is a weak solution in Ω and let $B(x_0, r) \Subset \Omega$. Since $C_0^\infty(B(x_0,r)) \subset C_0^\infty(\Omega)$, we have

$$\int_{B(x_0,r)} |\nabla u(x)|^{p-2}\nabla u(x) \cdot \nabla\varphi(x)\, dx = \int_\Omega |\nabla u(x)|^{p-2}\nabla u(x) \cdot \nabla\varphi(x)\, dx = 0$$

for every $\varphi \in C_0^\infty(B(x_0,r))$. The restriction of u belongs to $W^{1,p}(B(x_0,r))$, and thus u is a weak solution in $B(x_0, r)$. Assume then that for every $x \in \Omega$ there exists a ball $B(x, r_x) \Subset \Omega$, in which u is a weak solution. Observe that $u \in W^{1,p}_{\text{loc}}(\Omega)$ since $u \in W^{1,p}_{\text{loc}}(B(x, r_x))$ for every $x \in \Omega$. Let $\varphi \in C_0^\infty(\Omega)$. The collection $\{B(x, \frac{r_x}{2}) : x \in \operatorname{supp}\varphi\}$ is an open cover of $\operatorname{supp}\varphi$. Since $\operatorname{supp}\varphi$ is compact, there exists a finite number of points $x_i \in \operatorname{supp}\varphi$ and corresponding balls $B(x_i, \frac{r_i}{2})$ with $r_i = r_{x_i}$ for $i = 1, 2, \ldots, k$, such that $\operatorname{supp}\varphi \subset \bigcup_{i=1}^k B(x_i, \frac{r_i}{2})$. We consider a partition of unity $\psi_i \in C_0^\infty(B(x_i, r_i))$ with $0 \le \psi_i \le 1$ for every $i = 1, 2, \ldots, k$, and $\sum_{i=1}^k \psi_i(x) = 1$ for every $x \in \operatorname{supp}\varphi$. Then $\varphi\psi_i \in C_0^\infty(B(x_i, r_i))$ for every $i = 1, 2, \ldots, k$, and $\sum_{i=1}^k \nabla\psi_i(x) = 0$ for almost every $x \in \operatorname{supp}\varphi$ by Lemma 2.13. By summing up we obtain

$$\begin{aligned}
0 &= \sum_{i=1}^k \int_{B(x_i,r_i)} |\nabla u(x)|^{p-2}\nabla u(x) \cdot \nabla(\varphi(x)\psi_i(x))\, dx \\
&= \int_\Omega |\nabla u(x)|^{p-2}\nabla u(x) \cdot \sum_{i=1}^k \nabla(\varphi(x)\psi_i(x))\, dx \\
&= \int_\Omega |\nabla u(x)|^{p-2}\nabla u(x) \cdot \sum_{i=1}^k (\nabla\varphi(x)\psi_i(x) + \varphi(x)\nabla\psi_i(x))\, dx \\
&= \int_\Omega |\nabla u(x)|^{p-2}\nabla u(x) \cdot \nabla\varphi(x)\, dx.
\end{aligned}$$

This shows that u is a weak solution to the p-Laplace equation in Ω. $\square$

EXAMPLE 11.3. The function $\Phi\colon \mathbb{R}^n \setminus \{0\} \to \mathbb{R}$,

$$\Phi(x) = \begin{cases} |x|^{\frac{p-n}{p-1}}, & p \neq n, \\ \log|x|, & p = n, \end{cases}$$

is a classical solution to the p-Laplace equation in $\mathbb{R}^n \setminus \{0\}$ and it is called the fundamental solution of the p-Laplace equation. However, if $1 < p \leq n$, then Φ fails to be a weak solution to the p-Laplace equation in $\mathbb{R}^n$. Indeed, direct computations show that in this case $\Phi \in L^q_{\text{loc}}(\mathbb{R}^n)$ for every $1 \leq q < \frac{n(p-1)}{n-p}$, which in the case $p = n$ means that $\Phi \in L^q_{\text{loc}}(\mathbb{R}^n)$ for every $1 \leq q < \infty$. Moreover, $\nabla\Phi \in L^q_{\text{loc}}(\mathbb{R}^n)$ for every $1 \leq q < \frac{n(p-1)}{n-1}$, and thus $\Phi \in W^{1,q}(B(0,1))$ for every $1 \leq q < \frac{n(p-1)}{n-1}$. This upper bound is optimal since

$$\Phi \notin W^{1,\frac{n(p-1)}{n-1}}_{\text{loc}}(B(0,1))$$

for $1 < p \leq n$. Observe that $\frac{n(p-1)}{n-1} \leq p$ for $1 < p \leq n$. On the other hand, $\Phi \in W^{1,p}_{\text{loc}}(\mathbb{R}^n)$ if $p > n$.

The following example illustrates that a weak solution to the p-Laplace equation does not necessarily belong to $W^{1,p}(\Omega)$.

EXAMPLE 11.4. Consider the fundamental solution Φ in Example 11.3 with $1 < p < n$. Then $u = 1 - \Phi$ is a weak solution to the p-Laplace equation in $\Omega = B(0,1) \setminus \{0\}$. By Example 11.3 and Theorem 5.30 applied with $v = \min\{0, 1 - \Phi\}$, we have $u \in W^{1,q}_0(\Omega)$ for $1 < q < \frac{n(p-1)}{n-1}$, but $u \notin W^{1,q}(\Omega)$ for $q \geq \frac{n(p-1)}{n-1}$. In particular, $u \notin W^{1,p}(\Omega)$. On the other hand, also the constant function $u = 0$ is a weak solution to the p-Laplace equation in Ω with the zero boundary values. This shows that the p-Laplace equation with the boundary condition $u \in W^{1,q}_0(\Omega)$ for $q < p$ does not have a unique solution in general. Observe that in this case we cannot use u as a test function in the definition of a weak solution and thereby conclude that

$$\int_\Omega |\nabla u(x)|^p \, dx = \int_\Omega |\nabla u(x)|^{p-2} \nabla u(x) \cdot \nabla u(x) \, dx = 0,$$

which would imply that $\nabla u = 0$ almost everywhere in Ω and consequently $u = 0$ almost everywhere in Ω, by Lemma 3.15.

Next we show that

$$-\operatorname{div}(|\nabla\Phi|^{p-2}\nabla\Phi) = C(n,p)\delta_0$$

in the weak sense in $\mathbb{R}^n$, where δ_0 is Dirac's delta at the origin.

THEOREM 11.5. *Let $1 < p < \infty$ and let Φ be as in Example 11.3. There exists a constant $C = C(n,p) \neq 0$ such that*

$$\int_{\mathbb{R}^n} |\nabla\Phi(x)|^{p-2} \nabla\Phi(x) \cdot \nabla\varphi(x) \, dx = C\varphi(0)$$

for every $\varphi \in C^\infty_0(\mathbb{R}^n)$.

PROOF. Let $\varphi \in C_0^\infty(\mathbb{R}^n)$, let $B(0,R)$ be a ball with $\operatorname{supp}\varphi \Subset B(0,R)$, and let $0 < \varepsilon < R$. Then

$$\int_{\mathbb{R}^n} |\nabla\Phi(x)|^{p-2}\nabla\Phi(x)\cdot\nabla\varphi(x)\,dx = \int_{B(0,R)\setminus B(0,\varepsilon)} |\nabla\Phi(x)|^{p-2}\nabla\Phi(x)\cdot\nabla\varphi(x)\,dx + \int_{B(0,\varepsilon)} |\nabla\Phi(x)|^{p-2}\nabla\Phi(x)\cdot\nabla\varphi(x)\,dx.$$

For the second integral on the right-hand side we have

$$\begin{aligned}\left|\int_{B(0,\varepsilon)} |\nabla\Phi(x)|^{p-2}\nabla\Phi(x)\cdot\nabla\varphi(x)\,dx\right| &\le C(n,p)\int_{B(0,\varepsilon)} |x|^{1-n}|\nabla\varphi(x)|\,dx\\ &\le C(n,p)\|\nabla\varphi\|_{L^\infty(\mathbb{R}^n)}\int_0^\varepsilon r^{1-n}r^{n-1}\,dr\\ &= C(n,p)\varepsilon\|\nabla\varphi\|_{L^\infty(\mathbb{R}^n)} \xrightarrow{\varepsilon\to 0} 0.\end{aligned}$$

To estimate the other integral, we observe by a direct computation that

$$|\nabla\Phi(x)|^{p-2}\nabla\Phi(x) = C(n,p)\frac{x}{|x|^n}$$

for every $x \ne 0$. Since Φ is a classical solution to the p-Laplace equation in $\mathbb{R}^n \setminus \{0\}$, using the divergence theorem we obtain

$$\begin{aligned}&\int_{B(0,R)\setminus B(0,\varepsilon)} |\nabla\Phi(x)|^{p-2}\nabla\Phi(x)\cdot\nabla\varphi(x)\,dx\\ &\quad= \int_{B(0,R)\setminus B(0,\varepsilon)} \operatorname{div}(|\nabla\Phi(x)|^{p-2}\nabla\Phi(x)\varphi(x))\,dx\\ &\quad= -\int_{\partial B(0,\varepsilon)} |\nabla\Phi(x)|^{p-2}\nabla\Phi(x)\varphi(x)\cdot\frac{x}{|x|}\,d\sigma(x)\\ &\quad= C(n,p)\int_{\partial B(0,\varepsilon)} |x|^{1-n}\varphi(x)\,d\sigma(x)\\ &\quad= C(n,p)\varepsilon^{1-n}\int_{\partial B(0,\varepsilon)} \varphi(x)\,d\sigma(x)\\ &\quad= \frac{C(n,p)}{\sigma(\partial B(0,\varepsilon))}\int_{\partial B(0,\varepsilon)} \varphi(x)\,d\sigma(x) \xrightarrow{\varepsilon\to 0} C(n,p)\varphi(0).\end{aligned}$$

□

If $u \in W^{1,p}(\Omega)$ is a weak solution, then in Definition 11.1 the class of test functions can be taken to be the Sobolev space $W_0^{1,p}(\Omega)$ with zero boundary values.

LEMMA 11.6. *Let $1 < p < \infty$ and let $\Omega \subset \mathbb{R}^n$ be an open set. Assume that $u \in W^{1,p}(\Omega)$ is a weak solution to the p-Laplace equation in Ω. Then*

$$\int_\Omega |\nabla u(x)|^{p-2}\nabla u(x)\cdot\nabla\varphi(x)\,dx = 0$$

for every $\varphi \in W_0^{1,p}(\Omega)$.

PROOF. Let $\varphi_i \in C_0^\infty(\Omega)$, $i \in \mathbb{N}$, be such that $\varphi_i \to \varphi$ in $W^{1,p}(\Omega)$ as $i \to \infty$. By the Cauchy–Schwarz inequality and Hölder's inequality, we obtain

$$\begin{aligned}
&\left|\int_\Omega |\nabla u(x)|^{p-2}\nabla u(x)\cdot\nabla\varphi(x)\,dx\right| \\
&\quad = \left|\int_\Omega |\nabla u(x)|^{p-2}\nabla u(x)\cdot\nabla\varphi(x)\,dx - \int_\Omega |\nabla u(x)|^{p-2}\nabla u(x)\cdot\nabla\varphi_i(x)\,dx\right| \\
&\quad = \left|\int_\Omega |\nabla u(x)|^{p-2}\nabla u(x)\cdot(\nabla\varphi(x) - \nabla\varphi_i(x))\,dx\right| \\
&\quad \le \int_\Omega |\nabla u(x)|^{p-1}|\nabla\varphi(x) - \nabla\varphi_i(x)|\,dx \\
&\quad \le \left(\int_\Omega |\nabla u(x)|^p\,dx\right)^{\frac{p-1}{p}} \left(\int_\Omega |\nabla\varphi(x) - \nabla\varphi_i(x)|^p\,dx\right)^{\frac{1}{p}} \xrightarrow{i\to\infty} 0. \qquad \square
\end{aligned}$$

REMARK 11.7. Let $1 < p < \infty$ and assume that $u \in W^{1,p}_{\mathrm{loc}}(\Omega)$ is a weak solution to the p-Laplace equation in an open set $\Omega \subset \mathbb{R}^n$. Then

$$\int_\Omega |\nabla u(x)|^{p-2}\nabla u(x)\cdot\nabla\varphi(x)\,dx = 0$$

for every $\varphi \in W_0^{1,p}(\Omega)$ with $\operatorname{supp}\varphi \Subset \Omega$. Indeed, by convolution approximation we obtain a compact set $K \subset \Omega$ and functions $\varphi_i \in C_0^\infty(\Omega)$ such that $\operatorname{supp}\varphi \cup \operatorname{supp}\varphi_i \subset K$, for every $i \in \mathbb{N}$, and $\nabla\varphi_i \to \nabla\varphi$ in $L^p(\Omega)$ as $i \to \infty$. The rest of the proof is similar to the proof of Lemma 11.6 using the fact that $\int_K |\nabla u(x)|^p\,dx < \infty$. We remark that if $\varphi \ge 0$, then also $\varphi_i \ge 0$ for every $i \in \mathbb{N}$.

REMARK 11.8. Let $1 < p < \infty$ and assume that $u \in W^{1,p}(\mathbb{R}^n)$ is a weak solution to the p-Laplace equation in $\mathbb{R}^n$. By Lemma 11.6, we may use $\varphi \in W_0^{1,p}(\mathbb{R}^n)$ as a test function in (11.1). Since $u \in W^{1,p}(\mathbb{R}^n) = W_0^{1,p}(\mathbb{R}^n)$, we obtain

$$\int_{\mathbb{R}^n} |\nabla u(x)|^p\,dx = \int_{\mathbb{R}^n} |\nabla u(x)|^{p-2}\nabla u(x)\cdot\nabla u(x)\,dx = 0.$$

Thus $\nabla u = 0$ almost everywhere in $\mathbb{R}^n$, and by Lemma 3.15 we conclude that $u = 0$ almost everywhere in $\mathbb{R}^n$.

As an application of Lemma 11.6, we obtain a Sobolev space version of the comparison principle. In the proof we apply the following vectorial inequalities: if $2 \le p < \infty$, then

$$(|\zeta|^{p-2}\zeta - |\eta|^{p-2}\eta)\cdot(\zeta-\eta) \ge 2^{2-p}|\zeta-\eta|^p \tag{11.2}$$

for every $\zeta, \eta \in \mathbb{R}^n$; and if $1 < p \le 2$, then

$$(|\zeta|^{p-2}\zeta - |\eta|^{p-2}\eta)\cdot(\zeta-\eta) \ge (p-1)(1+|\zeta|+|\eta|)^{p-2}|\zeta-\eta|^2 \tag{11.3}$$

for every $\zeta, \eta \in \mathbb{R}^n$, $\zeta \ne \eta$. See Lindqvist [**280**, Chapter 12].

THEOREM 11.9. *Let $1 < p < \infty$ and let $\Omega \subset \mathbb{R}^n$ be an open set. Assume that $u, v \in W^{1,p}(\Omega)$ are weak solutions to the p-Laplace equation in Ω. If $\min\{v-u, 0\} \in W_0^{1,p}(\Omega)$, then $u \le v$ almost everywhere in Ω.*

PROOF. Since $u, v \in W^{1,p}(\Omega)$ are weak solutions, by Lemma 11.6 we have

$$\int_\Omega \big(|\nabla v(x)|^{p-2}\nabla v(x) - |\nabla u(x)|^{p-2}\nabla u(x)\big)\cdot\nabla\varphi(x)\,dx = 0$$

for every $\varphi \in W_0^{1,p}(\Omega)$. Let $\varphi = \min\{v-u,0\} \in W_0^{1,p}(\Omega)$. Assume first that $p \geq 2$. By Lemma 2.25 and (11.2) we obtain

$$\begin{aligned}\int_\Omega |\nabla \min\{v(x)-u(x),0\}|^p\,dx &= \int_{\{u>v\}} |\nabla v(x)-\nabla u(x)|^p\,dx \\ &\leq C(p)\int_{\{u>v\}} \left(|\nabla v(x)|^{p-2}\nabla v(x) - |\nabla u(x)|^{p-2}\nabla u(x)\right)\cdot(\nabla v(x)-\nabla u(x))\,dx \\ &= C(p)\int_\Omega \left(|\nabla v(x)|^{p-2}\nabla v(x) - |\nabla u(x)|^{p-2}\nabla u(x)\right)\cdot\nabla\varphi(x)\,dx = 0.\end{aligned}$$

Consequently $\nabla \min\{v-u,0\} = 0$ almost everywhere in Ω. Since $\min\{v-u,0\} \in W_0^{1,p}(\Omega)$, Lemma 3.15 implies that $\min\{v-u,0\} = 0$ almost everywhere Ω. This shows that $u \leq v$ almost everywhere in Ω.

For $1 < p < 2$, we use a similar reasoning as above but apply (11.3) instead of (11.2), and obtain

$$\begin{aligned}&\int_\Omega (1+|\nabla v(x)|+|\nabla u(x)|)^{p-2}|\nabla \min\{v(x)-u(x),0\}|^2\,dx \\ &\qquad = \int_{\{u>v\}} (1+|\nabla v(x)|+|\nabla u(x)|)^{p-2}|\nabla v(x)-\nabla u(x)|^2\,dx = 0.\end{aligned}$$

From this we conclude as above that $u \leq v$ almost everywhere in Ω. □

The comparison principle above implies that a weak solution to the p-Laplace equation with given Sobolev boundary values is unique whenever it exists.

COROLLARY 11.10. *Let $1 < p < \infty$ and let $\Omega \subset \mathbb{R}^n$ be an open set. Assume that $u, v \in W^{1,p}(\Omega)$ are weak solutions to the p-Laplace equation in Ω with $u - v \in W_0^{1,p}(\Omega)$. Then $v = u$ almost everywhere in Ω.*

PROOF. Theorem 2.30 implies $\min\{v-u,0\} \in W_0^{1,p}(\Omega)$ and $\min\{u-v,0\} \in W_0^{1,p}(\Omega)$. Theorem 11.9 shows that $u \leq v$ almost everywhere in Ω and $v \leq u$ almost everywhere in Ω. Thus $v = u$ almost everywhere in Ω. □

11.2. A variational approach

Next we consider a variational approach to the Dirichlet problem for the p-Laplace equation. We show that the p-Laplace equation is the Euler–Lagrange equation of the p-Dirichlet integral

$$I(u) = \int_\Omega |\nabla u(x)|^p\,dx.$$

In other words, a function with prescribed boundary values is a minimizer of the p-Dirichlet integral if and only if it is a weak solution to the p-Laplace equation.

THEOREM 11.11. *Let $1 < p < \infty$ and let $\Omega \subset \mathbb{R}^n$ be an open set. Assume that $u, g \in W^{1,p}(\Omega)$ and $u - g \in W_0^{1,p}(\Omega)$. Then*

$$I(u) = \inf\left\{I(v) : v \in W^{1,p}(\Omega),\ v - g \in W_0^{1,p}(\Omega)\right\}$$

if and only if

$$\int_\Omega |\nabla u(x)|^{p-2}\nabla u(x)\cdot\nabla\varphi(x)\,dx = 0$$

for every $\varphi \in C_0^\infty(\Omega)$.

PROOF. First assume that

$$I(u)=\inf\Big\{I(v):v\in W^{1,p}(\Omega),\ v-g\in W_0^{1,p}(\Omega)\Big\}.$$

Let $\varphi\in C_0^\infty(\Omega)$ and $0<\varepsilon\le 1$. Then $v=u+\varepsilon\varphi\in W^{1,p}(\Omega)$ and $v-g\in W_0^{1,p}(\Omega)$. Thus

$$\int_\Omega|\nabla u(x)|^p\,dx=I(u)\le I(v)=\int_\Omega|\nabla u(x)+\varepsilon\nabla\varphi(x)|^p\,dx$$

and so

$$\int_\Omega\frac{1}{\varepsilon}\big(|\nabla u(x)+\varepsilon\nabla\varphi(x)|^p-|\nabla u(x)|^p\big)\,dx\ge 0.$$

On the other hand, since $\nabla(|\zeta|^p)=p|\zeta|^{p-2}\zeta$ for $\zeta\in\mathbb{R}^n$, $\zeta\ne 0$, we have

$$\frac{1}{\varepsilon}\big(|\nabla u(x)+\varepsilon\nabla\varphi(x)|^p-|\nabla u(x)|^p\big)\xrightarrow{\varepsilon\to 0}p|\nabla u(x)|^{p-2}\nabla u(x)\cdot\nabla\varphi(x),$$

if $\nabla u(x)\ne 0$. By the mean value theorem for $\varepsilon\mapsto|\nabla u(x)+\varepsilon\nabla\varphi(x)|^p$ and Young's inequality, there is a constant $\gamma\in(0,1)$ such that

$$\begin{aligned}\frac{1}{\varepsilon}\big||\nabla u(x)+\varepsilon\nabla\varphi(x)|^p-|\nabla u(x)|^p\big|&\le p|\nabla u(x)+\gamma\varepsilon\nabla\varphi(x)|^{p-1}|\nabla\varphi(x)|\\&\le(p-1)|\nabla u(x)+\gamma\varepsilon\nabla\varphi(x)|^p+|\nabla\varphi(x)|^p\\&\le C(p)(|\nabla u(x)|^p+|\nabla\varphi(x)|^p)\end{aligned}$$

for every $x\in\mathbb{R}^n$ since the estimate above is trivial in the case $\nabla u(x)=0$. The dominated convergence theorem gives

$$\begin{aligned}p\int_\Omega|\nabla u(x)|^{p-2}\nabla u(x)\cdot\nabla\varphi(x)\,dx&=\int_\Omega\lim_{\varepsilon\to 0}\frac{1}{\varepsilon}\big(|\nabla u(x)+\varepsilon\nabla\varphi(x)|^p-|\nabla u(x)|^p\big)\,dx\\&=\lim_{\varepsilon\to 0}\int_\Omega\frac{1}{\varepsilon}\big(|\nabla u(x)+\varepsilon\nabla\varphi(x)|^p-|\nabla u(x)|^p\big)\,dx\\&\ge 0.\end{aligned}$$

Thus

$$\int_\Omega|\nabla u(x)|^{p-2}\nabla u(x)\cdot\nabla\varphi(x)\,dx\ge 0$$

for every $\varphi\in C_0^\infty(\Omega)$. The reverse inequality follows by replacing φ with $-\varphi$. This shows that u is a weak solution to the p-Laplace equation in Ω.

Then assume that $u\in W^{1,p}(\Omega)$ is a weak solution in Ω with $u-g\in W_0^{1,p}(\Omega)$, and let $v\in W^{1,p}(\Omega)$ with $v-g\in W_0^{1,p}(\Omega)$. We have $u-v=(u-g)-(v-g)\in W_0^{1,p}(\Omega)$ and by Lemma 11.6

$$\int_\Omega|\nabla u(x)|^{p-2}\nabla u(x)\cdot(\nabla u(x)-\nabla v(x))\,dx=0.$$

Young's inequality implies

$$\begin{aligned}\int_\Omega |\nabla u(x)|^p\,dx &= \int_\Omega |\nabla u(x)|^{p-2}\nabla u(x)\cdot\nabla u(x)\,dx\\ &= \int_\Omega |\nabla u(x)|^{p-2}\nabla u(x)\cdot\nabla v(x)\,dx\\ &\le \int_\Omega |\nabla u(x)|^{p-1}|\nabla v(x)|\,dx\\ &\le \frac{p-1}{p}\int_\Omega |\nabla u(x)|^p\,dx + \frac{1}{p}\int_\Omega |\nabla v(x)|^p\,dx,\end{aligned}$$

and by reorganizing the terms we obtain

$$\int_\Omega |\nabla u(x)|^p\,dx \le \int_\Omega |\nabla v(x)|^p\,dx.$$

This shows that

$$I(u) = \inf\Big\{I(v) : v \in W^{1,p}(\Omega), v-g \in W_0^{1,p}(\Omega)\Big\}.$$

□

REMARK 11.12. The proof of Theorem 11.11 also gives a local version of the theorem: a function $u \in W^{1,p}_{\mathrm{loc}}(\Omega)$ is a weak solution to the p-Laplace equation in Ω if and only if

$$\int_\Omega |\nabla u(x)|^p\,dx \le \int_\Omega |\nabla u(x)+\nabla\varphi(x)|^p\,dx$$

for every $\varphi \in C_0^\infty(\Omega)$.

Next we prove existence and uniqueness of a solution to the Dirichlet problem using the direct method in the calculus variations. This means that, instead of the partial differential equation, the argument uses the variational integral.

THEOREM 11.13. *Let $1 < p < \infty$, let $\Omega \subset \mathbb{R}^n$ be a bounded open set, and assume that $g \in W^{1,p}(\Omega)$. There exists a unique $u \in W^{1,p}(\Omega)$ such that $u-g \in W_0^{1,p}(\Omega)$ and*

$$\int_\Omega |\nabla u(x)|^p\,dx \le \int_\Omega |\nabla v(x)|^p\,dx$$

for every $v \in W^{1,p}(\Omega)$ with $v-g \in W_0^{1,p}(\Omega)$.

PROOF. Let

$$I = \inf\Big\{I(v) : v \in W^{1,p}(\Omega) \text{ and } v-g \in W_0^{1,p}(\Omega)\Big\}.$$

We note that I is finite since $I \le I(g) < \infty$. There exists a minimizing sequence $(u_i)_{i\in\mathbb{N}}$, with $u_i \in W^{1,p}(\Omega)$ and $u_i - g \in W_0^{1,p}(\Omega)$, such that $I(u_i) \to I$ as $i \to \infty$. The existence of the limit implies that the sequence $(I(u_i))_{i\in\mathbb{N}}$ is bounded. Thus there exists a constant $M < \infty$ such that

$$I(u_i) = \int_\Omega |\nabla u_i(x)|^p\,dx \le M$$

for every $i \in \mathbb{N}$. Using Corollary 3.7 with $q = p$, we obtain

$$\begin{aligned}
\|u_i\|^p_{W^{1,p}(\Omega)} &= \int_\Omega \big(|u_i(x)|^p + |\nabla u_i(x)|^p\big)\,dx \\
&\le 2^{p-1} \int_\Omega \big(|u_i(x) - g(x)|^p + |g(x)|^p + |\nabla u_i(x)|^p\big)\,dx \\
&\le C \operatorname{diam}(\Omega)^p \int_\Omega |\nabla u_i(x) - \nabla g(x)|^p\,dx + 2^{p-1} \int_\Omega \big(|g(x)|^p + |\nabla u_i(x)|^p\big)\,dx \\
&\le C(\operatorname{diam}(\Omega)^p + 1) \int_\Omega \big(|\nabla u_i(x)|^p + |g(x)|^p + |\nabla g(x)|^p\big)\,dx \\
&\le C(\operatorname{diam}(\Omega)^p + 1)\big(M + \|g\|^p_{W^{1,p}(\Omega)}\big) < \infty
\end{aligned}$$

for every $i \in \mathbb{N}$, with $C = C(n,p)$. This shows that $(u_i)_{i\in\mathbb{N}}$ is a bounded sequence in $W^{1,p}(\Omega)$. By Theorem 2.39, there exists a subsequence $(u_{i_k})_{k\in\mathbb{N}}$ and $u \in W^{1,p}(\Omega)$, with $u - g \in W_0^{1,p}(\Omega)$, such that $(u_{i_k}, \nabla u_{i_k}) \to (u, \nabla u)$ weakly in $L^p(\Omega;\mathbb{R}^{n+1})$ as $k \to \infty$. Lemma 2.34 gives

$$I(u) \le \liminf_{k\to\infty} I(u_{i_k}) = I$$

and consequently u is a minimizer.

To show the uniqueness, assume that $u_1, u_2 \in W^{1,p}(\Omega)$ are minimizers of the variational integral I with $u_1 - g \in W_0^{1,p}(\Omega)$ and $u_2 - g \in W_0^{1,p}(\Omega)$. For a contradiction, we assume that

$$|\{x \in \Omega : u_1(x) \ne u_2(x)\}| > 0.$$

By Corollary 3.7 with $q = p$, we have

$$0 < \int_\Omega |u_1(x) - u_2(x)|^p\,dx \le C(n,p) \operatorname{diam}(\Omega)^p \int_\Omega |\nabla u_1(x) - \nabla u_2(x)|^p\,dx$$

and thus

$$|\{x \in \Omega : \nabla u_1(x) \ne \nabla u_2(x)\}| > 0. \tag{11.4}$$

By Theorem 11.11, both u_1 and u_2 are weak solutions to the p-Laplace equation in Ω. Lemma 11.6 implies

$$\int_\Omega |\nabla u_i(x)|^{p-2} \nabla u_i(x) \cdot \nabla\varphi(x)\,dx = 0$$

for every $\varphi \in W_0^{1,p}(\Omega)$ and $i = 1, 2$. By subtracting these equations for $i = 1, 2$, choosing $\varphi = u_1 - u_2$, and applying (11.4), we obtain the contradiction

$$\begin{aligned}
0 &= \int_\Omega \big(|\nabla u_1(x)|^{p-2}\nabla u_1(x) - |\nabla u_2(x)|^{p-2}\nabla u_2(x)\big) \cdot \nabla\varphi(x)\,dx \\
&= \int_\Omega \big(|\nabla u_1(x)|^{p-2}\nabla u_1(x) - |\nabla u_2(x)|^{p-2}\nabla u_2(x)\big) \cdot (\nabla u_1(x) - \nabla u_2(x))\,dx > 0.
\end{aligned}$$

The final inequality follows from the fact that

$$\big(|\nabla u_1(x)|^{p-2}\nabla u_1(x) - |\nabla u_2(x)|^{p-2}\nabla u_2(x)\big) \cdot (\nabla u_1(x) - \nabla u_2(x)) > 0$$

in the set $\{x \in \Omega : \nabla u_1(x) \ne \nabla u_2(x)\}$, by (11.2) and (11.3). □

11.3. Weak super- and subsolutions

We consider not only weak solutions but also weak super- and subsolutions to the p-Laplace equation. As we shall see, weak super- and subsolutions are more flexible than weak solutions. Another advantage is that the properties of super- and subsolutions can be considered separately.

DEFINITION 11.14. Let $1 < p < \infty$ and let $\Omega \subset \mathbb{R}^n$ be an open set. A function $u \in W^{1,p}_{\mathrm{loc}}(\Omega)$ is a weak supersolution to the p-Laplace equation in Ω if

$$\int_\Omega |\nabla u(x)|^{p-2}\nabla u(x) \cdot \nabla \varphi(x)\, dx \geq 0$$

for every $\varphi \in C_0^\infty(\Omega)$ with $\varphi \geq 0$. Analogously, a function $u \in W^{1,p}_{\mathrm{loc}}(\Omega)$ is a weak subsolution to the p-Laplace equation in Ω if

$$\int_\Omega |\nabla u(x)|^{p-2}\nabla u(x) \cdot \nabla \varphi(x)\, dx \leq 0$$

for every $\varphi \in C_0^\infty(\Omega)$ with $\varphi \geq 0$.

REMARK 11.15. We can use functions $\varphi \in W_0^{1,p}(\Omega)$ satisfying $\operatorname{supp}\varphi \Subset \Omega$ and $\varphi \geq 0$ almost everywhere in Ω as test functions for weak super- and subsolutions in Ω. Indeed, by convolution approximation, we obtain a compact set $K \subset \Omega$ and nonnegative functions $\varphi_i \in C_0^\infty(\Omega)$ such that $\operatorname{supp}\varphi \cup \operatorname{supp}\varphi_i \subset K$, for every $i \in \mathbb{N}$, and $\nabla\varphi_i \to \nabla\varphi$ in $L^p(\Omega)$ as $i \to \infty$. The rest of the proof is similar to the proof of Lemma 11.6.

LEMMA 11.16. *Let $1 < p < \infty$ and let $\Omega \subset \mathbb{R}^n$ be an open set. A function $u \in W^{1,p}_{\mathrm{loc}}(\Omega)$ is a weak solution to the p-Laplace equation in Ω if and only if u is both weak super- and weak subsolution to the p-Laplace equation in Ω.*

PROOF. It is clear that a weak solution is both super- and subsolution. Assume then that $u \in W^{1,p}_{\mathrm{loc}}(\Omega)$ is a weak super- and subsolution in Ω. Let $\varphi \in C_0^\infty(\Omega)$. Theorem 2.30 implies that $\varphi = \varphi_+ - \varphi_-$ with $\varphi_+ = \max\{\varphi, 0\} \in W_0^{1,p}(\Omega)$ and $\varphi_- = -\min\{\varphi, 0\} \in W_0^{1,p}(\Omega)$. Since φ_+ and φ_- are compactly supported in Ω, by Remark 11.15 we obtain

$$\int_\Omega |\nabla u(x)|^{p-2}\nabla u(x) \cdot \nabla \varphi_+(x)\, dx \geq 0$$

and

$$\int_\Omega |\nabla u(x)|^{p-2}\nabla u(x) \cdot \nabla \varphi_-(x)\, dx \leq 0.$$

Thus

$$\begin{aligned}\int_\Omega &|\nabla u(x)|^{p-2}\nabla u(x) \cdot \nabla \varphi(x)\, dx \\ &= \int_\Omega |\nabla u(x)|^{p-2}\nabla u(x) \cdot \nabla \varphi_+(x)\, dx - \int_\Omega |\nabla u(x)|^{p-2}\nabla u(x) \cdot \nabla \varphi_-(x)\, dx \geq 0.\end{aligned}$$

The reverse inequality follows by replacing φ with $-\varphi$. This shows that u is a weak solution. □

It follows immediately from Definition 11.14 that $u \in W^{1,p}_{\mathrm{loc}}(\Omega)$ is a weak supersolution if and only if $-u$ is a weak subsolution in Ω. Next we discuss another symmetry property of super- and subsolutions.

LEMMA 11.17. *Let $1 < p < \infty$ and let $\Omega \subset \mathbb{R}^n$ be an open set. Let $\varepsilon > 0$ and assume that $u \in W^{1,p}_{\mathrm{loc}}(\Omega)$ is a weak supersolution to the p-Laplace equation in Ω with $u(x) \ge \varepsilon$ for every $x \in \Omega$. Then $\frac{1}{u}$ is a weak subsolution to the p-Laplace equation in Ω.*

PROOF. Let $\varphi \in C_0^\infty(\Omega)$ with $\varphi \ge 0$ and let $v = \frac{1}{u}$. Then $0 < \varepsilon^2 v \le u$ and by the chain rule $\nabla v = -u^{-2}\nabla u$ almost everywhere in Ω. Thus $\varepsilon^2|\nabla v| \le |\nabla u|$ almost everywhere in Ω and so $v \in W^{1,p}_{\mathrm{loc}}(\Omega)$. This implies

$$|\nabla v(x)|^{p-2}\nabla v(x) = -u^{2(1-p)}|\nabla u(x)|^{p-2}\nabla u(x) \tag{11.5}$$

for almost every $x \in \Omega$. Since $u \ge \varepsilon > 0$ and $\varphi u^{2(1-p)} \in W_0^{1,p}(\Omega)$ is compactly supported in Ω, by Remark 11.15 we may apply it as a test function. This, together with (11.5), gives

$$\begin{aligned}
0 &\le \int_\Omega |\nabla u(x)|^{p-2}\nabla u(x) \cdot \big(\nabla\varphi(x)u(x)^{2(1-p)} + 2(1-p)u(x)^{1-2p}\nabla u(x)\varphi(x)\big)\,dx \\
&= \int_\Omega u(x)^{2(1-p)}|\nabla u(x)|^{p-2}\nabla u(x)\cdot\nabla\varphi(x)\,dx \\
&\qquad + 2(1-p)\int_\Omega \varphi(x)u(x)^{1-2p}|\nabla u(x)|^p\,dx \\
&\le -\int_\Omega |\nabla v(x)|^{p-2}\nabla v(x)\cdot\nabla\varphi(x)\,dx
\end{aligned}$$

for every $\varphi \in C_0^\infty(\Omega)$ with $\varphi \ge 0$. Hence v is a weak subsolution in Ω. □

The class of weak subsolutions is closed with respect to truncation from below.

LEMMA 11.18. *Let $1 < p < \infty$ and let $\Omega \subset \mathbb{R}^n$ be an open set. Assume that $u \in W^{1,p}_{\mathrm{loc}}(\Omega)$ is a weak subsolution to the p-Laplace equation in Ω. Then $u_+ = \max\{u, 0\} \in W^{1,p}_{\mathrm{loc}}(\Omega)$ is a weak subsolution to the p-Laplace equation in Ω.*

PROOF. By Theorem 2.25, we have $u_+ \in W^{1,p}_{\mathrm{loc}}(\Omega)$. Let $\varphi \in C_0^\infty(\Omega)$ with $\varphi \ge 0$, and define $u_k = \min\{ku_+, 1\}$ for every $k \in \mathbb{N}$. Then $(u_k)_{k\in\mathbb{N}}$ is a pointwise increasing sequence in $W^{1,p}_{\mathrm{loc}}(\Omega)$, $0 \le u_k \le 1$ for every $k \in \mathbb{N}$ and

$$\lim_{k\to\infty} u_k(x) = \chi_{\{x\in\Omega:u(x)>0\}}(x)$$

for every $x \in \Omega$. Notice that $u_k\varphi \ge 0$ is compactly supported in Ω and

$$\nabla u_k = \begin{cases} k\nabla u & \text{a.e. in } \{x\in\Omega : 0 < ku(x) < 1\}, \\ 0 & \text{a.e. in } \{x\in\Omega : ku(x) \ge 1\} \cup \{x \in \Omega : u(x) \le 0\}. \end{cases}$$

By Remark 11.15 we can use $u_k\varphi$ as a test function, and the Leibniz rule (2.17) gives

$$\begin{aligned}
0 &\ge \int_\Omega |\nabla u(x)|^{p-2}\nabla u(x)\cdot\nabla(u_k\varphi)(x)\,dx \\
&= \int_\Omega \varphi(x)|\nabla u(x)|^{p-2}\nabla u(x)\cdot\nabla u_k(x)\,dx + \int_\Omega u_k(x)|\nabla u(x)|^{p-2}\nabla u(x)\cdot\nabla\varphi(x)\,dx \\
&= k\int_{\{x\in\Omega:0<u(x)<\frac{1}{k}\}} \varphi(x)|\nabla u(x)|^p\,dx + \int_\Omega u_k(x)|\nabla u(x)|^{p-2}\nabla u(x)\cdot\nabla\varphi(x)\,dx.
\end{aligned}$$

Since the last two integrals are finite, this implies

$$\int_\Omega u_k(x)|\nabla u(x)|^{p-2}\nabla u(x)\cdot\nabla\varphi(x)\,dx \le -k\int_{\{x\in\Omega:0<u(x)<\frac{1}{k}\}}\varphi(x)|\nabla u(x)|^p\,dx \le 0.$$

By Theorem 2.25 (c) and the dominated convergence theorem

$$\begin{aligned}\int_\Omega|\nabla u_+(x)|^{p-2}\nabla u_+(x)\cdot\nabla\varphi(x)\,dx &= \int_\Omega\lim_{k\to\infty}u_k(x)|\nabla u(x)|^{p-2}\nabla u(x)\cdot\nabla\varphi(x)\,dx\\ &= \lim_{k\to\infty}\int_\Omega u_k(x)|\nabla u(x)|^{p-2}\nabla u(x)\cdot\nabla\varphi(x)\,dx\\ &\le 0.\end{aligned}$$

This shows that u_+ is a weak subsolution in Ω. □

REMARK 11.19. The following versions of Lemma 11.18 can be proved in a similar manner. If u, v are weak subsolutions, then $\max\{u, v\}$ is a weak subsolution. In particular $\max\{u, k\}$ is a weak subsolution if u is a weak subsolution and $k\in\mathbb{Z}$. Similarly, if u, v are weak supersolutions, then $\min\{u, v\}$ is a weak supersolution. In particular $\min\{u, k\}$ is a weak supersolution if u is a weak supersolution and $k\in\mathbb{Z}$.

11.4. Energy estimates

The purpose of energy estimates, sometimes called Caccioppoli estimates, is to provide an upper bound for the gradient of a solution with respect to the function itself. In many cases the partial differential equation is only used to prove energy estimates and the rest of the argument applies to all functions that satisfy the corresponding estimate. This is a powerful method since it applies to a whole class of partial differential equations simultaneously.

THEOREM 11.20. *Let $1<p<\infty$ and let $\Omega\subset\mathbb{R}^n$ be an open set. Assume that $u\in W^{1,p}_{\mathrm{loc}}(\Omega)$ is a weak solution to the p-Laplace equation in Ω. There exists a constant $C=C(p)$ such that*

$$\int_\Omega\varphi(x)^p|\nabla u(x)|^p\,dx \le C\int_\Omega|u(x)|^p|\nabla\varphi(x)|^p\,dx$$

for every $\varphi\in C_0^\infty(\Omega)$ with $\varphi\ge 0$.

PROOF. Let $\varphi\in C_0^\infty(\Omega)$ with $\varphi\ge0$ and define $v=\varphi^p u\in W_0^{1,p}(\Omega)$. Then v is compactly supported in Ω and

$$\nabla v(x)=\varphi(x)^p\nabla u(x)+p\varphi(x)^{p-1}\nabla\varphi(x)u(x)$$

for almost every $x\in\Omega$. By Remark 11.7, we obtain

$$\begin{aligned}0&=\int_\Omega|\nabla u(x)|^{p-2}\nabla u(x)\cdot\nabla v(x)\,dx\\ &=\int_\Omega|\nabla u(x)|^p\varphi(x)^p\,dx+p\int_\Omega\varphi(x)^{p-1}u(x)|\nabla u(x)|^{p-2}\nabla u(x)\cdot\nabla\varphi(x)\,dx.\end{aligned}$$

The last integral is finite, and therefore by Young's inequality

$$\begin{aligned}\int_\Omega |\nabla u(x)|^p \varphi(x)^p\,dx &= -p\int_\Omega \varphi(x)^{p-1}u(x)|\nabla u(x)|^{p-2}\nabla u(x)\cdot\nabla\varphi(x)\,dx\\ &\le p\int_\Omega \varphi(x)^{p-1}|u(x)||\nabla u(x)|^{p-1}|\nabla\varphi(x)|\,dx\\ &\le \frac{1}{2}\int_\Omega |\nabla u(x)|^p\varphi(x)^p\,dx + C(p)\int_\Omega |u(x)|^p|\nabla\varphi(x)|^p\,dx.\end{aligned}$$

Both terms on the right-hand side are finite since $u \in W^{1,p}_{\text{loc}}(\Omega)$ and $\varphi \in C_0^\infty(\Omega)$. The claim follows by absorbing the first term on the right-hand side. □

REMARK 11.21. By a more careful application of Young's inequality or Hölder's inequality in the proof of Theorem 11.20, we obtain

$$\int_\Omega \varphi(x)^p|\nabla u(x)|^p\,dx \le p^p\int_\Omega |u(x)|^p|\nabla\varphi(x)|^p\,dx.$$

As a consequence of the energy estimate, we obtain the following reverse Poincaré inequality.

LEMMA 11.22. *Let $1 < p < \infty$ and let $\Omega \subset \mathbb{R}^n$ be an open set. Assume that $u \in W^{1,p}_{\text{loc}}(\Omega)$ is a weak solution to the p-Laplace equation in Ω. There exists a constant $C = C(p)$ such that*

$$\int_{B(z,r)} |\nabla u(x)|^p\,dx \le \frac{C}{r^p}\int_{B(z,2r)} |u(x)-u_{B(z,2r)}|^p\,dx$$

for every ball $B(z,r)$ with $B(z,2r) \Subset \Omega$.

PROOF. Let $\varphi \in C_0^\infty(B(z,2r))$ be such that $0 \le \varphi \le 1$, $\varphi = 1$ in $B(z,r)$ and $|\nabla\varphi| \le \frac{C}{r}$. Then $u - u_{B(z,2r)} \in W^{1,p}(B(z,2r))$ is a weak solution to the p-Laplace equation in $B(z,2r)$. Theorem 11.20 implies

$$\begin{aligned}\int_{B(z,r)} |\nabla u(x)|^p\,dx &\le \int_{B(z,2r)} \varphi(x)^p|\nabla(u-u_{B(z,2r)})(x)|^p\,dx\\ &\le C(p)\int_{B(z,2r)} |u(x)-u_{B(z,2r)}|^p|\nabla\varphi(x)|^p\,dx\\ &\le \frac{C(p)}{r^p}\int_{B(z,2r)} |u(x)-u_{B(z,2r)}|^p\,dx.\end{aligned}$$ □

We prove a logarithmic energy estimate for weak supersolutions. Observe that the right-hand side of (11.6) is independent of the solution.

THEOREM 11.23. *Let $1 < p < \infty$ and let $\Omega \subset \mathbb{R}^n$ be an open set. Let $\varepsilon > 0$ and assume that $u \in W^{1,p}_{\text{loc}}(\Omega)$ is a weak supersolution to the p-Laplace equation in Ω with $u(x) \ge \varepsilon$ for every $x \in \Omega$. There exists a constant $C = C(p)$ such that*

$$\int_\Omega |\nabla \log u(x)|^p\varphi(x)^p\,dx \le C\int_\Omega |\nabla\varphi(x)|^p\,dx \tag{11.6}$$

for every $\varphi \in C_0^\infty(\Omega)$ with $\varphi \ge 0$.

PROOF. Let $\varphi \in C_0^\infty(\Omega)$ with $\varphi \ge 0$. Since $0 \le u(x)^{1-p} \le \varepsilon^{1-p}$ and

$$|\nabla(u(x)^{1-p})| = |(1-p)u(x)^{-p}\nabla u(x)| \le (p-1)\varepsilon^{-p}|\nabla u(x)|$$

for almost every $x \in \Omega$, we have $u^{1-p} \in W^{1,p}_{\mathrm{loc}}(\Omega)$ and $u^{1-p}\varphi^p \in W^{1,p}_0(\Omega)$. Moreover, $u^{1-p}\varphi^p$ is compactly supported in Ω. By Remark 11.15, we may apply $u^{1-p}\varphi^p \geq 0$ as a test function. Thus

$$\begin{aligned}
0 &\leq \int_\Omega |\nabla u(x)|^{p-2}\nabla u(x) \cdot \nabla(u(x)^{1-p}\varphi(x)^p)\,dx \\
&= \int_\Omega \big((1-p)|\nabla u(x)|^p u(x)^{-p}\varphi(x)^p \\
&\qquad + p\varphi(x)^{p-1}u(x)^{1-p}|\nabla u(x)|^{p-2}\nabla u(x)\cdot\nabla\varphi(x)\big)\,dx \\
&= (1-p)\int_\Omega |\nabla \log u(x)|^p\varphi(x)^p\,dx \\
&\qquad + p\int_\Omega \varphi(x)^{p-1}u(x)^{1-p}|\nabla u(x)|^{p-2}\nabla u(x)\cdot\nabla\varphi(x)\,dx,
\end{aligned}$$

where the integrals are finite. By rearranging the terms and applying Young's inequality, we obtain

$$\begin{aligned}
\int_\Omega |\nabla \log u(x)|^p\varphi(x)^p\,dx &\leq \frac{p}{p-1}\int_\Omega |\nabla u(x)|^{p-1}u(x)^{-(p-1)}\varphi(x)^{p-1}|\nabla\varphi(x)|\,dx \\
&= \int_\Omega |\nabla \log u(x)|^{p-1}\varphi(x)^{p-1}\frac{p}{p-1}|\nabla\varphi(x)|\,dx \\
&\leq \frac{p-1}{p}\int_\Omega |\nabla \log u(x)|^p\varphi(x)^p\,dx + \frac{1}{p}\int_\Omega \Big(\frac{p}{p-1}\Big)^p|\nabla\varphi(x)|^p\,dx.
\end{aligned}$$

By absorbing the first term on the right hand-side, we have

$$\int_\Omega |\nabla \log u(x)|^p\varphi(x)^p\,dx \leq \Big(\frac{p}{p-1}\Big)^p\int_\Omega |\nabla\varphi(x)|^p\,dx.$$

The absorbed term is finite since $\varphi\nabla u \in L^p(\Omega)$ and

$$|\nabla \log u(x)|^p\varphi(x)^p = u(x)^{-p}|\nabla u(x)|^p\varphi(x)^p \leq \varepsilon^{-p}|\nabla u(x)|^p\varphi(x)^p$$

for almost every $x \in \Omega$. □

The next result shows that the logarithm of a positive weak solution to the p-Laplace equation is of bounded mean oscillation over balls that are far away from the boundary. This result comes with a uniform bound for the mean oscillation over all such weak solutions.

THEOREM 11.24. *Let $1 < p < \infty$ and let $\Omega \subset \mathbb{R}^n$ be an open set. Let $\varepsilon > 0$, assume that $u \in W^{1,p}_{\mathrm{loc}}(\Omega)$ is a weak supersolution to the p-Laplace equation in Ω with $u(x) \geq \varepsilon$ for every $x \in \Omega$, and define $v = \log u$. There exists a constant $C = C(n,p)$ such that*

$$\fint_{B(z,r)} |v(x) - v_{B(z,r)}|\,dx \leq C \tag{11.7}$$

for every ball $B(z,r)$ with $B(z,2r) \Subset \Omega$.

PROOF. Let $\varphi \in C_0^\infty(B(z,2r))$ be such that $0 \le \varphi \le 1$, $\varphi = 1$ in $B(z,r)$ and $|\nabla\varphi| \le \frac{C}{r}$. By Theorem 11.23, we obtain

$$\begin{aligned}\int_{B(z,r)} |\nabla v(x)|^p\,dx &\le \int_\Omega \varphi(x)^p |\nabla v(x)|^p\,dx \le C(p)\int_\Omega |\nabla\varphi(x)|^p\,dx\\ &\le \frac{C(p)}{r^p}\int_{B(z,2r)} 1\,dx = C(n,p)r^{n-p},\end{aligned}$$

and thus $|\nabla v| \in L^p_{\mathrm{loc}}(\Omega)$. On the other hand, since $u \ge \varepsilon$ and $u \in L^p_{\mathrm{loc}}(\Omega)$, we have $v \in L^p_{\mathrm{loc}}(\Omega)$, and so $v \in W^{1,p}_{\mathrm{loc}}(\Omega)$. By the $(1,p)$-Poincaré inequality, see Theorem 3.14, we conclude that

$$\begin{aligned}\fint_{B(z,r)} |v(x) - v_{B(z,r)}|\,dx &\le C(n,p)r\left(\fint_{B(z,r)} |\nabla v(x)|^p\,dx\right)^{\frac{1}{p}}\\ &\le C(n,p)r(r^{-n}r^{n-p})^{\frac{1}{p}} = C(n,p).\end{aligned}$$

□

The energy estimate in Theorem 11.20 can be generalized by incorporating suitable powers of the function u.

THEOREM 11.25. *Let $1 < p < \infty$ and let $\Omega \subset \mathbb{R}^n$ be an open set. Assume that $u \in W^{1,p}_{\mathrm{loc}}(\Omega)$ is a weak subsolution to the p-Laplace equation in Ω and let $\alpha > 0$. There exists a constant $C = C(p)$ such that*

$$\int_{\{u>0\}} u(x)^{\alpha-1}|\nabla u(x)|^p\varphi(x)^p\,dx \le \frac{C}{\alpha^p}\int_{\{u>0\}} u(x)^{p+\alpha-1}|\nabla\varphi(x)|^p\,dx$$

for every $\varphi \in C_0^\infty(\Omega)$ with $\varphi \ge 0$. Here $\{u > 0\} = \{x \in \Omega : u(x) > 0\}$.

PROOF. By Lemma 11.18 and Theorem 2.25, we may consider $u = u_+$. We would like to apply $u^\alpha\varphi^p$ as a test function, but it is not clear whether this function belongs to $W_0^{1,p}(\Omega)$. Thus we define $\psi_k = \varphi^p \min\{u^\alpha, ku\}$ for every $k \in \mathbb{N}$. Then

$$\nabla\psi_k = p\varphi^{p-1}\nabla\varphi\min\{u^\alpha, ku\} + \varphi^p\nabla\min\{u^\alpha, ku\} \tag{11.8}$$

almost everywhere in Ω. Let

$$\Omega_k = \{x \in \Omega : 0 < u^\alpha(x) \le ku(x)\}$$

for every $k \in \mathbb{N}$. Since $\nabla u = 0$ almost everywhere in the set $\{u = 0\}$, we have

$$\nabla\min\{u^\alpha, ku\} = \begin{cases}\alpha u^{\alpha-1}\nabla u & \text{a.e. in } \Omega_k,\\ k\nabla u & \text{a.e. in } \Omega\setminus\Omega_k.\end{cases} \tag{11.9}$$

It follows that $\psi_k \in W_0^{1,p}(\Omega)$ for every $k \in \mathbb{N}$. Observe that $u^{\alpha-1} \le k$ in Ω_k.

Fix $k \in \mathbb{N}$. Since u is a weak subsolution in Ω and $\psi_k \ge 0$ is compactly supported in Ω, we have by Remark 11.15 and (11.8) that

$$\begin{aligned}0 &\ge \int_\Omega |\nabla u(x)|^{p-2}\nabla u(x)\cdot\nabla\psi_k(x)\,dx\\ &= \int_\Omega \varphi(x)^p|\nabla u(x)|^{p-2}\nabla u(x)\cdot\nabla\min\{u(x)^\alpha, ku(x)\}\,dx \qquad (11.10)\\ &\quad + p\int_\Omega \varphi(x)^{p-1}\min\{u(x)^\alpha, ku(x)\}|\nabla u(x)|^{p-2}\nabla u(x)\cdot\nabla\varphi(x)\,dx.\end{aligned}$$

By (11.9), (11.10), and Young's inequality we obtain

$$\begin{aligned}
&\alpha \int_{\Omega_k} \varphi(x)^p u(x)^{\alpha-1} |\nabla u(x)|^p \, dx + k \int_{\Omega \setminus \Omega_k} \varphi(x)^p |\nabla u(x)|^p \, dx \\
&\quad = \int_\Omega \varphi(x)^p |\nabla u(x)|^{p-2} \nabla u(x) \cdot \nabla \min\{u(x)^\alpha, ku(x)\} \, dx \\
&\quad \le p \left| \int_\Omega \varphi(x)^{p-1} \min\{u(x)^\alpha, ku(x)\} |\nabla u(x)|^{p-2} \nabla u(x) \cdot \nabla \varphi(x) \, dx \right| \\
&\quad \le p \int_{\Omega_k} \varphi(x)^{p-1} u(x)^\alpha |\nabla u(x)|^{p-1} |\nabla \varphi(x)| \, dx \\
&\qquad + pk \int_{\Omega \setminus \Omega_k} \varphi(x)^{p-1} u(x) |\nabla u(x)|^{p-1} |\nabla \varphi(x)| \, dx \\
&\quad \le \frac{\alpha}{2} \int_{\Omega_k} \varphi(x)^p u(x)^{\alpha-1} |\nabla u(x)|^p \, dx + \frac{k}{2} \int_{\Omega \setminus \Omega_k} \varphi(x)^p |\nabla u(x)|^p \, dx \\
&\qquad + C(p) \alpha^{1-p} \int_{\Omega_k} u(x)^{p+\alpha-1} |\nabla \varphi(x)|^p \, dx + C(p) k \int_{\Omega \setminus \Omega_k} u(x)^p |\nabla \varphi(x)|^p \, dx.
\end{aligned}$$

Since $0 < u^\alpha \le ku$ in Ω_k and $u \in W^{1,p}_{\mathrm{loc}}(\Omega)$, we have

$$\int_{\Omega_k} \varphi(x)^p u(x)^{\alpha-1} |\nabla u(x)|^p \, dx \le k \int_\Omega \varphi(x)^p |\nabla u(x)|^p \, dx < \infty$$

and

$$\int_{\Omega \setminus \Omega_k} \varphi(x)^p |\nabla u(x)|^p \, dx \le \int_\Omega \varphi(x)^p |\nabla u(x)|^p \, dx < \infty.$$

Hence these terms can be absorbed into the left-hand side. This gives

$$\begin{aligned}
&\alpha \int_{\Omega_k} \varphi(x)^p u(x)^{\alpha-1} |\nabla u(x)|^p \, dx + k \int_{\Omega \setminus \Omega_k} \varphi(x)^p |\nabla u(x)|^p \, dx \\
&\quad \le C(p) \alpha^{1-p} \int_{\Omega_k} u(x)^{p+\alpha-1} |\nabla \varphi(x)|^p \, dx + C(p) k \int_{\Omega \setminus \Omega_k} u(x)^p |\nabla \varphi(x)|^p \, dx,
\end{aligned}$$

where

$$k \int_{\Omega \setminus \Omega_k} \varphi(x)^p |\nabla u(x)|^p \, dx \xrightarrow{k \to \infty} 0$$

and

$$k \int_{\Omega \setminus \Omega_k} u(x)^p |\nabla \varphi(x)|^p \, dx \le \int_{\Omega \setminus \Omega_k} u(x)^{p+\alpha-1} |\nabla \varphi(x)|^p \, dx \xrightarrow{k \to \infty} 0,$$

since $ku \le u^\alpha$ in $\Omega \setminus \Omega_k$. Here we may assume that $u^{p+\alpha-1} |\nabla \varphi|^p \in L^1(\Omega)$ since otherwise there is nothing to prove. By the monotone convergence theorem we conclude that

$$\begin{aligned}\int_{\{u>0\}} u(x)^{\alpha-1}|\nabla u(x)|^p\varphi(x)^p\,dx &= \lim_{k\to\infty}\int_{\Omega_k} u(x)^{\alpha-1}|\nabla u(x)|^p\varphi(x)^p\,dx\\
&= \lim_{k\to\infty}\left(\int_{\Omega_k}\varphi(x)^p u(x)^{\alpha-1}|\nabla u(x)|^p\,dx + \frac{k}{\alpha}\int_{\Omega\setminus\Omega_k}\varphi(x)^p|\nabla u(x)|^p\,dx\right)\\
&\le \lim_{k\to\infty}\left(\frac{C(p)}{\alpha^p}\int_{\Omega_k} u(x)^{p+\alpha-1}|\nabla\varphi(x)|^p\,dx + \frac{C(p)k}{\alpha}\int_{\Omega\setminus\Omega_k} u(x)^p|\nabla\varphi(x)|^p\,dx\right)\\
&= \frac{C(p)}{\alpha^p}\int_{\{u>0\}} u(x)^{p+\alpha-1}|\nabla\varphi(x)|^p\,dx. \qquad\square\end{aligned}$$

Next we prove for weak supersolutions a similar estimate as in Theorem 11.25.

THEOREM 11.26. *Let $1<p<\infty$ and let $\Omega\subset\mathbb{R}^n$ be an open set. Assume that $u\in W^{1,p}_{\mathrm{loc}}(\Omega)$ is a nonnegative weak supersolution to the p-Laplace equation in Ω and let $\alpha>0$. There exists $C=C(p)$ such that*

$$\int_\Omega |\nabla u(x)|^p u(x)^{-1-\alpha}\varphi(x)^p\,dx \le \frac{C}{\alpha^p}\int_\Omega u(x)^{p-1-\alpha}|\nabla\varphi(x)|^p\,dx$$

for every $\varphi\in C_0^\infty(\Omega)$ with $\varphi\ge 0$.

PROOF. Let $u_k=u+\frac{1}{k}$ for every $k\in\mathbb{N}$ and apply $\varphi^p u_k^{-\alpha}\in W_0^{1,p}(\Omega)$ as a test function. Observe that this is compactly supported in Ω. Since u is a weak supersolution, by Remark 11.15 we have

$$\begin{aligned}0&\le \int_\Omega |\nabla u(x)|^{p-2}\nabla u(x)\cdot\big(p\varphi(x)^{p-1}\nabla\varphi(x)u_k(x)^{-\alpha} - \alpha\varphi(x)^p u_k(x)^{-1-\alpha}\nabla u_k(x)\big)\,dx\\
&= p\int_\Omega |\nabla u_k(x)|^{p-2}\nabla u_k(x)\cdot\varphi(x)^{p-1}\nabla\varphi(x)u_k(x)^{-\alpha}\,dx\\
&\quad -\alpha\int_\Omega |\nabla u_k(x)|^p u_k(x)^{-1-\alpha}\varphi(x)^p\,dx,\end{aligned}$$

where the last term is finite. By Young's inequality

$$\begin{aligned}\int_\Omega |\nabla u_k(x)|^p u_k(x)^{-1-\alpha}\varphi(x)^p\,dx &\le \frac{p}{\alpha}\int_\Omega |\nabla u_k(x)|^{p-1}u_k(x)^{-\alpha}\varphi(x)^{p-1}|\nabla\varphi(x)|\,dx\\
&\le \frac{p-1}{p}\int_\Omega |\nabla u_k(x)|^p u_k(x)^{-1-\alpha}\varphi(x)^p\,dx\\
&\quad + p^{p-1}\alpha^{-p}\int_\Omega |\nabla\varphi(x)|^p u_k(x)^{p-1-\alpha}\,dx.\end{aligned}$$

The first term on the right-hand side is finite, and absorption of this term gives

$$\int_\Omega |\nabla u_k(x)|^p u_k(x)^{-1-\alpha}\varphi(x)^p\,dx \le \left(\frac{p}{\alpha}\right)^p\int_\Omega |\nabla\varphi(x)|^p u_k(x)^{p-1-\alpha}\,dx.$$

By Fatou's lemma, we obtain

$$\begin{aligned}\int_\Omega |\nabla u(x)|^p u(x)^{-1-\alpha}\varphi(x)^p\,dx &\le \liminf_{k\to\infty}\int_\Omega |\nabla u_k(x)|^p u_k(x)^{-1-\alpha}\varphi(x)^p\,dx\\
&\le \liminf_{k\to\infty}\left(\frac{p}{\alpha}\right)^p\int_\Omega |\nabla\varphi(x)|^p u_k(x)^{p-1-\alpha}\,dx\\
&= \left(\frac{p}{\alpha}\right)^p\int_\Omega |\nabla\varphi(x)|^p u(x)^{p-1-\alpha}\,dx,\end{aligned}$$

where the final step follows from the monotone convergence theorem if $p-1-\alpha \leq 0$, and from the dominated convergence theorem otherwise; in the latter case, we may clearly assume that the last integral is finite. $\square$

11.5. Local boundedness of weak solutions

For $p > n$, Morrey's inequality implies that every function $u \in W^{1,p}_{\mathrm{loc}}(\Omega)$ is continuous and thus locally bounded in Ω, see Remark 3.24. In this section we show that a weak solution $u \in W^{1,p}_{\mathrm{loc}}(\Omega)$ to the p-Laplace equation is locally bounded for every $1 < p < \infty$. Moreover, this result comes with a useful estimate that is sometimes called a weak maximum principle, see Theorem 11.30. First we show that a weak subsolution is locally bounded from above. The proof is based on the Moser iteration technique together with a Caccioppoli inequality and a Sobolev inequality. We will later extend this result for every $\beta > 0$, see Lemma 11.32.

THEOREM 11.27. *Let $1 < p < \infty$ and let $\Omega \subset \mathbb{R}^n$ be an open set. Assume that $u \in W^{1,p}_{\mathrm{loc}}(\Omega)$ is a weak subsolution to the p-Laplace equation in Ω and let $\beta > p-1$. There exist constants $C = C(n,p,\beta)$ and $\tau = \tau(n,p) > 0$ such that*

$$\operatorname*{ess\,sup}_{x\in B(z,r)} u_+(x) \leq C\Big(\Big(\frac{R}{R-r}\Big)^{\tau} \fint_{B(z,R)} u_+(x)^{\beta}\,dx\Big)^{\frac{1}{\beta}} \tag{11.11}$$

whenever $B(z,R) \Subset \Omega$ and $0 < r < R$.

REMARK 11.28. By choosing $r = \frac{R}{2}$ and $\beta = p$ in Theorem 11.27, we have

$$\begin{aligned}\operatorname*{ess\,sup}_{x\in B(z,r)} u(x) \leq \operatorname*{ess\,sup}_{x\in B(z,r)} u_+(x) &\leq C(n,p)\Big(\fint_{B(z,2r)} u_+(x)^p\,dx\Big)^{\frac{1}{p}}\\ &\leq C(n,p)\Big(\fint_{B(z,2r)} |u(x)|^p\,dx\Big)^{\frac{1}{p}} < \infty.\end{aligned}$$

In particular, weak subsolutions are locally essentially bounded from above in Ω.

PROOF OF THEOREM 11.27. Lemma 11.18 implies that $u_+ \in W^{1,p}_{\mathrm{loc}}(\Omega)$ is a weak subsolution in Ω. There are two stages in the proof. First we show that $u_+ \in L^{\beta}_{\mathrm{loc}}(\Omega)$ for every $0 < \beta < \infty$. As a byproduct, we obtain a reverse Hölder inequality for u_+. In the second stage, we prove inequality (11.11) by using the so-called Moser iteration technique, where the reverse Hölder inequality is applied recursively.

In the first stage, we begin by assuming that $u_+ \in L^{\beta}_{\mathrm{loc}}(\Omega)$ for some $p-1 < \beta < \infty$. Observe that for $\beta = p$, and hence also for $0 < \beta \leq p$, this follows from the assumption $u \in W^{1,p}_{\mathrm{loc}}(\Omega)$. Let $B(z,R) \Subset \Omega$ and $0 < r < R$, and let $\varphi \in C^{\infty}_0(B(z,R))$ be such that $\varphi = 1$ in $B(z,r)$, $0 \leq \varphi \leq 1$ and $|\nabla\varphi| \leq \frac{C}{R-r}$. Using Theorem 11.25 with $\alpha = \beta - (p-1) > 0$ and $\{u > 0\} = \{x \in \Omega : u(x) > 0\}$, we obtain

$$\begin{aligned}\int_{\Omega} \big|\varphi(x)\nabla\big(u_+(x)^{\frac{\beta}{p}}\big)\big|^p\,dx &= \int_{\{u>0\}} \varphi(x)^p\Big|\frac{\beta}{p}u(x)^{\frac{\beta}{p}-1}\nabla u(x)\Big|^p\,dx\\ &= \Big(\frac{\beta}{p}\Big)^p \int_{\{u>0\}} \varphi(x)^p u(x)^{\beta-p}|\nabla u(x)|^p\,dx\\ &\leq C(p)\Big(\frac{\beta}{\beta-(p-1)}\Big)^p \int_{\Omega} u_+(x)^{\beta}|\nabla\varphi(x)|^p\,dx.\end{aligned}$$

By the Leibniz rule,

$$\left|\nabla\left(\varphi(x)u_+(x)^{\frac{\beta}{p}}\right)\right| \le \left|\varphi(x)\nabla\left(u_+(x)^{\frac{\beta}{p}}\right)\right| + \left|u_+(x)^{\frac{\beta}{p}}\nabla\varphi(x)\right|,$$

and thus

$$\begin{aligned}
&\int_\Omega \left|\nabla\left(\varphi(x)u_+(x)^{\frac{\beta}{p}}\right)\right|^p dx \\
&\quad \le 2^{p-1}\left(\int_\Omega \left|\varphi(x)\nabla\left(u_+(x)^{\frac{\beta}{p}}\right)\right|^p dx + \int_\Omega \left|u_+(x)^{\frac{\beta}{p}}\nabla\varphi(x)\right|^p dx\right) \\
&\quad \le C(p)\left(\left(\frac{\beta}{\beta-(p-1)}\right)^p + 1\right)\int_\Omega \left|u_+(x)^{\frac{\beta}{p}}\nabla\varphi(x)\right|^p dx.
\end{aligned}$$

Since $p > 1$, we have

$$\left(\frac{\beta}{\beta-(p-1)}\right)^p + 1 = \frac{\beta^p + (\beta-(p-1))^p}{(\beta-(p-1))^p} \le \left(\frac{2\beta-p+1}{\beta-p+1}\right)^p,$$

and using the properties of φ we obtain

$$\begin{aligned}
\fint_{B(z,R)} \left|\nabla\left(\varphi(x)u_+(x)^{\frac{\beta}{p}}\right)\right|^p dx &\le C(p)\left(\frac{2\beta-p+1}{\beta-p+1}\right)^p \fint_{B(z,R)} \left|u_+(x)^{\frac{\beta}{p}}\nabla\varphi(x)\right|^p dx \\
&\le C(p)\left(\frac{2\beta-p+1}{\beta-p+1}\frac{1}{R-r}\right)^p \fint_{B(z,R)} u_+(x)^\beta\, dx.
\end{aligned}$$

By the estimates above, we find that $\varphi u_+^{\frac{\beta}{p}} \in W_0^{1,p}(B(z,R))$. The Sobolev inequality in Theorem 3.17 implies

$$\left(\fint_{B(z,R)} \left|\varphi(x)u_+(x)^{\frac{\beta}{p}}\right|^{\kappa p} dx\right)^{\frac{1}{\kappa p}} \le C(n,p)R\left(\fint_{B(z,R)} \left|\nabla\left(\varphi(x)u_+(x)^{\frac{\beta}{p}}\right)\right|^p dx\right)^{\frac{1}{p}},$$

where $\kappa = \kappa(n,p) > 1$. By combining the previous estimates and using the properties of φ, we obtain

$$\begin{aligned}
\left(\fint_{B(z,r)} u_+(x)^{\kappa\beta}\, dx\right)^{\frac{1}{\kappa\beta}} &\le \left(\left(\frac{R}{r}\right)^n \fint_{B(z,R)} \left|\varphi(x)u_+(x)^{\frac{\beta}{p}}\right|^{\kappa p} dx\right)^{\frac{1}{\kappa\beta}} \\
&\le C(n,p)^{\frac{p}{\beta}}\left(\frac{R}{r}\right)^{\frac{n}{\kappa\beta}}\left(R^p \fint_{B(z,R)} \left|\nabla\left(\varphi(x)u_+(x)^{\frac{\beta}{p}}\right)\right|^p dx\right)^{\frac{1}{\beta}} \\
&\le C(n,p)^{\frac{p}{\beta}}\left(\frac{R}{r}\right)^{\frac{n}{\beta}}\left(\left(\frac{2\beta-p+1}{\beta-p+1}\frac{R}{R-r}\right)^p \fint_{B(z,R)} u_+(x)^\beta\, dx\right)^{\frac{1}{\beta}}.
\end{aligned} \tag{11.12}$$

This is a reverse Hölder inequality, which holds whenever $B(z,R) \Subset \Omega$ and $0 < r < R$. From (11.12) and the assumption $u_+ \in L^\beta_{\mathrm{loc}}(\Omega)$ we conclude that $u_+ \in L^{\kappa\beta}_{\mathrm{loc}}(\Omega)$ with $\kappa > 1$. This observation makes it possible to increase the level of local integrability stepwise. In particular, since initially $u_+ \in L^p_{\mathrm{loc}}(\Omega)$ by the assumptions of the theorem, we may iterate (11.12) starting with $\beta = p$ and conclude that $u_+ \in L^\beta_{\mathrm{loc}}(\Omega)$ for every $0 < \beta < \infty$.

Next we fix $\beta_0 > p-1$ and show that (11.11) holds with this β_0. Note that if $\beta \ge \beta_0$, then

$$\frac{2\beta-p+1}{\beta-p+1} \le \frac{2\beta_0-p+1}{\beta_0-p+1} = C(p,\beta_0)$$

and $u_+ \in L^\beta_{\mathrm{loc}}(\Omega)$ by the first part of the proof. Using (11.12) we obtain

$$\left(\fint_{B(z,r)} u_+(x)^{\kappa\beta}\,dx\right)^{\frac{1}{\kappa\beta}} \le C(n,p,\beta_0)^{\frac{p}{\beta}}\left(\frac{R}{r}\right)^{\frac{n}{\beta}}\left(\frac{R}{R-r}\right)^{\frac{p}{\beta}}\left(\fint_{B(z,R)} u_+(x)^{\beta}\,dx\right)^{\frac{1}{\beta}}.$$

We apply this estimate recursively. Let $r_0 = R$ and $r_k = r + \frac{R-r}{2^k}$ for every $k \in \mathbb{N}$. Then $r < r_{k+1} < r_k \le R$, $\frac{r_k}{r_{k+1}} \le 2$ and $\frac{r_k}{r_k - r_{k+1}} \le \frac{2^{k+1}R}{R-r}$ for every $k \in \mathbb{N}_0$, and thus

$$\begin{aligned}&\left(\fint_{B(z,r_{k+1})} u_+(x)^{\kappa\beta}\,dx\right)^{\frac{1}{\kappa\beta}}\\ &\qquad \le C(n,p,\beta_0)^{\frac{p}{\beta}}\cdot 2^{\frac{n}{\beta}}\left(\frac{2^{k+1}R}{R-r}\right)^{\frac{p}{\beta}}\left(\fint_{B(z,r_k)} u_+(x)^{\beta}\,dx\right)^{\frac{1}{\beta}}\end{aligned} \tag{11.13}$$

for every $\beta \ge \beta_0$ and $k \in \mathbb{N}_0$. In the first step we have, by (11.13),

$$\begin{aligned}&\left(\fint_{B(z,r_1)} u_+(x)^{\kappa\beta_0}\,dx\right)^{\frac{1}{\kappa\beta_0}}\\ &\qquad \le C(n,p,\beta_0)^{\frac{p}{\beta_0}}\cdot 2^{\frac{n}{\beta_0}}\left(\frac{2R}{R-r}\right)^{\frac{p}{\beta_0}}\left(\fint_{B(z,r_0)} u_+(x)^{\beta_0}\,dx\right)^{\frac{1}{\beta_0}}.\end{aligned}$$

In the second step, by applying (11.13) twice,

$$\begin{aligned}&\left(\fint_{B(z,r_2)} u_+(x)^{\kappa^2\beta_0}\,dx\right)^{\frac{1}{\kappa^2\beta_0}}\\ &\qquad \le C(n,p,\beta_0)^{\frac{p}{\kappa\beta_0}}\cdot 2^{\frac{n}{\kappa\beta_0}}\left(\frac{2^2R}{R-r}\right)^{\frac{p}{\kappa\beta_0}}\left(\fint_{B(z,r_1)} u_+(x)^{\kappa\beta_0}\,dx\right)^{\frac{1}{\kappa\beta_0}}\\ &\qquad \le C(n,p,\beta_0)^{\frac{p}{\beta_0}+\frac{p}{\kappa\beta_0}}\cdot 2^{\frac{n}{\beta_0}+\frac{n}{\kappa\beta_0}}\\ &\qquad\qquad \cdot 2^{\frac{p}{\beta_0}+\frac{2p}{\kappa\beta_0}}\left(\frac{R}{R-r}\right)^{\frac{p}{\beta_0}+\frac{p}{\kappa\beta_0}}\left(\fint_{B(z,r_0)} u_+(x)^{\beta_0}\,dx\right)^{\frac{1}{\beta_0}}.\end{aligned}$$

By applying (11.13) recursively we obtain

$$\begin{aligned}\left(\fint_{B(z,r_k)} u_+(x)^{\kappa^k\beta_0}\,dx\right)^{\frac{1}{\kappa^k\beta_0}} &\le C(n,p,\beta_0)^{\frac{p}{\beta_0}\sum_{i=1}^k\frac{1}{\kappa^{i-1}}}\cdot 2^{\frac{n}{\beta_0}\sum_{i=1}^k\frac{1}{\kappa^{i-1}}}\\ &\quad \cdot 2^{\frac{p}{\beta_0}\sum_{i=1}^k\frac{i}{\kappa^{i-1}}}\left(\frac{R}{R-r}\right)^{\frac{p}{\beta_0}\sum_{i=1}^k\frac{1}{\kappa^{i-1}}}\left(\fint_{B(z,r_0)} u_+(x)^{\beta_0}\,dx\right)^{\frac{1}{\beta_0}}\end{aligned} \tag{11.14}$$

for every $k \in \mathbb{N}$. The sum of a geometric series gives

$$\sum_{i=1}^k \frac{1}{\kappa^{i-1}} \xrightarrow{k\to\infty} \sum_{i=1}^\infty \frac{1}{\kappa^{i-1}} = \frac{1}{1-\frac{1}{\kappa}} = \frac{\kappa}{\kappa-1},$$

and by recognizing the derivative of a geometric series we obtain

$$\sum_{i=1}^k \frac{i}{\kappa^{i-1}} \xrightarrow{k\to\infty} \sum_{i=1}^\infty \frac{i}{\kappa^{i-1}} = \frac{1}{(1-\frac{1}{\kappa})^2}.$$

Hence we conclude from (11.14) that

$$\begin{aligned}\limsup_{k\to\infty}\Big(\fint_{B(z,r)} u_+(x)^{\kappa^k\beta_0}\,dx\Big)^{\frac{1}{\kappa^k\beta_0}} &\le \limsup_{k\to\infty}\Big(\Big(\frac{r_k}{r}\Big)^n\fint_{B(z,r_k)} u_+(x)^{\kappa^k\beta_0}\,dx\Big)^{\frac{1}{\kappa^k\beta_0}}\\ &\le C(n,p,\beta_0)\Big(\frac{R}{R-r}\Big)^{\frac{p}{\beta_0}\frac{\kappa}{\kappa-1}}\Big(\fint_{B(z,r_0)} u_+(x)^{\beta_0}\,dx\Big)^{\frac{1}{\beta_0}}.\end{aligned}$$

Thus u is essentially bounded in the ball $B(z,r)$ and

$$\operatorname*{ess\,sup}_{x\in B(z,r)} u_+(x) \le C(n,p,\beta_0)\Big(\Big(\frac{R}{R-r}\Big)^{\frac{p\kappa}{\kappa-1}}\fint_{B(z,R)} u_+(x)^{\beta_0}\,dx\Big)^{\frac{1}{\beta_0}}.$$

The claim follows with $\tau=\frac{p\kappa}{\kappa-1}$. □

REMARK 11.29. For $1<p<n$, by the proof of Theorem 11.27, we may choose $\kappa=\frac{n}{n-p}$ and thus $\tau=\frac{p\kappa}{\kappa-1}=n$. When $p\ge n$, the proof gives $\tau=\frac{p\kappa}{\kappa-1}>p\ge n$.

Next we conclude that every weak solution is locally essentially bounded.

THEOREM 11.30. *Let $1<p<\infty$ and let $\Omega\subset\mathbb{R}^n$ be an open set. Assume that $u\in W^{1,p}_{\mathrm{loc}}(\Omega)$ is a weak solution to the p-Laplace equation in Ω and let $\beta>p-1$. There exist constants $C=C(n,p,\beta)$ and $\tau=\tau(n,p)>0$ such that*

$$\operatorname*{ess\,sup}_{x\in B(z,r)} |u(x)| \le C\Big(\Big(\frac{R}{R-r}\Big)^{\tau}\fint_{B(z,R)} |u(x)|^{\beta}\,dx\Big)^{\frac{1}{\beta}}$$

whenever $B(z,R)\Subset\Omega$ and $0<r<R$.

PROOF. As a weak solution, u is a weak subsolution the p-Laplace equation in Ω. By Lemma 11.18, $u_+\in W^{1,p}_{\mathrm{loc}}(\Omega)$ is a weak subsolution in Ω and thus by Theorem 11.27 we have $u_+\in L^\infty_{\mathrm{loc}}(\Omega)$ with

$$\operatorname*{ess\,sup}_{x\in B(z,r)} u_+(x) \le C(n,p,\beta)\Big(\Big(\frac{R}{R-r}\Big)^{\tau}\fint_{B(z,R)} u_+(x)^{\beta}\,dx\Big)^{\frac{1}{\beta}}.$$

On the other hand, since u is a weak solution $-u$ is a weak solution as well. Again by Lemma 11.18, $(-u)_+=u_-\in W^{1,p}_{\mathrm{loc}}(\Omega)$ is a weak subsolution, and by Theorem 11.27 we have $u_-\in L^\infty_{\mathrm{loc}}(\Omega)$ with

$$\operatorname*{ess\,sup}_{x\in B(z,r)} u_-(x) \le C(n,p,\beta)\Big(\Big(\frac{R}{R-r}\Big)^{\tau}\fint_{B(z,R)} u_-(x)^{\beta}\,dx\Big)^{\frac{1}{\beta}}.$$

This shows that $u=u_+-u_-\in L^\infty_{\mathrm{loc}}(\Omega)$. Moreover,

$$\begin{aligned}\operatorname*{ess\,sup}_{x\in B(z,r)} |u(x)| &= \operatorname*{ess\,sup}_{x\in B(z,r)} (u_+(x)+u_-(x)) \le \operatorname*{ess\,sup}_{x\in B(z,r)} u_+(x) + \operatorname*{ess\,sup}_{x\in B(z,r)} u_-(x)\\ &\le C(n,p,\beta)\Big(\Big(\frac{R}{R-r}\Big)^{\tau}\fint_{B(z,R)} u_+(x)^{\beta}\,dx\Big)^{\frac{1}{\beta}}\\ &\quad + C(n,p,\beta)\Big(\Big(\frac{R}{R-r}\Big)^{\tau}\fint_{B(z,R)} u_-(x)^{\beta}\,dx\Big)^{\frac{1}{\beta}}\\ &\le C(n,p,\beta)\Big(\Big(\frac{R}{R-r}\Big)^{\tau}\fint_{B(z,R)} |u(x)|^{\beta}\,dx\Big)^{\frac{1}{\beta}}.\end{aligned}$$

□

The following technical lemma will be used when proving that Theorem 11.27 holds for every $\beta > 0$.

LEMMA 11.31. *Let $\Psi\colon [0, R_0] \to \mathbb{R}$ be a nonnegative bounded function and assume that there exist $C_1 > 0$, $\alpha > 0$ and $0 < \varepsilon < 1$ such that*

$$\Psi(r) \leq \varepsilon\Psi(R) + C_1(R - r)^{-\alpha}$$

for every $0 \leq r < R \leq R_0$. Then there exists a constant $C = C(\alpha, \varepsilon)$ such that

$$\Psi(r) \leq C(R - r)^{-\alpha} C_1$$

for every $0 \leq r < R \leq R_0$.

PROOF. Let $0 < \tau < 1$ and $0 \leq r < R \leq R_0$. We define $t_0 = r$ and

$$t_{i+1} = t_i + (1 - \tau)\tau^i(R - r),$$

for every $i \in \mathbb{N}_0$. Then $r \leq t_i < t_{i+1} \leq R$ for every $i \in \mathbb{N}_0$ and

$$\begin{aligned}
\Psi(t_0) &\leq \varepsilon\Psi(t_1) + C_1(t_1 - t_0)^{-\alpha} \\
&= \varepsilon\Psi(t_1) + C_1(1 - \tau)^{-\alpha}(R - r)^{-\alpha} \\
&\leq \varepsilon(\varepsilon\Psi(t_2) + C_1(t_2 - t_1)^{-\alpha}) + C_1(1 - \tau)^{-\alpha}(R - r)^{-\alpha} \\
&= \varepsilon^2\Psi(t_2) + \varepsilon C_1\tau^{-\alpha}(1 - \tau)^{-\alpha}(R - r)^{-\alpha} + C_1(1 - \tau)^{-\alpha}(R - r)^{-\alpha} \\
&= \varepsilon^2\Psi(t_2) + C_1(1 - \tau)^{-\alpha}(R - r)^{-\alpha}(\varepsilon\tau^{-\alpha} + 1).
\end{aligned}$$

Recursively, we obtain

$$\Psi(r) = \Psi(t_0) \leq \varepsilon^k\Psi(t_k) + C_1(R - r)^{-\alpha}(1 - \tau)^{-\alpha}\sum_{i=0}^{k-1}\varepsilon^i\tau^{-i\alpha}$$

for every $k \in \mathbb{N}$. Since Ψ is bounded, here $\varepsilon^k\Psi(t_k) \to 0$ as $k \to \infty$. By choosing $\tau = \tau(\varepsilon, \alpha)$ with $\frac{\varepsilon}{\tau^\alpha} < 1$, we conclude that

$$\begin{aligned}
\Psi(r) &\leq \limsup_{k\to\infty}\left(\varepsilon^k\Psi(t_k) + C_1(R - r)^{-\alpha}(1 - \tau)^{-\alpha}\sum_{i=0}^{k-1}\varepsilon^i\tau^{-i\alpha}\right) \\
&= C(\alpha, \varepsilon)C_1(R - r)^{-\alpha}.
\end{aligned}$$

□

LEMMA 11.32. *Theorems 11.27 and 11.30 hold for every $\beta > 0$.*

PROOF. We may assume that $0 < \beta \leq p - 1$ since for $\beta > p - 1$ the results are covered by Theorem 11.27 and Theorem 11.30, respectively. By Remark 11.29 we may also assume that $\tau = \tau(n, p) \geq n$ in Theorem 11.27. Let $B(z, R_0) \Subset \Omega$, $0 < r < R \leq R_0$, and $0 < \varepsilon < 1$. Since $B(z, R) \Subset \Omega$, Theorem 11.27 and Young's inequality imply

$$\begin{aligned}
\operatorname*{ess\,sup}_{x\in B(z,r)} u_+(x) &\le C(n,p)\Big(\Big(\frac{R}{R-r}\Big)^{\tau} \fint_{B(z,R)} u_+(x)^p\,dx\Big)^{\frac{1}{p}}\\
&\le C(n,p)\Big(\Big(\frac{R}{R-r}\Big)^{\tau} \fint_{B(z,R)} u_+(x)^\beta \operatorname*{ess\,sup}_{x\in B(z,R)} u_+(x)^{p-\beta}\,dx\Big)^{\frac{1}{p}}\\
&= C(n,p)\Big(\Big(\frac{R}{R-r}\Big)^{\tau} \fint_{B(z,R)} u_+(x)^\beta\,dx\Big)^{\frac{1}{p}} \operatorname*{ess\,sup}_{x\in B(z,R)} u_+(x)^{1-\frac{\beta}{p}}\\
&\le \varepsilon \operatorname*{ess\,sup}_{x\in B(z,R)} u_+(x) + C(n,p,\varepsilon,\beta)\Big(\Big(\frac{R}{R-r}\Big)^{\tau} \fint_{B(z,R)} u_+(x)^\beta\,dx\Big)^{\frac{1}{\beta}}\\
&\le \varepsilon \operatorname*{ess\,sup}_{x\in B(z,R)} u_+(x) + C_1(R-r)^{-\frac{\tau}{\beta}},
\end{aligned}$$

where

$$C_1 = C(n,p,\varepsilon,\beta)R_0^{\frac{\tau-n}{\beta}}\Big(\int_{B(z,R_0)} u_+(x)^\beta\,dx\Big)^{\frac{1}{\beta}} < \infty.$$

Here we used the facts that $\tau \ge n$ and that by Theorem 11.27 we have $u_+ \in L^\infty_{\mathrm{loc}}(\Omega)$. It is important to use R_0 instead of R above since C_1 is not allowed to depend on R. Without loss of generality, we may assume that $C_1 > 0$. We let $\Psi(0) = 0$ and

$$\Psi(r) = \operatorname*{ess\,sup}_{x\in B(z,r)} u_+(x) \le \operatorname*{ess\,sup}_{x\in B(z,R_0)} u_+(x) < \infty$$

for every $0 < r \le R_0$. By Lemma 11.31 we then have

$$\begin{aligned}
\operatorname*{ess\,sup}_{x\in B(z,r)} u_+(x) &\le C(n,p,\varepsilon,\beta)(R-r)^{-\frac{\tau}{\beta}} R_0^{\frac{\tau-n}{\beta}}\Big(\int_{B(z,R_0)} u_+(x)^\beta\,dx\Big)^{\frac{1}{\beta}}\\
&\le C(n,p,\varepsilon,\beta)\Big(\Big(\frac{R_0}{R-r}\Big)^{\tau} \fint_{B(z,R_0)} u_+(x)^\beta\,dx\Big)^{\frac{1}{\beta}}
\end{aligned}$$

whenever $0 < r < R \le R_0$. By choosing $R = R_0$ we conclude that Theorem 11.27 holds for every $\beta > 0$. Finally, the proof of Theorem 11.30 together with the knowledge that Theorem 11.27 holds for every $\beta > 0$ shows that also Theorem 11.30 holds for every $\beta > 0$. □

Next we consider a version of Theorem 11.27 for nonnegative weak supersolutions.

LEMMA 11.33. *Let $1 < p < \infty$ and let $\Omega \subset \mathbb{R}^n$ be an open set. Assume that $u \in W^{1,p}_{\mathrm{loc}}(\Omega)$ is a nonnegative weak supersolution to the p-Laplace equation in Ω and let $\beta > 0$. There exist constants $C = C(n,p,\beta)$ and $\tau = \tau(n,p) > 0$ such that*

$$\Big(\Big(\frac{R}{R-r}\Big)^{\tau} \fint_{B(z,R)} u(x)^{-\beta}\,dx\Big)^{-\frac{1}{\beta}} \le C \operatorname*{ess\,inf}_{x\in B(z,r)} u(x) \tag{11.15}$$

whenever $B(z,R) \Subset \Omega$ and $0 < r < R$. Moreover, the constants C and τ are the same as in Theorem 11.27 and Lemma 11.32.

PROOF. We may assume that the left-hand side of (11.15) is not zero. Since the function $u_k = u + \frac{1}{k} \ge \frac{1}{k}$ is a weak supersolution in Ω for every $k \in \mathbb{N}$, Lemma 11.17

implies that $\frac{1}{u_k} = (\frac{1}{u_k})_+$ is a weak subsolution in Ω for every $k \in \mathbb{N}$. For a fixed $k \in \mathbb{N}$, Theorem 11.27 and Lemma 11.32 give

$$0 < \operatorname*{ess\,sup}_{x\in B(z,r)} \frac{1}{u_k(x)} \le C(n,p,\beta)\left(\left(\frac{R}{R-r}\right)^{\tau} \fint_{B(z,R)} \left(\frac{1}{u_k(x)}\right)^{\beta} dx\right)^{\frac{1}{\beta}}$$

and equivalently

$$\begin{aligned}
0 < \left(\left(\frac{R}{R-r}\right)^{\tau} \fint_{B(z,R)} u_k(x)^{-\beta}\, dx\right)^{-\frac{1}{\beta}} &\le C(n,p,\beta)\left(\operatorname*{ess\,sup}_{x\in B(z,r)} \frac{1}{u_k(x)}\right)^{-1} \\
&= C(n,p,\beta) \operatorname*{ess\,inf}_{x\in B(z,r)} u_k(x) \\
&= C(n,p,\beta)\left(\operatorname*{ess\,inf}_{x\in B(z,r)} u(x) + \frac{1}{k}\right).
\end{aligned}$$

The claim follows from the monotone convergence theorem by letting $k \to \infty$ since

$$\begin{aligned}
0 < \left(\fint_{B(z,R)} u(x)^{-\beta}\, dx\right)^{-\frac{1}{\beta}} &= \left(\lim_{k\to\infty} \fint_{B(z,R)} u_k(x)^{-\beta}\, dx\right)^{-\frac{1}{\beta}} \\
&= \lim_{k\to\infty} \left(\fint_{B(z,R)} u_k(x)^{-\beta}\, dx\right)^{-\frac{1}{\beta}}. \qquad \square
\end{aligned}$$

Another way to prove Lemma 11.33 is to apply the Moser iteration technique as in the proof of Theorem 11.27 using Theorem 11.26 for weak supersolutions. This alternative approach avoids the use of Lemma 11.17.

11.6. Harnack's inequality

Harnack's inequality states that locally the supremum of a nonnegative solution is bounded by a constant times the infimum of the solution. However, since a function $u \in W^{1,p}_{\mathrm{loc}}(\Omega)$ is defined only up to a set of measure zero, we consider the essential supremum and infimum. For a continuous function, these can be replaced by the standard supremum and infimum. Harnack's inequality can be seen as a quantitative version of the maximum principle. We begin with Harnack's inequality for $p > n$. In this case, Harnack's inequality holds for nonnegative weak supersolutions.

THEOREM 11.34. *Let $n < p < \infty$ and let $\Omega \subset \mathbb{R}^n$ be an open set. Assume that $u \in W^{1,p}_{\mathrm{loc}}(\Omega)$ is a nonnegative weak supersolution to the p-Laplace equation in Ω. There exists a constant $C = C(n,p)$ such that*

$$\operatorname*{ess\,sup}_{x\in B(z,r)} u(x) \le C \operatorname*{ess\,inf}_{x\in B(z,r)} u(x)$$

whenever $B(z,r)$ is a ball with $B(z,2r) \Subset \Omega$.

PROOF. The function $u_k = u + \frac{1}{k} \ge \frac{1}{k}$ is weak supersolution in Ω for every $k \in \mathbb{N}$. As in the proof of Theorem 11.24, we have $\log u_k \in W^{1,p}_{\mathrm{loc}}(\Omega)$ for every $k \in \mathbb{N}$. Let $k \in \mathbb{N}$ and let $B(z,r)$ be a ball with $B(z,2r) \Subset \Omega$. Since $p > n$, by Morrey's inequality, see Remark 3.24, there is a set $E_k \subset B(z,r)$ such that $|E_k| = 0$ and

$$\begin{aligned}
|\log u_k(x) - \log u_k(y)| &\le C(n,p)|x-y|^{1-\frac{n}{p}} \|\nabla \log u_k\|_{L^p(B(z,r))} \\
&\le C(n,p) r^{1-\frac{n}{p}} \|\nabla \log u_k\|_{L^p(B(z,r))}
\end{aligned}$$

for every $x, y \in B(z,r) \setminus E_k$. Theorem 11.23 implies

$$\int_\Omega |\nabla \log u_k(x)|^p \varphi(x)^p \, dx \le C(p) \int_\Omega |\nabla \varphi(x)|^p \, dx$$

for every $\varphi \in C_0^\infty(\Omega)$ with $\varphi \ge 0$. By taking $\varphi \in C_0^\infty(B(z,2r))$ such that $0 \le \varphi \le 1$, $\varphi = 1$ in $B(z,r)$ and $|\nabla \varphi| \le \frac{C}{r}$, we obtain

$$\int_{B(z,r)} |\nabla \log u_k(x)|^p \, dx \le C(p) \int_{B(z,2r)} |\nabla \varphi(x)|^p \, dx \le C(n,p) r^{n-p},$$

and thus

$$\|\nabla \log u_k\|_{L^p(B(z,r))} \le C(n,p) r^{\frac{n}{p}-1}.$$

By the estimates above,

$$\begin{aligned}\left|\log \frac{u_k(x)}{u_k(y)}\right| &= \left|\log u_k(x) - \log u_k(y)\right| \le C(n,p) r^{1-\frac{n}{p}} \|\nabla \log u_k\|_{L^p(B(z,r))} \\ &\le C(n,p) r^{1-\frac{n}{p}} r^{\frac{n}{p}-1} = C(n,p)\end{aligned}$$

and consequently

$$u_k(x) \le e^{C(n,p)} u_k(y)$$

for every $x, y \in B(z,r) \setminus E_k$. Finally, by taking the limit $k \to \infty$, we see that

$$u(x) \le e^{C(n,p)} u(y)$$

for every $x, y \in B(z,r) \setminus E$, where $E = \bigcup_{k \in \mathbb{N}} E_k$ satisfies $|E| = 0$. This proves Harnack's inequality

$$\operatorname*{ess\,sup}_{x \in B(z,r)} u(x) \le e^{C(n,p)} \operatorname*{ess\,inf}_{x \in B(z,r)} u(x). \qquad \square$$

REMARK 11.35. In the case $p = n$, there is a direct proof for Harnack's inequality for nonnegative weak solutions to the n-Laplace equation, see Granlund, Lindqvist and Martio [**161**, Theorem 4.15].

Next we consider Harnack's inequality in the case $1 < p < \infty$. In the proof we apply BMO and the John–Nirenberg inequality, see Section 9.7. Since it was convenient to use cubes in Section 9.7, we also use cubes in the proof below. However, this is just a technical point and we could work either with cubes or balls throughout.

THEOREM 11.36. *Let $1 < p < \infty$ and let $\Omega \subset \mathbb{R}^n$ be an open set. Assume that $u \in W^{1,p}_{\mathrm{loc}}(\Omega)$ is a nonnegative weak solution to the p-Laplace equation in Ω. There exists a constant $C = C(n,p)$ such that*

$$\operatorname*{ess\,sup}_{x \in B(z,r)} u(x) \le C \operatorname*{ess\,inf}_{x \in B(z,r)} u(x)$$

for every ball $B(z,r)$ with $B(z, 6\sqrt{n} r) \Subset \Omega$.

PROOF. Let $B(z,r)$ be a ball with $B(z, 6\sqrt{n} r) \Subset \Omega$. Since $u = u_+$ is a weak sub- and supersolution, Lemma 11.32 and Lemma 11.33 respectively imply

$$\operatorname*{ess\,sup}_{x \in B(z,r)} u(x) \le C(n,p,\beta) \left(\fint_{B(z,2r)} u(x)^\beta \, dx \right)^{\frac{1}{\beta}}$$

and

$$\left(\fint_{B(z,2r)} u(x)^{-\beta}\,dx\right)^{-\frac{1}{\beta}} \le C(n,p,\beta)\operatorname*{ess\,inf}_{x\in B(z,r)} u(x),$$

for every $\beta > 0$. It suffices to show that there exists $\beta = \beta(n,p) > 0$ satisfying the reverse Hölder inequality

$$\left(\fint_{B(z,2r)} u(x)^{\beta}\,dx\right)^{\frac{1}{\beta}} \le C(n,p)\left(\fint_{B(z,2r)} u(x)^{-\beta}\,dx\right)^{-\frac{1}{\beta}}, \tag{11.16}$$

since this, together with the estimates above, gives

$$\begin{aligned}\operatorname*{ess\,sup}_{x\in B(z,r)} u(x) &\le C(n,p)\left(\fint_{B(z,2r)} u(x)^{\beta}\,dx\right)^{\frac{1}{\beta}}\\ &\le C(n,p)\left(\fint_{B(z,2r)} u(x)^{-\beta}\,dx\right)^{-\frac{1}{\beta}} \le C(n,p)\operatorname*{ess\,inf}_{x\in B(z,r)} u(x).\end{aligned}$$

For $k \in \mathbb{N}$, let

$$v_k = \log\left(u + \tfrac{1}{k}\right),$$

and let $Q' = Q(z',r') \subset Q(z,2r)$. Since $B(z,6\sqrt{n}r) \Subset \Omega$ and $u + \frac{1}{k} \ge \frac{1}{k} > 0$ is a weak supersolution in Ω for every $k \in \mathbb{N}$, Theorem 11.24 implies

$$\fint_{Q'} |v_k(x) - (v_k)_{Q'}|\,dx \le C(n)\fint_{B'} \left|v_k(x) - (v_k)_{B'}\right|dx \le C(n,p),$$

where $B' = B(z',\sqrt{n}r')$. It follows that v_k is of bounded mean oscillation over such cubes Q', and thus $v_k \in \mathrm{BMO}^d(Q(z,2r))$ with

$$\|v_k\|_{\mathrm{BMO}^d(Q(z,2r))} \le C(n,p).$$

By Theorem 9.41, there exists $\beta = \beta(n,p) > 0$, independent of $k \in \mathbb{N}$, such that

$$\fint_{Q(z,2r)} e^{\beta|v_k(x)-(v_k)_{Q(z,2r)}|}\,dx \le C(n).$$

Hence,

$$\begin{aligned}&\fint_{B(z,2r)} e^{\beta v_k(x)}\,dx \fint_{B(z,2r)} e^{-\beta v_k(x)}\,dx \le C(n)\fint_{Q(z,2r)} e^{\beta v_k(x)}\,dx \fint_{Q(z,2r)} e^{-\beta v_k(x)}\,dx\\ &\quad= C(n)\fint_{Q(z,2r)} e^{\beta(v_k(x)-(v_k)_{Q(z,2r)})}\,dx \fint_{Q(z,2r)} e^{-\beta(v_k(x)-(v_k)_{Q(z,2r)})}\,dx\\ &\quad\le C(n)\left(\fint_{Q(z,2r)} e^{\beta|v_k(x)-(v_k)_{Q(z,2r)}|}\,dx\right)^2 \le C(n).\end{aligned}$$

We emphasize that the constants in this estimate are independent of $k \in \mathbb{N}$. Since $u+1 \in L^{\beta}_{\mathrm{loc}}(\Omega)$, we can apply both dominated and monotone convergence theorems to conclude that

$$\begin{aligned}\left(\fint_{B(z,2r)} u(x)^\beta \, dx\right)^{\frac{1}{\beta}} &= \lim_{k\to\infty} \left(\fint_{B(z,2r)} \left(u(x)+\tfrac{1}{k}\right)^\beta dx\right)^{\frac{1}{\beta}} \\ &= \lim_{k\to\infty} \left(\fint_{B(z,2r)} e^{\beta v_k(x)} \, dx\right)^{\frac{1}{\beta}} \\ &\le C(n,\beta) \lim_{k\to\infty} \left(\fint_{B(z,2r)} e^{-\beta v_k(x)} \, dx\right)^{-\frac{1}{\beta}} \\ &= C(n,\beta) \lim_{k\to\infty} \left(\fint_{B(z,2r)} \left(u(x)+\tfrac{1}{k}\right)^{-\beta} dx\right)^{-\frac{1}{\beta}} \\ &= C(n,\beta) \left(\fint_{B(z,2r)} u(x)^{-\beta} \, dx\right)^{-\frac{1}{\beta}}.\end{aligned}$$

Since $\beta = \beta(n,p)$, the desired reverse Hölder inequality (11.16) follows, and the proof is complete. □

Harnack's inequality implies the following strong maximum principle for weak solutions. An analogous argument gives a strong minimum principle as well.

THEOREM 11.37. *Let $1 < p < \infty$ and let $\Omega \subset \mathbb{R}^n$ be a connected open set. Assume that $u \in W^{1,p}_{\mathrm{loc}}(\Omega)$ is a continuous representative of a weak solution to the p-Laplace equation in Ω. If u attains its maximum in Ω, then u is constant in Ω.*

PROOF. Assume that there exists $x_0 \in \Omega$ such that

$$u(x_0) = \max_{x\in\Omega} u(x).$$

Then the function $x \mapsto u(x_0) - u(x)$ is a nonnegative weak solution in Ω. Let $r_0 > 0$ be such that $B(x_0, \sqrt{6n}r_0) \Subset \Omega$. By Harnack's inequality in Theorem 11.36, we have

$$\sup_{x\in B(x_0,r_0)} (u(x_0)-u(x)) \le C(n,p) \min_{x\in B(x_0,r_0)} (u(x_0)-u(x)) = 0.$$

Thus $u(x) = u(x_0)$ for every $x \in B(x_0, r_0)$.

Let then $x \in \Omega$. Since Ω is connected, there exists a finite chain of balls $B(x_i, r_i)$, $i = 0, \dots, k$, such that $B(x_i, 6\sqrt{n}r_i) \Subset \Omega$, $x_k = x$ and $B(x_i, r_i) \cap B(x_{i+1}, r_{i+1}) \neq \emptyset$, for every $i = 0, 1, \dots, k-1$. By repeating the first part of the proof in each of these balls we conclude that u is constant in each of the balls, and thus $u(x) = u(x_0)$ for every $x \in \Omega$. □

REMARK 11.38. The strong maximum principle implies the standard maximum principle: if $1 < p < \infty$ and $u \in C(\overline{\Omega}) \cap W^{1,p}_{\mathrm{loc}}(\Omega)$ is a weak solution to the p-Laplace equation in a bounded open set $\Omega \subset \mathbb{R}^n$, then

$$\max_{\overline{\Omega}} u = \max_{\partial\Omega} u.$$

REMARK 11.39. Let $1 < p < \infty$ and let $\Omega \subset \mathbb{R}^n$ be a connected open set. Assume that $u \in C(\Omega) \cap W^{1,p}_{\mathrm{loc}}(\Omega)$ is a nonnegative weak solution to the p-Laplace equation in Ω and let $\Omega' \Subset \Omega$. A chaining argument and Harnack's inequality in Theorem 11.36 give the pointwise estimate

$$\frac{1}{C} u(y) \le u(x) \le C u(y)$$

for every $x, y \in \Omega'$, with a constant $C = C(n, p, \Omega', \Omega) \geq 1$. In particular, if $u(x_0) = 0$ for some $x_0 \in \Omega$, then $u(x) = 0$ for every $x \in \Omega$.

11.7. Local Hölder continuity

For $p > n$, a weak solution $u \in W^{1,p}_{\mathrm{loc}}(\Omega)$ to the p-Laplace equation in $\Omega \subset \mathbb{R}^n$ has a locally Hölder continuous representative solely based on the Sobolev regularity, see Remark 3.24. For $1 < p \leq n$, a function $u \in W^{1,p}_{\mathrm{loc}}(\Omega)$ is not necessarily even continuous. However, in this section we show that if u is a weak solution to the p-Laplace equation for any $1 < p < \infty$, then u has a locally Hölder continuous representative. We begin with the case $p = n$.

LEMMA 11.40. *Let $\Omega \subset \mathbb{R}^n$ be an open set. Assume that $u \in W^{1,n}_{\mathrm{loc}}(\Omega)$ is a weak solution to the n-Laplace equation in Ω. Then u has a locally Hölder continuous representative in Ω.*

PROOF. First assume that $B(x, 2R) \Subset \Omega$. Let $0 < r < R$, let $\varphi \in C_0^\infty(B(x, 2r))$ be such that $0 \leq \varphi \leq 1$, $\varphi = 1$ in $B(x, r)$ and $|\nabla \varphi| \leq \frac{C}{r}$, and define $A(x, r) = B(x, 2r) \setminus B(x, r)$. Then $u - u_{A(x,2r)} \in W^{1,n}_{\mathrm{loc}}(\Omega)$ is a weak solution to the n-Laplace equation in Ω. Theorem 11.20 implies

$$\begin{aligned}\int_{B(x,r)} |\nabla u(y)|^n \, dy &\leq \int_\Omega \varphi(y)^n |\nabla(u - u_{A(x,r)})(y)|^n \, dy \\ &\leq C(n) \int_\Omega |u(y) - u_{A(x,r)}|^n |\nabla \varphi(y)|^n \, dy \\ &\leq \frac{C(n)}{r^n} \int_{A(x,r)} |u(y) - u_{A(x,r)}|^n \, dy.\end{aligned}$$

It is straightforward to verify that $A(x, r)$ satisfies the Boman chain condition with $\tau = \tau(n)$, see Definition 9.2. By Theorem 9.32 and Hölder's inequality we obtain

$$\int_{A(x,r)} |u(y) - u_{A(x,r)}|^n \, dy \leq C(n) r^n \int_{A(x,r)} |\nabla u(y)|^n \, dy.$$

Thus

$$\int_{B(x,r)} |\nabla u(y)|^n \, dy \leq C_1 \int_{A(x,r)} |\nabla u(y)|^n \, dy,$$

where $C_1 = C(n)$. This implies

$$\begin{aligned}(1 + C_1) \int_{B(x,r)} |\nabla u(y)|^n \, dy &\leq C_1 \int_{B(x,r)} |\nabla u(y)|^n \, dy + C_1 \int_{A(x,r)} |\nabla u(y)|^n \, dy \\ &= C_1 \int_{B(x,2r)} |\nabla u(y)|^n \, dy,\end{aligned}$$

and consequently

$$\int_{B(x,r)} |\nabla u(y)|^n \, dy \leq \frac{C_1}{1 + C_1} \int_{B(x,2r)} |\nabla u(y)|^n \, dy.$$

Let

$$\Psi(r) = \int_{B(x,r)} |\nabla u(y)|^n \, dy \quad \text{and} \quad \gamma = \frac{C_1}{1 + C_1}.$$

Then $0 < \gamma < 1$ and $\Psi(r) \leq \gamma \Psi(2r)$. By iteration

$$\Psi(r) \leq \gamma^k \Psi(2^k r)$$

for every $k \in \mathbb{N}$ with $2^{k-1}r < R$. Let $k \in \mathbb{N}$ with $2^{k-1}r < R \le 2^k r$. Then

$$\Psi(r) \le \gamma^k \Psi(2^k r) \le \left(\frac{r}{R}\right)^{\delta} \Psi(2R) < \infty,$$

where $\delta = -\log_2 \gamma > 0$. Moreover, by considering $\min\{\delta, 1\}$ instead of δ, we may assume that $0 < \delta \le 1$. By Hölder's inequality,

$$\begin{aligned} \fint_{B(x,r)} |\nabla u(y)|\,dy &\le \left(\fint_{B(x,r)} |\nabla u(y)|^n\,dy\right)^{\frac{1}{n}} = C(n) r^{-1} \Psi(r)^{\frac{1}{n}} \\ &\le C(n)\big(R^{-\delta}\Psi(2R)\big)^{\frac{1}{n}} r^{\frac{\delta}{n}-1}. \end{aligned} \tag{11.17}$$

Next we use the fractional maximal function with $\alpha = 1 - \frac{\delta}{n} > 0$, see Definition 1.38. By taking supremum over $0 < r < R$ in (11.17), we obtain

$$M_{\alpha,R}|\nabla u|(x) \le C(n)\left(R^{-\delta} \int_{B(x,2R)} |\nabla u(y)|^n\,dy\right)^{\frac{1}{n}} < \infty,$$

which holds for every $x \in \Omega$ and $R > 0$ satisfying $B(x, 2R) \Subset \Omega$. Let $B(z, 3R_0) \Subset \Omega$ and let $x, x' \in B(z, \frac{R_0}{8})$. Then $B(x, 2R_0) \Subset \Omega$ and thus

$$\begin{aligned} M_{\alpha,4|x-x'|}|\nabla u|(x) \le M_{\alpha,R_0}|\nabla u|(x) &\le C(n)\left(R_0^{-\delta} \int_{B(x,2R_0)} |\nabla u(y)|^n\,dy\right)^{\frac{1}{n}} \\ &\le C(n)\left(R_0^{-\delta} \int_{B(z,3R_0)} |\nabla u(y)|^n\,dy\right)^{\frac{1}{n}} < \infty. \end{aligned}$$

A corresponding estimate holds for $M_{\alpha,R_0}|\nabla u|(x')$ as well. By a localized version of Theorem 4.17, there exists $E \subset \mathbb{R}^n$ such that $|E| = 0$ and

$$\begin{aligned} |u(x) - u(x')| &\le C(n,\alpha)|x - x'|^{1-\alpha}\big(M_{\alpha,4|x-x'|}|\nabla u|(x) + M_{\alpha,4|x-x'|}|\nabla u|(x')\big) \\ &\le C(n,\alpha)|x - x'|^{1-\alpha}\left(R_0^{-\delta} \int_{B(z,3R_0)} |\nabla u(y)|^n\,dy\right)^{\frac{1}{n}} \end{aligned}$$

for every $x, x' \in B(z, \frac{R_0}{8}) \setminus E$. This shows that u has a locally Hölder continuous representative in Ω. □

The following theorem shows that a weak solution to the p-Laplace equation has a locally Hölder continuous representative for every $1 < p < \infty$. The proof is based on Harnack's inequality and an iteration scheme.

THEOREM 11.41. *Let $1 < p < \infty$ and let $\Omega \subset \mathbb{R}^n$ be an open set. Assume that $u \in W^{1,p}_{\mathrm{loc}}(\Omega)$ is a weak solution to the p-Laplace equation in Ω. Then u has a locally Hölder continuous representative in Ω.*

PROOF. Assume that $B(x, 2R) \Subset \Omega$. For $0 < r \le R$, we define

$$m(x,r) = \operatorname*{ess\,inf}_{y\in B(x,r)} u(y) \quad \text{and} \quad M(x,r) = \operatorname*{ess\,sup}_{y\in B(x,r)} u(y).$$

The oscillation of u in $B(x, r)$ is then

$$\operatorname*{osc}_{B(x,r)} u = M(x,r) - m(x,r).$$

By Theorem 11.30, we have $-\infty < m(x, r) \le M(x, r) < \infty$ for every $0 < r \le R$.

Let $b = 7\sqrt{n}$ and fix $0 < r < \frac{R}{b}$. Then functions $y \mapsto u(y) - m(x, br)$ and $y \mapsto M(x, br) - u(y)$ are weak solutions to the p-Laplace equation in $B(x, br)$ and $u - m(x, br) \geq 0$ and $M(x, br) - u \geq 0$ almost everywhere in $B(x, br)$. By Harnack's inequality in Theorem 11.36, applied in the open set $B(x, br)$, we have

$$\begin{aligned} M(x,r) - m(x,br) &= \operatorname*{ess\,sup}_{y \in B(x,r)} (u(y) - m(x,br)) \\ &\leq C(n,p) \operatorname*{ess\,inf}_{y \in B(x,r)} (u(y) - m(x,br)) \\ &= C(n,p)(m(x,r) - m(x,br)). \end{aligned}$$

A similar argument gives

$$\begin{aligned} M(x,br) - m(x,r) &= \operatorname*{ess\,sup}_{y \in B(x,r)} (M(x,br) - u(y)) \\ &\leq C(n,p) \operatorname*{ess\,inf}_{y \in B(x,r)} (M(x,br) - u(y)) \\ &= C(n,p)(M(x,br) - M(x,r)). \end{aligned}$$

By combining these estimates we obtain

$$\begin{aligned} &M(x,r) - m(x,br) + M(x,br) - m(x,r) \\ &\qquad \leq C(n,p)\big(m(x,r) - m(x,br) + M(x,br) - M(x,r)\big). \end{aligned}$$

Reorganization of the terms gives

$$M(x,r) - m(x,r) \leq \frac{C(n,p) - 1}{C(n,p) + 1}(M(x,br) - m(x,br)),$$

that is,

$$\operatorname*{osc}_{B(x,r)} u \leq \gamma \operatorname*{osc}_{B(x,br)} u, \qquad \text{with } \gamma = \frac{C(n,p) - 1}{C(n,p) + 1} \in (0,1). \tag{11.18}$$

Let $k \in \mathbb{N}$ with $b^k r < R \leq b^{k+1} r$. By iterating (11.18), we obtain

$$\operatorname*{osc}_{B(x,r)} u \leq \gamma^k \operatorname*{osc}_{B(x,b^k r)} u \leq \frac{1}{\gamma}\left(\frac{r}{R}\right)^{\alpha} \operatorname*{osc}_{B(x,R)} u, \tag{11.19}$$

where $\alpha = -\log_b \gamma$. This estimate holds whenever $B(x, 2R) \Subset \Omega$ and $0 < r < \frac{R}{b}$ with $b = 7\sqrt{n}$.

Let $B(z, 3R_0) \Subset \Omega$ and let $x, x' \in B(z, \frac{R_0}{4b})$, $x \neq x'$, be Lebesgue points of u. Then $B(x, 2R_0) \Subset \Omega$ and $2|x - x'| < \frac{R_0}{b}$. Since x and x' are Lebesgue points of u, Theorem 1.21 and (11.19) imply

$$\begin{aligned} |u(x) - u(x')| &\leq \operatorname*{osc}_{B(x,2|x-x'|)} u \leq C(n,p)\left(\frac{2|x - x'|}{R_0}\right)^{\alpha} \operatorname*{osc}_{B(x,R_0)} u \\ &\leq C(n,p) R_0^{-\alpha} |x - x'|^{\alpha} \operatorname*{ess\,sup}_{y \in B(x,R_0)} |u(y)| \\ &\leq C(n,p) R_0^{-\alpha} |x - x'|^{\alpha} \operatorname*{ess\,sup}_{y \in B(z,2R_0)} |u(y)|. \end{aligned}$$

Here

$$\operatorname*{ess\,sup}_{y \in B(z,2R_0)} |u(y)| < \infty$$

by Theorem 11.30, and therefore the function $u \in W^{1,p}_{\mathrm{loc}}(\Omega)$ has a locally Hölder continuous representative in Ω. $\square$

REMARK 11.42. Let $1 < p < \infty$ and let $u \in W^{1,p}_{\mathrm{loc}}(\Omega)$ be a weak solution to the p-Laplace equation in an open set $\Omega \subset \mathbb{R}^n$. Theorem 11.41 shows that u is locally Hölder continuous in Ω, that is, $u \in C^{0,\alpha}_{\mathrm{loc}}(\Omega)$. By regularity theory even the gradient ∇u is locally Hölder continuous in Ω, that is, $u \in C^{1,\alpha}_{\mathrm{loc}}(\Omega)$. See DiBenedetto [**101**], Evans [**120**], Lewis [**270**], Tolksdorf [**382**], Uhlenbeck [**389**] and Ural′tseva [**390**].

REMARK 11.43. Let $1 < p < \infty$, let $\Omega \subset \mathbb{R}^n$ be a bounded open set, and assume that $g \in W^{1,p}(\Omega)$. By Theorem 11.13 and Theorem 11.11 there exists a unique weak solution $u \in W^{1,p}(\Omega)$ to the p-Laplace equation with $u - g \in W^{1,p}_0(\Omega)$. Recall from Remark 6.21 that a point $y \in \partial\Omega$ satisfies the Wiener criterion if

$$\int_0^1 \left(\frac{\operatorname{cap}_p(\Omega^c \cap \overline{B}(y,r), B(y,2r))}{\operatorname{cap}_p(\overline{B}(y,r), B(y,2r))}\right)^{\frac{1}{p-1}} \frac{dr}{r} = \infty.$$

This is a sufficient condition the for regularity of a boundary point for the Dirichlet problem for the p-Laplace equation. That is, if $y \in \partial\Omega$ satisfies the Wiener criterion, then for every $g \in C(\partial\Omega)$ the solution u to the Dirichlet problem with boundary values g satisfies

$$\lim_{x\to y} u(x) = g(y).$$

This result was obtained by Maz′ya [**317**]. See also Gariepy and Ziemer [**140, 141**]. Note that if the complement Ω^c satisfies the p-capacity density condition in Definition 6.17, then the Wiener criterion is satisfied at every point of the boundary.

The Wiener criterion in fact characterizes regular boundary points. The necessity was proved by Lindqvist and Martio [**281**] for $p > n - 1$ and by Kilpeläinen and Malý [**222**] for $p > 1$. See also Adams and Hedberg [**4**], Heinonen, Kilpeläinen and Martio [**187**], Malý and Ziemer [**303**] and Mikkonen [**327**]. For the Laplace equation such a characterization was obtained by Wiener [**396**], for uniformly elliptic linear equations by Littman, Stampacchia and Weinberger [**282**] and for linear equations in a weighted case by Fabes, Kenig and Serapioni [**123**].

REMARK 11.44. Let $1 < p < \infty$ and let $\Omega \subset \mathbb{R}^n$ be a bounded open set. Assume that $g \in W^{1,p}(\Omega) \cap C(\overline{\Omega})$ and that u is the weak solution to p-Laplace equation in Ω with $u - g \in W^{1,p}_0(\Omega)$. There exists a constant $C = C(n,p)$ such that

$$\begin{aligned}\operatorname*{osc}_{\Omega\cap B(y,r_1)} u \le &\operatorname*{osc}_{\partial\Omega\cap \overline{B}(y,2r_2)} g\\ &+ \operatorname*{osc}_{\partial\Omega} g \exp\left(-C\int_{r_1}^{r_2} \left(\frac{\operatorname{cap}_p(\Omega^c \cap \overline{B}(y,r), B(y,2r))}{\operatorname{cap}_p(\overline{B}(y,r), B(y,2r))}\right)^{\frac{1}{p-1}} \frac{dr}{r}\right)\end{aligned}$$

for every $y \in \partial\Omega$ and $0 < r_1 \le r_2 < \infty$. Sufficiency of the Wiener criterion for the boundary regularity follows form this estimate, see Björn [**34**], Björn and Björn [**31**, Chapter 11], Björn, MacManus and Shanmugalingam [**36**], DiBenedetto and Gianazza [**102**], Heinonen, Kilpeläinen and Martio [**187**, Theorem 6.18, Theorem 6.27], Maz′ya [**316, 317**], Stredulinsky [**374**] and Ziemer [**406**].

In particular, if Ω^c satisfies the p-capacity density condition in Definition 6.17 with a constant C_1, then

$$\operatorname*{osc}_{\Omega\cap B(y,r_1)} u \le \operatorname*{osc}_{\partial\Omega\cap \overline{B}(y,r_2)} g + \operatorname*{osc}_{\partial\Omega} g \left(\frac{r_1}{r_2}\right)^{\delta}$$

for every $y \in \partial\Omega$ and $0 < r_1 \le r_2 < \infty$ with $\delta = \delta(n,p,C_1) > 0$. Here the oscillation of a function is as in the proof of Theorem 11.41. This estimate implies that if Ω^c

satisfies the p-capacity density condition, then a solution to the Dirichlet problem for Hölder continuous boundary values is Hölder continuous in $\overline{\Omega}$, but possibly with a different Hölder continuity exponent. For further details, we refer to Heinonen, Kilpeläinen and Martio [**187**, Theorem 6.44].

11.8. Notes

Our presentation of the theory of the p-Laplace equation is based on Heinonen, Kilpeläinen and Martio [**187**], Lindqvist [**280**] and Malý and Ziemer [**303**]. For the corresponding theory on metric measure spaces, we refer to Björn and Björn [**31**].

The Moser iteration scheme is based on the works of Moser [**331, 332**], Serrin [**359**] and Trudinger [**386**]. It can be found in Björn and Björn [**31**, Chapter 8], Björn and Marola [**33**], Chen and Wu [**82**, Chapter 4], Fabes, Kenig and Serapioni [**124**], Giaquinta [**148**, Section 5.2], Giaquinta and Martinazzi [**149**, Section 8.3], Gilbarg and Trudinger [**152**, Chapter 8], Han and Lin [**178**], Heinonen, Kilpeläinen and Martio [**187**, Chapter 3], Lindqvist [**280**, Chapter 3], Malý and Ziemer [**303**, Section 2.3.5] and Stredulinsky [**374**, Chapter 3].

It is possible to avoid the use of the John–Nirenberg inequality in the proof of the Harnack inequality, see Bombieri and Giusti [**54**], Han and Lin [**178**, Section 4.4], Saloff-Coste [**357**, Chapter 2] and Trudinger [**387**]. For the parabolic case we refer to Ivert, Marola and Masson [**205**], Kinnunen and Kuusi [**228**] and Marola and Masson [**307**]. The hole filling technique in the proof of Theorem 11.40 is from Widman [**395**]. See also Granlund, Lindqvist and Martio [**161**] and Lindqvist [**280**].

De Giorgi's method [**97**] gives an alternative approach to Hölder continuity and Harnack inequalities. See also DiBenedetto and Trudinger [**103**], Giaquinta [**148**, Chapter 5], Giaquinta and Martinazzi [**149**, Section 8.3], Giusti [**153**, Chapter 7], Kinnunen and Shanmugalingam [**238**], Ladyzhenskaya and Ural′tseva [**253**] and Malý and Ziemer [**303**, Section 2.3.4].

CHAPTER 12

Stability Results for the p-Laplace Equation

In this chapter we apply results and techniques from the previous chapters of the book for three themes related to Sobolev space properties of weak solutions to the p-Laplace equation. These themes are higher integrability of the gradient of a weak solution, stability of weak solutions with given boundary values with respect to a varying p, and very weak solutions, which are assumed to belong to a Sobolev space below the natural exponent p. In particular, these results show that for a solution to the p-Laplace equation, we can reach a Sobolev space higher than the natural one from a Sobolev space below than the natural one. The capacity density condition for the complement, see Definition 6.17, turns out to be essentially optimal for the global higher integrability and stability results, but these are nontrivial already for sets with smooth boundaries.

12.1. Higher integrability of the gradient

We begin with a local higher integrability result for the gradient of a weak solution, showing that a weak solution $u \in W^{1,p}_{\mathrm{loc}}(\Omega)$ to the p-Laplace equation belongs to a slightly higher Sobolev space, that is, $u \in W^{1,p+\delta}_{\mathrm{loc}}(\Omega)$ for some $\delta > 0$. The proof is based on the energy estimate in Lemma 11.22 and a Sobolev–Poincaré inequality in Theorem 3.14, which give a reverse Hölder inequality for the weak gradient. Local higher integrability then follows from the self-improvement property of reverse Hölder inequalities, see Theorem 8.43.

THEOREM 12.1. *Let $1 < p < \infty$ and let $\Omega \subset \mathbb{R}^n$ be an open set. Assume that $u \in W^{1,p}_{\mathrm{loc}}(\Omega)$ is a weak solution to the p-Laplace equation in Ω. There exists $\delta = \delta(n,p) > 0$ such that $|\nabla u| \in L^{p+\delta}_{\mathrm{loc}}(\Omega)$. Moreover, there exists a constant $C = C(n,p)$ such that*

$$\left(\fint_{B(z,r)} |\nabla u(x)|^{p+\delta}\,dx\right)^{\frac{1}{p+\delta}} \le C\left(\fint_{B(z,2r)} |\nabla u(x)|^{p}\,dx\right)^{\frac{1}{p}}$$

whenever $B(z,2r) \Subset \Omega$.

PROOF. Let

$$q = \begin{cases} \max\{\frac{np}{n+p}, 1\}, & 1 < p < n, \\ \frac{p}{2}, & p \ge n, \end{cases}$$

and

$$q^* = \begin{cases} \frac{nq}{n-q}, & 1 \le q < n, \\ 2q, & q \ge n. \end{cases}$$

Then $1 \le q < p \le q^* < \infty$ and $u \in W^{1,p}_{\mathrm{loc}}(\Omega) \subset W^{1,q}_{\mathrm{loc}}(\Omega)$. By Theorem 3.14, we have

$$\left(\fint_{B(z,2r)} |u(x) - u_{B(z,2r)}|^{q^*}\,dx\right)^{\frac{1}{q^*}} \le C(n,q)r\left(\fint_{B(z,2r)} |\nabla u(x)|^q\,dx\right)^{\frac{1}{q}} \tag{12.1}$$

whenever $B(z,2r) \Subset \Omega$. Since $q = q(n,p)$, we conclude that the constant C in the Sobolev–Poincaré inequality above only depends on n and p. By Lemma 11.22, Hölder's inequality and (12.1), we obtain

$$\begin{aligned}\left(\fint_{B(z,r)} |\nabla u(x)|^p\,dx\right)^{\frac{1}{p}} &\le \frac{C(n,p)}{r}\left(\fint_{B(z,2r)} |u(x) - u_{B(z,2r)}|^p\,dx\right)^{\frac{1}{p}}\\ &\le \frac{C(n,p)}{r}\left(\fint_{B(z,2r)} |u(x) - u_{B(z,2r)}|^{q^*}\,dx\right)^{\frac{1}{q^*}}\\ &\le C(n,p)\left(\fint_{B(z,2r)} |\nabla u(x)|^q\,dx\right)^{\frac{1}{q}}.\end{aligned}$$

Let $f = |\nabla u|^q$. The estimate above can be rewritten as

$$\left(\fint_{B(z,r)} f(x)^{\frac{p}{q}}\,dx\right)^{\frac{q}{p}} \le C(n,p)\fint_{B(z,2r)} f(x)\,dx$$

for every $B(z,2r) \Subset \Omega$. The claim follows from Remark 8.44. □

REMARK 12.2. It is known that the gradient of a weak solution to the p-Laplace equation is locally Hölder continuous and thus $\nabla u \in L^\infty_{\mathrm{loc}}(\Omega)$, see Remark 11.42. However, Theorem 12.1 comes with a reverse Hölder inequality for the weak gradient. Moreover, a weak solution to a more general partial differential equation of the p-Laplacian type is not more regular than locally Hölder continuous and the argument also applies in this case. An example in Meyers [**324**] and Serrin [**360**] shows that the higher integrability exponent in Theorem 12.1 is not very large and depends on the structural constants for general elliptic elliptic equations. See also Chen and Wu [**82**, p. 189] and Giaquinta [**147**, p. 157].

COROLLARY 12.3. *Let $1 < p < \infty$ and let $\Omega \subset \mathbb{R}^n$ be an open set. Assume that $u \in W^{1,p}_{\mathrm{loc}}(\Omega)$ is a weak solution to the p-Laplace equation in Ω. There exists $\delta = \delta(n,p) > 0$ such that $u \in W^{1,p+\delta}_{\mathrm{loc}}(\Omega)$.*

PROOF. By Theorem 3.14, we obtain $q = q(n,p) > p$ such that

$$\left(\fint_{B(z,r)} |u(x) - u_{B(z,r)}|^q\,dx\right)^{\frac{1}{q}} \le C(n,q)r\left(\fint_{B(z,r)} |\nabla u(x)|^p\,dx\right)^{\frac{1}{p}}$$

for every $B(z,r) \Subset \Omega$, which implies

$$\begin{aligned}\left(\fint_{B(z,r)} |u(x)|^q\,dx\right)^{\frac{1}{q}} &\le \left(\fint_{B(z,r)} |u(x) - u_{B(z,r)}|^q\,dx\right)^{\frac{1}{q}} + |u_{B(z,r)}|\\ &\le C(n,q)r\left(\fint_{B(z,r)} |\nabla u(x)|^p\,dx\right)^{\frac{1}{p}} + |u|_{B(z,r)} < \infty.\end{aligned}$$

This shows that $u \in L^q_{\mathrm{loc}}(\Omega)$ for some $q = q(n,p) > p$. Together with Theorem 12.1 this implies that $u \in W^{1,p+\delta}_{\mathrm{loc}}(\Omega)$ for some $\delta = \delta(n,p) > 0$. □

REMARK 12.4. Theorem 11.41 asserts that a weak solution to the p-Laplace equation is continuous, and thus $u \in L^\infty_{\mathrm{loc}}(\Omega)$. This fact together with Theorem 12.1 can be used to give an alternative proof of the previous corollary.

Next we consider a global higher integrability result over the entire open set Ω. In the argument we need the following variant of the energy estimate given in Theorem 11.20. Observe that the support of the test function φ in the following lemma need not be a compact subset of Ω.

LEMMA 12.5. *Let $1 < p < \infty$ and let $\Omega \subset \mathbb{R}^n$ be an open set. Assume that $g \in W^{1,p}(\Omega)$ and that $u \in W^{1,p}(\Omega)$ is a weak solution to the p-Laplace equation in Ω with $u - g \in W_0^{1,p}(\Omega)$. There exists a constant $C = C(p)$ such that*

$$\int_\Omega \varphi(x)^p |\nabla u(x)|^p \, dx \le C\left(\int_\Omega |u(x) - g(x)|^p |\nabla \varphi(x)|^p \, dx + \int_\Omega \varphi(x)^p |\nabla g(x)|^p \, dx \right)$$

for every $\varphi \in C_0^\infty(\mathbb{R}^n)$ with $\varphi \ge 0$.

PROOF. Let $\varphi \in C_0^\infty(\mathbb{R}^n)$ with $\varphi \ge 0$ and let $v = \varphi^p(u - g)$. Since $u - g \in W_0^{1,p}(\Omega)$, Corollary 5.29 implies $v \in W_0^{1,p}(\Omega)$. By Leibniz's rule

$$\nabla v(x) = \varphi(x)^p(\nabla u(x) - \nabla g(x)) + p\varphi(x)^{p-1}\nabla\varphi(x)(u(x) - g(x))$$

for almost every $x \in \Omega$. Lemma 11.6 gives

$$\begin{aligned}
0 &= \int_\Omega |\nabla u(x)|^{p-2} \nabla u(x) \cdot \nabla v(x) \, dx \\
&= \int_\Omega \varphi(x)^p |\nabla u(x)|^{p-2} \nabla u(x) \cdot (\nabla u(x) - \nabla g(x)) \, dx \\
&\qquad + p \int_\Omega \varphi(x)^{p-1} (u(x) - g(x)) |\nabla u(x)|^{p-2} \nabla u(x) \cdot \nabla \varphi(x) \, dx \\
&= \int_\Omega \varphi(x)^p |\nabla u(x)|^p \, dx - \int_\Omega \varphi(x)^p |\nabla u(x)|^{p-2} \nabla u(x) \cdot \nabla g(x) \, dx \\
&\qquad + p \int_\Omega \varphi(x)^{p-1} (u(x) - g(x)) |\nabla u(x)|^{p-2} \nabla u(x) \cdot \nabla \varphi(x) \, dx
\end{aligned}$$

and thus

$$\begin{aligned}
\int_\Omega \varphi(x)^p |\nabla u(x)|^p \, dx &\le \int_\Omega \varphi(x)^p |\nabla u(x)|^{p-1} |\nabla g(x)| \, dx \\
&\qquad + p \int_\Omega \varphi(x)^{p-1} |u(x) - g(x)| |\nabla u(x)|^{p-1} |\nabla \varphi(x)| \, dx,
\end{aligned}$$

where we also used the fact that the two terms on the right-hand side are finite. By Young's inequality

$$\begin{aligned}
\int_\Omega \varphi(x)^p |\nabla u(x)|^{p-1} |\nabla g(x)| \, dx &= \int_\Omega \varphi(x)^{p-1} |\nabla u(x)|^{p-1} \varphi(x) |\nabla g(x)| \, dx \\
&\le \frac{1}{4} \int_\Omega \varphi(x)^p |\nabla u(x)|^p \, dx + C(p) \int_\Omega \varphi(x)^p |\nabla g(x)|^p \, dx
\end{aligned}$$

and

$$\begin{aligned}
p &\int_\Omega \varphi(x)^{p-1} |u(x) - g(x)| |\nabla u(x)|^{p-1} |\nabla \varphi(x)| \, dx \\
&\le \frac{1}{4} \int_\Omega \varphi(x)^p |\nabla u(x)|^p \, dx + C(p) \int_\Omega |u(x) - g(x)|^p |\nabla \varphi(x)|^p \, dx.
\end{aligned}$$

This implies

$$\int_\Omega \varphi(x)^p|\nabla u(x)|^p\,dx \le \frac{1}{2}\int_\Omega \varphi(x)^p|\nabla u(x)|^p\,dx + C(p)\int_\Omega \varphi(x)^p|\nabla g(x)|^p\,dx$$
$$+ C(p)\int_\Omega |u(x)-g(x)|^p|\nabla\varphi(x)|^p\,dx.$$

The first term on the right-hand side is finite by the assumption $u \in W^{1,p}(\Omega)$, and the claim follows by absorbing this term. $\square$

We state a global higher integrability result on open sets whose complement satisfies the capacity density condition in Definition 6.17. The self-improvement of the capacity density condition, see Theorem 7.21, is applied in the proof.

THEOREM 12.6. *Let $1 < p < \infty$ and let $\Omega \subset \mathbb{R}^n$ be a bounded open set such that the complement Ω^c satisfies the p-capacity density condition with a constant C_1. Assume that $g \in W^{1,s}(\Omega)$ for some $s > p$ and that $u \in W^{1,p}(\Omega)$ is a weak solution to the p-Laplace equation in Ω with $u - g \in W_0^{1,p}(\Omega)$. There exists $\delta = \delta(n,p,s,C_1) > 0$ with $p + \delta \le s$ such that $|\nabla u| \in L^{p+\delta}(\Omega)$. Moreover, there exists a constant $C = C(n,p,s,C_1,\Omega)$ such that*

$$\left(\int_\Omega |\nabla u(x)|^{p+\delta}\,dx\right)^{\frac{1}{p+\delta}} \le C\left[\left(\int_\Omega |\nabla u(x)|^p\,dx\right)^{\frac{1}{p}} + \left(\int_\Omega |\nabla g(x)|^{p+\delta}\,dx\right)^{\frac{1}{p+\delta}}\right].$$

PROOF. Let $B(z,r)$ be a ball with $\overline{B}(z,2r)\cap\Omega^c \ne \emptyset$ and let $\varphi \in C_0^\infty(B(z,2r))$ be such that $\varphi = 1$ in $B(z,r)$, $0 \le \varphi \le 1$ and $|\nabla\varphi| \le \frac{C}{r}$. Since Ω is bounded, we have $u - g \in W_0^{1,p}(\Omega)$ and $g \in W^{1,s}(\Omega) \subset W^{1,p}(\Omega)$. Lemma 12.5 implies

$$\begin{aligned}\int_{B(z,r)\cap\Omega} |\nabla u(x)|^p\,dx &\le \int_\Omega \varphi(x)^p|\nabla u(x)|^p\,dx \\ &\le \frac{C(p)}{r^p}\int_{B(z,2r)\cap\Omega}|u(x)-g(x)|^p\,dx + C(p)\int_{B(z,2r)\cap\Omega}|\nabla g(x)|^p\,dx.\end{aligned} \tag{12.2}$$

Theorem 7.21 gives an exponent $1 < q_0 = q_0(n,p,C_1) < p$ such that Ω^c satisfies the q_0-capacity density condition with a constant $C = C(n,p,C_1)$. By increasing the exponent q in the proof of Theorem 12.1, if necessary, we obtain exponents $q_0 \le q = q(n,p,C_1) < p$ and $p \le q^* = q^*(n,p,C_1)$ with $1 \le q^* \le \frac{nq}{n-q}$ for $1 < q < n$ and $1 \le q^* < \infty$ for $n \le q < \infty$. Since $q_0 \le q$, Ω^c satisfies the q-capacity density condition with a constant $C = C(n,p,C_1)$, see Remark 6.20.

By Hölder's inequality $u - g \in W_0^{1,p}(\Omega) \subset W_0^{1,q}(\Omega)$. By considering a q-quasicontinuous representative $v \in W^{1,q}(\mathbb{R}^n)$ of $u - g \in W_0^{1,q}(\Omega)$, given by Corollary 5.29, we have $v = 0$ q-quasieverywhere in Ω^c. Since $\overline{B}(z,2r)\cap\Omega^c \ne \emptyset$, there exists a point $y \in \overline{B}(z,2r)\cap\Omega^c$. Let $A = \{x \in \overline{B}(y,4r) : v(x) = 0\}$. By Theorem 5.34, Theorem 5.38 and Lemma 5.35

$$\begin{aligned}\operatorname{cap}_q(A, B(y,8r)) &\ge \operatorname{cap}_q(\overline{B}(y,4r)\cap\Omega^c, B(y,8r)) \\ &\ge C(n,p,C_1)\operatorname{cap}_q(\overline{B}(y,4r), B(y,8r)) \ge C(n,p,C_1)r^{n-q}.\end{aligned}$$

Theorem 5.47 gives

$$\begin{aligned}\Big(\fint_{B(y,8r)} |v(x)|^{q^*}\,dx\Big)^{\frac{1}{q^*}} &\le C(n,p,C_1)\Big(\frac{1}{\operatorname{cap}_q(A,B(y,8r))}\int_{B(y,8r)}|\nabla v(x)|^q\,dx\Big)^{\frac{1}{q}}\\ &\le C(n,p,C_1)r\Big(\fint_{B(y,8r)}|\nabla v(x)|^q\,dx\Big)^{\frac{1}{q}}.\end{aligned}$$

Since $v=u-g$ almost everywhere in Ω and $v=0$ almost everywhere in Ω^c, it follows that

$$\begin{aligned}\frac{1}{r}\Big(\frac{1}{|B(z,r)|}&\int_{B(z,2r)\cap\Omega}|u(x)-g(x)|^p\,dx\Big)^{\frac{1}{p}}\\ &\le \frac{C}{r}\Big(\fint_{B(y,8r)}|v(x)|^{q^*}\,dx\Big)^{\frac{1}{q^*}} \le C\Big(\fint_{B(y,8r)}|\nabla v(x)|^q\,dx\Big)^{\frac{1}{q}}\\ &= C\Big(\frac{1}{|B(y,8r)|}\int_{B(y,8r)\cap\Omega}|\nabla u(x)-\nabla g(x)|^q\,dx\Big)^{\frac{1}{q}}\\ &\le C\Big(\frac{1}{|B(z,10r)|}\int_{B(z,10r)\cap\Omega}|\nabla u(x)|^q\,dx\Big)^{\frac{1}{q}}\\ &\qquad + C\Big(\frac{1}{|B(z,10r)|}\int_{B(z,10r)\cap\Omega}|\nabla g(x)|^p\,dx\Big)^{\frac{1}{p}},\end{aligned}$$

with $C=C(n,p,C_1)$. By (12.2) and the estimates above we have

$$\begin{aligned}\Big(\frac{1}{|B(z,r)|}&\int_{B(z,r)\cap\Omega}|\nabla u(x)|^p\,dx\Big)^{\frac{1}{p}}\\ &\le C\bigg[\Big(\frac{1}{|B(z,10r)|}\int_{B(z,10r)\cap\Omega}|\nabla u(x)|^q\,dx\Big)^{\frac{1}{q}}\\ &\qquad + \Big(\frac{1}{|B(z,10r)|}\int_{B(z,10r)\cap\Omega}|\nabla g(x)|^p\,dx\Big)^{\frac{1}{p}}\bigg],\end{aligned}\tag{12.3}$$

with $C=C(n,p,C_1)$, whenever $\overline{B}(z,2r)\cap\Omega^c\neq\emptyset$. By the proof of Theorem 12.1, we conclude that (12.3) is also valid when $\overline{B}(z,2r)\cap\Omega^c=\emptyset$.

Let $f=|\nabla u|^q\chi_\Omega$ and $h=|\nabla g|^q\chi_\Omega$. Then (12.3) implies

$$\begin{aligned}\Big(\fint_{B(z,r)} f(x)^{\frac{p}{q}}\,dx\Big)^{\frac{q}{p}} \le C(n,p,C_1)\bigg[\fint_{B(z,10r)} f(x)\,dx + \Big(\fint_{B(z,10r)} h(x)^{\frac{p}{q}}\,dx\Big)^{\frac{q}{p}}\bigg]\end{aligned}$$

for every $B(z,r) \subset \mathbb{R}^n$. By Remark 8.44, there exists $\delta = \delta(n,p,s,C_1) > 0$ with $p+\delta \le s$ and $C = C(n,p,C_1)$ such that

$$\begin{aligned}\left(\frac{1}{|B(z,r)|}\int_{B(z,r)\cap\Omega}|\nabla u(x)|^{p+\delta}\,dx\right)^{\frac{1}{p+\delta}} &\le C\Bigg[\left(\frac{1}{|B(z,2r)|}\int_{B(z,2r)\cap\Omega}|\nabla u(x)|^{p}\,dx\right)^{\frac{1}{p}}\\ &\quad+\left(\frac{1}{|B(z,2r)|}\int_{B(z,2r)\cap\Omega}|\nabla g(x)|^{p+\delta}\,dx\right)^{\frac{1}{p+\delta}}\Bigg]\end{aligned}$$

for every $B(z,r) \subset \mathbb{R}^n$. The claim follows by considering a ball $B(z,r)$ with $z \in \partial\Omega$ and $r = \operatorname{diam}(\Omega)$. □

REMARK 12.7. Let $1 < p_0 < \infty$. Assume that $\Omega \subset \mathbb{R}^n$ is a bounded open set such that the complement Ω^c satisfies the p_0-capacity density condition with a constant C_1 and that $g \in W^{1,s}(\Omega)$ for some $s > p_0$. There exists $\delta = \delta(n,p_0,s,C_1) > 0$ such that the conclusion of Theorem 12.6 holds for every $p_0 - \delta < p < p_0 + \delta$ and for every weak solution $u \in W^{1,p}(\Omega)$ to the p-Laplace equation in Ω with $u - g \in W_0^{1,p}(\Omega)$, with uniform constants 2δ and $C = C(n,p_0,s,C_1,\Omega)$. This follows by carefully tracking the constants in the proof of Theorem 12.6.

COROLLARY 12.8. *Let $1 < p < \infty$ and let $\Omega \subset \mathbb{R}^n$ be a bounded open set such that the complement Ω^c satisfies the p-capacity density condition with a constant C_1. Assume that $g \in W^{1,s}(\Omega)$ for some $s > p$ and that $u \in W^{1,p}(\Omega)$ is a weak solution to the p-Laplace equation in Ω with $u-g \in W_0^{1,p}(\Omega)$. There exists $\delta = \delta(n,p,s,C_1) > 0$ with $p+\delta \le s$ such that $u \in W^{1,p+\delta}(\Omega)$.*

PROOF. Since $u - g \in W_0^{1,p}(\Omega)$, Corollary 3.7 implies that there exists $q = q(n,p,s) > p$ with $q \le s$ such that

$$\left(\int_\Omega |u(x)-g(x)|^q\,dx\right)^{\frac{1}{q}} \le C(n,p,s,\Omega)\left(\int_\Omega |\nabla u(x)-\nabla g(x)|^p\,dx\right)^{\frac{1}{p}} < \infty.$$

It follows that

$$\left(\int_\Omega |u(x)|^q\,dx\right)^{\frac{1}{q}} \le \left(\int_\Omega |u(x)-g(x)|^q\,dx\right)^{\frac{1}{q}} + \left(\int_\Omega |g(x)|^q\,dx\right)^{\frac{1}{q}} < \infty$$

and thus $u \in L^{p+\delta}(\Omega)$ for some $\delta = \delta(n,p,s) > 0$. Together with Theorem 12.6 this implies that $u \in W^{1,p+\delta}(\Omega)$ for some $\delta = \delta(n,p,s,C_1) > 0$. □

Next example shows that the capacity density condition in Definition 6.17 is essentially the weakest possible condition for the global higher integrability in Theorem 12.6.

EXAMPLE 12.9. Let $1 < p < n$. By Example 6.28 and Theorem 5.38 there exists a compact set $K \subset B(0,1) \subset \mathbb{R}^n$ such that

$$\operatorname{cap}_p(K, B(0,2)) > 0 \quad\text{and}\quad \dim_{\mathrm{H}}(K) = n-p.$$

Without loss of generality, we may assume that $0 \in K$. Theorem 5.42 and Theorem 5.38 show that

$$\operatorname{cap}_q(K, B(0,2)) = 0 \tag{12.4}$$

for every $1 < q < p$. Let $\Omega = B(0,2) \setminus K$. Then Ω^c cannot satisfy the p-capacity density condition since otherwise Theorem 7.21 would imply the existence of $1 < q < p$ and $C > 0$ such that

$$\operatorname{cap}_q(K, B(0,2)) = \operatorname{cap}_q\big(\Omega^c \cap \overline{B}(0,1), B(0,2)\big) \geq C > 0,$$

contradicting (12.4).

Let $g \in C_0^\infty(B(0,2))$ with $g = 1$ on K. By Theorem 11.13 and Theorem 11.11, there exists a unique weak solution $u \in W^{1,p}(\Omega)$ to the p-Laplace equation in Ω with $u - g \in W_0^{1,p}(\Omega)$. That is, u is the weak solution to the Dirichlet problem for the p-Laplace equation in Ω with the Sobolev boundary values zero on $\partial B(0,2)$ and one on K. By Theorem 11.41, we may assume that u is continuous in Ω. Corollary 5.29 implies the existence of a p-quasicontinuous function $v \in W^{1,p}(\mathbb{R}^n)$ such that $v = u - g$ almost everywhere in Ω and $v = 0$ p-quasieverywhere in $\mathbb{R}^n \setminus \Omega$. Thus $v = 0$ p-quasieverywhere in K and consequently $v + g = 1$ p-quasieverywhere in K. By Theorem 5.46 we may use the p-quasicontinuous function $v+g \in W_0^{1,p}(B(0,2))$ as a test function for the capacity $\operatorname{cap}_p(K, B(0,2))$. This gives

$$0 < \operatorname{cap}_p(K, B(0,2)) \leq \int_{B(0,2)} |\nabla v(x) + \nabla g(x)|^p \, dx = \int_{B(0,2)} |\nabla u(x)|^p \, dx.$$

The final equality holds since $v + g = u$ almost everywhere in Ω and $|K| = 0$. We conclude that u is not constant. By Theorem 11.9 we have $0 \leq u \leq 1$ everywhere in Ω. Since $\dim_{\mathrm{H}}(K) = n - p < n - 1$, the set Ω is connected and Theorem 11.37 implies that $0 < u(x) < 1$ for every $y \in \Omega$.

We claim that $\nabla u \notin L^{p+\delta}(\Omega)$ for any $\delta > 0$. For a contradiction, assume that

$$\int_\Omega |\nabla u(x)|^{p+\delta} \, dx < \infty \tag{12.5}$$

for some $\delta > 0$ and let $q = \frac{p+\delta}{1+\delta} < p$. We prove that $w = v + g \in W_0^{1,p}(B(0,2))$ is a weak solution to the p-Laplace equation in $B(0,2)$. Since $w = u$ almost everywhere in Ω, w is a p-quasicontinuous extension of u to $B(0,2)$. Recall from (12.4) that $\operatorname{cap}_q(K, B(0,2)) = 0$. Hence, by Theorem 5.44 there exist functions $\psi_j \in \operatorname{Lip}_0(B(0,2))$, $j \in \mathbb{N}$, such that $0 \leq \psi_j \leq 1$, $\psi_j = 1$ in K for every $j \in \mathbb{N}$, and

$$\int_{B(0,2)} |\nabla \psi_j(x)|^q \, dx \xrightarrow{j \to \infty} 0. \tag{12.6}$$

Let $\varphi \in C_0^\infty(B(0,2))$ and fix $j \in \mathbb{N}$. Then $\varphi(1 - \psi_j) \in \operatorname{Lip}_0(\Omega) \subset W_0^{1,p}(\Omega)$. Since $u \in W^{1,p}(\Omega)$ is a weak solution in Ω and $w = u$ almost everywhere in Ω, Lemma 11.6 implies

$$\int_\Omega |\nabla w(x)|^{p-2} \nabla w(x) \cdot \nabla((1 - \psi_j(x))\varphi(x)) \, dx = 0.$$

Consequently, since $|K| = 0$,

$$\begin{aligned}
&\int_{B(0,2)} |\nabla w(x)|^{p-2} \nabla w(x) \cdot \nabla \varphi(x)\, dx \\
&\quad = \int_{B(0,2)} |\nabla w(x)|^{p-2} \nabla w(x) \cdot \nabla((1-\psi_j(x))\varphi(x))\, dx \\
&\qquad + \int_{B(0,2)} |\nabla w(x)|^{p-2} \nabla w(x) \cdot \nabla(\psi_j(x)\varphi(x))\, dx \\
&\quad = \int_{\Omega} |\nabla w(x)|^{p-2} \nabla w(x) \cdot \nabla((1-\psi_j(x))\varphi(x))\, dx \\
&\qquad + \int_{B(0,2)} |\nabla w(x)|^{p-2} \nabla w(x) \cdot \nabla(\psi_j(x)\varphi(x))\, dx \\
&\quad = \int_{B(0,2)} |\nabla w(x)|^{p-2} \nabla w(x) \cdot \nabla(\psi_j(x)\varphi(x))\, dx.
\end{aligned}$$

By Hölder's inequality, we have

$$\begin{aligned}
&\left| \int_{B(0,2)} |\nabla w(x)|^{p-2} \nabla w(x) \cdot \nabla \varphi(x)\, dx \right| \\
&\quad = \left| \int_{B(0,2)} |\nabla w(x)|^{p-2} \nabla w(x) \cdot \nabla(\psi_j(x)\varphi(x))\, dx \right| \\
&\quad \le \int_{B(0,2)} |\nabla w(x)|^{p-1} |\nabla(\psi_j(x)\varphi(x))|\, dx \\
&\quad \le \left(\int_{B(0,2)} |\nabla w(x)|^{p+\delta}\, dx \right)^{\frac{p-1}{p+\delta}} \left(\int_{B(0,2)} |\nabla(\psi_j(x)\varphi(x))|^{\frac{p+\delta}{1+\delta}}\, dx \right)^{\frac{1+\delta}{p+\delta}}.
\end{aligned} \tag{12.7}$$

The first integral on the right-hand side of (12.7) is finite by (12.5) since $w = u$ almost everywhere in Ω and $|K| = 0$. To estimate the second integral on the right-hand side of (12.7), we recall that $q = \frac{p+\delta}{1+\delta}$. Theorem 3.17 implies

$$\|\psi_j\|_{L^q(B(0,2))} \le C(n,p,\delta) \|\nabla \psi_j\|_{L^q(B(0,2))},$$

and thus, by (12.6),

$$\begin{aligned}
&\left(\int_{B(0,2)} |\nabla(\psi_j(x)\varphi(x))|^{\frac{p+\delta}{1+\delta}}\, dx \right)^{\frac{1+\delta}{p+\delta}} \le \|\varphi \nabla \psi_j\|_{L^q(B(0,2))} + \|\psi_j \nabla \varphi\|_{L^q(B(0,2))} \\
&\quad \le \|\varphi\|_{L^\infty(B(0,2))} \|\nabla \psi_j\|_{L^q(B(0,2))} + \|\nabla \varphi\|_{L^\infty(B(0,2))} \|\psi_j\|_{L^q(B(0,2))} \\
&\quad \le C(n,p,\delta) \big(\|\varphi\|_{L^\infty(B(0,2))} + \|\nabla \varphi\|_{L^\infty(B(0,2))} \big) \|\nabla \psi_j\|_{L^q(B(0,2))} \xrightarrow{j\to\infty} 0.
\end{aligned}$$

We conclude that $w \in W_0^{1,p}(B(0,2))$ is a weak solution to the p-Laplace equation in $B(0,2)$ with the property that $0 < w(x) < 1$ for almost every $y \in \Omega$. This is however a contradiction, since by Theorem 11.13 the zero function is the unique solution to the p-Laplace equation in $B(0,2)$ that belongs to $W_0^{1,p}(B(0,2))$. Thus (12.5) cannot be true. In conclusion, $u \in W^{1,p}(\Omega)$ is a weak solution to the p-Laplace equation in Ω with a smooth boundary function g and $\nabla u \notin L^{p+\delta}(\Omega)$ for any $\delta > 0$. This shows that we can not omit in Theorem 12.6 the assumption that Ω^c satisfies the p-capacity density condition. Observe from Lemma 6.19 that this construction is not possible for $p > n$.

12.2. Stability with respect to the exponent

This section discusses a stability result, with respect to a varying exponent p, for weak solutions to the p-Laplace equation with given boundary values. This turns out to be a rather delicate problem. The main challenge is that the underlying Sobolev space changes as p varies and the associated energies are not necessarily finite. We show that under suitable assumptions solutions with varying exponent converge to the solution of the limit problem in the Sobolev space. The argument is based on the variational approach, and the equivalence in Theorem 11.11 between weak solutions to the p-Laplace equation and minimizers of the p-Dirichlet integral, see Theorem 11.13, is used throughout. The global higher integrability in Theorem 12.6 serves as a starting point for the proof of the stability.

THEOREM 12.10. *Let $1 < p < \infty$ and assume that $\Omega \subset \mathbb{R}^n$ is a nonempty bounded open set such that Ω^c satisfies the p-capacity density condition with a constant C_1. Let $(p_i)_{i\in\mathbb{N}}$ be a sequence with $1 < p_i < \infty$ for every $i \in \mathbb{N}$ and $p_i \to p$ as $i \to \infty$. Assume that $g \in W^{1,s}(\Omega)$ for some $s > p$ and let $u_i \in W^{1,p_i}(\Omega)$ be a weak solution to the p_i-Laplace equation in Ω with $u_i - g \in W_0^{1,p_i}(\Omega)$ for every $i \in \mathbb{N}$. Then there exist a unique weak solution $u \in W^{1,p}(\Omega)$ to the p-Laplace equation in Ω with $u - g \in W_0^{1,p}(\Omega)$ and a subsequence of $(u_i)_{i\in\mathbb{N}}$ that converges to u locally uniformly in Ω as $i \to \infty$.*

The lengthy proof of Theorem 12.10 is divided into several lemmas, which are of independent interest. We shall pass to a subsequence of $(u_i)_{i\in\mathbb{N}}$ several times and for simplicity the obtained subsequences will be denoted again by $(u_i)_{i\in\mathbb{N}}$.

LEMMA 12.11. *Under the assumptions in Theorem 12.10, there exist a subsequence of $(u_i)_{i\in\mathbb{N}}$ denoted by $(u_i)_{i\in\mathbb{N}}$, $0 < \delta \le s - p$, and a function $u \in W^{1,p+\delta}(\Omega)$ such that*

(a) *$(u_i)_{i\in\mathbb{N}}$ is a bounded sequence in $W^{1,p+\delta}(\Omega)$,*
(b) *$\nabla u_i \to \nabla u$ weakly in $L^{p+\delta}(\Omega;\mathbb{R}^n)$,*
(c) *$u_i \to u$ locally uniformly in Ω as $i \to \infty$.*

PROOF. First we show that $(u_i - g)_{i\in\mathbb{N}}$ is a bounded sequence in $W_0^{1,p+\delta}(\Omega)$ for some $\delta > 0$. By Theorem 12.6 and Remark 12.7, there exist $\delta = \delta(n,p,s,C_1) > 0$ and $C = C(n,p,s,C_1,\Omega)$ such that, for all sufficiently large $i \in \mathbb{N}$, the exponents $q = p + \delta$ and $p - \delta$ satisfy the assumptions of Corollary 3.7,

$$1 < p - \delta < p_i < p + \delta < p_i + 2\delta \le s \tag{12.8}$$

and

$$\begin{aligned} &\left(\int_\Omega |\nabla u_i(x)|^{p_i+2\delta}\,dx\right)^{\frac{1}{p_i+2\delta}} \\ &\quad\le C\left[\left(\int_\Omega |\nabla u_i(x)|^{p_i}\,dx\right)^{\frac{1}{p_i}} + \left(\int_\Omega |\nabla g(x)|^{p_i+2\delta}\,dx\right)^{\frac{1}{p_i+2\delta}}\right]. \end{aligned} \tag{12.9}$$

By passing to a subsequence, we may assume that both (12.8) and (12.9) hold for every $i \in \mathbb{N}$. Since $u_i - g \in W_0^{1,p_i}(\Omega) \subset W_0^{1,p-\delta}(\Omega)$ for every $i \in \mathbb{N}$, Corollary 3.7

and Hölder's inequality imply that

$$\begin{aligned}\left(\int_\Omega |u_i(x)-g(x)|^{p+\delta}\,dx\right)^{\frac{1}{p+\delta}} &\le C\left(\int_\Omega |\nabla u_i(x)-\nabla g(x)|^{p-\delta}\,dx\right)^{\frac{1}{p-\delta}}\\ &\le C\left(\int_\Omega |\nabla u_i(x)-\nabla g(x)|^{p_i}\,dx\right)^{\frac{1}{p_i}}\end{aligned} \tag{12.10}$$

with $C = C(n, p, s, C_1, \Omega)$.

By (12.8), we have $g \in W^{1,s}(\Omega) \subset W^{1,p_i}(\Omega)$ for every $i \in \mathbb{N}$. Theorem 11.11 asserts that u_i is a minimizer of the p_i-Dirichlet integral with $u_i - g \in W_0^{1,p_i}(\Omega)$ and thus

$$\left(\int_\Omega |\nabla u_i(x)|^{p_i}\,dx\right)^{\frac{1}{p_i}} \le \left(\int_\Omega |\nabla g(x)|^{p_i}\,dx\right)^{\frac{1}{p_i}}, \tag{12.11}$$

for every $i \in \mathbb{N}$. By Hölder's inequality and (12.9) we obtain

$$\begin{aligned}\left(\int_\Omega |\nabla u_i(x)|^{p+\delta}\,dx\right)^{\frac{1}{p+\delta}} &\le C\left(\int_\Omega |\nabla u_i(x)|^{p_i+2\delta}\,dx\right)^{\frac{1}{p_i+2\delta}}\\ &\le C\left[\left(\int_\Omega |\nabla u_i(x)|^{p_i}\,dx\right)^{\frac{1}{p_i}} + \left(\int_\Omega |\nabla g(x)|^{p_i+2\delta}\,dx\right)^{\frac{1}{p_i+2\delta}}\right]\\ &\le C\left(\int_\Omega |\nabla g(x)|^{s}\,dx\right)^{\frac{1}{s}}\end{aligned} \tag{12.12}$$

for every $i \in \mathbb{N}$ with $C = C(n, p, s, C_1, \Omega)$. This shows that $(\nabla u_i)_{i\in\mathbb{N}}$ is a bounded sequence in $L^{p+\delta}(\Omega; \mathbb{R}^n)$. It follows that

$$\begin{aligned}&\left(\int_\Omega |\nabla u_i(x)-\nabla g(x)|^{p+\delta}\,dx\right)^{\frac{1}{p+\delta}}\\ &\le \left(\int_\Omega |\nabla u_i(x)|^{p+\delta}\,dx\right)^{\frac{1}{p+\delta}} + \left(\int_\Omega |\nabla g(x)|^{p+\delta}\,dx\right)^{\frac{1}{p+\delta}}\\ &\le C\left(\int_\Omega |\nabla g(x)|^{s}\,dx\right)^{\frac{1}{s}}\end{aligned} \tag{12.13}$$

for every $i \in \mathbb{N}$ with $C = C(n, p, s, C_1, \Omega)$.

Using (12.10) and (12.13) we obtain

$$\begin{aligned}\left(\int_\Omega |u_i(x)-g(x)|^{p+\delta}\,dx\right)^{\frac{1}{p+\delta}} &\le C\left(\int_\Omega |\nabla u_i(x)-\nabla g(x)|^{p_i}\,dx\right)^{\frac{1}{p_i}}\\ &\le C\left(\int_\Omega |\nabla u_i(x)-\nabla g(x)|^{p+\delta}\,dx\right)^{\frac{1}{p+\delta}}\\ &\le C\left(\int_\Omega |\nabla g(x)|^{s}\,dx\right)^{\frac{1}{s}}\end{aligned}$$

for every $i \in \mathbb{N}$ with $C = C(n, p, s, C_1, \Omega)$. Thus $(u_i - g)_{i\in\mathbb{N}}$ is a bounded sequence in $L^{p+\delta}(\Omega)$. Together with (12.13) and the fact that $g \in W^{1,s}(\Omega) \subset W^{1,p+\delta}(\Omega)$ we conclude that $(u_i)_{i\in\mathbb{N}}$ is a bounded sequence in $W^{1,p+\delta}(\Omega)$. By Theorem 2.39 there exists $u \in W^{1,p+\delta}(\Omega)$ and a subsequence $(u_i)_{i\in\mathbb{N}}$ such that $u_i \to u$ weakly in $L^{p+\delta}(\Omega)$ and $\nabla u_i \to \nabla u$ weakly in $L^{p+\delta}(\Omega; \mathbb{R}^n)$ as $i \to \infty$.

By Theorem 11.41 we may assume that u_i is continuous in Ω, for every $i \in \mathbb{N}$. Let $\Omega' \Subset \Omega$ be an open set. By Theorem 11.30 there exists a constant $C = C(n,p,s,C_1)$ such that

$$\sup_{x\in B(z,r)} |u_i(x)| \le C\biggl(\fint_{B(z,2r)} |u_i(x)|^{p+\delta}\,dx\biggr)^{\frac{1}{p+\delta}}$$

for every $i \in \mathbb{N}$ whenever $B(z,2r) \Subset \Omega$. By covering $\overline{\Omega'}$ with finitely many such balls and using the fact that $(u_i)_{i\in\mathbb{N}}$ is a bounded sequence in $L^{p+\delta}(\Omega)$, we conclude that

$$\sup_{x\in\overline{\Omega'}} |u_i(x)| \le C \sup_{i\in\mathbb{N}}\biggl(\int_\Omega |u_i(x)|^{p+\delta}\,dx\biggr)^{\frac{1}{p+\delta}} < \infty, \tag{12.14}$$

for every $i \in \mathbb{N}$, with $C = C(n,p,s,C_1,\Omega')$. Assuming $B(z,3r) \Subset \Omega$, the proof of Theorem 11.41 and Theorem 11.30 show that there exist an exponent $\alpha = \alpha(n,p,s,C_1) > 0$ and a constant $C = C(n,p,s,C_1)$ such that

$$\begin{aligned} |u_i(x) - u_i(x')| &\le Cr^{-\alpha}|x-x'|^\alpha \sup_{y\in B(z,2r)} |u_i(y)| \\ &\le Cr^{-\alpha}|x-x'|^\alpha \biggl(\fint_{B(z,3r)} |u_i(x)|^{p+\delta}\,dx\biggr)^{\frac{1}{p+\delta}} \end{aligned} \tag{12.15}$$

for every $i \in \mathbb{N}$ and every $x, x' \in B\bigl(z, \frac{r}{28\sqrt{n}}\bigr)$. A covering argument using (12.15) and (12.14) gives

$$\sup_{x,x'\in\overline{\Omega'}} |u_i(x) - u_i(x')| \le C|x-x'|^\alpha \sup_{i\in\mathbb{N}}\biggl(\int_\Omega |u_i(x)|^{p+\delta}\,dx\biggr)^{\frac{1}{p+\delta}} < \infty \tag{12.16}$$

for all $i \in \mathbb{N}$, with $C = C(n,p,s,C_1,\Omega')$.

From (12.14) and (12.16) it follows that the sequence $(u_i)_{i\in\mathbb{N}}$ is both uniformly bounded and equicontinuous in $\overline{\Omega'}$. The Arzela–Ascoli theorem gives a subsequence $(u_i)_{i\in\mathbb{N}}$ such that $u_i \to u$ uniformly in $\Omega' \subset \overline{\Omega'}$ as $i \to \infty$. Here we modify u on a set of measure zero, if necessary, and identify the limit u by using the fact that $u_i \to u$ weakly in $L^{p+\delta}(\Omega')$.

Finally, we exhaust Ω by considering the sets $\Omega_j = \{x \in \Omega : d(x,\partial\Omega) > \frac{1}{j}\} \Subset \Omega$ for every $j \in \mathbb{N}$. Then $\Omega = \bigcup_{j=1}^\infty \Omega_j$. We take a subsequence of $(u_i)_{i\in\mathbb{N}}$ that converges uniformly to u in Ω_1 and then proceed recursively: in the jth step we choose from the previously chosen subsequence a new subsequence that converges uniformly in Ω_j to u. At the end, we take the diagonal subsequence. This diagonal subsequence converges uniformly to u in all compact subsets $K \subset \Omega$ since each such K is contained in Ω_j for some $j \in \mathbb{N}$. Clearly the diagonal subsequence also satisfies the conditions (a) and (b). □

REMARK 12.12. Let Ω and $(u_i)_{i\in\mathbb{N}}$ be as in Theorem 12.10. By the regularity theory $u_i \in C^{1,\alpha}_{\mathrm{loc}}(\Omega)$ for every $i \in \mathbb{N}$, see Remark 11.42. A careful analysis of constants in the regularity theory shows that $(u_i)_{i\in\mathbb{N}}$ and $(\nabla u_i)_{i\in\mathbb{N}}$ are locally equicontinuous and, in addition, locally uniformly bounded. Thus the Arzela–Ascoli theorem implies the existence of a subsequence $(u_i)_{i\in\mathbb{N}}$ such that $u_i \to u$ and $\nabla u_i \to \nabla u$ locally uniformly in Ω as $i \to \infty$. If the boundary function g is Hölder continuous on $\partial\Omega$, then $u_i \in C^{1,\alpha}(\Omega)$ for every $i \in \mathbb{N}$ since Ω^c satisfies the

p-capacity density condition, see Remark 11.44. Moreover, we have $u_i \to u$ and $\nabla u_i \to \nabla u$ uniformly in Ω as $i \to \infty$.

The limit function u in Lemma 12.11 has the correct boundary values. The p-capacity density condition is applied in the proof below.

LEMMA 12.13. *Under the assumptions in Theorem* 12.10 *and notation as in Lemma* 12.11, *we have* $u - g \in W_0^{1,p}(\Omega)$.

PROOF. Let $0 < \varepsilon < p - 1$ and let $(u_i)_{i\in\mathbb{N}}$ be the subsequence given by Lemma 12.11. For sufficiently large $i \in \mathbb{N}$ we have $p_i > p - \varepsilon$ and $u_i - g \in W_0^{1,p_i}(\Omega)$. It follows that $u_i - g \in W_0^{1,p-\varepsilon}(\Omega)$ for every such $i \in \mathbb{N}$, and by taking a further subsequence we may assume that this holds for every $i \in \mathbb{N}$. By Hölder's inequality, Lemma 12.11 (a), and the fact that $g \in W^{1,s}(\Omega)$ with $p - \varepsilon < p + \delta \le s$, we find that $(u_i - g)_{i\in\mathbb{N}}$ is a bounded sequence in $W_0^{1,p-\varepsilon}(\Omega)$. Lemma 12.11 (c) asserts that $u_i - g \to u - g$ locally uniformly and thus almost everywhere in Ω as $i \to \infty$. Hence Theorem 2.40 implies that $u - g \in W_0^{1,p-\varepsilon}(\Omega)$, and Lemma 12.11 gives $u - g \in W^{1,p}(\Omega)$. According to Remark 6.33 and the assumption that Ω^c satisfies the p-capacity density condition, we obtain

$$W^{1,p}(\Omega) \cap \bigcap_{0<\varepsilon<p-1} W_0^{1,p-\varepsilon}(\Omega) = W_0^{1,p}(\Omega),$$

from which we conclude that $u - g \in W_0^{1,p}(\Omega)$. $\square$

It remains to show that u is a minimizer of the p-Dirichlet integral with the boundary values g. We being with the following elementary inequality.

LEMMA 12.14. *Let* $a, b, \varepsilon > 0$. *Then*

$$\big||\zeta|^a - |\zeta|^b\big| \le \Big(\frac{1}{\varepsilon}|\zeta|^{\max\{a,b\}+\varepsilon} + \frac{1}{e}\Big(\frac{1}{a} + \frac{1}{b}\Big)\Big)|a - b| \tag{12.17}$$

for every $\zeta \in \mathbb{R}^n$, *where* e *is Napier's constant.*

PROOF. Clearly, we may assume that $\zeta \ne 0$. An application of the mean value theorem shows that

$$\begin{aligned}\big||\zeta|^a - |\zeta|^b\big| &= \big|\exp(a\log|\zeta|) - \exp(b\log|\zeta|)\big| \\ &\le \max_{x\in[a,b]}\Big|\frac{\partial}{\partial x}\big(\exp(x\log|\zeta|)\big)\Big||a - b| \le \big|\log|\zeta|\big|(|\zeta|^a + |\zeta|^b)|a - b|.\end{aligned}$$

By elementary calculus,

$$0 \le \frac{\log(t)}{t^s} \le \frac{1}{e}\cdot\frac{1}{s} \tag{12.18}$$

whenever $1 \le t < \infty$ and $s > 0$. If $|\zeta| \ge 1$, then (12.18) implies

$$\big|\log|\zeta|\big|(|\zeta|^a + |\zeta|^b)|a - b| \le \frac{2}{e}\cdot\frac{1}{\varepsilon}\cdot|\zeta|^{\max\{a,b\}+\varepsilon}|a - b| \le \frac{1}{\varepsilon}\cdot|\zeta|^{\max\{a,b\}+\varepsilon}|a - b|.$$

On the other hand, if $0 < |\zeta| \le 1$, then (12.18) with $t = \frac{1}{|\zeta|}$ gives

$$\big|\log|\zeta|\big|(|\zeta|^a + |\zeta|^b)|a - b| \le \frac{1}{e}\Big(\frac{1}{a} + \frac{1}{b}\Big)|a - b|.$$

The claim follows by combining the estimates above. $\square$

Next we prove a lower semicontinuity result in the varying exponent case.

LEMMA 12.15. *Under the assumptions in Theorem* 12.10 *and notation as in Lemma* 12.11, *we have*

$$\int_{\Omega'} |\nabla u(x)|^p \, dx \le \liminf_{i\to\infty} \int_{\Omega'} |\nabla u_i(x)|^{p_i} \, dx$$

whenever $\Omega' \subset \Omega$ *is an open set.*

PROOF. We consider the subsequence of $(u_i)_{i\in\mathbb{N}}$ given by Lemma 12.11. By the proof of Lemma 12.11, we may assume that the inequalities in (12.8) hold for every $i \in \mathbb{N}$. Let $0 < \varepsilon < \frac{p-1}{2}$. For sufficiently large $i \in \mathbb{N}$, we have $p_i > p - \varepsilon$. By taking a further subsequence, we may assume that this holds for every $i \in \mathbb{N}$.

Observe that

$$\left(\int_{\Omega'} |\nabla u_i(x)|^{p_i} \, dx\right) \le \left(\int_{\Omega'} |\nabla u_i(x)|^{p+\delta} \, dx\right)^{\frac{p_i}{p+\delta}} |\Omega'|^{1-\frac{p_i}{p+\delta}}.$$

Since $(u_i)_{i\in\mathbb{N}}$ is a bounded sequence in $L^{p+\delta}(\Omega)$ and $\Omega' \subset \Omega$ has a finite measure, the left-hand side is uniformly bounded over $i \in \mathbb{N}$. Inequality (12.17) implies

$$\begin{aligned} &\left|\left(\int_{\Omega'} |\nabla u_i(x)|^{p_i} \, dx\right)^{\frac{p-\varepsilon}{p_i}} - \left(\int_{\Omega'} |\nabla u_i(x)|^{p_i} \, dx\right)^{\frac{p-\varepsilon}{p}}\right| \\ &\quad \le \left[\left(\int_{\Omega'} |\nabla u_i(x)|^{p_i} \, dx\right)^{q_i} + \frac{1}{e}\left(\frac{p_i}{p-\varepsilon} + \frac{p}{p-\varepsilon}\right)\right] \left|\frac{p-\varepsilon}{p_i} - \frac{p-\varepsilon}{p}\right| \\ &\quad \xrightarrow{i\to\infty} 0, \end{aligned} \tag{12.19}$$

where $q_i = \max\{\frac{p-\varepsilon}{p_i}, \frac{p-\varepsilon}{p}\} + 1$. Since $1 < p - \varepsilon < p + \delta$ and $\Omega' \subset \Omega$, Lemma 12.11 and the fact that Ω is bounded implies that $\nabla u_i \to \nabla u$ weakly in $L^{p-\varepsilon}(\Omega'; \mathbb{R}^n)$ as $i \to \infty$. Using Lemma 2.34, Hölder's inequality and (12.19), we obtain

$$\begin{aligned} \int_{\Omega'} |\nabla u(x)|^{p-\varepsilon} \, dx &\le \liminf_{i\to\infty} \int_{\Omega'} |\nabla u_i(x)|^{p-\varepsilon} \, dx \\ &\le \liminf_{i\to\infty} \left[\left(\int_{\Omega'} |\nabla u_i(x)|^{p_i} \, dx\right)^{\frac{p-\varepsilon}{p_i}} |\Omega'|^{1-\frac{p-\varepsilon}{p_i}}\right] \\ &= \liminf_{i\to\infty} \left(\int_{\Omega'} |\nabla u_i(x)|^{p_i} \, dx\right)^{\frac{p-\varepsilon}{p}} |\Omega'|^{\frac{\varepsilon}{p}} \\ &= \left(\liminf_{i\to\infty} \int_{\Omega'} |\nabla u_i(x)|^{p_i} \, dx\right)^{\frac{p-\varepsilon}{p}} |\Omega'|^{\frac{\varepsilon}{p}}. \end{aligned} \tag{12.20}$$

We apply (12.17) with $\delta > 0$, $\zeta = \nabla u(x)$, $a = p - \varepsilon$ and $b = p$, and obtain

$$\begin{aligned} &\int_{\Omega'} \left||\nabla u(x)|^{p-\varepsilon} - |\nabla u(x)|^p\right| dx \\ &\quad \le |\varepsilon| \left(\frac{1}{\delta} \int_{\Omega'} |\nabla u(x)|^{p+\delta} \, dx + \frac{1}{e}|\Omega'| \left|\frac{1}{p-\varepsilon} + \frac{1}{p}\right|\right) \xrightarrow{\varepsilon\to 0} 0. \end{aligned}$$

Here we also used the fact that by Lemma 12.11, we have

$$\int_{\Omega'} |\nabla u(x)|^{p+\delta} \, dx < \infty.$$

From this, together with (12.20), we conclude that

$$\int_{\Omega'} |\nabla u(x)|^p \, dx = \lim_{\varepsilon \to 0} \int_{\Omega'} |\nabla u(x)|^{p-\varepsilon} \, dx \le \liminf_{i\to\infty} \int_{\Omega'} |\nabla u_i(x)|^{p_i} \, dx. \qquad \square$$

We obtain a local uniform integrability estimate for the gradients.

LEMMA 12.16. *Let K be a compact subset of Ω and let $K_\varepsilon = \{x \in \Omega : d(x,K) < \varepsilon\}$ for $0 < \varepsilon < d(K, \partial\Omega)$. Under the assumptions in Theorem 12.10 and notation as in Lemma 12.11, we have*

$$\limsup_{i\to\infty} \int_{K_\varepsilon} |\nabla u_i(x)|^{p_i} \, dx \le 3^p \int_{K_\varepsilon} |\nabla u(x)|^p \, dx$$

for a dense subset of ε with $0 < \varepsilon < d(K, \partial\Omega)$.

PROOF. We consider $\delta > 0$, $u \in W^{1,p+\delta}(\Omega)$ and the subsequence of $(u_i)_{i\in\mathbb{N}}$ given by Lemma 12.11. Taking a further subsequence, if necessary, we may assume that

$$1 < p - \delta < p_i \le p + \frac{\delta}{2} < p + \delta \le s \tag{12.21}$$

for every $i \in \mathbb{N}$. Let $0 < \varepsilon' < \varepsilon < d(K, \partial\Omega)$ and let $\psi \in \operatorname{Lip}_0(K_\varepsilon)$ be a cutoff function with $0 \le \psi \le 1$ and $\psi = 1$ in $K_{\varepsilon'}$. For every $i \in \mathbb{N}$, we define

$$v_i = u_i + \psi(u - u_i) = \psi u + (1 - \psi) u_i.$$

Let $i \in \mathbb{N}$. By (12.21), $v_i \in W^{1,p_i}(\Omega)$ with $v_i - g \in W_0^{1,p_i}(\Omega)$. By the minimizing property of u_i, see Theorem 11.11, and properties of v_i, we have

$$\begin{aligned}\int_{\Omega\setminus K_\varepsilon} |\nabla u_i(x)|^{p_i} \, dx + \int_{K_\varepsilon} |\nabla u_i(x)|^{p_i} \, dx &= \int_\Omega |\nabla u_i(x)|^{p_i} \, dx \le \int_\Omega |\nabla v_i(x)|^{p_i} \, dx \\ &= \int_{\Omega\setminus K_\varepsilon} |\nabla u_i(x)|^{p_i} \, dx + \int_{K_\varepsilon} |\nabla v_i(x)|^{p_i} \, dx\end{aligned}$$

and therefore

$$\int_{K_\varepsilon} |\nabla u_i(x)|^{p_i} \, dx \le \int_{K_\varepsilon} |\nabla v_i(x)|^{p_i} \, dx.$$

This implies

$$\begin{aligned}\int_{K_{\varepsilon'}} |\nabla u_i(x)|^{p_i} \, dx &\le \int_{K_\varepsilon} |\nabla u_i(x)|^{p_i} \, dx \le \int_{K_\varepsilon} |\nabla v_i(x)|^{p_i} \, dx \\ &= \int_{K_\varepsilon} \big|(1 - \psi(x))\nabla u_i(x) + (u(x) - u_i(x))\nabla\psi(x) + \psi(x)\nabla u(x)\big|^{p_i} \, dx \\ &\le 3^{p_i} \bigg[\int_{K_\varepsilon} (1 - \psi(x))^{p_i} |\nabla u_i(x)|^{p_i} \, dx + \int_{K_\varepsilon} \psi(x)^{p_i} |\nabla u(x)|^{p_i} \, dx \\ &\qquad + \int_{K_\varepsilon} |u_i(x) - u(x)|^{p_i} |\nabla\psi(x)|^{p_i} \, dx\bigg].\end{aligned}$$

Next we apply a hole filling technique. Using the facts that $\psi = 1$ in $K_{\varepsilon'} \subset K_\varepsilon$ and $0 \le \psi \le 1$, and by adding the term

$$3^{p_i} \int_{K_{\varepsilon'}} |\nabla u_i(x)|^{p_i} \, dx$$

to the both sides, we obtain

$$\begin{aligned}(1+3^{p_i})\int_{K_{\varepsilon'}}|\nabla u_i(x)|^{p_i}\,dx \le 3^{p_i}\Big[\int_{K_\varepsilon}|\nabla u_i(x)|^{p_i}\,dx + \int_{K_\varepsilon}\psi(x)^{p_i}|\nabla u(x)|^{p_i}\,dx\\ + \int_{K_\varepsilon}|u_i(x)-u(x)|^{p_i}|\nabla\psi(x)|^{p_i}\,dx\Big].\end{aligned}$$

This, together with $0\le\psi\le 1$, implies

$$\begin{aligned}\int_{K_{\varepsilon'}}|\nabla u_i(x)|^{p_i}\,dx \le \frac{3^{p_i}}{1+3^{p_i}}\Big[\int_{K_\varepsilon}|\nabla u_i(x)|^{p_i}\,dx + \int_{K_\varepsilon}|\nabla u(x)|^{p_i}\,dx\\ + \int_{K_\varepsilon}|u_i(x)-u(x)|^{p_i}|\nabla\psi(x)|^{p_i}\,dx\Big].\end{aligned}\tag{12.22}$$

Since $p_i\le p+\delta$ and $u_i\to u$ locally uniformly in Ω, see Lemma 12.11, we have

$$\begin{aligned}\int_{K_\varepsilon}|u_i(x)-u(x)|^{p_i}|\nabla\psi(x)|^{p_i}\,dx &\le \|\nabla\psi\|^{p_i}_{L^\infty(\mathbb{R}^n)}\int_{K_\varepsilon}|u_i(x)-u(x)|^{p_i}\,dx\\ &\le \|\nabla\psi\|^{p_i}_{L^\infty(\mathbb{R}^n)}\Big(\int_{K_\varepsilon}|u_i(x)-u(x)|^{p+\delta}\,dx\Big)^{\frac{p_i}{p+\delta}}|K_\varepsilon|^{1-\frac{p_i}{p+\delta}} \xrightarrow{i\to\infty} 0.\end{aligned}$$

By taking limes superior on both sides of (12.22), we obtain

$$\begin{aligned}&\limsup_{i\to\infty}\int_{K_{\varepsilon'}}|\nabla u_i(x)|^{p_i}\,dx\\ &\quad\le \frac{3^p}{1+3^p}\Big[\limsup_{i\to\infty}\int_{K_\varepsilon}|\nabla u_i(x)|^{p_i}\,dx + \limsup_{i\to\infty}\int_{K_\varepsilon}|\nabla u(x)|^{p_i}\,dx\Big].\end{aligned}$$

Let

$$\Psi(\varepsilon)=\limsup_{i\to\infty}\int_{K_\varepsilon}|\nabla u_i(x)|^{p_i}\,dx,$$

for $0<\varepsilon<d(K,\partial\Omega)$. Estimate (12.11) and inequality (12.21) give

$$\Big(\int_{K_\varepsilon}|\nabla u_i(x)|^{p_i}\,dx\Big)^{\frac{1}{p_i}} \le \Big(\int_\Omega|\nabla u_i(x)|^{p_i}\,dx\Big)^{\frac{1}{p_i}} \le C\Big(\int_\Omega|\nabla g(x)|^s\,dx\Big)^{\frac1s}<\infty$$

for all $i\in\mathbb{N}$, with $C=C(p,s,\delta,\Omega)$. This implies that Ψ is a bounded function of ε. Since Ψ is also nondecreasing, the set of points where Ψ is discontinuous is at most countable.

Let ε be a point at which Ψ is continuous. Then

$$\Psi(\varepsilon)=\lim_{\varepsilon'\to\varepsilon_-}\Psi(\varepsilon')\le\frac{3^p}{1+3^p}\Big[\Psi(\varepsilon)+\limsup_{i\to\infty}\int_{K_\varepsilon}|\nabla u(x)|^{p_i}\,dx\Big].\tag{12.23}$$

By absorbing the first term on the right-hand side of (12.23) we obtain

$$\limsup_{i\to\infty}\int_{K_\varepsilon}|\nabla u_i(x)|^{p_i}\,dx=\Psi(\varepsilon)\le 3^p\limsup_{i\to\infty}\int_{K_\varepsilon}|\nabla u(x)|^{p_i}\,dx.$$

It suffices to show that

$$\lim_{i\to\infty}\int_{K_\varepsilon}|\nabla u(x)|^{p_i}\,dx=\int_{K_\varepsilon}|\nabla u(x)|^p\,dx.\tag{12.24}$$

By (12.21), there exists $\delta' > 0$ such that $\max\{p_i, p\} + \delta' < p + \delta$ for every $i \in \mathbb{N}$. Lemma 12.11 implies

$$\begin{aligned}
&\limsup_{i\to\infty} \int_{K_\varepsilon} |\nabla u(x)|^{\max\{p_i,p\}+\delta'}\,dx \\
&\quad\le \limsup_{i\to\infty}\left[\left(\int_{K_\varepsilon} |\nabla u(x)|^{p+\delta}\,dx\right)^{\frac{\max\{p_i,p\}+\delta'}{p+\delta}} |K_\varepsilon|^{1-\frac{\max\{p_i,p\}+\delta'}{p+\delta}}\right] \\
&\quad= \left(\int_{K_\varepsilon} |\nabla u(x)|^{p+\delta}\,dx\right)^{\frac{p+\delta'}{p+\delta}} |K_\varepsilon|^{1-\frac{p+\delta'}{p+\delta}} < \infty.
\end{aligned}$$

By (12.17) with $\zeta = \nabla u(x)$, $a = p_i$ and $b = p$, we conclude that

$$\begin{aligned}
&\int_{K_\varepsilon} \big||\nabla u(x)|^{p_i} - |\nabla u(x)|^p\big|\,dx \\
&\quad\le |p_i - p|\left(\frac{1}{\delta'}\int_{K_\varepsilon} |\nabla u(x)|^{\max\{p_i,p\}+\delta'}\,dx + \frac{1}{e}|K_\varepsilon|\left|\frac{1}{p_i} + \frac{1}{p}\right|\right) \xrightarrow{i\to\infty} 0.
\end{aligned} \tag{12.25}$$

Thus (12.24) holds and this proves the claim for a dense set of ε with $0 < \varepsilon < d(K, \partial\Omega)$. □

We show that u is a minimizer of the p-Dirichlet integral among compactly supported smooth variations.

LEMMA 12.17. *Under the assumptions in Theorem* 12.10 *and notation as in Lemma* 12.11, *we have*

$$\int_\Omega |\nabla u(x)|^p\,dx \le \int_\Omega |\nabla u(x) + \nabla\varphi(x)|^p\,dx$$

for every $\varphi \in C_0^\infty(\Omega)$.

PROOF. We consider $\delta > 0$, $u \in W^{1,p+\delta}(\Omega)$ and the subsequence of $(u_i)_{i\in\mathbb{N}}$ given by Lemma 12.11. Without loss of generality, we may assume that (12.21) holds for every $i \in \mathbb{N}$. Fix $\varepsilon > 0$. Let $\varphi \in C_0^\infty(\Omega)$ and let $\Omega' \Subset \Omega'' \Subset \Omega$ be open sets such that $\operatorname{supp}\varphi \subset \Omega'$ and

$$\int_{\Omega\setminus\Omega'} |\nabla u(x)|^p\,dx < \frac{\varepsilon}{2\cdot 3^p\cdot 3^p}.$$

Let $\psi \in C_0^\infty(\Omega'')$ be a cutoff function with $0 \le \psi \le 1$ and $\psi = 1$ in Ω', and let

$$v_i = \varphi + \psi(u - u_i)$$

for every $i \in \mathbb{N}$. By Lemma 12.11 we have $u_i \in W^{1,p+\delta}(\Omega)$ for every $i \in \mathbb{N}$ and $u \in W^{1,p+\delta}(\Omega)$. From this and the fact that $p_i \le p + \delta$ for every $i \in \mathbb{N}$ we conclude that $v_i \in W_0^{1,p_i}(\Omega'')$ for every $i \in \mathbb{N}$. Since $\varphi = 0$ in $\Omega'' \setminus \Omega'$ and $\psi = 1$ in Ω', we have

$$u_i + v_i = \begin{cases} u + \varphi & \text{in } \Omega', \\ u_i + \psi(u - u_i) & \text{in } \Omega'' \setminus \Omega'. \end{cases}$$

The minimizing property of u_i in Ω'', see Theorem 11.11, implies

$$
\begin{aligned}
\int_{\Omega''} |\nabla u_i(x)|^{p_i}\,dx &\le \int_{\Omega''} |\nabla u_i(x) + \nabla v_i(x)|^{p_i}\,dx \\
&= \int_{\Omega'} |\nabla u_i(x) + \nabla v_i(x)|^{p_i}\,dx + \int_{\Omega''\setminus\Omega'} |\nabla u_i(x) + \nabla v_i(x)|^{p_i}\,dx \\
&= \int_{\Omega'} |\nabla u(x) + \nabla \varphi(x)|^{p_i}\,dx \\
&\quad + \int_{\Omega''\setminus\Omega'} \big|\nabla u_i(x) + \nabla\big(\psi(x)(u(x)-u_i(x))\big)\big|^{p_i}\,dx.
\end{aligned}
\tag{12.26}
$$

For the second integral on the right-hand side of (12.26), we have

$$
\begin{aligned}
&\int_{\Omega''\setminus\Omega'} \big|\nabla u_i(x) + \nabla\big(\psi(x)(u(x)-u_i(x))\big)\big|^{p_i}\,dx \\
&\quad \le 3^{p_i}\bigg[\int_{\Omega''\setminus\Omega'} (1-\psi(x))^{p_i}|\nabla u_i(x)|^{p_i}\,dx + \int_{\Omega''\setminus\Omega'} \psi(x)^{p_i}|\nabla u(x)|^{p_i}\,dx \\
&\qquad + \int_{\Omega''\setminus\Omega'} |\nabla\psi(x)|^{p_i}|u(x)-u_i(x)|^{p_i}\,dx\bigg].
\end{aligned}
\tag{12.27}
$$

We take limit superior on both sides of (12.27) and estimate the terms separately.

Let $K = \overline{\Omega''}\setminus\Omega'$ and $K_{\varepsilon'} = \{x\in\Omega : d(x,K) < \varepsilon'\}$ with $0<\varepsilon'<d(K,\partial\Omega)$ as in Lemma 12.16. By Lemma 12.16 there exists a dense set of ε' satisfying

$$
\begin{aligned}
\limsup_{i\to\infty} \int_{\Omega''\setminus\Omega'} (1-\psi(x))^{p_i}|\nabla u_i(x)|^{p_i}\,dx &\le \limsup_{i\to\infty} \int_{K_{\varepsilon'}} |\nabla u_i(x)|^{p_i}\,dx \\
&\le 3^p \int_{K_{\varepsilon'}} |\nabla u(x)|^p\,dx.
\end{aligned}
$$

By letting $\varepsilon'\to 0$ along such a dense set, and using dominated convergence theorem, we obtain

$$
\limsup_{i\to\infty} \int_{\Omega''\setminus\Omega'} (1-\psi(x))^{p_i}|\nabla u_i(x)|^{p_i}\,dx \le 3^p \int_K |\nabla u(x)|^p\,dx < \frac{\varepsilon}{2\cdot 3^p}.
$$

For the second integral on the right-hand side of (12.27) we apply (12.17) as in the proof of Lemma 12.16 and obtain

$$
\begin{aligned}
\limsup_{i\to\infty} \int_{\Omega''\setminus\Omega'} \psi(x)^{p_i}|\nabla u(x)|^{p_i}\,dx &\le \limsup_{i\to\infty} \int_{\Omega\setminus\Omega'} |\nabla u(x)|^{p_i}\,dx \\
&\le \int_{\Omega\setminus\Omega'} |\nabla u(x)|^p\,dx < \frac{\varepsilon}{2\cdot 3^p\cdot 3^p}.
\end{aligned}
$$

For the third integral on the right-hand side of (12.27), we have

$$
\begin{aligned}
\int_{\Omega''\setminus\Omega'} |\nabla\psi(x)|^{p_i}|u(x)-u_i(x)|^{p_i}\,dx &\le \|\nabla\psi\|_{L^\infty(\mathbb{R}^n)}^{p_i} \int_{\Omega''\setminus\Omega'} |u(x)-u_i(x)|^{p_i}\,dx \\
&\le \|\nabla\psi\|_{L^\infty(\mathbb{R}^n)}^{p_i} \bigg(\int_{\Omega''\setminus\Omega'} |u(x)-u_i(x)|^{p+\delta}\,dx\bigg)^{\frac{p_i}{p+\delta}} |\Omega''\setminus\Omega'|^{1-\frac{p_i}{p+\delta}} \xrightarrow{i\to\infty} 0,
\end{aligned}
$$

since $u_i \to u$ locally uniformly in Ω by Lemma 12.11. Thus we conclude from (12.27) that

$$\limsup_{i\to\infty} \int_{\Omega''\setminus\Omega'} |\nabla u_i(x) + \nabla\big(\psi(x)(u(x) - u_i(x))\big)|^{p_i}\,dx < \varepsilon.$$

Together with (12.26) and again (12.17) as in the proof of Lemma 12.16, this implies

$$\begin{aligned}\limsup_{i\to\infty} \int_{\Omega''} |\nabla u_i(x)|^{p_i}\,dx &\le \limsup_{i\to\infty} \int_{\Omega'} |\nabla u(x) + \nabla\varphi(x)|^{p_i}\,dx + \varepsilon \\ &= \int_{\Omega'} |\nabla u(x) + \nabla\varphi(x)|^{p}\,dx + \varepsilon.\end{aligned} \tag{12.28}$$

Finally, by (12.28) and Lemma 12.15 applied in Ω',

$$\begin{aligned}\int_{\Omega} |\nabla u(x)|^p\,dx &= \int_{\Omega\setminus\Omega'} |\nabla u(x)|^p\,dx + \int_{\Omega'} |\nabla u(x)|^p\,dx \\ &\le \int_{\Omega\setminus\Omega'} |\nabla u(x)|^p\,dx + \liminf_{i\to\infty} \int_{\Omega'} |\nabla u_i(x)|^{p_i}\,dx \\ &\le \int_{\Omega\setminus\Omega'} |\nabla u(x)|^p\,dx + \limsup_{i\to\infty} \int_{\Omega''} |\nabla u_i(x)|^{p_i}\,dx \\ &\le \int_{\Omega\setminus\Omega'} |\nabla u(x)|^p\,dx + \int_{\Omega'} |\nabla u(x) + \nabla\varphi(x)|^p\,dx + \varepsilon \\ &= \int_{\Omega} |\nabla u(x) + \nabla\varphi(x)|^p\,dx + \varepsilon,\end{aligned}$$

where the final equality holds since $\operatorname{supp}\varphi \subset \Omega'$. The claim follows by letting $\varepsilon \to 0$. □

The function u is a minimizer of the p-Dirichlet integral also in sense of Theorem 11.13.

LEMMA 12.18. *Under the assumptions in Theorem* 12.10 *and notation as in Lemma* 12.11, *we have*

$$\int_{\Omega} |\nabla u(x)|^p\,dx \le \int_{\Omega} |\nabla v(x)|^p\,dx$$

for every $v \in W^{1,p}(\Omega)$ *with* $v - g \in W_0^{1,p}(\Omega)$.

PROOF. Let $v \in W^{1,p}(\Omega)$ with $v - g \in W_0^{1,p}(\Omega)$. By Lemma 12.13 we have $v - u = (v - g) + (g - u) \in W_0^{1,p}(\Omega)$. There exist functions $\varphi_j \in C_0^\infty(\Omega)$, $j \in \mathbb{N}$, such that $\varphi_j \to v - u$ in $W^{1,p}(\Omega)$ as $j \to \infty$. Lemma 12.17 implies

$$\int_{\Omega} |\nabla u(x)|^p\,dx \le \int_{\Omega} |\nabla u(x) + \nabla\varphi_j(x)|^p\,dx$$

for every $j \in \mathbb{N}$, and thus

$$\begin{aligned}\int_{\Omega} |\nabla u(x)|^p\,dx &\le \lim_{j\to\infty} \int_{\Omega} |\nabla u(x) + \nabla\varphi_j(x)|^p\,dx \\ &= \int_{\Omega} |\nabla u(x) + (\nabla v(x) - \nabla u(x))|^p\,dx = \int_{\Omega} |\nabla v(x)|^p\,dx.\end{aligned}$$ □

We are ready to prove the main result of this section.

PROOF OF THEOREM 12.10. By Lemma 12.11 there exists $u \in W^{1,p}(\Omega)$ such that a subsequence of $(u_i)_{i\in\mathbb{N}}$ converges to u locally uniformly in Ω as $i \to \infty$. Lemma 12.13 asserts that $u - g \in W_0^{1,p}(\Omega)$. From Lemma 12.18 we conclude that u is a minimizer of the p-Dirichlet integral with the boundary values g in Ω and Theorem 11.11 implies that u is a weak solution to the p-Laplace equation in Ω. $\square$

We have also Sobolev convergence in the setting of Theorem 12.10.

LEMMA 12.19. *Under the assumptions in Theorem* 12.10 *and notation as in Lemma* 12.11, *we have* $u_i \to u$ *in* $W^{1,p}_{\mathrm{loc}}(\Omega)$ *as* $i \to \infty$.

PROOF. We use same notation as in the proof of Lemma 12.17. Let $\varepsilon > 0$. By Lemma 12.15 with Ω'' and (12.28) with $\varphi = 0$ we have

$$\begin{aligned}\int_{\Omega''} |\nabla u(x)|^p\,dx &\le \liminf_{i\to\infty} \int_{\Omega''} |\nabla u_i(x)|^{p_i}\,dx \\ &\le \limsup_{i\to\infty} \int_{\Omega''} |\nabla u_i(x)|^{p_i}\,dx \le \int_{\Omega''} |\nabla u(x)|^p\,dx + \varepsilon\end{aligned}$$

and thus, by letting $\varepsilon \to 0$, we conclude that

$$\lim_{i\to\infty} \int_{\Omega''} |\nabla u_i(x)|^{p_i}\,dx = \int_{\Omega''} |\nabla u(x)|^p\,dx.$$

Since $(u_i)_{i\in\mathbb{N}}$ is a bounded sequence in $W^{1,p+\delta}(\Omega)$, an application of (12.17) as in connection with (12.25) gives

$$\lim_{i\to\infty} \int_{\Omega''} |\nabla u_i(x)|^{p}\,dx = \lim_{i\to\infty} \int_{\Omega''} |\nabla u_i(x)|^{p_i}\,dx = \int_{\Omega''} |\nabla u(x)|^p\,dx.$$

Hence,

$$\|\nabla u_i\|_{L^p(\Omega'';\mathbb{R}^n)} \xrightarrow{i\to\infty} \|\nabla u\|_{L^p(\Omega'';\mathbb{R}^n)}.$$

Moreover, it follows from Lemma 12.11 that $\nabla u_i \to \nabla u$ weakly in $L^p(\Omega'';\mathbb{R}^n)$ as $i \to \infty$. By Lemma 2.37, we obtain

$$\|\nabla u_i - \nabla u\|_{L^p(\Omega'';\mathbb{R}^n)} \xrightarrow{i\to\infty} 0,$$

and consequently $\nabla u_i \to \nabla u$ in $L^p_{\mathrm{loc}}(\Omega;\mathbb{R}^n)$ as $i \to \infty$. Since $u_i \to u$ locally uniformly in Ω and thus in $L^p_{\mathrm{loc}}(\Omega)$, we conclude that $u_i \to u$ in $W^{1,p}_{\mathrm{loc}}(\Omega)$ as $i \to \infty$. $\square$

Next we show that the capacity density condition in Definition 6.17 is essentially the weakest possible condition for the stability result in Theorem 12.10. If the complement does not satisfy the p-capacity density condition, it may happen that a sequence of solutions to the p_i-Laplace equation, even with smooth boundary values, converges to a wrong solution as $p_i \to p$.

EXAMPLE 12.20. We modify Example 12.9 to the varying exponent case. Let K and $\Omega = B(0,2) \setminus K$ be as in Example 12.9 and let $(p_i)_{i\in\mathbb{N}}$ be an increasing sequence with $1 < p_i < p_{i+1} < p$ for every $i \in \mathbb{N}$ and $p_i \to p$ as $i \to \infty$. For every $i \in \mathbb{N}$, let $u_i \in W^{1,p_i}(\Omega)$ be the unique continuous weak solution to the p_i-Laplace equation in Ω with $u_i - g \in W_0^{1,p_i}(\Omega)$, where $g \in C_0^\infty(B(0,2))$ with $g = 1$ in K. By Corollary 5.29 there exists a p_i-quasicontinuous function $v_i \in W^{1,p_i}(\mathbb{R}^n)$ such that $v_i = u_i - g$ almost everywhere in Ω and $v_i = 0$ p_i-quasieverywhere in $\mathbb{R}^n \setminus \Omega$. This implies that $v_i = 0$ p_i-quasieverywhere in $B(0,2)^c$. By Corollary 5.29 we conclude that $v_i \in W_0^{1,p_i}(B(0,2))$ and consequently $v_i + g \in W_0^{1,p_i}(B(0,2))$ for every $i \in \mathbb{N}$. We claim that $w_i = v_i + g$ is a weak solution to the p_i-Laplace equation in $B(0,2)$ for every $i \in \mathbb{N}$. Observe that $w_i = u_i$ almost everywhere in Ω and thus w_i is an extension of u_i to $B(0,2)$ for every $i \in \mathbb{N}$.

Fix $i \in \mathbb{N}$. Let $(\psi_j)_{j\in\mathbb{N}}$ be as in (12.6) for $q = p_i < p$ and let $\varphi \in C_0^\infty(B(0,2))$. Then $\varphi(1-\psi_j) \in \operatorname{Lip}_0(\Omega) \subset W_0^{1,p_i}(\Omega)$ for every $j \in \mathbb{N}$. Since $u_i \in W^{1,p_i}(\Omega)$ is a weak solution to the p_i-Laplace equation in Ω and $w_i = u_i$ almost everywhere in Ω, Lemma 11.6 implies

$$\int_\Omega |\nabla w_i(x)|^{p_i-2}\nabla w_i(x)\cdot\nabla((1-\psi_j(x))\varphi(x))\,dx = 0.$$

As in (12.7), also using the fact that $|K| = 0$, we have

$$\begin{aligned}
&\left|\int_{B(0,2)} |\nabla w_i(x)|^{p_i-2}\nabla w_i(x)\cdot\nabla\varphi(x)\,dx\right| \\
&\quad= \left|\int_{B(0,2)} |\nabla w_i(x)|^{p_i-2}\nabla w_i(x)\cdot\nabla(\psi_j(x)\varphi(x))\,dx\right| \\
&\quad\le \int_{B(0,2)} |\nabla w_i(x)|^{p_i-1}|\nabla(\psi_j(x)\varphi(x))|\,dx \\
&\quad\le \left(\int_{B(0,2)} |\nabla w_i(x)|^{p_i}\,dx\right)^{\frac{p_i-1}{p_i}}\left(\int_{B(0,2)} |\nabla(\psi_j(x)\varphi(x))|^{p_i}\,dx\right)^{\frac{1}{p_i}}.
\end{aligned} \tag{12.29}$$

The first term on the right-hand side of (12.29) is finite. By (12.6) with $q = p_i < p$ we obtain

$$\begin{aligned}
\left(\int_{B(0,2)} |\nabla(\psi_j(x)\varphi(x))|^{p_i}\,dx\right)^{\frac{1}{p_i}} &\le \|\varphi\nabla\psi_j\|_{L^q(B(0,2))} + \|\psi_j\nabla\varphi\|_{L^q(B(0,2))} \\
&\le \|\varphi\|_{L^\infty(B(0,2))}\|\nabla\psi_j\|_{L^q(B(0,2))} + \|\nabla\varphi\|_{L^\infty(B(0,2))}\|\psi_j\|_{L^q(B(0,2))} \\
&\le C(n,p_i)\big(\|\varphi\|_{L^\infty(B(0,2))} + \|\nabla\varphi\|_{L^\infty(B(0,2))}\big)\|\nabla\psi_j\|_{L^q(B(0,2))} \xrightarrow{j\to\infty} 0.
\end{aligned}$$

Thus it follows from (12.29) that $w_i \in W_0^{1,p_i}(B(0,2))$ is a weak solution to the p_i-Laplace equation in $B(0,2)$. On the other hand, by Theorem 11.13 the zero function is the unique solution to the p_i-Laplace equation in $B(0,2)$ that belongs to $W_0^{1,p_i}(B(0,2))$. Hence $w_i = 0$ in $B(0,2)$ and consequently $u_i = 0$ in Ω for every $i \in \mathbb{N}$.

Finally, let $u \in W^{1,p}(\Omega)$ be the unique weak solution to the p-Laplace equation in Ω, with $u - g \in W_0^{1,p}(\Omega)$. Arguing as in Example 12.9, we find that $0 < u(x) < 1$ for every $x \in \Omega$, and thus there is no subsequence of $(u_i)_{i\in\mathbb{N}}$ that would converge to u pointwise almost everywhere in Ω. In conclusion, the claim in Theorem 12.10 does not hold in this case since the solutions u_i converge to the solution of a wrong limit problem. This is possible since Ω^c does not satisfy the p-capacity density condition.

12.3. Very weak solutions

In this section we take another look at the Sobolev space assumption in the definition of a weak solution to the p-Laplace equation. Let $1 < p < \infty$ and let $\Omega \subset \mathbb{R}^n$ be an open set. In Definition 11.1 we assumed that $u \in W_{\mathrm{loc}}^{1,p}(\Omega)$. However, the integral in (11.1) makes sense under the weaker assumption that $u \in W^{1,q}_{\mathrm{loc}}(\Omega)$ for some q with $\max\{1,p-1\} \le q < p$ since

$$\begin{aligned}\int_\Omega \big||\nabla u(x)|^{p-2}\nabla u(x)\cdot\nabla\varphi(x)\big|\,dx &\le \int_\Omega |\nabla u(x)|^{p-1}|\nabla\varphi(x)|\,dx\\ &\le \bigg(\int_{\operatorname{supp}\varphi}|\nabla u(x)|^q\,dx\bigg)^{\frac{p-1}{q}}\bigg(\int_\Omega|\nabla\varphi(x)|^{\frac{q}{q-p+1}}\,dx\bigg)^{\frac{q-p+1}{q}} < \infty\end{aligned}$$

for every $\varphi \in C_0^\infty(\Omega)$. Here the last integral is interpreted to be $\|\nabla\varphi\|_{L^\infty(\Omega)}$ if $q = p-1$.

DEFINITION 12.21. Let $1 < p < \infty$ and let $\Omega \subset \mathbb{R}^n$ be an open set. A function $u \in W^{1,q}_{\mathrm{loc}}(\Omega)$, with $\max\{1,p-1\} \le q < p$, is a very weak solution to the p-Laplace equation in Ω, if

$$\int_\Omega |\nabla u(x)|^{p-2}\nabla u(x)\cdot\nabla\varphi(x)\,dx = 0 \tag{12.30}$$

for every $\varphi \in C_0^\infty(\Omega)$.

Theorem 12.23 below shows that, for q close enough to p, a very weak solution to the p-Laplace equation in Ω belongs to the natural Sobolev space $W^{1,p}_{\mathrm{loc}}(\Omega)$ and thus is a weak solution in the sense of Definition 11.1. Moreover, Corollary 12.3 implies that $u \in W^{1,p+\delta}_{\mathrm{loc}}(\Omega)$ for some $\delta > 0$. The proof of Theorem 12.23 is based on localization, Lipschitz truncation and weighted norm inequalities. Also the following version of Lemma 11.6 is applied in the proof.

LEMMA 12.22. *Let $1 < p < \infty$ and $\max\{1,p-1\} < q < p$, and let $\Omega \subset \mathbb{R}^n$ be an open set. Assume that $u \in W^{1,q}_{\mathrm{loc}}(\Omega)$ satisfies (12.30) for every $\varphi \in C_0^\infty(\Omega)$. Then (12.30) holds for every $\varphi \in \operatorname{Lip}_c(\Omega)$.*

PROOF. Let $\varphi \in \operatorname{Lip}_c(\Omega)$. By convolution approximation, we obtain a compact set $K \subset \Omega$ and functions $\varphi_i \in C_0^\infty(\Omega)$ such that $\operatorname{supp}\varphi \cup \operatorname{supp}\varphi_i \subset K$ for every $i \in \mathbb{N}$ and $\nabla\varphi_i \to \nabla\varphi$ in $L^{\frac{q}{q-p+1}}(\Omega)$ as $i \to \infty$. As in the proof of Lemma 11.6, see also Remark 11.7, Hölder's inequality implies

$$\begin{aligned}\bigg|\int_\Omega |\nabla u(x)|^{p-2}\nabla u(x)\cdot\nabla\varphi(x)\,dx\bigg| &= \bigg|\int_\Omega |\nabla u(x)|^{p-2}\nabla u(x)\cdot(\nabla\varphi(x)-\nabla\varphi_i(x))\,dx\bigg|\\ &\le \int_K |\nabla u(x)|^{p-1}|\nabla\varphi(x)-\nabla\varphi_i(x)|\,dx\\ &\le \bigg(\int_K |\nabla u(x)|^q\,dx\bigg)^{\frac{p-1}{q}}\bigg(\int_\Omega|\nabla\varphi(x)-\nabla\varphi_i(x)|^{\frac{q}{q-p+1}}\,dx\bigg)^{\frac{q-p+1}{q}} \xrightarrow{i\to\infty} 0.\end{aligned}$$

□

THEOREM 12.23. *Let $1 < p < \infty$ and let $\Omega \subset \mathbb{R}^n$ be an open set. There exists $0 < \varepsilon = \varepsilon(n,p) < p-1$ such that if $u \in W^{1,p-\varepsilon}_{\mathrm{loc}}(\Omega)$ satisfies (12.30) for every $\varphi \in C_0^\infty(\Omega)$, then $u \in W^{1,p}_{\mathrm{loc}}(\Omega)$.*

PROOF. Let ε with

$$0 < \varepsilon < \min\bigg\{\frac{1}{2}, \frac{p-1}{2}, \frac{1}{2(p'-1)}\bigg\} \tag{12.31}$$

to be chosen later. Here $p' = \frac{p}{p-1}$. Assume that $u \in W^{1,p-\varepsilon}_{\mathrm{loc}}(\Omega)$ satisfies (12.30) for every $\varphi \in C_0^\infty(\Omega)$. Let $B(z,r)$ be a ball with $B(z,4r) \subset \Omega$, and let $\psi \in$

$\mathrm{Lip}_0(B(z,2r))$ be such that $0 \le \psi \le 1$, $\psi = 1$ in $B(z,r)$ and $|\nabla\psi| \le \frac{1}{r}$ almost everywhere in $\mathbb{R}^n$. Let

$$v = (u - u_{B(z,2r)})\psi$$

and

$$E_t = \{x \in \mathbb{R}^n : M|\nabla v|(x) > t\},$$

for $t > 0$. Then $v \in W^{1,1}(\mathbb{R}^n)$, and by Theorem 4.2 and Theorem 2.7 there exists a $C(n)t$-Lipschitz function $v_t\colon \mathbb{R}^n \to \mathbb{R}$ such that $v_t = v$ almost everywhere in

$$\mathbb{R}^n \setminus E_t = \{x \in \mathbb{R}^n : M|\nabla v|(x) \le t\}.$$

Since $v = 0$ in $\mathbb{R}^n \setminus B(z,2r)$, we have

$$M|\nabla v|(x) = \sup_{\rho>0} \fint_{B(x,\rho)} |\nabla v(y)|\,dy \le C(n) \fint_{B(z,2r)} |\nabla v(y)|\,dy = t_0$$

for every $x \in \mathbb{R}^n \setminus B(z,3r)$. It follows that if $t \ge t_0$, then $v_t = v = 0$ in $\mathbb{R}^n \setminus B(z,3r)$, and thus v_t is compactly supported in Ω for $t \ge t_0$.

By Lemma 12.22 we may use $v_t \in \mathrm{Lip}_c(\Omega)$ as a test function in (12.30) for every $t \ge t_0$. Thus

$$\begin{aligned}
&\int_{B(z,2r)\setminus E_t} |\nabla u(x)|^{p-2}\nabla u(x)\cdot\nabla v(x)\,dx \\
&\quad= \int_{\Omega\setminus E_t} |\nabla u(x)|^{p-2}\nabla u(x)\cdot\nabla v(x)\,dx \\
&\quad= \int_{\Omega\setminus E_t} |\nabla u(x)|^{p-2}\nabla u(x)\cdot\nabla v_t(x)\,dx \\
&\quad= -\int_{\Omega\cap E_t} |\nabla u(x)|^{p-2}\nabla u(x)\cdot\nabla v_t(x)\,dx \\
&\quad\le C(n)t\int_{B(z,4r)\cap E_t} |\nabla u(x)|^{p-1}\,dx
\end{aligned} \tag{12.32}$$

for every $t \ge t_0$, where the final inequality holds since $|\nabla v_t| \le C(n)t$ almost everywhere in $\mathbb{R}^n$ and $\nabla v_t = 0$ in $\Omega \setminus B(z,4r)$. We multiply (12.32) by $t^{-1-\varepsilon}$, integrate over (t_0,∞), and change the order of integration on the right-hand side. This and the assumption $0 < \varepsilon < \frac{1}{2}$ from (12.31) give

$$\begin{aligned}
&\int_{t_0}^{\infty} t^{-1-\varepsilon}\int_{B(z,2r)\setminus E_t} |\nabla u(x)|^{p-2}\nabla u(x)\cdot\nabla v(x)\,dx\,dt \\
&\quad\le C(n)\int_{t_0}^{\infty} t^{-\varepsilon}\int_{B(z,4r)\cap E_t} |\nabla u(x)|^{p-1}\,dx\,dt \\
&\quad\le C(n)\int_{0}^{\infty} t^{-\varepsilon}\int_{B(z,4r)\cap E_t} |\nabla u(x)|^{p-1}\,dx\,dt \\
&\quad= \frac{C(n)}{1-\varepsilon}\int_{B(z,4r)} (M|\nabla v|(x))^{1-\varepsilon}|\nabla u(x)|^{p-1}\,dx \\
&\quad\le C(n)\int_{B(z,4r)} (M|\nabla v|(x))^{1-\varepsilon}|\nabla u(x)|^{p-1}\,dx.
\end{aligned} \tag{12.33}$$

On the other hand,

$$\begin{aligned}\varepsilon\int_{t_0}^{\infty} t^{-1-\varepsilon}&\int_{B(z,2r)\setminus E_t}|\nabla u(x)|^{p-2}\nabla u(x)\cdot\nabla v(x)\,dx\,dt\\&=\int_{B(z,2r)\cap E_{t_0}}(M|\nabla v|(x))^{-\varepsilon}|\nabla u(x)|^{p-2}\nabla u(x)\cdot\nabla v(x)\,dx\\&\quad+t_0^{-\varepsilon}\int_{B(z,2r)\setminus E_{t_0}}|\nabla u(x)|^{p-2}\nabla u(x)\cdot\nabla v(x)\,dx.\end{aligned}$$

By substituting this equality to (12.33), we obtain

$$\begin{aligned}I_1=t_0^{-\varepsilon}&\int_{B(z,2r)\setminus E_{t_0}}|\nabla u(x)|^{p-2}\nabla u(x)\cdot\nabla v(x)\,dx\\&+\int_{B(z,2r)\cap E_{t_0}}(M|\nabla v|(x))^{-\varepsilon}|\nabla u(x)|^{p-2}\nabla u(x)\cdot\nabla v(x)\,dx\\\le C(n)\varepsilon&\int_{B(z,4r)}(M|\nabla v|(x))^{1-\varepsilon}|\nabla u(x)|^{p-1}\,dx=\varepsilon I_2.\end{aligned}\tag{12.34}$$

By the properties of ψ and Theorem 3.5, we have

$$|\nabla v(x)|\le|\nabla u(x)|\psi(x)+|u(x)-u_{B(z,2r)}||\nabla\psi(x)|\le C(n)M\big(|\nabla u|\chi_{B(z,4r)}\big)(x)$$

for almost every $x\in\mathbb{R}^n$. Recall from (12.31) that $0<\varepsilon<\min\{\frac{1}{2},\frac{p-1}{2}\}$. Hence,

$$5^n2^{p-\varepsilon}\frac{p-\varepsilon}{p-\varepsilon-1}\le5^n2^p\frac{p}{\frac{1}{2}(p-1)}=5^n2^{p+1}\frac{p}{p-1}=C_1(n,p)>1.\tag{12.35}$$

Thus we can apply Theorem 1.15 (c) with $p-\varepsilon>1$ and this constant $C_1(n,p)$ which is independent of ε; see the end of the proof of Theorem 1.15. Using also Young's inequality, with exponents $\frac{p-\varepsilon}{1-\varepsilon}>1$ and $\frac{p-\varepsilon}{p-1}>1$, we obtain

$$\begin{aligned}I_2&\le C(n)\int_{B(z,4r)}\big(M\big(M\big(|\nabla u|\chi_{B(z,4r)}\big)\big)(x)\big)^{1-\varepsilon}|\nabla u(x)|^{p-1}\,dx\\&\le C(n)\int_{B(z,4r)}|\nabla u(x)|^{p-\varepsilon}\,dx\\&\quad+C(n)\int_{B(z,4r)}\big(M\big(M\big(|\nabla u|\chi_{B(z,4r)}\big)\big)(x)\big)^{p-\varepsilon}\,dx\\&\le C(n,p)\int_{B(z,4r)}|\nabla u(x)|^{p-\varepsilon}\,dx.\end{aligned}\tag{12.36}$$

By combining (12.34) and (12.36), we have

$$I_1\le\varepsilon I_2\le C(n,p)\varepsilon\int_{B(z,4r)}|\nabla u(x)|^{p-\varepsilon}\,dx.\tag{12.37}$$

Since $M|\nabla v|(x) \le t_0$ for every $x \in B(z,2r) \setminus E_{t_0}$, we estimate I_1 from below as follows

$$
\begin{aligned}
I_1 &= t_0^{-\varepsilon} \int_{B(z,2r)\setminus E_{t_0}} |\nabla u(x)|^{p-2}\nabla u(x)\cdot \nabla v(x)\,dx \\
&\quad + \int_{B(z,2r)\cap E_{t_0}} (M|\nabla v|(x))^{-\varepsilon}|\nabla u(x)|^{p-2}\nabla u(x)\cdot\nabla v(x)\,dx \\
&\ge -t_0^{-\varepsilon}\int_{B(z,2r)\setminus E_{t_0}} |\nabla u(x)|^{p-1}|\nabla v(x)|\,dx \\
&\quad + \int_{B(z,2r)} (M|\nabla v|(x))^{-\varepsilon}|\nabla u(x)|^{p-2}\nabla u(x)\cdot\nabla v(x)\,dx \\
&\quad - \int_{B(z,2r)\setminus E_{t_0}} (M|\nabla v|(x))^{-\varepsilon}|\nabla u(x)|^{p-2}\nabla u(x)\cdot\nabla v(x)\,dx \\
&\ge \int_{B(z,2r)} (M|\nabla v|(x))^{-\varepsilon}|\nabla u(x)|^{p-2}\nabla u(x)\cdot\nabla v(x)\,dx \\
&\quad - 2\int_{B(z,2r)\setminus E_{t_0}} (M|\nabla v|(x))^{-\varepsilon}|\nabla u(x)|^{p-1}|\nabla v(x)|\,dx \\
&= J_1 - J_2.
\end{aligned}
$$

To estimate J_1, we consider

$$
D_1 = \big\{x \in B(z,2r)\setminus B(z,r) : M|\nabla v|(x) \le \varepsilon M\big(|\nabla u|\chi_{B(z,2r)}\big)(x)\big\}
$$

and $D_2 = (B(z,2r)\setminus B(z,r))\setminus D_1$. Then

$$
\begin{aligned}
J_1 &= \int_{B(z,r)} (M|\nabla v|(x))^{-\varepsilon}|\nabla u(x)|^{p-2}\nabla u(x)\cdot\nabla v(x)\,dx \\
&\quad + \int_{B(z,2r)\setminus B(z,r)} (M|\nabla v|(x))^{-\varepsilon}|\nabla u(x)|^{p-2}\nabla u(x)\cdot\nabla v(x)\,dx \\
&= \int_{B(z,r)} (M|\nabla v|(x))^{-\varepsilon}|\nabla u(x)|^{p}\,dx \\
&\quad + \int_{D_1} (M|\nabla v|(x))^{-\varepsilon}|\nabla u(x)|^{p-2}\nabla u(x)\cdot\nabla v(x)\,dx \\
&\quad + \int_{D_2} (M|\nabla v|(x))^{-\varepsilon}|\nabla u(x)|^{p-2}\nabla u(x)\cdot\nabla u(x)\psi(x)\,dx \\
&\quad + \int_{D_2} (M|\nabla v|(x))^{-\varepsilon}|\nabla u(x)|^{p-2}\nabla u(x)\cdot\nabla\psi(x)\big(u(x)-u_{B(z,2r)}\big)\,dx \\
&\ge \int_{B(z,r)} (M|\nabla v|(x))^{-\varepsilon}|\nabla u(x)|^{p}\,dx \\
&\quad - \int_{D_1} (M|\nabla v|(x))^{-\varepsilon}|\nabla u(x)|^{p-1}|\nabla v(x)|\,dx \\
&\quad - \frac{1}{r}\int_{D_2} (M|\nabla v|(x))^{-\varepsilon}|\nabla u(x)|^{p-1}|u(x)-u_{B(z,2r)}|\,dx \\
&= I_3 - I_4 - I_5.
\end{aligned}
$$

To estimate I_3, we recall from (12.31) that $0 < \varepsilon < \frac{1}{2(p'-1)}$, where $p' = \frac{p}{p-1}$. Theorem 8.33 implies $\sigma = (M|\nabla v|)^{\varepsilon(p'-1)} \in A_1$ with $[\sigma]_{A_1} \le C(n,p)$; observe that $\sigma(x) > 0$ for every $x \in \mathbb{R}^n$ unless $|\nabla v| = 0$ almost everywhere in $\mathbb{R}^n$, in which case the claim of the theorem is trivial. Thus $\sigma \in A_{p'}$ and $[\sigma]_{A_{p'}} \le [\sigma]_{A_1}$, and by the A_p duality in Lemma 8.13 (b),

$$w = \big(M|\nabla v|\big)^{-\varepsilon} = \big((M|\nabla v|)^{\varepsilon(p'-1)}\big)^{-\frac{1}{p'-1}} = \sigma^{1-p} \in A_p$$

with $[w]_{A_p} \le C(n,p)$. The weighted norm inequality for the Hardy–Littlewood maximal operator with a Muckenhoupt A_p weight in Theorem 8.28 implies

$$\begin{aligned} I_3 &= \int_{B(z,r)} |\nabla u(x)|^p \, w(x)\, dx \ge C(n,p) \int_{B(z,r)} \big(M(|\nabla u|\chi_{B(z,r)})(x)\big)^p \, w(x)\, dx \\ &= C(n,p) \int_{B(z,r)} \big(M|\nabla v|(x)\big)^{-\varepsilon} \big(M(|\nabla u|\chi_{B(z,r)})(x)\big)^p \, dx. \end{aligned}$$

Since $\nabla u = \nabla v$ almost everywhere in $B(z,r)$ and $|\nabla v| = 0$ almost everywhere in $\mathbb{R}^n \setminus B(z,2r)$, we have

$$\begin{aligned} M|\nabla v|(x) &\le \sup_{0<\rho\le r-|x-z|} \fint_{B(x,\rho)} |\nabla v(y)|\, dy + \sup_{\rho>r-|x-z|} \fint_{B(x,\rho)} |\nabla v(y)|\, dy \\ &\le \sup_{0<\rho\le r-|x-z|} \fint_{B(x,\rho)} |\nabla u(y)|\chi_{B(z,r)}(y)\, dy \\ &\qquad + C(n) \fint_{B(z,2r)} |\nabla v(y)|\, dy \\ &\le M\big(|\nabla u|\chi_{B(z,r)}\big)(x) + C(n) \fint_{B(z,2r)} |\nabla u(y)|\, dy \\ &\qquad + \frac{C(n)}{r} \fint_{B(z,2r)} |u(y) - u_{B(z,2r)}|\, dy \\ &\le M\big(|\nabla u|\chi_{B(z,r)}\big)(x) + C(n) \fint_{B(z,2r)} |\nabla u(y)|\, dy, \end{aligned} \tag{12.38}$$

for every $x \in B(z, \frac{r}{2})$. In the final step we applied the Poincaré inequality in Theorem 3.14. Let

$$G = \Big\{ x \in B(z, \tfrac{r}{2}) : M\big(|\nabla u|\chi_{B(z,r)}\big)(x) \ge C(n) \fint_{B(z,2r)} |\nabla u(y)|\, dy \Big\}.$$

By (12.38), we have $M|\nabla v|(x) \le 2M(|\nabla u|\chi_{B(z,r)})(x)$ for every $x \in G$. It follows that

$$\begin{aligned} I_3 &\ge C(n,p) \int_G \big(M\big(|\nabla u|\chi_{B(z,r)}\big)(x)\big)^{p-\varepsilon}\, dx \\ &= C(n,p) \int_{B(z,\frac{r}{2})} \big(M\big(|\nabla u|\chi_{B(z,r)}\big)(x)\big)^{p-\varepsilon}\, dx \\ &\qquad - C(n,p) \int_{B(z,\frac{r}{2})\setminus G} \big(M\big(|\nabla u|\chi_{B(z,r)}\big)(x)\big)^{p-\varepsilon}\, dx \\ &\ge C(n,p) \int_{B(z,\frac{r}{2})} |\nabla u(x)|^{p-\varepsilon}\, dx - C(n,p) r^n \Big(\fint_{B(z,2r)} |\nabla u(y)|\, dy \Big)^{p-\varepsilon}. \end{aligned} \tag{12.39}$$

On the other hand, by the definition of D_1 and Theorem 1.15, we obtain

$$\begin{aligned}
I_4 &= \int_{D_1} (M|\nabla v|(x))^{-\varepsilon}|\nabla u(x)|^{p-1}|\nabla v(x)|\,dx \\
&\le \int_{D_1} (M|\nabla v|(x))^{1-\varepsilon}|\nabla u(x)|^{p-1}\,dx \\
&\le \varepsilon^{1-\varepsilon}\int_{B(z,2r)} \big(M(|\nabla u|\chi_{B(z,2r)})(x)\big)^{1-\varepsilon}|\nabla u(x)|^{p-1}\,dx \\
&\le \varepsilon^{1-\varepsilon}\int_{B(z,2r)} \big(M(|\nabla u|\chi_{B(z,2r)})(x)\big)^{p-\varepsilon}\,dx \\
&\le \varepsilon^{\frac{1}{2}}C(n,p)\int_{B(z,2r)} |\nabla u(x)|^{p-\varepsilon}\,dx.
\end{aligned} \tag{12.40}$$

Write $B = B(z,2r)$. To estimate the somewhat critical term I_5, we use the fact that there exists $1 \le q = q(p) < p$ such that

$$\left(\fint_B |u(x)-u_B|^p\,dx\right)^{\frac{1}{p}} \le C(n,p,q)r\left(\fint_B |\nabla u(x)|^q\,dx\right)^{\frac{1}{q}}, \tag{12.41}$$

see Theorem 3.14. Moreover, we may assume that $0 < \varepsilon < p - q$. Fix $0 < \delta < 1$ to be chosen later. The definition of D_2, Young's inequality followed by Hölder's inequality, estimate (12.41) and Theorem 1.15 imply

$$\begin{aligned}
I_5 &= \frac{1}{r}\int_{D_2} (M|\nabla v|(x))^{-\varepsilon}|\nabla u(x)|^{p-1}|u(x)-u_B|\,dx \\
&\le \frac{\varepsilon^{-\varepsilon}}{r}\int_B \big(M(|\nabla u|\chi_B)(x)\big)^{-\varepsilon}|\nabla u(x)|^{p-1}|u(x)-u_B|\,dx \\
&\le \frac{\varepsilon^{-\varepsilon}}{r}\int_B \big(M(|\nabla u|\chi_B)(x)\big)^{p-1-\varepsilon}|u(x)-u_B|\,dx \\
&\le \delta\int_B \big(M(|\nabla u|\chi_B)(x)\big)^{p-\varepsilon}\,dx + \frac{C(p,\delta)}{r^{p-\varepsilon}}\int_B |u(x)-u_B|^{p-\varepsilon}\,dx \\
&\le \delta\int_B \big(M(|\nabla u|\chi_B)(x)\big)^{p-\varepsilon}\,dx + C(n,p,q,\delta)r^n\left(\fint_B |\nabla u(x)|^q\,dx\right)^{\frac{p-\varepsilon}{q}} \\
&\le \delta C(n,p)\int_B |\nabla u(x)|^{p-\varepsilon}\,dx + C(n,p,q,\delta)r^n\left(\fint_B |\nabla u(x)|^q\,dx\right)^{\frac{p-\varepsilon}{q}}.
\end{aligned} \tag{12.42}$$

Here we also used the fact that $\varepsilon^{-\varepsilon} = e^{-\varepsilon\log\varepsilon}$ is uniformly bounded for $0 < \varepsilon < \frac{1}{2}$. By combining (12.39), (12.40) and (12.42), using Hölder's inequality, and recalling that $B = B(z,2r)$, we conclude that

$$\begin{aligned}
J_1 \ge I_3 - I_4 - I_5 &\ge C(n,p)\int_{B(z,\frac{r}{2})} |\nabla u(x)|^{p-\varepsilon}\,dx \\
&\quad - C(n,p)(\varepsilon^{\frac{1}{2}}+\delta)\int_{B(z,2r)} |\nabla u(x)|^{p-\varepsilon}\,dx \\
&\quad - C(n,p,q,\delta)r^n\left(\fint_{B(z,2r)} |\nabla u(x)|^q\,dx\right)^{\frac{p-\varepsilon}{q}}.
\end{aligned} \tag{12.43}$$

To estimate

$$J_2 = 2\int_{B\setminus E_{t_0}} (M|\nabla v|(x))^{-\varepsilon}|\nabla u(x)|^{p-1}|\nabla v(x)|\,dx,$$

we introduce one more absorption parameter $0<\gamma<1$ that will be chosen later. Let $x\in B\setminus E_{t_0}$ be a Lebesgue point of $|\nabla v|$. Then $|\nabla v(x)|\le M|\nabla v|(x)\le t_0$ by the proof of Corollary 1.23. If $|\nabla u(x)|\ge \frac{t_0}{\gamma}$, then

$$\begin{aligned}(M|\nabla v|(x))^{-\varepsilon}|\nabla u(x)|^{p-1}|\nabla v(x)| &\le (M|\nabla v|(x))^{1-\varepsilon}|\nabla u(x)|^{p-1}\\ &\le t_0^{1-\varepsilon}|\nabla u(x)|^{p-1}\le \gamma^{1-\varepsilon}|\nabla u(x)|^{p-\varepsilon}.\end{aligned}$$

Then assume that $|\nabla u(x)|<\frac{t_0}{\gamma}$. Since $B=B(z,2r)\subset B(x,4r)$, we have

$$t_0 = C(n)\fint_B |\nabla v(y)|\,dy \le C(n)M|\nabla v|(x),$$

and hence

$$\begin{aligned}(M|\nabla v|(x))^{-\varepsilon}|\nabla u(x)|^{p-1}|\nabla v(x)| &\le C(n)t_0^{-\varepsilon}t_0^{p-1}\gamma^{1-p}|\nabla v(x)|\\ &\le C(n,p)\gamma^{1-p}\Big(\fint_B|\nabla v(y)|\,dy\Big)^{p-1-\varepsilon}|\nabla v(x)|.\end{aligned}$$

The definition of v, 1-Poincaré inequality, and Hölder's inequality imply

$$\begin{aligned}\fint_B|\nabla v(y)|\,dy &\le \fint_B|\nabla u(y)|\,dy+\frac{1}{r}\fint_B|u(y)-u_B|\,dy\\ &\le C(n)\fint_B|\nabla u(y)|\,dy\le C(n)\Big(\fint_B|\nabla u(y)|^q\,dy\Big)^{\frac1q}.\end{aligned}$$

By Hölder's inequality and the estimates above, we obtain

$$\begin{aligned}J_2 &\le 2\gamma^{1-\varepsilon}\int_B|\nabla u(x)|^{p-\varepsilon}\,dx\\ &\quad + C(n,p)\gamma^{1-p}\Big(\fint_B|\nabla v(y)|\,dy\Big)^{p-1-\varepsilon}\int_B|\nabla v(x)|\,dx\\ &= 2\gamma^{1-\varepsilon}\int_B|\nabla u(x)|^{p-\varepsilon}\,dx + C(n,p)\gamma^{1-p}r^n\Big(\fint_B|\nabla v(y)|\,dy\Big)^{p-\varepsilon}\\ &\le 2\gamma^{\frac12}\int_B|\nabla u(x)|^{p-\varepsilon}\,dx + C(n,p,\gamma)r^n\Big(\fint_B|\nabla u(y)|^q\,dy\Big)^{\frac{p-\varepsilon}{q}}.\end{aligned}\tag{12.44}$$

Estimates (12.43) and (12.44) give for I_1 the lower bound

$$\begin{aligned}I_1\ge J_1-J_2 &\ge C(n,p)\int_{B(z,\frac r2)}|\nabla u(x)|^{p-\varepsilon}\,dx\\ &\quad - C(n,p)\big(\varepsilon^{\frac12}+\delta+\gamma^{\frac12}\big)\int_{B(z,2r)}|\nabla u(x)|^{p-\varepsilon}\,dx\\ &\quad - C(n,p,q,\delta,\gamma)r^n\Big(\fint_{B(z,2r)}|\nabla u(x)|^q\,dx\Big)^{\frac{p-\varepsilon}{q}}.\end{aligned}$$

By combining this with the upper bound from (12.37), we arrive at

$$\begin{aligned}\left(\fint_{B(z,\frac{r}{2})}|\nabla u(x)|^{p-\varepsilon}\,dx\right)^{\frac{q}{p-\varepsilon}} &\le C_1\fint_{B(z,4r)}|\nabla u(x)|^q\,dx\\ &\quad+C(n,p)\big(\varepsilon+\varepsilon^{\frac12}+\delta+\gamma^{\frac12}\big)^{\frac{q}{p-\varepsilon}}\left(\fint_{B(z,4r)}|\nabla u(x)|^{p-\varepsilon}\,dx\right)^{\frac{q}{p-\varepsilon}},\end{aligned}\tag{12.45}$$

where $C_1=C_1(n,p,q,\delta,\gamma)$. This estimate holds for all sufficiently small $\varepsilon>0$ and for all balls $B(z,r)$ satisfying $B(z,4r)\subset\Omega$.

To conclude the proof, assume that $\theta=\theta(n,\frac{p}{q})>0$ is given and choose $0<\varepsilon_1,\delta,\gamma<1$, only depending on n and p, such that

$$C(n,p)\big(\varepsilon+\varepsilon^{\frac12}+\delta+\gamma^{\frac12}\big)^{\frac{q}{p-\varepsilon}}\le\theta$$

for every $0<\varepsilon<\varepsilon_1$. Observe that the constant $C_1=C_1(n,p,q,\delta,\gamma)=C_1(n,p)$ in (12.45) is independent of ε. Remark 8.45, with exponent $1<\frac{p}{q}<\infty$ and functions $f=|\nabla u|^q$ and $g=0$, gives $\varepsilon=\varepsilon(n,p)>0$ and $C>0$ such that

$$\left(\fint_{B(z,r)}|\nabla u(x)|^p\,dx\right)^{\frac{q}{p}}\le C\left(\fint_{B(z,2r)}|\nabla u(x)|^{p-\varepsilon}\,dx\right)^{\frac{q}{p-\varepsilon}}$$

whenever $B(z,2r)\subset\Omega$. Since $u\in W^{1,p-\varepsilon}_{\mathrm{loc}}(\Omega)$, it follows that $|\nabla u|\in L^p_{\mathrm{loc}}(\Omega)$. Finally, inequality (12.41) implies that also $u\in L^p_{\mathrm{loc}}(\Omega)$, and thus $u\in W^{1,p}_{\mathrm{loc}}(\Omega)$. □

REMARK 12.24. The proof above gives some exponent q with $p-1\le q<p$ such that a very weak solution $u\in W^{1,q}_{\mathrm{loc}}(\Omega)$ to a general partial differential equation of p-Laplacian type belongs to $W^{1,p}_{\mathrm{loc}}(\Omega)$, see Heinonen, Kilpeläinen and Martio [**187**] and Malý and Ziemer [**303**]. For this kind of equations the exponent q depends on structural conditions, see Meyers [**324**] and Serrin [**360**].

12.4. Notes

The local higher integrability result in Theorem 12.1 has been studied by Bojarski [**47**], Meyers [**324**] and Meyers and Elcrat [**325**]. See also Chen and Wu [**82**, Section 12.4], Giaquinta [**147**, Chapter 5], [**148**, Section 6.2], Giaquinta and Martinazzi [**149**, Section 9.2.2], Giaquinta and Modica [**150**], Giusti [**153**, Section 6.4], Stredulinsky [**374**, Section 3.3.0] and Zhikov [**403**]. For the global higher integrability result in Theorem 12.6, see Granlund [**160**] in the case of the measure density condition and Kilpeläinen and Koskela [**220**] in the case of the capacity density condition. See also Giaquinta [**147**, p. 151–154], Giusti [**153**, Section 6.5] and Maasalo and Zatorska-Goldstein [**295**]. Example 12.9 is from Kilpeläinen and Koskela [**220**, Remark 3.3]. For local higher integrability results for parabolic problems, see Antontsev and Zhikov [**17**], Arkhipova [**18**], Bögelein [**37**], Bögelein and Duzaar [**39**], Bögelein, Duzaar, Kinnunen and Scheven [**40**], Bögelein, Duzaar, Korte and Scheven [**41**], Bögelein, Duzaar and Scheven [**43**], Gianazza and Schwarzacher [**146,146**], Giaquinta and Struwe [**151**], Habermann [**163**], Kinnunen and Lewis [**230**], Masson, Miranda, Paronetto and Parviainen [**311**], Zhikov and Pastukhova [**404**] and Wieser [**398**]. For global higher integrability in the parabolic case, see Bögelein and Parviainen [**45**], Fujishima and Habermann [**137**], Habermann [**162**], Masson and Parviainen [**312**], Moring, Scheven, Schwarzacher and Singer [**330**] and Parviainen [**342,343**].

For stability results, we refer to Lindqvist [**278, 279**] and Maasalo and Zatorska-Goldstein [**295**]. Example 12.20 is from Lindqvist [**279**]. Related stability results are studied by Kinnunen, Marola and Martio [**233**] and Li and Martio [**273–276**]. Stability results for parabolic problems can be found in Fujishima, Habermann, Kinnunen and Masson [**138**] and Kinnunen and Parviainen [**236**].

Very weak solutions have been studied by Bojarski [**47**], Meyers [**324**], Iwaniec and Sbordone [**209**] with the Hodge decomposition and Lewis [**272**] by the Lipschitz truncation. Our presentation is based on Lewis [**272**] and Mikkonen [**327**]. See also Bögelein and Zatorska-Goldstein [**46**] and Zatorska-Goldstein [**402**]. For very weak solutions in the parabolic case, see Adimurthi and Byun [**6**], Bögelein [**38**], Bögelein and Li [**44**], Kinnunen and Lewis [**231**] and Li [**277**]. The Lipschitz truncation method has been applied not only in connection with very weak solutions but also in other questions in partial differential equation and the calculus of variations. For example, see Acerbi and Fusco [**2**], Breit, Diening and Schwarzacher [**55**], Bulíček, Burczak and Schwarzacher [**66**], Bulíček, Diening and Schwarzacher [**67**], Bulíček and Schwarzacher [**68**], Bögelein, Duzaar and Mingione [**42**], Diening, Růžička and Wolf [**105**], Diening, Schwarzacher, Stroffolini and Verde [**106**] and Duzaar and Mingione [**110**].

Bibliography

[1] Daniel Aalto and Juha Kinnunen, *The discrete maximal operator in metric spaces*, J. Anal. Math. **111** (2010), 369–390. MR2747071

[2] Emilio Acerbi and Nicola Fusco, *Semicontinuity problems in the calculus of variations*, Arch. Rational Mech. Anal. **86** (1984), no. 2, 125–145. MR751305

[3] David R. Adams, *Lectures on L^p-potential theory*, Technical Report 2, University of Umeå, 1981.

[4] David R. Adams and Lars Inge Hedberg, *Function spaces and potential theory*, Grundlehren der Mathematischen Wissenschaften [Fundamental Principles of Mathematical Sciences], vol. 314, Springer-Verlag, Berlin, 1996. MR1411441

[5] Robert A. Adams and John J. F. Fournier, *Sobolev spaces*, Second edition, Pure and Applied Mathematics (Amsterdam), vol. 140, Elsevier/Academic Press, Amsterdam, 2003. MR2424078

[6] Karthik Adimurthi and Sun-Sig Byun, *Boundary higher integrability for very weak solutions of quasilinear parabolic equations*, J. Math. Pures Appl. (9) **121** (2019), 244–285. MR3906171

[7] Hiroaki Aikawa, *Quasiadditivity of Riesz capacity*, Math. Scand. **69** (1991), no. 1, 15–30. MR1143471

[8] Hiroaki Aikawa and Matts Essén, *Potential theory—selected topics*, Lecture Notes in Mathematics, vol. 1633, Springer-Verlag, Berlin, 1996. MR1439503

[9] H. Aimar, M. Carena, R. Durán, and M. Toschi, *Powers of distances to lower dimensional sets as Muckenhoupt weights*, Acta Math. Hungar. **143** (2014), no. 1, 119–137. MR3215609

[10] J. M. Aldaz, *Remarks on the Hardy-Littlewood maximal function*, Proc. Roy. Soc. Edinburgh Sect. A **128** (1998), no. 1, 1–9. MR1606325

[11] J. M. Aldaz, L. Colzani, and J. Pérez Lázaro, *Optimal bounds on the modulus of continuity of the uncentered Hardy-Littlewood maximal function*, J. Geom. Anal. **22** (2012), no. 1, 132–167. MR2868961

[12] J. M. Aldaz and J. Pérez Lázaro, *Functions of bounded variation, the derivative of the one dimensional maximal function, and applications to inequalities*, Trans. Amer. Math. Soc. **359** (2007), no. 5, 2443–2461. MR2276629

[13] J. M. Aldaz and J. Pérez Lázaro, *Boundedness and unboundedness results for some maximal operators on functions of bounded variation*, J. Math. Anal. Appl. **337** (2008), no. 1, 130–143. MR2356061

[14] Alano Ancona, *On strong barriers and an inequality of Hardy for domains in $\mathbf{R}^n$*, J. London Math. Soc. (2) **34** (1986), no. 2, 274–290. MR856511

[15] Kenneth F. Andersen and Wo-Sang Young, *On the reverse weak type inequality for the Hardy maximal function and the weighted classes $L(\log L)^k$*, Pacific J. Math. **112** (1984), no. 2, 257–264. MR743983

[16] Kenneth F. Andersen and Wo-Sang Young, *Weighted reverse weak type inequalities for the ergodic maximal function and the classes $L\log^+ L$*, Proc. Amer. Math. Soc. **95** (1985), no. 2, 275–279. MR801338

[17] Stanislav Antontsev and Vasilii Zhikov, *Higher integrability for parabolic equations of $p(x,t)$-Laplacian type*, Adv. Differential Equations **10** (2005), no. 9, 1053–1080. MR2161759

[18] A. A. Arkhipova, *Reverse Hölder inequalities with boundary integrals and L_p-estimates for solutions of nonlinear elliptic and parabolic boundary-value problems*, Nonlinear evolution equations, Amer. Math. Soc. Transl. Ser. 2, vol. 164, Amer. Math. Soc., Providence, RI 1995, pp. 15–42. MR1334137

[19] Patrice Assouad, *Étude d'une dimension métrique liée à la possibilité de plongements dans* $\mathbf{R}^n$, C. R. Acad. Sci. Paris Sér. A-B **288** (1979), no. 15, A731–A734. MR532401

[20] Patrice Assouad, *Plongements lipschitziens dans* $\mathbf{R}^n$, Bull. Soc. Math. France **111** (1983), no. 4, 429–448. MR763553

[21] D. Bakry, T. Coulhon, M. Ledoux, and L. Saloff-Coste, *Sobolev inequalities in disguise*, Indiana Univ. Math. J. **44** (1995), no. 4, 1033–1074. MR1386760

[22] Alexander A. Balinsky, W. Desmond Evans, and Roger T. Lewis, *The analysis and geometry of Hardy's inequality*, Universitext, Springer, Cham, 2015. MR3408787

[23] G. Barbatis, S. Filippas, and A. Tertikas, *Refined geometric* L^p *Hardy inequalities*, Commun. Contemp. Math. **5** (2003), no. 6, 869–881. MR2030561

[24] G. Barbatis, S. Filippas, and A. Tertikas, *Series expansion for* L^p *Hardy inequalities*, Indiana Univ. Math. J. **52** (2003), no. 1, 171–190. MR1970026

[25] G. Barbatis, S. Filippas, and A. Tertikas, *A unified approach to improved* L^p *Hardy inequalities with best constants*, Trans. Amer. Math. Soc. **356** (2004), no. 6, 2169–2196. MR2048514

[26] David Beltran, João Pedro Ramos, and Olli Saari, *Regularity of fractional maximal functions through Fourier multipliers*, J. Funct. Anal. **276** (2019), no. 6, 1875–1892. MR3912794

[27] Colin Bennett, Ronald A. DeVore, and Robert Sharpley, *Weak-*L^∞ *and BMO*, Ann. of Math. (2) **113** (1981), no. 3, 601–611. MR621018

[28] Colin Bennett and Robert Sharpley, *Interpolation of operators*, Pure and Applied Mathematics, vol. 129, Academic Press, Inc., Boston, MA, 1988. MR928802

[29] Antonio Bernal, *A note on the one-dimensional maximal function*, Proc. Roy. Soc. Edinburgh Sect. A **111** (1989), no. 3-4, 325–328. MR1007529

[30] Marco Biroli and Umberto Mosco, *Sobolev and isoperimetric inequalities for Dirichlet forms on homogeneous spaces*, Atti Accad. Naz. Lincei Cl. Sci. Fis. Mat. Natur. Rend. Lincei (9) Mat. Appl. **6** (1995), no. 1, 37–44. MR1340280

[31] Anders Björn and Jana Björn, *Nonlinear potential theory on metric spaces*, EMS Tracts in Mathematics, vol. 17, European Mathematical Society (EMS), Zürich, 2011. MR2867756

[32] Anders Björn, Jana Björn, and Juha Lehrbäck, *Sharp capacity estimates for annuli in weighted* $\mathbf{R}^n$ *and in metric spaces*, Math. Z. **286** (2017), no. 3-4, 1173–1215. MR3671573

[33] Anders Björn and Niko Marola, *Moser iteration for (quasi)minimizers on metric spaces*, Manuscripta Math. **121** (2006), no. 3, 339–366. MR2267657

[34] Jana Björn, *Sharp exponents and a Wiener type condition for boundary regularity of quasiminimizers*, Adv. Math. **301** (2016), 804–819. MR3539390

[35] Jana Björn, Stephen M. Buckley, and Stephen Keith, *Admissible measures in one dimension*, Proc. Amer. Math. Soc. **134** (2006), no. 3, 703–705. MR2180887

[36] Jana Björn, Paul MacManus, and Nageswari Shanmugalingam, *Fat sets and pointwise boundary estimates for p-harmonic functions in metric spaces*, J. Anal. Math. **85** (2001), 339–369. MR1869615 (2002j:31017)

[37] Verena Bögelein, *Higher integrability for weak solutions of higher order degenerate parabolic systems*, Ann. Acad. Sci. Fenn. Math. **33** (2008), no. 2, 387–412. MR2431372

[38] Verena Bögelein, *Very weak solutions of higher-order degenerate parabolic systems*, Adv. Differential Equations **14** (2009), no. 1-2, 121–200. MR2478931

[39] Verena Bögelein and Frank Duzaar, *Higher integrability for parabolic systems with non-standard growth and degenerate diffusions*, Publ. Mat. **55** (2011), no. 1, 201–250. MR2779582

[40] Verena Bögelein, Frank Duzaar, Juha Kinnunen, and Christoph Scheven, *Higher integrability for doubly nonlinear parabolic systems*, J. Math. Pures Appl. (9) **143** (2020), 31–72. MR4163123

[41] Verena Bögelein, Frank Duzaar, Riikka Korte, and Christoph Scheven, *The higher integrability of weak solutions of porous medium systems*, Adv. Nonlinear Anal. **8** (2019), no. 1, 1004–1034. MR3918415

[42] Verena Bögelein, Frank Duzaar, and Giuseppe Mingione, *The regularity of general parabolic systems with degenerate diffusion*, Mem. Amer. Math. Soc. **221** (2013), no. 1041, vi+143. MR3059060

[43] Verena Bögelein, Frank Duzaar, and Christoph Scheven, *Higher integrability for the singular porous medium system*, J. Reine Angew. Math. **767** (2020), 203–230. MR4160307

[44] Verena Bögelein and Qifan Li, *Very weak solutions of degenerate parabolic systems with non-standard* $p(x,t)$*-growth*, Nonlinear Anal. **98** (2014), 190–225. MR3158453

[45] Verena Bögelein and Mikko Parviainen, *Self-improving property of nonlinear higher order parabolic systems near the boundary*, NoDEA Nonlinear Differential Equations Appl. **17** (2010), no. 1, 21–54. MR2596493

[46] Verena Bögelein and Anna Zatorska-Goldstein, *Higher integrability of very weak solutions of systems of $p(x)$-Laplacean type*, J. Math. Anal. Appl. **336** (2007), no. 1, 480–497. MR2348520

[47] Bogdan Bojarski, *Generalized solutions of a system of differential equations of first order and of elliptic type with discontinuous coefficients*, Mat. Sb. N.S. **43(85)** (1957), 451–503. MR0106324

[48] Bogdan Bojarski, *Sharp maximal operator of fractional order and Sobolev imbedding inequalities*, Bull. Polish Acad. Sci. Math. **33** (1985), no. 1-2, 7–16. MR798722

[49] Bogdan Bojarski, *Remarks on Sobolev imbedding inequalities*, Complex analysis, Joensuu 1987, Lecture Notes in Math., vol. 1351, Springer, Berlin 1988, pp. 52–68. MR982072

[50] Bogdan Bojarski and Piotr Hajłasz, *Pointwise inequalities for Sobolev functions and some applications*, Studia Math. **106** (1993), no. 1, 77–92. MR1226425

[51] Bogdan Bojarski, Piotr Hajłasz, and Paweł Strzelecki, *Improved $C^{k,\lambda}$ approximation of higher order Sobolev functions in norm and capacity*, Indiana Univ. Math. J. **51** (2002), no. 3, 507–540. MR1911044

[52] Bogdan Bojarski and Tadeusz Iwaniec, *Analytical foundations of the theory of quasiconformal mappings in $\mathbf{R}^n$*, Ann. Acad. Sci. Fenn. Ser. A I Math. **8** (1983), no. 2, 257–324. MR731786

[53] Jan Boman, *L^p-estimates for very strongly elliptic systems*, Technical Report 29, Department of Mathematics, Stockholm University, 1982.

[54] E. Bombieri and E. Giusti, *Harnack's inequality for elliptic differential equations on minimal surfaces*, Invent. Math. **15** (1972), 24–46. MR308945

[55] D. Breit, L. Diening, and S. Schwarzacher, *Solenoidal Lipschitz truncation for parabolic PDEs*, Math. Models Methods Appl. Sci. **23** (2013), no. 14, 2671–2700. MR3119635

[56] Haim Brezis, *Functional analysis, Sobolev spaces and partial differential equations*, Universitext, Springer, New York, 2011. MR2759829

[57] Haim Brezis and Moshe Marcus, *Hardy's inequalities revisited*, Ann. Scuola Norm. Sup. Pisa Cl. Sci. (4) **25** (1997), no. 1-2, 217–237 (1998). Dedicated to Ennio De Giorgi. MR1655516

[58] Haim Brezis, Moshe Marcus, and Itai Shafrir, *Extremal functions for Hardy's inequality with weight*, J. Funct. Anal. **171** (2000), no. 1, 177–191. MR1742864

[59] Alexander Brudnyi and Yuri Brudnyi, *Methods of geometric analysis in extension and trace problems. Volume 1*, Monographs in Mathematics, vol. 102, Birkhäuser/Springer Basel AG, Basel, 2012. MR2882877

[60] Alexander Brudnyi and Yuri Brudnyi, *Methods of geometric analysis in extension and trace problems. Volume 2*, Monographs in Mathematics, vol. 103, Birkhäuser/Springer Basel AG, Basel, 2012. MR2868143

[61] Stephen M. Buckley, *Inequalities of John-Nirenberg type in doubling spaces*, J. Anal. Math. **79** (1999), 215–240. MR1749313

[62] Stephen M. Buckley, *Is the maximal function of a Lipschitz function continuous?*, Ann. Acad. Sci. Fenn. Math. **24** (1999), no. 2, 519–528. MR1724375

[63] Stephen M. Buckley and Pekka Koskela, *Sobolev-Poincaré implies John*, Math. Res. Lett. **2** (1995), no. 5, 577–593. MR1359964

[64] Stephen M. Buckley and Pekka Koskela, *Orlicz-Hardy inequalities*, Illinois J. Math. **48** (2004), no. 3, 787–802. MR2114252 (2005k:26050)

[65] Stephen M. Buckley, Pekka Koskela, and Guozhen Lu, *Boman equals John*, XVIth Rolf Nevanlinna Colloquium (Joensuu, 1995), de Gruyter, Berlin 1996, pp. 91–99. MR1427074

[66] M. Bulíček, J. Burczak, and S. Schwarzacher, *A unified theory for some non-Newtonian fluids under singular forcing*, SIAM J. Math. Anal. **48** (2016), no. 6, 4241–4267. MR3582412

[67] Miroslav Bulíček, Lars Diening, and Sebastian Schwarzacher, *Existence, uniqueness and optimal regularity results for very weak solutions to nonlinear elliptic systems*, Anal. PDE **9** (2016), no. 5, 1115–1151. MR3531368

[68] Miroslav Bulíček and Sebastian Schwarzacher, *Existence of very weak solutions to elliptic systems of p-Laplacian type*, Calc. Var. Partial Differential Equations **55** (2016), no. 3, Art. 52, 14. MR3500290

[69] Alberto Pedro Calderón, *Estimates for singular integral operators in terms of maximal functions*, Studia Math. **44** (1972), 563–582. Collection of articles honoring the completion by Antoni Zygmund of 50 years of scientific activity, VI. MR0348555

[70] Alberto Pedro Calderón and Ridgway Scott, *Sobolev type inequalities for $p > 0$*, Studia Math. **62** (1978), no. 1, 75–92. MR0487419

[71] Alberto Pedro Calderón and Antoni Zygmund, *Local properties of solutions of elliptic partial differential equations*, Studia Math. **20** (1961), 171–225. MR0136849

[72] Javier Canto and Carlos Pérez, *Extensions of the John-Nirenberg theorem and applications*, Proc. Amer. Math. Soc. **149** (2021), no. 4, 1507–1525. MR4242308

[73] Luca Capogna, Donatella Danielli, and Nicola Garofalo, *The geometric Sobolev embedding for vector fields and the isoperimetric inequality*, Comm. Anal. Geom. **2** (1994), no. 2, 203–215. MR1312686

[74] Emanuel Carneiro, Renan Finder, and Mateus Sousa, *On the variation of maximal operators of convolution type II*, Rev. Mat. Iberoam. **34** (2018), no. 2, 739–766. MR3809456

[75] Emanuel Carneiro and José Madrid, *Derivative bounds for fractional maximal functions*, Trans. Amer. Math. Soc. **369** (2017), no. 6, 4063–4092. MR3624402

[76] Emanuel Carneiro, José Madrid, and Lillian B. Pierce, *Endpoint Sobolev and BV continuity for maximal operators*, J. Funct. Anal. **273** (2017), no. 10, 3262–3294. MR3695894

[77] Emanuel Carneiro and Diego Moreira, *On the regularity of maximal operators*, Proc. Amer. Math. Soc. **136** (2008), no. 12, 4395–4404. MR2431055

[78] Emanuel Carneiro and Benar F. Svaiter, *On the variation of maximal operators of convolution type*, J. Funct. Anal. **265** (2013), no. 5, 837–865. MR3063097

[79] S.-Y. A. Chang, J. M. Wilson, and T. H. Wolff, *Some weighted norm inequalities concerning the Schrödinger operators*, Comment. Math. Helv. **60** (1985), no. 2, 217–246. MR800004

[80] Sagun Chanillo, Jan-Olov Strömberg, and Richard L. Wheeden, *Norm inequalities for potential-type operators*, Rev. Mat. Iberoamericana **3** (1987), no. 3-4, 311–335. MR996820

[81] Sagun Chanillo and Richard L. Wheeden, *Weighted Poincaré and Sobolev inequalities and estimates for weighted Peano maximal functions*, Amer. J. Math. **107** (1985), no. 5, 1191–1226. MR805809

[82] Ya-Zhe Chen and Lan-Cheng Wu, *Second order elliptic equations and elliptic systems*, Translations of Mathematical Monographs, vol. 174, American Mathematical Society, Providence, RI, 1998. Translated from the 1991 Chinese original by Bei Hu. MR1616087

[83] Filippo Chiarenza and Michele Frasca, *A note on a weighted Sobolev inequality*, Proc. Amer. Math. Soc. **93** (1985), no. 4, 703–704. MR776206

[84] Filippo Chiarenza and Michele Frasca, *Morrey spaces and Hardy-Littlewood maximal function*, Rend. Mat. Appl. (7) **7** (1987), no. 3-4, 273–279 (1988). MR985999

[85] Seng-Kee Chua, *Weighted Sobolev inequalities on domains satisfying the chain condition*, Proc. Amer. Math. Soc. **117** (1993), no. 2, 449–457. MR1140667

[86] Andrea Cianchi, *Hardy inequalities in Orlicz spaces*, Trans. Amer. Math. Soc. **351** (1999), no. 6, 2459–2478. MR1433113

[87] Ronald R. Coifman and Charles L. Fefferman, *Weighted norm inequalities for maximal functions and singular integrals*, Studia Math. **51** (1974), 241–250. MR0358205

[88] Ronald R. Coifman, Peter W. Jones, and José L. Rubio de Francia, *Constructive decomposition of BMO functions and factorization of A_p weights*, Proc. Amer. Math. Soc. **87** (1983), no. 4, 675–676. MR687639

[89] Ronald R. Coifman and Richard Rochberg, *Another characterization of BMO*, Proc. Amer. Math. Soc. **79** (1980), no. 2, 249–254. MR565349

[90] Thierry Coulhon, *Espaces de Lipschitz et inégalités de Poincaré*, J. Funct. Anal. **136** (1996), no. 1, 81–113. MR1375154

[91] David V. Cruz-Uribe, *Two weight inequalities for fractional integral operators and commutators*, Advanced courses of mathematical analysis VI, World Sci. Publ., Hackensack, NJ 2017, pp. 25–85. MR3642364

[92] David V. Cruz-Uribe, José Maria Martell, and Carlos Pérez, *Weights, extrapolation and the theory of Rubio de Francia*, Operator Theory: Advances and Applications, vol. 215, Birkhäuser/Springer Basel AG, Basel, 2011. MR2797562

[93] David V. Cruz-Uribe and Kabe Moen, *A fractional Muckenhoupt-Wheeden theorem and its consequences*, Integral Equations Operator Theory **76** (2013), no. 3, 421–446. MR3065302

[94] David V. Cruz-Uribe, Kabe Moen, and Scott Rodney, *Regularity results for weak solutions of elliptic PDEs below the natural exponent*, Ann. Mat. Pura Appl. (4) **195** (2016), no. 3, 725–740. MR3500302

[95] Donatella Danielli, Nicola Garofalo, and Nguyen Cong Phuc, *Inequalities of Hardy-Sobolev type in Carnot-Carathéodory spaces*, Sobolev spaces in mathematics. I, Int. Math. Ser. (N. Y.), vol. 8, Springer, New York 2009, pp. 117–151. MR2508841

[96] E. B. Davies, *The Hardy constant*, Quart. J. Math. Oxford Ser. (2) **46** (1995), no. 184, 417–431. MR1366614

[97] Ennio De Giorgi, *Sulla differenziabilità e l'analiticità delle estremali degli integrali multipli regolari*, Mem. Accad. Sci. Torino. Cl. Sci. Fis. Mat. Nat. (3) **3** (1957), 25–43. MR0093649

[98] Miguel de Guzmán, *Differentiation of integrals in R^n*, Lecture Notes in Mathematics, Vol. 481, Springer-Verlag, Berlin-New York, 1975. With appendices by Antonio Córdoba, and Robert Fefferman, and two by Roberto Moriyón. MR0457661

[99] Miguel de Guzmán, *Real variable methods in Fourier analysis*, North-Holland Mathematics Studies, vol. 46, North-Holland Publishing Co., Amsterdam-New York, 1981. Notas de Matemática [Mathematical Notes], 75. MR596037

[100] Ronald A. DeVore and Robert C. Sharpley, *Maximal functions measuring smoothness*, Mem. Amer. Math. Soc. **47** (1984), no. 293, viii+115. MR727820

[101] Emmanuele DiBenedetto, *$C^{1+\alpha}$ local regularity of weak solutions of degenerate elliptic equations*, Nonlinear Anal. **7** (1983), no. 8, 827–850. MR709038

[102] Emmanuele DiBenedetto and Ugo Gianazza, *A Wiener-type condition for boundary continuity of quasi-minima of variational integrals*, Manuscripta Math. **149** (2016), no. 3-4, 339–346. MR3458172

[103] Emmanuele DiBenedetto and Neil S. Trudinger, *Harnack inequalities for quasiminima of variational integrals*, Ann. Inst. H. Poincaré Anal. Non Linéaire **1** (1984), no. 4, 295–308. MR778976

[104] Lars Diening, Petteri Harjulehto, Peter Hästö, and Michael Růžička, *Lebesgue and Sobolev spaces with variable exponents*, Lecture Notes in Mathematics, vol. 2017, Springer, Heidelberg, 2011. MR2790542

[105] Lars Diening, Michael Růžička, and Jörg Wolf, *Existence of weak solutions for unsteady motions of generalized Newtonian fluids*, Ann. Sc. Norm. Super. Pisa Cl. Sci. (5) **9** (2010), no. 1, 1–46. MR2668872

[106] Lars Diening, Sebastian Schwarzacher, Bianca Stroffolini, and Anna Verde, *Parabolic Lipschitz truncation and caloric approximation*, Calc. Var. Partial Differential Equations **56** (2017), no. 4, Art. 120, 27. MR3672391

[107] Javier Duoandikoetxea, *Fourier analysis*, Graduate Studies in Mathematics, vol. 29, American Mathematical Society, Providence, RI, 2001. Translated and revised from the 1995 Spanish original by D. V. Cruz-Uribe. MR1800316

[108] Javier Duoandikoetxea, *Forty years of Muckenhoupt weights*, Function Spaces and Inequalities, Lecture Notes Paseky nad Jizerou 2013, Matfyzpress, Praga 2013, pp. 23–75.

[109] Ricardo G. Durán and Fernando López García, *Solutions of the divergence and analysis of the Stokes equations in planar Hölder-α domains*, Math. Models Methods Appl. Sci. **20** (2010), no. 1, 95–120. MR2606245

[110] Frank Duzaar and Giuseppe Mingione, *The p-harmonic approximation and the regularity of p-harmonic maps*, Calc. Var. Partial Differential Equations **20** (2004), no. 3, 235–256. MR2062943

[111] Bartłomiej Dyda, Lizaveta Ihnatsyeva, Juha Lehrbäck, Heli Tuominen, and Antti V. Vähäkangas, *Muckenhoupt A_p-properties of distance functions and applications to Hardy-Sobolev–type inequalities*, Potential Anal. **50** (2019), no. 1, 83–105. MR3900847

[112] Bartłomiej Dyda, Lizaveta Ihnatsyeva, and Antti V. Vähäkangas, *On improved fractional Sobolev-Poincaré inequalities*, Ark. Mat. **54** (2016), no. 2, 437–454. MR3546360

[113] Bartłomiej Dyda and Antti V. Vähäkangas, *A framework for fractional Hardy inequalities*, Ann. Acad. Sci. Fenn. Math. **39** (2014), no. 2, 675–689. MR3237044

[114] Bartłomiej Dyda and Antti V. Vähäkangas, *Characterizations for fractional Hardy inequality*, Adv. Calc. Var. **8** (2015), no. 2, 173–182. MR3331699

[115] David E. Edmunds and W. Desmond Evans, *Hardy operators, function spaces and embeddings*, Springer Monographs in Mathematics, Springer-Verlag, Berlin, 2004. MR2091115

[116] David E. Edmunds and W. Desmond Evans, *Spectral theory and differential operators*, Second edition, Oxford Mathematical Monographs, Oxford University Press, Oxford, 2018. MR3823299

[117] David E. Edmunds, Ritva Hurri-Syrjänen, and Antti V. Vähäkangas, *Fractional Hardy-type inequalities in domains with uniformly fat complement*, Proc. Amer. Math. Soc. **142** (2014), no. 3, 897–907. MR3148524

[118] Moritz Egert, Robert Haller-Dintelmann, and Joachim Rehberg, *Hardy's inequality for functions vanishing on a part of the boundary*, Potential Anal. **43** (2015), no. 1, 49–78. MR3361789

[119] Sylvester Eriksson-Bique and Antti V. Vähäkangas, *Self-improvement of pointwise Hardy inequality*, Trans. Amer. Math. Soc. **372** (2019), no. 3, 2235–2250. MR3976590

[120] Lawrence C. Evans, *A new proof of local $C^{1,\alpha}$ regularity for solutions of certain degenerate elliptic p.d.e*, J. Differential Equations **45** (1982), no. 3, 356–373. MR672713

[121] Lawrence C. Evans, *Partial differential equations*, Second edition, Graduate Studies in Mathematics, vol. 19, American Mathematical Society, Providence, RI, 2010. MR2597943

[122] Lawrence C. Evans and Ronald F. Gariepy, *Measure theory and fine properties of functions*, Studies in Advanced Mathematics, CRC Press, Boca Raton, FL, 1992. MR1158660

[123] Eugene B. Fabes, D. Jerison, and Carlos E. Kenig, *The Wiener test for degenerate elliptic equations*, Ann. Inst. Fourier (Grenoble) **32** (1982), no. 3, vi, 151–182. MR688024

[124] Eugene B. Fabes, Carlos E. Kenig, and Raul P. Serapioni, *The local regularity of solutions of degenerate elliptic equations*, Comm. Partial Differential Equations **7** (1982), no. 1, 77–116. MR643158

[125] Kenneth Falconer, *Fractal geometry. Mathematical foundations and applications*, John Wiley & Sons, Ltd., Chichester, 1990. MR1102677

[126] Herbert Federer and William P. Ziemer, *The Lebesgue set of a function whose distribution derivatives are p-th power summable*, Indiana Univ. Math. J. **22** (1972/73), 139–158. MR0435361

[127] Charles L. Fefferman, *The uncertainty principle*, Bull. Amer. Math. Soc. (N.S.) **9** (1983), no. 2, 129–206. MR707957

[128] Charles L. Fefferman and Elias M. Stein, *Some maximal inequalities*, Amer. J. Math. **93** (1971), 107–115. MR0284802

[129] Alberto Fiorenza and Miroslav Krbec, *On the domain and range of the maximal operator*, Nagoya Math. J. **158** (2000), 43–61. MR1766576

[130] Wendell H. Fleming, *Functions whose partial derivatives are measures*, Illinois J. Math. **4** (1960), 452–478. MR0130338

[131] Bruno Franchi, Cristian E. Gutiérrez, and Richard L. Wheeden, *Weighted Sobolev-Poincaré inequalities for Grushin type operators*, Comm. Partial Differential Equations **19** (1994), no. 3-4, 523–604. MR1265808

[132] Bruno Franchi, Piotr Hajłasz, and Pekka Koskela, *Definitions of Sobolev classes on metric spaces*, Ann. Inst. Fourier (Grenoble) **49** (1999), no. 6, 1903–1924. MR1738070

[133] Bruno Franchi, Guozhen Lu, and Richard L. Wheeden, *Representation formulas and weighted Poincaré inequalities for Hörmander vector fields*, Ann. Inst. Fourier (Grenoble) **45** (1995), no. 2, 577–604. MR1343563

[134] Bruno Franchi, Guozhen Lu, and Richard L. Wheeden, *A relationship between Poincaré-type inequalities and representation formulas in spaces of homogeneous type*, Internat. Math. Res. Notices **1** (1996), 1–14. MR1383947

[135] Bruno Franchi and Richard L. Wheeden, *Some remarks about Poincaré type inequalities and representation formulas in metric spaces of homogeneous type*, J. Inequal. Appl. **3** (1999), no. 1, 65–89. MR1731670

[136] Jonathan M. Fraser, *Assouad Dimension in Fractal Geometry: theory, variations, and applications*, 2020. Cambridge Tracts in Mathematics 222, Cambridge University Press, Cambridge.

[137] Yohei Fujishima and Jens Habermann, *Global higher integrability for non-quadratic parabolic quasi-minimizers on metric measure spaces*, Adv. Calc. Var. **10** (2017), no. 3, 267–301. MR3667049

[138] Yohei Fujishima, Jens Habermann, Juha Kinnunen, and Mathias Masson, *Stability for parabolic quasiminimizers*, Potential Anal. **41** (2014), no. 3, 983–1004. MR3264831

[139] José García-Cuerva and José L. Rubio de Francia, *Weighted norm inequalities and related topics*, North-Holland Mathematics Studies, vol. 116, North-Holland Publishing Co., Amsterdam, 1985. Notas de Matemática [Mathematical Notes], 104. MR807149

[140] Ronald Gariepy and William P. Ziemer, *Behavior at the boundary of solutions of quasilinear elliptic equations*, Arch. Rational Mech. Anal. **56** (1974/75), 372–384. MR355332

[141] Ronald Gariepy and William P. Ziemer, *A regularity condition at the boundary for solutions of quasilinear elliptic equations*, Arch. Rational Mech. Anal. **67** (1977), no. 1, 25–39. MR492836

[142] John B. Garnett, *Bounded analytic functions*, first edition, Graduate Texts in Mathematics, vol. 236, Springer, New York, 2007. MR2261424

[143] Nicola Garofalo and Duy-Minh Nhieu, *Lipschitz continuity, global smooth approximations and extension theorems for Sobolev functions in Carnot-Carathéodory spaces*, J. Anal. Math. **74** (1998), 67–97. MR1631642

[144] F. W. Gehring, *The L^p-integrability of the partial derivatives of a quasiconformal mapping*, Acta Math. **130** (1973), 265–277. MR402038

[145] Ioseb Genebashvili, Amiran Gogatishvili, Vakhtang Kokilashvili, and Miroslav Krbec, *Weight theory for integral transforms on spaces of homogeneous type*, Pitman Monographs and Surveys in Pure and Applied Mathematics, vol. 92, Longman, Harlow, 1998. MR1791462

[146] Ugo Gianazza and Sebastian Schwarzacher, *Self-improving property of degenerate parabolic equations of porous medium-type*, Amer. J. Math. **141** (2019), no. 2, 399–446. MR3928041

[147] Mariano Giaquinta, *Multiple integrals in the calculus of variations and nonlinear elliptic systems*, Annals of Mathematics Studies, vol. 105, Princeton University Press, Princeton, NJ, 1983. MR717034

[148] Mariano Giaquinta, *Introduction to regularity theory for nonlinear elliptic systems*, Lectures in Mathematics ETH Zürich, Birkhäuser Verlag, Basel, 1993. MR1239172

[149] Mariano Giaquinta and Luca Martinazzi, *An introduction to the regularity theory for elliptic systems, harmonic maps and minimal graphs*, Second edition, Appunti. Scuola Normale Superiore di Pisa (Nuova Serie) [Lecture Notes. Scuola Normale Superiore di Pisa (New Series)], vol. 11, Edizioni della Normale, Pisa, 2012. MR3099262

[150] Mariano Giaquinta and Giuseppe Modica, *Regularity results for some classes of higher order nonlinear elliptic systems*, J. Reine Angew. Math. **311(312)** (1979), 145–169. MR549962

[151] Mariano Giaquinta and Michael Struwe, *On the partial regularity of weak solutions of nonlinear parabolic systems*, Math. Z. **179** (1982), no. 4, 437–451. MR652852

[152] David Gilbarg and Neil S. Trudinger, *Elliptic partial differential equations of second order*, Classics in Mathematics, Springer-Verlag, Berlin, 2001. Reprint of the 1998 edition. MR1814364

[153] Enrico Giusti, *Direct methods in the calculus of variations*, World Scientific Publishing Co., Inc., River Edge, NJ, 2003. MR1962933

[154] V. M. Gol′dshteĭn and Yu. G. Reshetnyak, *Quasiconformal mappings and Sobolev spaces*, Mathematics and its Applications (Soviet Series), vol. 54, Kluwer Academic Publishers Group, Dordrecht, 1990. Translated and revised from the 1983 Russian original, Translated by O. Korneeva. MR1136035

[155] Jasun Gong, *Rigidity of derivations in the plane and in metric measure spaces*, Illinois J. Math. **56** (2012), no. 4, 1109–1147. MR3231476

[156] Loukas Grafakos, *Classical Fourier analysis*, Second edition, Graduate Texts in Mathematics, vol. 249, Springer, New York, 2008. MR2445437

[157] Loukas Grafakos, *Modern Fourier analysis*, Second edition, Graduate Texts in Mathematics, vol. 250, Springer, New York, 2009. MR2463316

[158] Loukas Grafakos and Juha Kinnunen, *Sharp inequalities for maximal functions associated with general measures*, Proc. Roy. Soc. Edinburgh Sect. A **128** (1998), no. 4, 717–723. MR1635412

[159] Loukas Grafakos and Stephen Montgomery-Smith, *Best constants for uncentred maximal functions*, Bull. London Math. Soc. **29** (1997), no. 1, 60–64. MR1416408

[160] Seppo Granlund, *An L^p-estimate for the gradient of extremals*, Math. Scand. **50** (1982), no. 1, 66–72. MR664508

[161] Seppo Granlund, Peter Lindqvist, and Olli Martio, *Conformally invariant variational integrals*, Trans. Amer. Math. Soc. **277** (1983), no. 1, 43–73. MR690040

[162] Jens Habermann, *Global gradient estimates for non-quadratic vector-valued parabolic quasi-minimizers*, Nonlinear Anal. **114** (2015), 42–73. MR3300784

[163] Jens Habermann, *Higher integrability for vector-valued parabolic quasi-minimizers on metric measure spaces*, Ark. Mat. **54** (2016), no. 1, 85–123. MR3475819

[164] Piotr Hajłasz, *Geometric approach to Sobolev spaces and badly degenerated elliptic equations*, Nonlinear analysis and applications (Warsaw, 1994), GAKUTO Internat. Ser. Math. Sci. Appl., vol. 7, Gakkotosho, Tokyo 1996, pp. 141–168. MR1422932

[165] Piotr Hajłasz, *Sobolev spaces on an arbitrary metric space*, Potential Anal. **5** (1996), no. 4, 403–415. MR1401074

[166] Piotr Hajłasz, *Pointwise Hardy inequalities*, Proc. Amer. Math. Soc. **127** (1999), no. 2, 417–423. MR1458875

[167] Piotr Hajłasz, *Sobolev inequalities, truncation method, and John domains*, Papers on analysis, Rep. Univ. Jyväskylä Dep. Math. Stat., vol. 83, Univ. Jyväskylä, Jyväskylä 2001, pp. 109–126. MR1886617

[168] Piotr Hajłasz, *A new characterization of the Sobolev space*, Studia Math. **159** (2003), no. 2, 263–275. MR2052222

[169] Piotr Hajłasz, *Sobolev spaces on metric-measure spaces*, Heat kernels and analysis on manifolds, graphs, and metric spaces (Paris, 2002), Contemp. Math., vol. 338, Amer. Math. Soc., Providence, RI 2003, pp. 173–218. MR2039955

[170] Piotr Hajłasz and Juha Kinnunen, *Hölder quasicontinuity of Sobolev functions on metric spaces*, Rev. Mat. Iberoamericana **14** (1998), no. 3, 601–622. MR1681586

[171] Piotr Hajłasz and Pekka Koskela, *Sobolev met Poincaré*, Mem. Amer. Math. Soc. **145** (2000), no. 688, x+101. MR1683160

[172] Piotr Hajłasz and Zhuomin Liu, *Sobolev spaces, Lebesgue points and maximal functions*, J. Fixed Point Theory Appl. **13** (2013), no. 1, 259–269. MR3071953

[173] Piotr Hajłasz and Zhuomin Liu, *Maximal potentials, maximal singular integrals, and the spherical maximal function*, Proc. Amer. Math. Soc. **142** (2014), no. 11, 3965–3974. MR3251736

[174] Piotr Hajłasz and Jan Malý, *On approximate differentiability of the maximal function*, Proc. Amer. Math. Soc. **138** (2010), no. 1, 165–174. MR2550181

[175] Piotr Hajłasz and Olli Martio, *Traces of Sobolev functions on fractal type sets and characterization of extension domains*, J. Funct. Anal. **143** (1997), no. 1, 221–246. MR1428124

[176] Piotr Hajłasz and Jani Onninen, *On boundedness of maximal functions in Sobolev spaces*, Ann. Acad. Sci. Fenn. Math. **29** (2004), no. 1, 167–176. MR2041705

[177] Heikki Hakkarainen and Juha Kinnunen, *The BV-capacity in metric spaces*, Manuscripta Math. **132** (2010), no. 1-2, 51–73. MR2609288

[178] Qing Han and Fanghua Lin, *Elliptic partial differential equations*, Second edition, Courant Lecture Notes in Mathematics, vol. 1, Courant Institute of Mathematical Sciences, New York; American Mathematical Society, Providence, RI, 2011. MR2777537

[179] G. H. Hardy and J. E. Littlewood, *A maximal theorem with function-theoretic applications*, Acta Math. **54** (1930), no. 1, 81–116. MR1555303

[180] G. H. Hardy, J. E. Littlewood, and G. Pólya, *Inequalities*, Second edition, Cambridge, at the University Press, 1952. MR0046395

[181] Lars Inge Hedberg, *On certain convolution inequalities*, Proc. Amer. Math. Soc. **36** (1972), 505–510. MR0312232

[182] Lars Inge Hedberg and Tero Kilpeläinen, *On the stability of Sobolev spaces with zero boundary values*, Math. Scand. **85** (1999), no. 2, 245–258. MR1724237

[183] Toni Heikkinen, Juha Kinnunen, Janne Korvenpää, and Heli Tuominen, *Regularity of the local fractional maximal function*, Ark. Mat. **53** (2015), no. 1, 127–154. MR3319617

[184] Toni Heikkinen and Heli Tuominen, *Smoothing properties of the discrete fractional maximal operator on Besov and Triebel-Lizorkin spaces*, Publ. Mat. **58** (2014), no. 2, 379–399. MR3264503

[185] Juha Heinonen, *Lectures on analysis on metric spaces*, Universitext, Springer-Verlag, New York, 2001. MR1800917

[186] Juha Heinonen, *Lectures on Lipschitz analysis*, Report. University of Jyväskylä Department of Mathematics and Statistics, vol. 100, University of Jyväskylä, Jyväskylä, 2005. MR2177410

[187] Juha Heinonen, Tero Kilpeläinen, and Olli Martio, *Nonlinear potential theory of degenerate elliptic equations*, Dover Publications, Inc., Mineola, NY, 2006. Unabridged republication of the 1993 original. MR2305115

[188] Juha Heinonen and Pekka Koskela, *Quasiconformal maps in metric spaces with controlled geometry*, Acta Math. **181** (1998), no. 1, 1–61. MR1654771

[189] Juha Heinonen and Pekka Koskela, *A note on Lipschitz functions, upper gradients, and the Poincaré inequality*, New Zealand J. Math. **28** (1999), no. 1, 37–42. MR1691958

[190] Juha Heinonen, Pekka Koskela, Nageswari Shanmugalingam, and Jeremy T. Tyson, *Sobolev spaces on metric measure spaces. An approach based on upper gradients*, New Mathematical Monographs, vol. 27, Cambridge University Press, Cambridge, 2015. MR3363168

[191] Edwin Hewitt and Karl Stromberg, *Real and abstract analysis. A modern treatment of the theory of functions of a real variable, third printing*, Graduate Texts in Mathematics, vol. 25, Springer-Verlag, New York-Heidelberg, 1975. MR0367121

[192] Toshio Horiuchi, *The imbedding theorems for weighted Sobolev spaces*, J. Math. Kyoto Univ. **29** (1989), no. 3, 365–403. MR1021144

[193] Toshio Horiuchi, *The imbedding theorems for weighted Sobolev spaces. II*, Bull. Fac. Sci. Ibaraki Univ. Ser. A **23** (1991), 11–37. MR1118940

[194] Ritva Hurri, *The weighted Poincaré inequalities*, Math. Scand. **67** (1990), no. 1, 145–160. MR1081294

[195] Ritva Hurri-Syrjänen, *Unbounded Poincaré domains*, Ann. Acad. Sci. Fenn. Ser. A I Math. **17** (1992), no. 2, 409–423. MR1190332

[196] Ritva Hurri-Syrjänen, *The John-Nirenberg inequality and a Sobolev inequality in general domains*, J. Math. Anal. Appl. **175** (1993), no. 2, 579–587. MR1219196

[197] Ritva Hurri-Syrjänen, Niko Marola, and Antti V. Vähäkangas, *Aspects of local-to-global results*, Bull. Lond. Math. Soc. **46** (2014), no. 5, 1032–1042. MR3262204

[198] Ritva Hurri-Syrjänen and Antti V. Vähäkangas, *Fractional Sobolev-Poincaré and fractional Hardy inequalities in unbounded John domains*, Mathematika **61** (2015), no. 2, 385–401. MR3343059

[199] Tuomas Hytönen, *A framework for non-homogeneous analysis on metric spaces, and the RBMO space of Tolsa*, Publ. Mat. **54** (2010), no. 2, 485–504. MR2675934

[200] Tuomas Hytönen, Jan van Neerven, Mark Veraar, and Lutz Weis, *Analysis in Banach spaces. Vol. I. Martingales and Littlewood-Paley theory*, Ergebnisse der Mathematik und ihrer Grenzgebiete. 3. Folge. A Series of Modern Surveys in Mathematics [Results in Mathematics and Related Areas. 3rd Series. A Series of Modern Surveys in Mathematics], vol. 63, Springer, Cham, 2016. MR3617205

[201] Lizaveta Ihnatsyeva, Juha Lehrbäck, Heli Tuominen, and Antti V. Vähäkangas, *Fractional Hardy inequalities and visibility of the boundary*, Studia Math. **224** (2014), no. 1, 47–80. MR3277052

[202] Lizaveta Ihnatsyeva and Antti V. Vähäkangas, *Characterization of traces of smooth functions on Ahlfors regular sets*, J. Funct. Anal. **265** (2013), no. 9, 1870–1915. MR3084491

[203] Lizaveta Ihnatsyeva and Antti V. Vähäkangas, *Hardy inequalities in Triebel-Lizorkin spaces*, Indiana Univ. Math. J. **62** (2013), no. 6, 1785–1807. MR3205532

[204] Lizaveta Ihnatsyeva and Antti V. Vähäkangas, *Hardy inequalities in Triebel-Lizorkin spaces II. Aikawa dimension*, Ann. Mat. Pura Appl. (4) **194** (2015), no. 2, 479–493. MR3322431

[205] Per-Anders Ivert, Niko Marola, and Mathias Masson, *Energy estimates for variational minimizers of a parabolic doubly nonlinear equation on metric measure spaces*, Ann. Acad. Sci. Fenn. Math. **39** (2014), no. 2, 711–719. MR3237046

[206] Tadeusz Iwaniec, *On L^p-integrability in PDEs and quasiregular mappings for large exponents*, Ann. Acad. Sci. Fenn. Ser. A I Math. **7** (1982), no. 2, 301–322. MR686647

[207] Tadeusz Iwaniec and Gaven Martin, *Geometric function theory and non-linear analysis*, Oxford Mathematical Monographs, The Clarendon Press, Oxford University Press, New York, 2001. MR1859913

[208] Tadeusz Iwaniec and Craig A. Nolder, *Hardy-Littlewood inequality for quasiregular mappings in certain domains in $\mathbf{R}^n$*, Ann. Acad. Sci. Fenn. Ser. A I Math. **10** (1985), 267–282. MR802488

[209] Tadeusz Iwaniec and Carlo Sbordone, *Weak minima of variational integrals*, J. Reine Angew. Math. **454** (1994), 143–161. MR1288682

[210] Marian Jakszto, *Another proof that L^p-bounded pointwise convergence implies weak convergence*, Real Anal. Exchange **36** (2010/11), no. 2, 479–481. MR3016731

[211] Pentti Järvi and Matti Vuorinen, *Uniformly perfect sets and quasiregular mappings*, J. London Math. Soc. (2) **54** (1996), no. 3, 515–529. MR1413895

[212] Peter W. Jones, *Extension theorems for BMO*, Indiana Univ. Math. J. **29** (1980), no. 1, 41–66. MR554817

[213] Peter W. Jones, *Factorization of A_p weights*, Ann. of Math. (2) **111** (1980), no. 3, 511–530. MR577135

[214] Peter W. Jones, *Quasiconformal mappings and extendability of functions in Sobolev spaces*, Acta Math. **147** (1981), no. 1-2, 71–88. MR631089

[215] Alf Jonsson and Hans Wallin, *Function spaces on subsets of $\mathbf{R}^n$*, Math. Rep. **2** (1984), no. 1, xiv+221. MR820626

[216] Jean-Lin Journé, *Calderón-Zygmund operators, pseudodifferential operators and the Cauchy integral of Calderón*, Lecture Notes in Mathematics, vol. 994, Springer-Verlag, Berlin, 1983. MR706075

[217] Antti Käenmäki, Juha Lehrbäck, and Matti Vuorinen, *Dimensions, Whitney covers, and tubular neighborhoods*, Indiana Univ. Math. J. **62** (2013), no. 6, 1861–1889. MR3205534

[218] Tero Kilpeläinen, *A remark on the uniqueness of quasi continuous functions*, Ann. Acad. Sci. Fenn. Math. **23** (1998), no. 1, 261–262. MR1601887

[219] Tero Kilpeläinen, Juha Kinnunen, and Olli Martio, *Sobolev spaces with zero boundary values on metric spaces*, Potential Anal. **12** (2000), no. 3, 233–247. MR1752853

[220] Tero Kilpeläinen and Pekka Koskela, *Global integrability of the gradients of solutions to partial differential equations*, Nonlinear Anal. **23** (1994), no. 7, 899–909. MR1302151

[221] Tero Kilpeläinen, Pekka Koskela, and Hiroaki Masaoka, *Lattice property of p-admissible weights*, Proc. Amer. Math. Soc. **143** (2015), no. 6, 2427–2437. MR3326025

[222] Tero Kilpeläinen and Jan Malý, *The Wiener test and potential estimates for quasilinear elliptic equations*, Acta Math. **172** (1994), no. 1, 137–161. MR1264000

[223] Juha Kinnunen, *Higher integrability with weights*, Ann. Acad. Sci. Fenn. Ser. A I Math. **19** (1994), no. 2, 355–366. MR1274087

[224] Juha Kinnunen, *The Hardy-Littlewood maximal function of a Sobolev function*, Israel J. Math. **100** (1997), 117–124. MR1469106

[225] Juha Kinnunen and Riikka Korte, *Characterizations of Sobolev inequalities on metric spaces*, J. Math. Anal. Appl. **344** (2008), no. 2, 1093–1104. MR2426336

[226] Juha Kinnunen and Riikka Korte, *Characterizations for the Hardy inequality*, Around the research of Vladimir Maz′ya. I, Int. Math. Ser. (N. Y.), vol. 11, Springer, New York 2010, pp. 239–254. MR2723821

[227] Juha Kinnunen, Riikka Korte, Nageswari Shanmugalingam, and Heli Tuominen, *Lebesgue points and capacities via the boxing inequality in metric spaces*, Indiana Univ. Math. J. **57** (2008), no. 1, 401–430. MR2400262

[228] Juha Kinnunen and Tuomo Kuusi, *Local behaviour of solutions to doubly nonlinear parabolic equations*, Math. Ann. **337** (2007), no. 3, 705–728. MR2274548

[229] Juha Kinnunen and Visa Latvala, *Lebesgue points for Sobolev functions on metric spaces*, Rev. Mat. Iberoamericana **18** (2002), no. 3, 685–700. MR1954868

[230] Juha Kinnunen and John L. Lewis, *Higher integrability for parabolic systems of p-Laplacian type*, Duke Math. J. **102** (2000), no. 2, 253–271. MR1749438

[231] Juha Kinnunen and John L. Lewis, *Very weak solutions of parabolic systems of p-Laplacian type*, Ark. Mat. **40** (2002), no. 1, 105–132. MR1948889

[232] Juha Kinnunen and Peter Lindqvist, *The derivative of the maximal function*, J. Reine Angew. Math. **503** (1998), 161–167. MR1650343

[233] Juha Kinnunen, Niko Marola, and Olli Martio, *Harnack's principle for quasiminimizers*, Ric. Mat. **56** (2007), no. 1, 73–88. MR2330352

[234] Juha Kinnunen and Olli Martio, *The Sobolev capacity on metric spaces*, Ann. Acad. Sci. Fenn. Math. **21** (1996), no. 2, 367–382. MR1404091

[235] Juha Kinnunen and Olli Martio, *Hardy's inequalities for Sobolev functions*, Math. Res. Lett. **4** (1997), no. 4, 489–500. MR1470421

[236] Juha Kinnunen and Mikko Parviainen, *Stability for degenerate parabolic equations*, Adv. Calc. Var. **3** (2010), no. 1, 29–48. MR2604616

[237] Juha Kinnunen and Eero Saksman, *Regularity of the fractional maximal function*, Bull. London Math. Soc. **35** (2003), no. 4, 529–535. MR1979008

[238] Juha Kinnunen and Nageswari Shanmugalingam, *Regularity of quasi-minimizers on metric spaces*, Manuscripta Math. **105** (2001), no. 3, 401–423. MR1856619

[239] Soulaymane Korry, *Boundedness of Hardy-Littlewood maximal operator in the framework of Lizorkin-Triebel spaces*, Rev. Mat. Complut. **15** (2002), no. 2, 401–416. MR1951818

[240] Soulaymane Korry, *A class of bounded operators on Sobolev spaces*, Arch. Math. (Basel) **82** (2004), no. 1, 40–50. MR2034469

[241] Riikka Korte, Juha Lehrbäck, and Heli Tuominen, *The equivalence between pointwise Hardy inequalities and uniform fatness*, Math. Ann. **351** (2011), no. 3, 711–731. MR2854110 (2012k:31012)

[242] Riikka Korte and Nageswari Shanmugalingam, *Equivalence and self-improvement of p-fatness and Hardy's inequality, and association with uniform perfectness*, Math. Z. **264** (2010), no. 1, 99–110. MR2564934

[243] Pekka Koskela and Juha Lehrbäck, *Weighted pointwise Hardy inequalities*, J. Lond. Math. Soc. (2) **79** (2009), no. 3, 757–779. MR2506697

[244] Pekka Koskela and Paul MacManus, *Quasiconformal mappings and Sobolev spaces*, Studia Math. **131** (1998), no. 1, 1–17. MR1628655

[245] Pekka Koskela and Eero Saksman, *Pointwise characterizations of Hardy-Sobolev functions*, Math. Res. Lett. **15** (2008), no. 4, 727–744. MR2424909

[246] Pekka Koskela and Xiao Zhong, *Hardy's inequality and the boundary size*, Proc. Amer. Math. Soc. **131** (2003), no. 4, 1151–1158 (electronic). MR1948106 (2004e:26021)

[247] Manfred Kronz, *Some function spaces on spaces of homogeneous type*, Manuscripta Math. **106** (2001), no. 2, 219–248. MR1865566

[248] Alois Kufner, *Weighted Sobolev spaces*, A Wiley-Interscience Publication, John Wiley & Sons, Inc., New York, 1985. Translated from the Czech. MR802206

[249] Alois Kufner, Lech Maligranda, and Lars-Erik Persson, *The prehistory of the Hardy inequality*, Amer. Math. Monthly **113** (2006), no. 8, 715–732. MR2256532

[250] Alois Kufner and Lars-Erik Persson, *Weighted inequalities of Hardy type*, World Scientific Publishing Co., Inc., River Edge, NJ, 2003. MR1982932

[251] Ondřej Kurka, *On the variation of the Hardy-Littlewood maximal function*, Ann. Acad. Sci. Fenn. Math. **40** (2015), no. 1, 109–133. MR3310075

[252] Emma-Karoliina Kurki and Antti V. Vähäkangas, *Weighted norm inequalities in a bounded domain by the sparse domination method*, Rev. Mat. Complut. **34** (2021), no. 2, 435–467. MR4254077

[253] Olga A. Ladyzhenskaya and Nina N. Ural′tseva, *Linear and quasilinear elliptic equations*, Translated from the Russian by Scripta Technica, Inc. Translation editor: Leon Ehrenpreis, Academic Press, New York-London, 1968. MR0244627

[254] N. S. Landkof, *Foundations of modern potential theory*, Springer-Verlag, New York-Heidelberg, 1972. Translated from the Russian by A. P. Doohovskoy, Die Grundlehren der mathematischen Wissenschaften, Band 180. MR0350027

[255] D. G. Larman, *A new theory of dimension*, Proc. London Math. Soc. (3) **17** (1967), 178–192. MR0203691

[256] Juha Lehrbäck, *Pointwise Hardy inequalities and uniformly fat sets*, Proc. Amer. Math. Soc. **136** (2008), no. 6, 2193–2200. MR2383525

[257] Juha Lehrbäck, *Self-improving properties of weighted Hardy inequalities*, Adv. Calc. Var. **1** (2008), no. 2, 193–203. MR2427452

[258] Juha Lehrbäck, *Weighted Hardy inequalities and the size of the boundary*, Manuscripta Math. **127** (2008), no. 2, 249–273. MR2442898

[259] Juha Lehrbäck, *Necessary conditions for weighted pointwise Hardy inequalities*, Ann. Acad. Sci. Fenn. Math. **34** (2009), no. 2, 437–446. MR2553805

[260] Juha Lehrbäck, *Weighted Hardy inequalities beyond Lipschitz domains*, Proc. Amer. Math. Soc. **142** (2014), no. 5, 1705–1715. MR3168477

[261] Juha Lehrbäck, *Hardy inequalities and Assouad dimensions*, J. Anal. Math. **131** (2017), 367–398. MR3631460

[262] Juha Lehrbäck and Nageswari Shanmugalingam, *Quasiadditivity of variational capacity*, Potential Anal. **40** (2014), no. 3, 247–265. MR3189220

[263] Juha Lehrbäck and Heli Tuominen, *A note on the dimensions of Assouad and Aikawa*, J. Math. Soc. Japan **65** (2013), no. 2, 343–356. MR3055588

[264] Juha Lehrbäck, Heli Tuominen, and Antti V. Vähäkangas, *Self-improvement of uniform fatness revisited*, Math. Ann. **368** (2017), no. 3-4, 1439–1464. MR3673660

[265] Juha Lehrbäck and Antti V. Vähäkangas, *In between the inequalities of Sobolev and Hardy*, J. Funct. Anal. **271** (2016), no. 2, 330–364. MR3501850

[266] Giovanni Leoni, *A first course in Sobolev spaces*, Second edition, Graduate Studies in Mathematics, vol. 181, American Mathematical Society, Providence, RI, 2017. MR3726909

[267] Andrei K. Lerner, *An elementary approach to several results on the Hardy-Littlewood maximal operator*, Proc. Amer. Math. Soc. **136** (2008), no. 8, 2829–2833. MR2399047

[268] Andrei K. Lerner, *A "local mean oscillation" decomposition and some of its applications*, Function spaces, approximation, inequalities and lineability. Lectures of the spring school in analysis, Matfyzpres, Prague 2011, pp. 71–106.

[269] Andrei K. Lerner, Sheldy Ombrosi, and Israel P. Rivera-Ríos, *On pointwise and weighted estimates for commutators of Calderón-Zygmund operators*, Adv. Math. **319** (2017), 153–181. MR3695871

[270] John L. Lewis, *Regularity of the derivatives of solutions to certain degenerate elliptic equations*, Indiana Univ. Math. J. **32** (1983), no. 6, 849–858. MR721568

[271] John L. Lewis, *Uniformly fat sets*, Trans. Amer. Math. Soc. **308** (1988), no. 1, 177–196. MR946438 (89e:31012)

[272] John L. Lewis, *On very weak solutions of certain elliptic systems*, Comm. Partial Differential Equations **18** (1993), no. 9-10, 1515–1537. MR1239922

[273] Gongbao Li and Olli Martio, *Stability in obstacle problems*, Math. Scand. **75** (1994), no. 1, 87–100. MR1308940

[274] Gongbao Li and Olli Martio, *Convergence properties of supersolutions and a-superharmonic functions*, Nonlinear Anal. **28** (1997), no. 3, 453–462. MR1420792

[275] Gongbao Li and Olli Martio, *Stability of solutions of varying degenerate elliptic equations*, Indiana Univ. Math. J. **47** (1998), no. 3, 873–891. MR1665792

[276] Gongbao Li and Olli Martio, *Stability and higher integrability of derivatives of solutions in double obstacle problems*, J. Math. Anal. Appl. **272** (2002), no. 1, 19–29. MR1930700

[277] Qifan Li, *Very weak solutions of subquadratic parabolic systems with non-standard $p(x,t)$-growth*, Nonlinear Anal. **156** (2017), 17–41. MR3634766

[278] Peter Lindqvist, *Stability for the solutions of* $\operatorname{div}(|\nabla u|^{p-2}\nabla u) = f$ *with varying* p, J. Math. Anal. Appl. **127** (1987), no. 1, 93–102. MR904212

[279] Peter Lindqvist, *On nonlinear Rayleigh quotients*, Potential Anal. **2** (1993), no. 3, 199–218. MR1245239

[280] Peter Lindqvist, *Notes on the stationary p-Laplace equation*, SpringerBriefs in Mathematics, Springer, Cham, 2019. MR3931688

[281] Peter Lindqvist and Olli Martio, *Two theorems of N. Wiener for solutions of quasilinear elliptic equations*, Acta Math. **155** (1985), no. 3-4, 153–171. MR806413

[282] W. Littman, G. Stampacchia, and H. F. Weinberger, *Regular points for elliptic equations with discontinuous coefficients*, Ann. Scuola Norm. Sup. Pisa Cl. Sci. (3) **17** (1963), 43–77. MR161019

[283] Feng Liu, Ting Chen, and Huoxiong Wu, *A note on the endpoint regularity of the Hardy-Littlewood maximal functions*, Bull. Aust. Math. Soc. **94** (2016), no. 1, 121–130. MR3539328

[284] Fon Che Liu, *A Luzin type property of Sobolev functions*, Indiana Univ. Math. J. **26** (1977), no. 4, 645–651. MR0450488

[285] Rui Lin Long and Fu Sheng Nie, *Weighted Sobolev inequality and eigenvalue estimates of Schrödinger operators*, Harmonic analysis (Tianjin, 1988), Lecture Notes in Math., vol. 1494, Springer, Berlin 1991, pp. 131–141. MR1187073

[286] Shanzhen Lu, Yong Ding, and Dunyan Yan, *Singular integrals and related topics*, World Scientific Publishing Co. Pte. Ltd., Hackensack, NJ, 2007. MR2354214

[287] Hannes Luiro, *Continuity of the maximal operator in Sobolev spaces*, Proc. Amer. Math. Soc. **135** (2007), no. 1, 243–251. MR2280193

[288] Hannes Luiro, *On the regularity of the Hardy-Littlewood maximal operator on subdomains of* $\mathbb{R}^n$, Proc. Edinb. Math. Soc. (2) **53** (2010), no. 1, 211–237. MR2579688

[289] Hannes Luiro, *On the continuous and discontinuous maximal operators*, Nonlinear Anal. **172** (2018), 36–58. MR3790366

[290] Hannes Luiro, *The variation of the maximal function of a radial function*, Ark. Mat. **56** (2018), no. 1, 147–161. MR3800463

[291] Hannes Luiro and Antti V. Vähäkangas, *Local maximal operators on fractional Sobolev spaces*, J. Math. Soc. Japan **68** (2016), no. 3, 1357–1368. MR3523550

[292] Hannes Luiro and Antti V. Vähäkangas, *Beyond local maximal operators*, Potential Anal. **46** (2017), no. 2, 201–226. MR3605166

[293] Jouni Luukkainen, *Assouad dimension: antifractal metrization, porous sets, and homogeneous measures*, J. Korean Math. Soc. **35** (1998), no. 1, 23–76. MR1608518

[294] Outi Elina Maasalo, *Global integrability of p-superharmonic functions on metric spaces*, J. Anal. Math. **106** (2008), 191–207. MR2448985

[295] Outi Elina Maasalo and Anna Zatorska-Goldstein, *Stability of quasiminimizers of the p-Dirichlet integral with varying p on metric spaces*, J. Lond. Math. Soc. (2) **77** (2008), no. 3, 771–788. MR2418304

[296] Roberto A. Macías and Carlos Segovia, *Lipschitz functions on spaces of homogeneous type*, Adv. in Math. **33** (1979), no. 3, 257–270. MR546295

[297] P. Maheux and L. Saloff-Coste, *Analyse sur les boules d'un opérateur sous-elliptique*, Math. Ann. **303** (1995), no. 4, 713–740. MR1359957

[298] Tero Mäkäläinen, *Adams inequality on metric measure spaces*, Rev. Mat. Iberoam. **25** (2009), no. 2, 533–558. MR2569546

[299] Jan Malý, *Hölder type quasicontinuity*, Potential Anal. **2** (1993), no. 3, 249–254. MR1245242

[300] Jan Malý, *Coarea integration in metric spaces*, NAFSA 7—Nonlinear analysis, function spaces and applications. Vol. 7, Czech. Acad. Sci., Prague 2003, pp. 148–192. MR2657115

[301] Jan Malý, *Coarea properties of Sobolev functions*, Function spaces, differential operators and nonlinear analysis (Teistungen, 2001), Birkhäuser, Basel 2003, pp. 371–381. MR1984185

[302] Jan Malý, David Swanson, and William P. Ziemer, *The co-area formula for Sobolev mappings*, Trans. Amer. Math. Soc. **355** (2003), no. 2, 477–492. MR1932709

[303] Jan Malý and William P. Ziemer, *Fine regularity of solutions of elliptic partial differential equations*, Mathematical Surveys and Monographs, vol. 51, American Mathematical Society, Providence, RI, 1997. MR1461542

[304] Moshe Marcus and Victor J. Mizel, *Every superposition operator mapping one Sobolev space into another is continuous*, J. Funct. Anal. **33** (1979), no. 2, 217–229. MR546508

[305] Moshe Marcus, Victor J. Mizel, and Yehuda Pinchover, *On the best constant for Hardy's inequality in* $\mathbb{R}^n$, Trans. Amer. Math. Soc. **350** (1998), no. 8, 3237–3255. MR1458330

[306] Moshe Marcus and Itai Shafrir, *An eigenvalue problem related to Hardy's* L^p *inequality*, Ann. Scuola Norm. Sup. Pisa Cl. Sci. (4) **29** (2000), no. 3, 581–604. MR1817710

[307] Niko Marola and Mathias Masson, *On the Harnack inequality for parabolic minimizers in metric measure spaces*, Tohoku Math. J. (2) **65** (2013), no. 4, 569–589. MR3161434

[308] Niko Marola and Olli Saari, *Local to global results for spaces of BMO type*, Math. Z. **282** (2016), no. 1-2, 473–484. MR3448391

[309] Joaquim Martín and Mario Milman, *Gehring's lemma for nondoubling measures*, Michigan Math. J. **47** (2000), no. 3, 559–573. MR1813544

[310] Olli Martio, *John domains, bi-Lipschitz balls and Poincaré inequality*, Rev. Roumaine Math. Pures Appl. **33** (1988), no. 1-2, 107–112. MR948443

[311] Mathias Masson, Michele Miranda Jr., Fabio Paronetto, and Mikko Parviainen, *Local higher integrability for parabolic quasiminimizers in metric spaces*, Ric. Mat. **62** (2013), no. 2, 279–305. MR3120053

[312] Mathias Masson and Mikko Parviainen, *Global higher integrability for parabolic quasiminimizers in metric measure spaces*, J. Anal. Math. **126** (2015), 307–339. MR3358035

[313] J. Mateu, P. Mattila, A. Nicolau, and J. Orobitg, *BMO for nondoubling measures*, Duke Math. J. **102** (2000), no. 3, 533–565. MR1756109

[314] Tanya Matskewich and Pavel E. Sobolevskii, *The best possible constant in generalized Hardy's inequality for convex domain in* $\mathbf{R}^n$, Nonlinear Anal. **28** (1997), no. 9, 1601–1610. MR1431208

[315] Pertti Mattila, *Geometry of sets and measures in Euclidean spaces. Fractals and rectifiability*, Cambridge Studies in Advanced Mathematics, vol. 44, Cambridge University Press, Cambridge, 1995. MR1333890

[316] Vladimir G. Maz′ya, *On the boundary regularity of solutions of elliptic equations and of a conformal mapping*, Dokl. Akad. Nauk SSSR **152** (1963), 1297–1300. MR0163053

[317] Vladimir G. Maz′ya, *The continuity at a boundary point of the solutions of quasi-linear elliptic equations*, Vestnik Leningrad. Univ. **25** (1970), no. 13, 42–55. MR0274948

[318] Vladimir G. Maz′ya, *Sobolev spaces*, Springer Series in Soviet Mathematics, Springer-Verlag, Berlin, 1985. Translated from the Russian by T. O. Shaposhnikova. MR817985 (87g:46056)

[319] Vladimir G. Maz′ya, *Lectures on isoperimetric and isocapacitary inequalities in the theory of Sobolev spaces*, Heat kernels and analysis on manifolds, graphs, and metric spaces (Paris, 2002), Contemp. Math., vol. 338, Amer. Math. Soc., Providence, RI 2003, pp. 307–340. MR2039959

[320] Vladimir G. Maz′ya, *Conductor and capacitary inequalities for functions on topological spaces and their applications to Sobolev-type imbeddings*, J. Funct. Anal. **224** (2005), no. 2, 408–430. MR2146047

[321] Vladimir G. Maz′ya, *Sobolev spaces with applications to elliptic partial differential equations*, augmented edition, Grundlehren der Mathematischen Wissenschaften [Fundamental Principles of Mathematical Sciences], vol. 342, Springer, Heidelberg, 2011. MR2777530

[322] Vladimir G. Maz′ya and Victor P. Havin, *A nonlinear potential theory*, Uspehi Mat. Nauk **27** (1972), no. 6, 67–138. MR0409858

[323] Antonios D. Melas, *The best constant for the centered Hardy-Littlewood maximal inequality*, Ann. of Math. (2) **157** (2003), no. 2, 647–688. MR1973058

[324] Norman G. Meyers, *An L^p-estimate for the gradient of solutions of second order elliptic divergence equations*, Ann. Scuola Norm. Sup. Pisa Cl. Sci. (3) **17** (1963), 189–206. MR159110

[325] Norman G. Meyers and Alan Elcrat, *Some results on regularity for solutions of non-linear elliptic systems and quasi-regular functions*, Duke Math. J. **42** (1975), 121–136. MR417568

[326] J. H. Michael and William P. Ziemer, *A Lusin type approximation of Sobolev functions by smooth functions*, Classical real analysis (Madison, Wis., 1982), Contemp. Math., vol. 42, Amer. Math. Soc., Providence, RI 1985, pp. 135–167. MR807986

[327] Pasi Mikkonen, *On the Wolff potential and quasilinear elliptic equations involving measures*, Ann. Acad. Sci. Fenn. Math. Diss. **104** (1996), 71. MR1386213 (97e:35069)

[328] Akihiko Miyachi, *Hardy-Sobolev spaces and maximal functions*, J. Math. Soc. Japan **42** (1990), no. 1, 73–90. MR1027541

[329] Yoshihiro Mizuta, *Potential theory in Euclidean spaces*, GAKUTO International Series. Mathematical Sciences and Applications, vol. 6, Gakkotosho Co., Ltd., Tokyo, 1996. MR1428685

[330] Kristian Moring, Christoph Scheven, Sebastian Schwarzacher, and Thomas Singer, *Global higher integrability of weak solutions of porous medium systems*, Commun. Pure Appl. Anal. **19** (2020), no. 3, 1697–1745. MR4064049

[331] Jürgen Moser, *A new proof of De Giorgi's theorem concerning the regularity problem for elliptic differential equations*, Comm. Pure Appl. Math. **13** (1960), 457–468. MR170091

[332] Jürgen Moser, *On Harnack's theorem for elliptic differential equations*, Comm. Pure Appl. Math. **14** (1961), 577–591. MR159138

[333] Benjamin Muckenhoupt, *Weighted norm inequalities for the Hardy maximal function*, Trans. Amer. Math. Soc. **165** (1972), 207–226. MR0293384

[334] Benjamin Muckenhoupt, *The equivalence of two conditions for weight functions*, Studia Math. **49** (1973/74), 101–106. MR350297

[335] Benjamin Muckenhoupt and Richard L. Wheeden, *Weighted norm inequalities for fractional integrals*, Trans. Amer. Math. Soc. **192** (1974), 261–274. MR0340523

[336] Benjamin Muckenhoupt and Richard L. Wheeden, *Weighted bounded mean oscillation and the Hilbert transform*, Studia Math. **54** (1975/76), no. 3, 221–237. MR399741

[337] Jindřich Nečas, *Sur une méthode pour résoudre les équations aux dérivées partielles du type elliptique, voisine de la variationnelle*, Ann. Scuola Norm. Sup. Pisa (3) **16** (1962), 305–326. MR0163054

[338] Makoto Ohtsuka, *Extremal length and precise functions*, GAKUTO International Series. Mathematical Sciences and Applications, vol. 19, Gakkotosho Co., Ltd., Tokyo, 2003. With a preface by Fumi-Yuki Maeda. MR2068961

[339] Sheldy Ombrosi, Carlos Pérez, Ezequiel Rela, and Israel P. Rivera-Ríos, *A note on generalized Fujii-Wilson conditions and BMO spaces*, Israel J. Math. **238** (2020), no. 2, 571–591. MR4145810

[340] Bohumír Opic and Alois Kufner, *Hardy-type inequalities*, Pitman Research Notes in Mathematics Series, vol. 219, Longman Scientific & Technical, Harlow, 1990. MR1069756

[341] Joan Orobitg and Joan Verdera, *Choquet integrals, Hausdorff content and the Hardy-Littlewood maximal operator*, Bull. London Math. Soc. **30** (1998), no. 2, 145–150. MR1489325

[342] Mikko Parviainen, *Global higher integrability for parabolic quasiminimizers in nonsmooth domains*, Calc. Var. Partial Differential Equations **31** (2008), no. 1, 75–98. MR2342615

[343] Mikko Parviainen, *Reverse Hölder inequalities for singular parabolic equations near the boundary*, J. Differential Equations **246** (2009), no. 2, 512–540. MR2468726

[344] Carlos Pérez, *Two weighted norm inequalities for Riesz potentials and uniform L^p-weighted Sobolev inequalities*, Indiana Univ. Math. J. **39** (1990), no. 1, 31–44. MR1052009

[345] Carlos Pérez, *Two weighted inequalities for potential and fractional type maximal operators*, Indiana Univ. Math. J. **43** (1994), no. 2, 663–683. MR1291534

[346] Carlos Pérez, Tiago Picon, Olli Saari, and Mateus Sousa, *Regularity of maximal functions on Hardy-Sobolev spaces*, Bull. Lond. Math. Soc. **50** (2018), no. 6, 1007–1015. MR3891939

[347] Yehuda Pinchover and Kyril Tintarev, *Ground state alternative for p-Laplacian with potential term*, Calc. Var. Partial Differential Equations **28** (2007), no. 2, 179–201. MR2284565

[348] Yehuda Pinchover and Kyril Tintarev, *On positive solutions of minimal growth for singular p-Laplacian with potential term*, Adv. Nonlinear Stud. **8** (2008), no. 2, 213–234. MR2404043

[349] Yehuda Pinchover and Kyril Tintarev, *On the Hardy-Sobolev-Maz′ya inequality and its generalizations*, Sobolev spaces in mathematics. I, Int. Math. Ser. (N. Y.), vol. 8, Springer, New York 2009, pp. 281–297. MR2508846

[350] Augusto C. Ponce, *Elliptic PDEs, measures and capacities. From the Poisson equations to nonlinear Thomas-Fermi problems*, EMS Tracts in Mathematics, vol. 23, European Mathematical Society (EMS), Zürich, 2016. MR3675703

[351] Tapio Rajala and Markku Vilppolainen, *Weakly controlled Moran constructions and iterated functions systems in metric spaces*, Illinois J. Math. **55** (2011), no. 3, 1015–1051 (2013). MR3069294

[352] Hans Martin Reimann and Thomas Rychener, *Funktionen beschränkter mittlerer Oszillation*, Lecture Notes in Mathematics, Vol. 487, Springer-Verlag, Berlin-New York, 1975. MR0511997

[353] Yu. G. Reshetnyak, *Integral representations of differentiable functions in domains with a nonsmooth boundary*, Sibirsk. Mat. Zh. **21** (1980), no. 6, 108–116, 221. MR601195

[354] Yu. G. Reshetnyak, *Space mappings with bounded distortion*, Translations of Mathematical Monographs, vol. 73, American Mathematical Society, Providence, RI, 1989. Translated from the Russian by H. H. McFaden. MR994644

[355] Marcel Riesz, *L'intégrale de Riemann-Liouville et le problème de Cauchy*, Acta Math. **81** (1949), 1–223. MR0030102

[356] Cora Sadosky, *Interpolation of operators and singular integrals. An introduction to harmonic analysis*, Monographs and Textbooks in Pure and Applied Math., vol. 53, Marcel Dekker, Inc., New York, 1979. MR551747

[357] Laurent Saloff-Coste, *Aspects of Sobolev-type inequalities*, London Mathematical Society Lecture Note Series, vol. 289, Cambridge University Press, Cambridge, 2002. MR1872526

[358] Stephen Semmes, *Finding curves on general spaces through quantitative topology, with applications to Sobolev and Poincaré inequalities*, Selecta Math. (N.S.) **2** (1996), no. 2, 155–295. MR1414889

[359] James Serrin, *Local behavior of solutions of quasi-linear equations*, Acta Math. **111** (1964), 247–302. MR170096

[360] James Serrin, *Pathological solutions of elliptic differential equations*, Ann. Scuola Norm. Sup. Pisa Cl. Sci. (3) **18** (1964), 385–387. MR170094

[361] Pavel Shvartsman, *Local approximations and intrinsic characterization of spaces of smooth functions on regular subsets of $\mathbb{R}^n$*, Math. Nachr. **279** (2006), no. 11, 1212–1241. MR2247585

[362] Peter Sjögren, *A remark on the maximal function for measures in* $\mathbf{R}^n$, Amer. J. Math. **105** (1983), no. 5, 1231–1233. MR714775

[363] Mark A. Smith and Barry Turett, *Rotundity in Lebesgue-Bochner function spaces*, Trans. Amer. Math. Soc. **257** (1980), no. 1, 105–118. MR549157

[364] Wayne Smith and David A. Stegenga, *Hölder domains and Poincaré domains*, Trans. Amer. Math. Soc. **319** (1990), no. 1, 67–100. MR978378

[365] S. L. Sobolev, *Some applications of functional analysis in mathematical physics*, Translations of Mathematical Monographs, vol. 90, American Mathematical Society, Providence,

RI, 1991. Translated from the third Russian edition by Harold H. McFaden, With comments by V. P. Palamodov. MR1125990

[366] Susan G. Staples, *L^p-averaging domains and the Poincaré inequality*, Ann. Acad. Sci. Fenn. Ser. A I Math. **14** (1989), no. 1, 103–127. MR997974

[367] Susan G. Staples, *L^p-averaging domains in homogeneous spaces*, J. Math. Anal. Appl. **317** (2006), no. 2, 550–564. MR2209578

[368] Elias M. Stein, *Note on the class L log L*, Studia Math. **32** (1969), 305–310. MR0247534

[369] Elias M. Stein, *Singular integrals and differentiability properties of functions*, Princeton Mathematical Series, No. 30, Princeton University Press, Princeton, N.J., 1970. MR0290095

[370] Elias M. Stein, *Some results in harmonic analysis in $\mathbf{R}^n$, for $n \to \infty$*, Bull. Amer. Math. Soc. (N.S.) **9** (1983), no. 1, 71–73. MR699317

[371] Elias M. Stein, *Harmonic analysis: real-variable methods, orthogonality, and oscillatory integrals*, Princeton Mathematical Series, vol. 43, Princeton University Press, Princeton, NJ, 1993. With the assistance of Timothy S. Murphy, Monographs in Harmonic Analysis, III. MR1232192

[372] Elias M. Stein and Jan-Olov Strömberg, *Behavior of maximal functions in $\mathbf{R}^n$ for large n*, Ark. Mat. **21** (1983), no. 2, 259–269. MR727348

[373] Edward W. Stredulinsky, *Higher integrability from reverse Hölder inequalities*, Indiana Univ. Math. J. **29** (1980), no. 3, 407–413. MR570689

[374] Edward W. Stredulinsky, *Weighted inequalities and degenerate elliptic partial differential equations*, Lecture Notes in Mathematics, vol. 1074, Springer-Verlag, Berlin, 1984. MR757718

[375] Jan-Olov Strömberg and Alberto Torchinsky, *Weights, sharp maximal functions and Hardy spaces*, Bull. Amer. Math. Soc. (N.S.) **3** (1980), no. 3, 1053–1056. MR585189

[376] Jan-Olov Strömberg and Alberto Torchinsky, *Weighted Hardy spaces*, Lecture Notes in Mathematics, vol. 1381, Springer-Verlag, Berlin, 1989. MR1011673

[377] Toshiyuki Sugawa, *Uniformly perfect sets: analytic and geometric aspects [translation of Sūgaku* **53** *(2001), no. 4, 387–402*, Sugaku Expositions **16** (2003), no. 2, 225–242. MR2019172

[378] Hitoshi Tanaka, *A remark on the derivative of the one-dimensional Hardy-Littlewood maximal function*, Bull. Austral. Math. Soc. **65** (2002), no. 2, 253–258. MR1898539

[379] Luc Tartar, *Imbedding theorems of Sobolev spaces into Lorentz spaces*, Boll. Unione Mat. Ital. Sez. B Artic. Ric. Mat. (8) **1** (1998), no. 3, 479–500. MR1662313

[380] Jesper Tidblom, *A geometrical version of Hardy's inequality for $\overset{\circ}{W}{}^{1,p}(\Omega)$*, Proc. Amer. Math. Soc. **132** (2004), no. 8, 2265–2271. MR2052402

[381] Jesper Tidblom, *A Hardy inequality in the half-space*, J. Funct. Anal. **221** (2005), no. 2, 482–495. MR2124873

[382] Peter Tolksdorf, *Regularity for a more general class of quasilinear elliptic equations*, J. Differential Equations **51** (1984), no. 1, 126–150. MR727034

[383] Alberto Torchinsky, *Real-variable methods in harmonic analysis*, Pure and Applied Mathematics, vol. 123, Academic Press, Inc., Orlando, FL, 1986. MR869816

[384] Hans Triebel, *Function spaces and wavelets on domains*, EMS Tracts in Mathematics, vol. 7, European Mathematical Society (EMS), Zürich, 2008. MR2455724

[385] Hans Triebel, *Fractals and spectra. Related to Fourier analysis and function spaces*, Modern Birkhäuser Classics, Birkhäuser Verlag, Basel, 2011. MR2732650

[386] Neil S. Trudinger, *On Harnack type inequalities and their application to quasilinear elliptic equations*, Comm. Pure Appl. Math. **20** (1967), 721–747. MR226198

[387] Neil S. Trudinger, *On the regularity of generalized solutions of linear, non-uniformly elliptic equations*, Arch. Rational Mech. Anal. **42** (1971), 50–62. MR344656

[388] Bengt Ove Turesson, *Nonlinear potential theory and weighted Sobolev spaces*, Lecture Notes in Mathematics, vol. 1736, Springer-Verlag, Berlin, 2000. MR1774162

[389] Karen Uhlenbeck, *Regularity for a class of non-linear elliptic systems*, Acta Math. **138** (1977), no. 3-4, 219–240. MR474389

[390] Nina N. Ural′tseva, *Degenerate quasilinear elliptic systems*, Zap. Naučn. Sem. Leningrad. Otdel. Mat. Inst. Steklov. (LOMI) **7** (1968), 184–222. MR0244628

[391] Nicholas Th. Varopoulos, *Distances associées aux opérateurs sous-elliptiques du second ordre*, C. R. Acad. Sci. Paris Sér. I Math. **309** (1989), no. 10, 663–667. MR1054273

[392] S. K. Vodop′yanov, *Monotone functions and quasiconformal mappings on Carnot groups*, Sibirsk. Mat. Zh. **37** (1996), no. 6, 1269–1295, ii. MR1440383

[393] Andreas Wannebo, *Hardy inequalities*, Proc. Amer. Math. Soc. **109** (1990), no. 1, 85–95. MR1010807 (90h:26025)

[394] Andreas Wannebo, *Hardy and Hardy PDO type inequalities in domains. Part I*, 2004. Unpublished manuscript, arXiv:math/0401253.

[395] Kjell-Ove Widman, *Hölder continuity of solutions of elliptic systems*, Manuscripta Math. **5** (1971), 299–308. MR296484

[396] Norbert Wiener, *The Dirichlet problem*, J. Math. and Physics **3** (1924), 127–146.

[397] Norbert Wiener, *The ergodic theorem*, Duke Math. J. **5** (1939), no. 1, 1–18. MR1546100

[398] Wilfried Wieser, *Parabolic Q-minima and minimal solutions to variational flow*, Manuscripta Math. **59** (1987), no. 1, 63–107. MR901250

[399] Ingemar Wik, *A comparison of the integrability of f and Mf with that of $f^\sharp$*, Technical Report 2, University of Umeå, 1983.

[400] P. Wojtaszczyk, *Banach spaces for analysts*, Cambridge Studies in Advanced Mathematics, vol. 25, Cambridge University Press, Cambridge, 1991. MR1144277

[401] Kōsaku Yosida, *Functional analysis*, Classics in Mathematics, Springer-Verlag, Berlin, 1995. Reprint of the sixth (1980) edition. MR1336382

[402] Anna Zatorska-Goldstein, *Very weak solutions of nonlinear subelliptic equations*, Ann. Acad. Sci. Fenn. Math. **30** (2005), no. 2, 407–436. MR2173373

[403] Vasiliĭ V. Zhikov, *On some variational problems*, Russian J. Math. Phys. **5** (1997), no. 1, 105–116 (1998). MR1486765

[404] Vasiliĭ V. Zhikov and Svetlana E. Pastukhova, *On the property of higher integrability for parabolic systems of variable order of nonlinearity*, Mat. Zametki **87** (2010), no. 2, 179–200. MR2731471

[405] Xiao Zhong, *On nonhomogeneous quasilinear elliptic equations*, Ann. Acad. Sci. Fenn. Math. Diss. **117** (1998), 46. Dissertation, University of Jyväskylä, Jyväskylä, 1998. MR1648847

[406] William P. Ziemer, *Boundary regularity for quasiminima*, Arch. Rational Mech. Anal. **92** (1986), no. 4, 371–382. MR823124

[407] William P. Ziemer, *Weakly differentiable functions. Sobolev spaces and functions of bounded variation*, Graduate Texts in Mathematics, vol. 120, Springer-Verlag, New York, 1989. MR1014685

Index

Selected Published Titles in This Series

257 **Juha Kinnunen, Juha Lehrbäck, and Antti Vähäkangas,** Maximal Function Methods for Sobolev Spaces, 2021

256 **Michio Jimbo, Tetsuji Miwa, and Fedor Smirnov,** Local Operators in Integrable Models I, 2021

255 **Alexandre Boritchev and Sergei Kuksin,** One-Dimensional Turbulence and the Stochastic Burgers Equation, 2021

254 **Karim Belabas and Henri Cohen,** Numerical Algorithms for Number Theory, 2021

252 **Julie Déserti,** The Cremona Group and Its Subgroups, 2021

251 **David Hoff,** Linear and Quasilinear Parabolic Systems, 2020

250 **Bachir Bekka and Pierre de la Harpe,** Unitary Representations of Groups, Duals, and Characters, 2020

249 **Nikolai M. Adrianov, Fedor Pakovich, and Alexander K. Zvonkin,** Davenport–Zannier Polynomials and Dessins d'Enfants, 2020

248 **Paul B. Larson and Jindrich Zapletal,** Geometric Set Theory, 2020

247 **István Heckenberger and Hans-Jürgen Schneider,** Hopf Algebras and Root Systems, 2020

246 **Matheus C. Bortolan, Alexandre N. Carvalho, and José A. Langa,** Attractors Under Autonomous and Non-autonomous Perturbations, 2020

245 **Aiping Wang and Anton Zettl,** Ordinary Differential Operators, 2019

244 **Nabile Boussaïd and Andrew Comech,** Nonlinear Dirac Equation, 2019

243 **José M. Isidro,** Jordan Triple Systems in Complex and Functional Analysis, 2019

242 **Bhargav Bhatt, Ana Caraiani, Kiran S. Kedlaya, Peter Scholze, and Jared Weinstein,** Perfectoid Spaces, 2019

241 **Dana P. Williams,** A Tool Kit for Groupoid C^*-Algebras, 2019

240 **Antonio Fernández López,** Jordan Structures in Lie Algebras, 2019

239 **Nicola Arcozzi, Richard Rochberg, Eric T. Sawyer, and Brett D. Wick,** The Dirichlet Space and Related Function Spaces, 2019

238 **Michael Tsfasman, Serge Vlăduţ, and Dmitry Nogin,** Algebraic Geometry Codes: Advanced Chapters, 2019

237 **Dusa McDuff, Mohammad Tehrani, Kenji Fukaya, and Dominic Joyce,** Virtual Fundamental Cycles in Symplectic Topology, 2019

236 **Bernard Host and Bryna Kra,** Nilpotent Structures in Ergodic Theory, 2018

235 **Habib Ammari, Brian Fitzpatrick, Hyeonbae Kang, Matias Ruiz, Sanghyeon Yu, and Hai Zhang,** Mathematical and Computational Methods in Photonics and Phononics, 2018

234 **Vladimir I. Bogachev,** Weak Convergence of Measures, 2018

233 **N. V. Krylov,** Sobolev and Viscosity Solutions for Fully Nonlinear Elliptic and Parabolic Equations, 2018

232 **Dmitry Khavinson and Erik Lundberg,** Linear Holomorphic Partial Differential Equations and Classical Potential Theory, 2018

231 **Eberhard Kaniuth and Anthony To-Ming Lau,** Fourier and Fourier-Stieltjes Algebras on Locally Compact Groups, 2018

230 **Stephen D. Smith,** Applying the Classification of Finite Simple Groups, 2018

229 **Alexander Molev,** Sugawara Operators for Classical Lie Algebras, 2018

228 **Zhenbo Qin,** Hilbert Schemes of Points and Infinite Dimensional Lie Algebras, 2018

227 **Roberto Frigerio,** Bounded Cohomology of Discrete Groups, 2017

226 **Marcelo Aguiar and Swapneel Mahajan,** Topics in Hyperplane Arrangements, 2017

For a complete list of titles in this series, visit the
AMS Bookstore at **www.ams.org/bookstore/survseries/**.